FIFTH EDITION

Microbes and Society

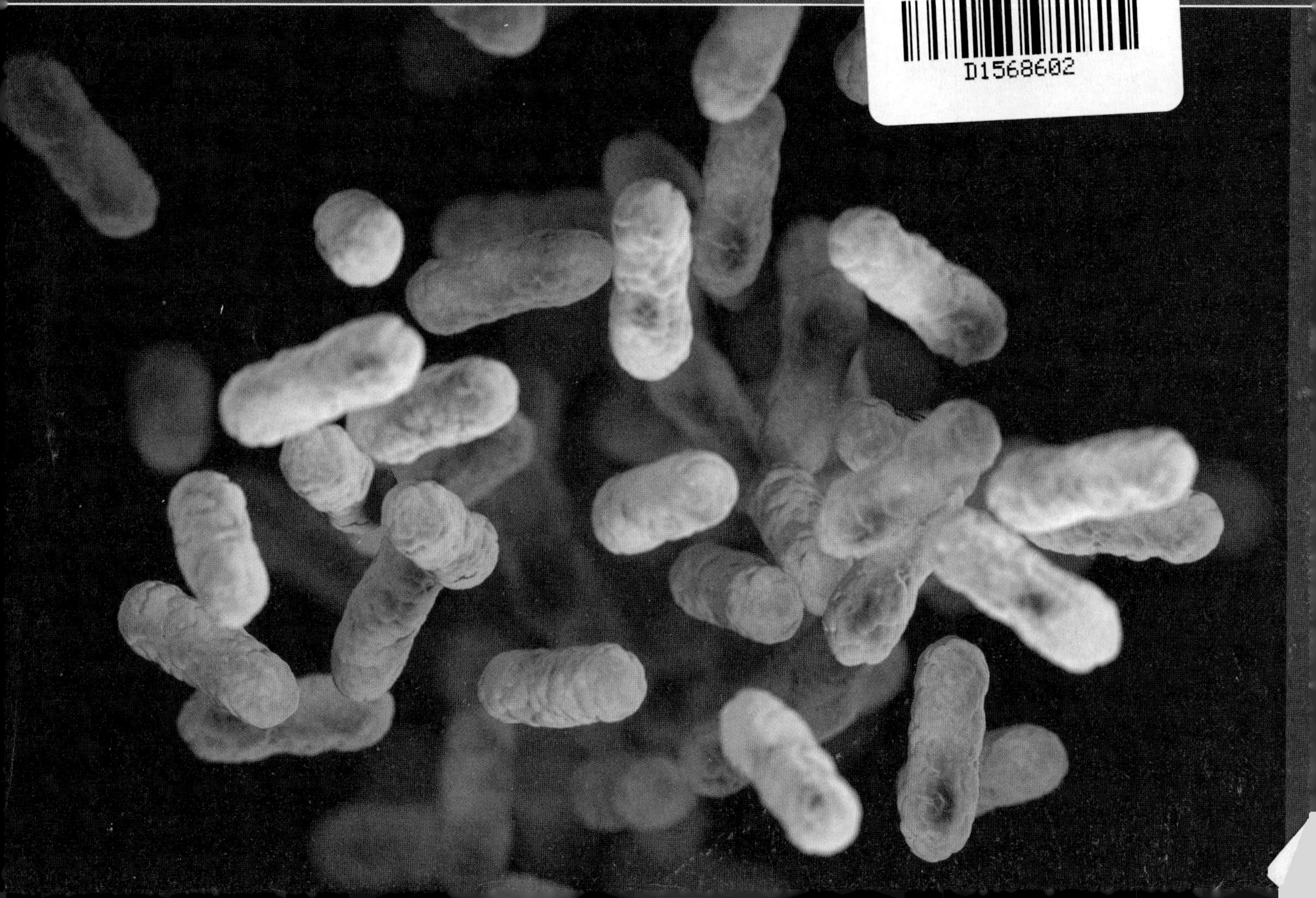

FIFTH EDITION

Microbes and Society

Jeffrey C. Pommerville, PhD
Emeritus Professor of Biology and Microbiology
Glendale Community College
Department of Biology
Glendale, AZ

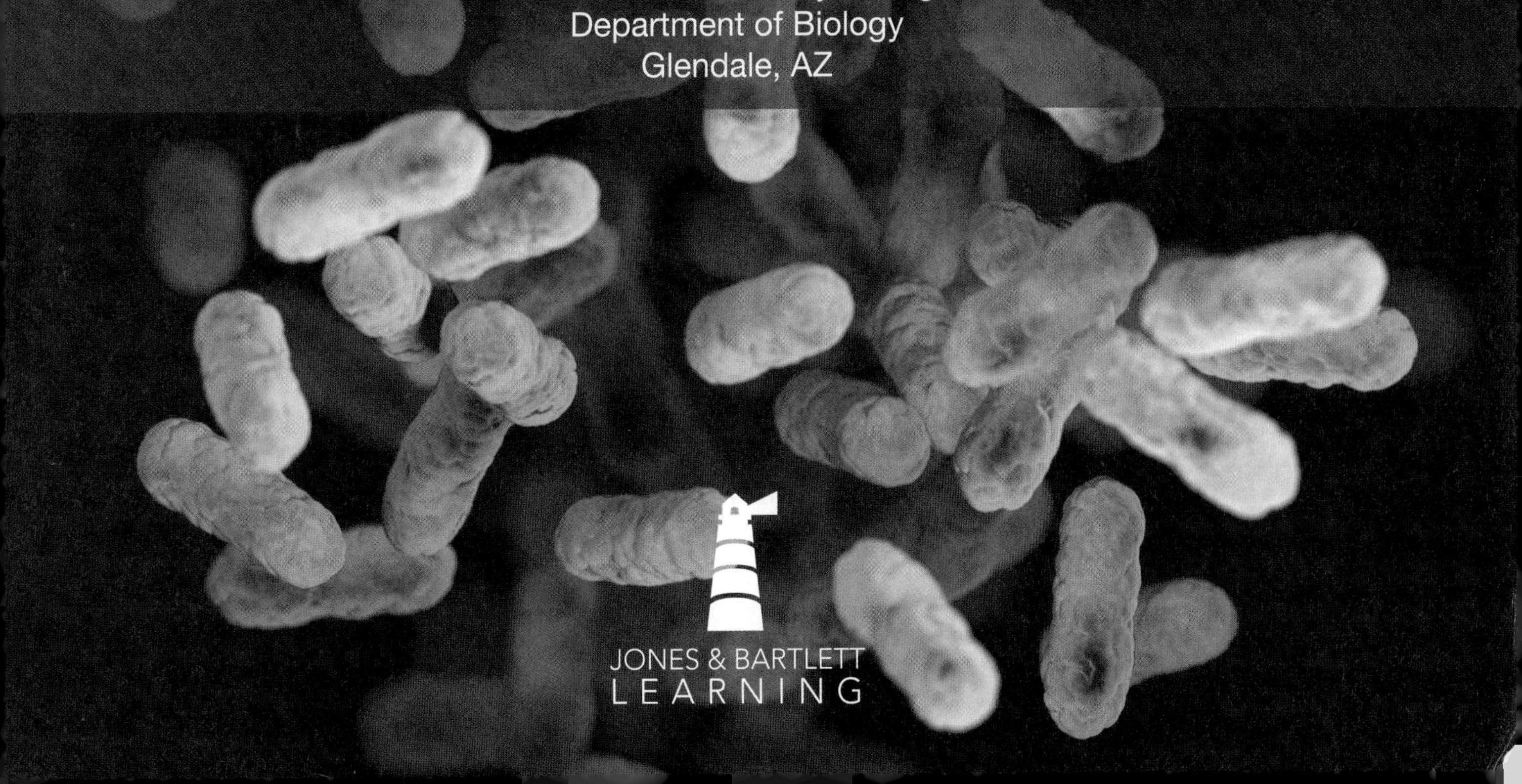

JONES & BARTLETT
LEARNING

World Headquarters
Jones & Bartlett Learning
5 Wall Street
Burlington, MA 01803
978-443-5000
info@jblearning.com
www.jblearning.com

Jones & Bartlett Learning books and products are available through most bookstores and online booksellers. To contact Jones & Bartlett Learning directly, call 800-832-0034, fax 978-443-8000, or visit our website, www.jblearning.com.

Substantial discounts on bulk quantities of Jones & Bartlett Learning publications are available to corporations, professional associations, and other qualified organizations. For details and specific discount information, contact the special sales department at Jones & Bartlett Learning via the above contact information or send an email to specialsales@jblearning.com.

17217-1

Production Credits
VP, Product Management: Amanda Martin
Director of Product Management: Laura Pagluica
Product Specialist: Audrey Schwinn
Product Coordinator: Paula-Yuan Gregory
Senior Project Specialist: Dan Stone
Digital Project Specialist: Rachel Reyes
Marketing Manager: Suzy Balk
Manufacturing and Inventory Control Supervisor: Amy Bacus
Composition: codeMantra U.S. LLC
Cover Design: Michael O'Donnell
Text Design: Michael O'Donnell
Media Development Editor: Troy Liston
Rights & Media Specialist: John Rusk
Cover Image TOP: ©NYCstocker/iStock/Thinkstock;
BOTTOM: Courtesy of CDC/Sarah Bailey Cutchin.
Illustrator Meredith Newlove.
Printing and Binding: LSC Communications
Cover Printing: LSC Communications

Library of Congress Cataloging-in-Publication Data
Library of Congress Cataloging-in-Publication Data unavailable at time of printing.

LCCN: 2019947448

6048

Printed in the United States of America

23 22 21 20 19   10 9 8 7 6 5 4 3 2 1

©NYCstocker/iStock/Thinkstock

Brief Contents

©NYCstocker/iStock/Thinkstock

Contents

©NYCstocker/iStock/Thinkstock

Preface

I am excited about the new fifth edition of *Microbes and Society*. Why? This new edition reflects the current state of the science of microbiology and stresses more than ever the roles microorganisms play in society and in human health. Enthusiastic responses from students and faculty have helped shape this edition into a form that fits the evolving needs of today's students. It emphasizes how microorganisms are connected to our lives every day, and the textbook highlights many microbial interactions that most of us would not realize occur. It shows the connections of science with other fields, including the social sciences and humanities.

Both our own health and the health of the planet depend on the interactions with microorganisms. When we see or hear a report on a disease outbreak, the discovery of some new and often exotic microorganism, or a news story about how microorganisms influence our good health and the environment, many students and the public in general often do not have the background to properly evaluate many of the statements or claims made. After reading *Microbes and Society,* all that will change.

Studies have reported that the use of more readable textbooks in college often is correlated with better grades. Therefore, the chapters of *Microbes and Society, Fifth Edition*, have been either completely rewritten or heavily edited to limit the volume of information. In addition, unfamiliar science vocabulary can be challenging, as the number of technical terms can be overwhelming. To be a science-literate student in today's society, some mastery of scientific jargon is necessary. However, the mass of technical terms has been reduced by almost 10 percent. Consequently, *Microbes and Society*, *Fifth Edition* presents a balanced approach by providing a sufficient volume of information and introducing enough vocabulary to promote effective learning and communication. In the end, students will be more science and biology literate, which is essential in today's world.

So, welcome to the world of microbiology! I hope you find the journey exciting, fun, informative, and fascinating.

Audience

Microbes and Society, Fifth Edition is written for the nonscience undergraduate and inquiring citizen of the 21st century. The textbook assumes little or no science background, and it should accommodate a one-quarter or one-semester course. Students will find that understanding microbes will help them do well in such fields as business, sociology, food science, pharmaceutical and health sciences, economics, and agriculture. It discusses such topics as the place of microbes in ecology and the environment, the use of microbes in biotechnology, the role of microbes in food production, and the numerous other ways that microbes contribute to the quality of our lives and our well-being. The book also examines the problem of antibiotic resistance, discusses the importance of vaccines, describes the intimate association of microorganisms in our gut, and surveys several microbial diseases of historical and contemporary times.

Objectives

The 21st century is destined to be the Century of Biology. In future decades we can anticipate new products of biotechnology, new ways of preserving and protecting our environment, new methods in agriculture, new practices to maintain human health, and new technologies not yet even in the idea stage. Microbes are at the center of all of these happenings. They are the hammers and nails of genetic engineering, the worker bees for purifying polluted water, the sources of imaginative insecticides and pesticides, our internal guardians protecting us from disease and helping keep us healthy, and the jumping off points for futuristic technologies. Knowing the microbes is essential to knowing the future. Helping you to know the microbes is the first major objective of this book.

Rarely does a day go by when we do not enjoy a "microbial food;" each time we put out the garbage, we assume that microbes will break it down; whenever we take a breath, we inhale oxygen that microbes have put into the atmosphere; and each time we cover a sneeze, we try to stop the spread of microbes. Helping you to understand the places that microbes occupy in our day-to-day existence is this book's second major objective.

But what would the present and future be without the past? The third major objective is to provide some examples of how microbes have had a significant impact on history. We shall study, for example, how plague changed the course of Western civilization, how malaria influenced the building of the Panama Canal, how microbes influenced the way human cultures arose, and how microbes made much of the current work in biotechnology possible. Few groups of organisms have played such a rich and powerful role in history and society.

I hope you will enjoy your education in microbiology and come to understand the influence of microbes on our society today, in the past, and in the future.

Organization

Microbes and Society, Fifth Edition contains two parts. Part I introduces the microbial world over the span of 11 chapters. Individual chapters explore the bacteria, viruses, fungi, and protists. Other chapters describe how these microbes grow and reproduce, the unique genetic patterns they display, and the methods used to control them.

Part II moves to the practical applications of microbiology. We visit a restaurant for a microbial meal, we wander through a research facility and see microbes at work, we stop at various locations in the environment and observe microbes acting on our behalf, and we examine their place in disease. The bottom line is that microbes are relevant.

Must the chapters be studied in sequence? Absolutely not. Time constraints often prevent courses from using the entire book, so instructors and students are invited to "cherry pick" those topics that best fit their situation. To encourage flexibility, each chapter has been written independently of the others, and each section in a chapter stands alone. Instructors may, therefore, design their own approach to microbiology according to their students' needs. Instructors seeking additional information on creating a custom textbook for their course should contact their Jones & Bartlett Learning Account Manager at go.jblearning.com/findmyrep.

The Student Experience

Approaching a course in microbiology can be an anxious undertaking. There are new insights to learn, new concepts to understand, and an entirely new vocabulary to master. To help smooth over the bumps, this book incorporates several features that should help increase students' comfort level as they read *Microbes and Society, Fifth Edition.*

Students might note that all chapters are about the same length. This was done purposefully to provide a symmetrical framework in which students can learn. Each chapter has several sections and numerous smaller subsections to accommodate limited study times. The ultimate goal has been to provide a thorough and balanced presentation of microbiology within an enjoyable context.

Each chapter begins with an engaging story that helps set a tone for the pages and topics that follow. A section titled **"Looking Ahead"** indicates to students what they should take away from the chapter.

LOOKING AHEAD

After reading and completing this chapter, you will be able to:

2.1 Compare the contributions to microbiology that were made by each of the four pioneers.
2.2 Write correctly an organism's scientific name.
2.3 Explain the differences between (a) the light microscope and the electron microscope (EM) and (b) the transmission EM and scanning EM.
2.4 Identify the general characteristics of the microbes and viruses.
2.5 Describe the tree of life, and assess its significance to microbiology and the evolution of life.

Special topics boxes in each chapter (**"A Closer Look"**) encourage a moment of relief from the rigors of study, and present a historical insight, an interesting aside, or a health issue. Most figures are presented in full color, and special attention has been given to setting them close to their text reference. In the chapter margins of each page, definitions to some scientific terms and organism pronunciations (sometimes appearing to be tongue twisting names) are presented. In addition, a pronunciation guide has been added as an Appendix. The chapter topics conclude with **A Final Thought**

At the end of each chapter are the **Chapter Discussion Questions**. This includes **What Was He Thinking** that has the student think about the most important information presented in the chapter. The **Key Terms** again appear and provide the vocabulary to answer the **Questions For Discussion**.

Even many of the paragraphs are about the same size (there should be a rhythm in reading).

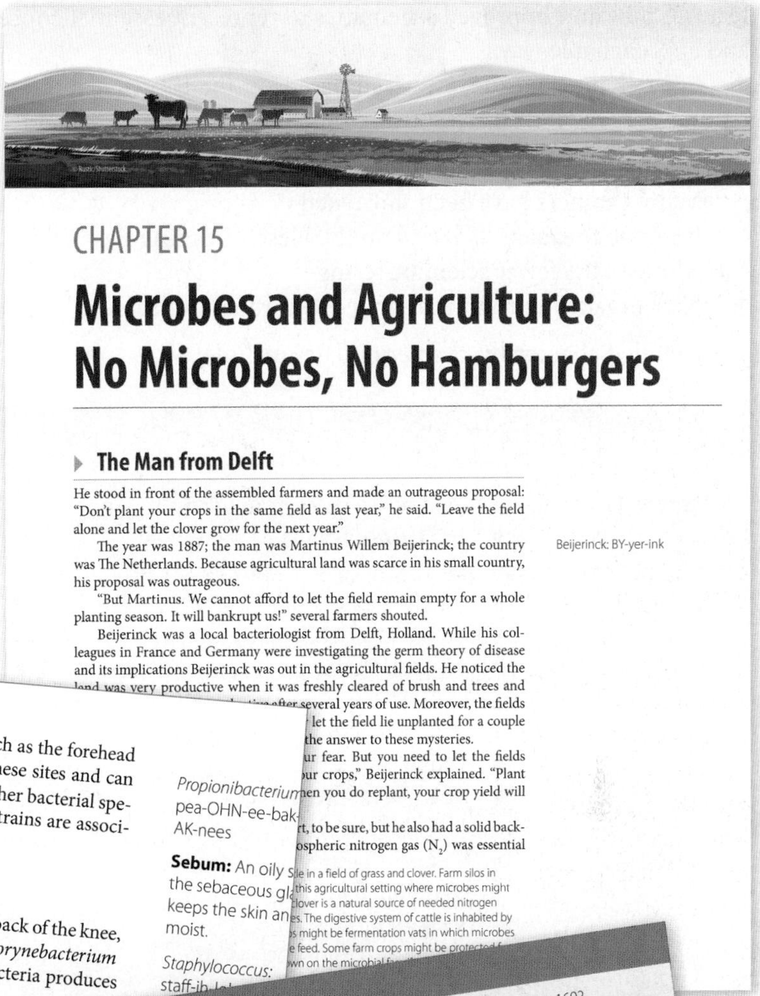
CHAPTER 15

Microbes and Agriculture: No Microbes, No Hamburgers

The Man from Delft

He stood in front of the assembled farmers and made an outrageous proposal: "Don't plant your crops in the same field as last year," he said. "Leave the field alone and let the clover grow for the next year."

The year was 1887; the man was Martinus Willem Beijerinck; the country was The Netherlands. Because agricultural land was scarce in his small country, his proposal was outrageous.

"But Martinus. We cannot afford to let the field remain empty for a whole planting season. It will bankrupt us!" several farmers shouted.

Beijerinck was a local bacteriologist from Delft, Holland. While his colleagues in France and Germany were investigating the germ theory of disease and its implications Beijerinck was out in the agricultural fields. He noticed the land was very productive when it was freshly cleared of brush and trees and

Beijerinck: BY-yer-ink

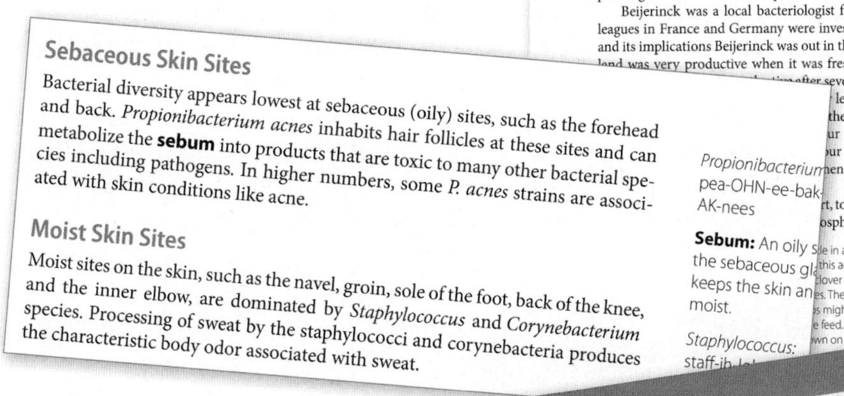
Sebaceous Skin Sites

Bacterial diversity appears lowest at sebaceous (oily) sites, such as the forehead and back. *Propionibacterium acnes* inhabits hair follicles at these sites and can metabolize the **sebum** into products that are toxic to many other bacterial species including pathogens. In higher numbers, some *P. acnes* strains are associated with skin conditions like acne.

Moist Skin Sites

Moist sites on the skin, such as the navel, groin, sole of the foot, back of the knee, and the inner elbow, are dominated by *Staphylococcus* and *Corynebacterium* species. Processing of sweat by the staphylococci and corynebacteria produces the characteristic body odor associated with sweat.

Propionibacterium: pea-OHN-ee-bak-AK-nees

Sebum: An oily

the sebaceous gl

keeps the skin an

moist.

Staphylococcus:

A CLOSER LOOK 5.4

"The Work of the Devil"

As an undergraduate, Linda Caporael was missing a critical history course for graduation. Little did she know that through this class she was about to provide a possible answer for one of the biggest mysteries of early American history: the cause of the Salem Witch Trials. These trials in 1692 led to the execution of 20 people who had been accused of being witches in Salem, Massachusetts (see **FIGURE A**).

Caporael, now a behavioral psychologist at New York's Rensselaer Polytechnic Institute, in preparation of a paper for her history course had read a book where the author could not explain the hallucinations among the people in Salem during the witchcraft trials. Caporael made a connection between the "Salem witches" and a French story of ergot poisoning in 1951. In Pont-Saint-Esprit, a small village in the south of France, more than 50 villagers went insane for a month after eating ergotized rye flour. Some people had fits of insomnia, others had hallucinogenic visions, and still others were reported to have fits of hysterical laughing or crying. In the end, three people died.

Caporael noticed a link between these bizarre

FIGURE A A witch trial in Massachusetts, 1692.

sensations on the skin, and a host of other symptoms—all of which, Linda Caporael found in the records of the Salem witchcraft trials. Ergot thrives in warm, damp, rainy springs and summers, which were the exact conditions Caporael says existed in Salem in 1691. In addition, parts of Salem village consisted of swampy meadows that would be the perfect environment for fungal growth. And because rye was the staple grain of Salem, it is not a stretch to suggest that the rye crop consumed in the winter of 1691–1692

What's New in This Edition?

Microbiology is a dynamic science and so this fifth edition reflects many of these advances as they pertain to humans, the environment, and society. The textbook also

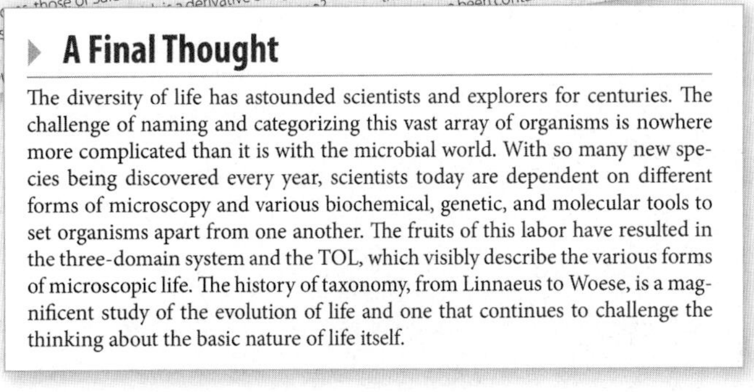
A Final Thought

The diversity of life has astounded scientists and explorers for centuries. The challenge of naming and categorizing this vast array of organisms is nowhere more complicated than it is with the microbial world. With so many new species being discovered every year, scientists today are dependent on different forms of microscopy and various biochemical, genetic, and molecular tools to set organisms apart from one another. The fruits of this labor have resulted in the three-domain system and the TOL, which visibly describe the various forms of microscopic life. The history of taxonomy, from Linnaeus to Woese, is a magnificent study of the evolution of life and one that continues to challenge the thinking about the basic nature of life itself.

features new and improved schematics to represent essential concepts and patterns of thought in a more accurate and approachable way.

Each chapter has been revised based on the most current and significant knowledge available. New to this edition:

- All chapters have been extensively edited and rewritten
- Many chapters have been shortened
- Improved readability based on the Flesch Reading Ease Test and the Flesch-Kincaid Grade Level Test
- Almost 10% fewer scientific terms
- New organism pronunciation guide throughout the text

Detailed chapter-by-chapter updates include:

Chapter	New Features and Content
Chapter 1	■ Chapter completely rewritten to emphasize the connections between microbes and society ■ 16 new figures
Chapter 2	■ Chapter restructured for information flow ■ New end-of-chapter questions ■ 14 new figures
Chapter 3	■ New end-of-chapter questions ■ 1 modified figure ■ New "A Closer Look"
Chapter 4	■ Former Chapter 5 in previous edition ■ New chapter opener ■ New end-of-chapter questions ■ 3 modified figures ■ 11 new figures
Chapter 5	■ Chapter 5 combines (and condenses) Chapters 7 and 8 ■ New end-of-chapter questions ■ 4 modified figures ■ 16 new figures
Chapter 6	■ Chapter placement re-sequenced for better understanding after the discussion of prokaryotes and eukaryotes (Chapters 4 and 5) ■ New chapter opener ■ New end-of-chapter questions ■ 5 modified figures
Chapter 7	■ Chapter rewritten for better clarity and understanding ■ New end-of-chapter questions ■ 6 new figures ■ 6 modified figures
Chapter 8	■ Former Chapter 4 in previous edition ■ New chapter introduction ■ New end-of-chapter questions ■ New A Closer Look ■ 7 new figures ■ 3 modified figures ■ 1 new table

Chapter 9	■ Former Chapter 10 in previous edition ■ 4 new figures ■ Extensive revision of 7 figures ■ New end-of-chapter questions
Chapter 10	■ Former Chapter 11 in previous edition ■ 6 new figures ■ Revisions to 5 figures ■ 1 new table ■ New end-of-chapter questions
Chapter 11	■ New chapter covering microbial social behavior through biofilms and the human microbiome. ■ 18 new figures ■ 2 new A Closer Look boxed features ■ 9 new end-of-chapter questions
Chapter 12	■ Former Chapter 13 in previous edition ■ 11 new figures ■ 1 new table ■ 8 new end-of-chapter questions
Chapter 13	■ Former Chapter 12 in previous edition ■ New chapter opener ■ 7 new figures ■ 3 new A Closer Look boxed features ■ New end-of-chapter questions and modifications
Chapter 14	■ New chapter opener ■ 6 new figures ■ Revisions to 7 figures ■ 1 new A Closer Look boxed feature ■ New end-of-chapter questions and modifications
Chapter 15	■ 14 new figures ■ Revisions to 2 figures ■ 1 new A Closer Look boxed feature ■ New end-of-chapter questions and modifications
Chapter 16	■ 14 new figures ■ New chapter opener ■ New end-of-chapter questions and modifications
Chapter 17	Chapter has been completely rewritten and immune system material is much more approachable for students. Vaccines and vaccine safety have been added to this chapter. ■ 9 new figures ■ Revisions to 5 figures ■ New end-of-chapter questions and modifications
Chapter 18	Information on viral diseases updated ■ 8 new figures ■ Information on Zika virus disease added
Chapter 19	■ 12 new figures ■ Revisions to 4 figures ■ 2 new A Closer Look boxed features ■ New end-of-chapter questions and modifications

Teaching Tools

The following resources are available to instructors to assist with course preparation:

- Sample Syllabus
- In-depth Lecture Presentations in PowerPoint format that feature art from the book
- Test Banks
- Image Bank of figures from the book
- Answers to the end-of-chapter questions from the book

Qualified instructors can obtain this material by contacting their Jones & Bartlett Learning Account Manager or by going to www.jblearning.com.

For instructors who offer a lab component with their course we have the following resources:

- *Laboratory Fundamentals of Microbiology, Eleventh Edition* is the completely modernized lab manual revolution built for today's learners, focusing on the student's experience in the lab. Lab activities in this new, full-color resource have been expanded and reorganized into new sections, such as "Laboratory Safety," "Population Growth" and "Immunology." Access to over 100 minutes of 34 instructor-chosen, high-quality videos of actual students performing the most common lab skills, procedures, and techniques is included with each new copy.
- Access to the *Fundamentals of Microbiology Laboratory Video Series* is also available for purchase as a subscription if your students do not require a printed lab manual.

Contact your Jones & Bartlett Learning Account Manager at go.jblearning.com/findmyrep for more information on bundle pricing if you plan to use more than one of these resources in your course.

Note to Instructors

Microbiology embodies the beautiful and ugly, the simple and complex, and the big and the small of life, and in this regard is a fascinating, useful, and approachable topic for non-science majors to learn. Because of the real value of microbes to the quality of human life, the environment, and society, as well as their ability to cause disease, I saw the need for a course in microbiology for the non-science major. *Microbes and Society, Fifth Edition* is written and designed to support such a course; I hope you find it to be a useful tool in your pedagogical mission.

Please feel free to email me (jeffpommervillephd@gmail.com) anytime with questions, comments, ideas, and/or suggestions that you believe could strengthen this text and make it an even more exciting learning experience for students.

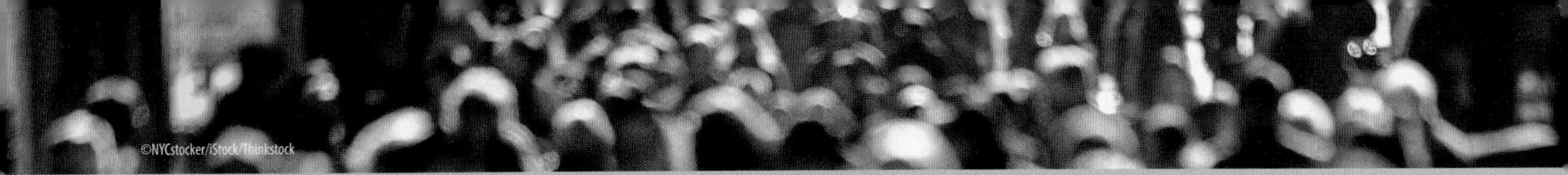

Acknowledgments

Putting together a new edition of a textbook always requires the input of a whole team and so I must recognize those at Jones & Bartlett Learning who helped put together this fifth edition of *Microbes and Society.* I want to thank my editors Matt Kane and Laura Pagluica for giving me the opportunity to revise and update *Alcamo's Microbes and Society, Fourth Edition,* and for their support and encouragement throughout the process; and Audrey Schwinn (new mom!) and Loren-Marie Durr for managing the process expertly, guiding the revisions, and keeping me on a timetable. I also want to thank Dan Stone for his expert work at the production end of the publication process, and to photo researcher John Rusk. Thanks also to Elizabeth Hamblin for copyediting the manuscript and to Jill Perri for proofreading the text. It takes a team of talented and dedicated professionals to put together a text and the *Microbes and Society* team is stellar. I salute everyone on the team and am honored to work with you.

The author and the publisher would also like to thank the following individuals for their services as reviewers of this edition.

Jeremiah J. Davie, Ph.D.
Jennifer B. Ellington, Ph.D., Belmont Abbey College
Julia K. Laverty, M.S., MLS (ASCP) ^{CM}*, Cecil College*
Joanna A. Miller, Ph.D., Drew University
Veronica L. Mittak, DHEd, MS, MPH, NY Chiropractic College and Cayuga Community College
Professor Natalie Osterhoudt, M.S. Zoology, Broward College
Christopher S. Registad. Ph.D., Concordia University Chicago

About the Author

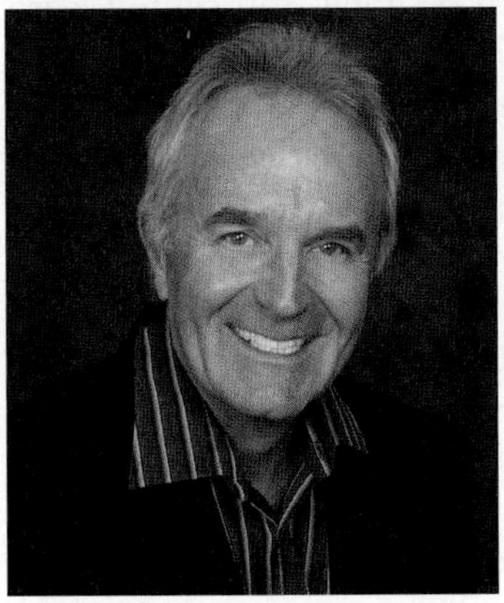

Today, I am a Emeritus professor of biology and microbiologist emeritus (at Glendale Community College in Glendale, AZ), researcher, and science educator. My plans did not start with that intent. While in high school in Santa Barbara, California, I wanted to play professional baseball, study the stars, and own a '66 Corvette. None of these desires would come true—my batting average was miserable (but I was a good defensive third baseman), I hated the astronomy correspondence course I took, and I have yet to buy that Corvette.

I found an interest in biology at Santa Barbara City College. After squeaking through college calculus, I transferred to the University of California at Santa Barbara (UCSB), where I received a BS in biology and stayed on to pursue a PhD degree studying cell communication and sexual pheromones in a water fungus in the lab of Ian Ross. After receiving my doctorate in cell and organismal biology, my graduation was written up in the local newspaper as a native son who was a "fungal sex biologist"—an image that was not lost on my three older brothers!

While in graduate school at UCSB, I rescued a secretary in distress from being licked to death by a German shepherd. Within a year, we were married (the secretary and I). When I finished my doctoral thesis, I spent several years as a postdoctoral fellow at the University of Georgia. Worried that I was involved in too many research projects, a faculty member told me something I will never forget. He said, "Jeff, it's when you can't think of a project or what to do that you need to worry." Well, I have never had to worry!

I then moved on to Texas A&M University, where I spent eight years in teaching and research—and telling Aggie jokes. Toward the end of this time, after publishing over 30 peer-reviewed papers in national and international research journals, I realized I had a real interest in teaching and education. Leaving the sex biologist career behind, I headed farther west to Arizona to join the biology faculty at Glendale Community College, where I continued to teach introductory biology and microbiology until my retirement in 2018.

I have been lucky to be part of several educational research projects and have been honored, two of my colleagues, with a Team Innovation of the Year Award by the League of Innovation in the Community Colleges. In 2000, I became project director and lead principal investigator for a National Science Foundation grant to improve student outcomes in science through changes in curriculum and pedagogy. I had a fascinating three years coordinating more than 60 science faculty members (who at times were harder to manage than students) in designing and field testing 18 interdisciplinary science units. This culminated with me being honored in 2003 with the Gustav Ohaus Award (College Division) for Innovations in Science Teaching from the National Science Teachers Association.

I was the Perspectives Editor for the *Journal of Microbiology* and *Biology Education*, the science education research journal of the American Society for Microbiology (ASM) and in 2004 was co-chair for the ASM Conference for Undergraduate Educators. From 2006 to 2007, I was the chair of Undergraduate Education Division of ASM. In 2006, I was selected as one of four outstanding instructors at Glendale Community College. The culmination of my teaching career came in 2008 when I was recognized nationally by being awarded the Carski Foundation Distinguished Undergraduate Teaching Award for distinguished teaching of microbiology to undergraduate students and encouraging them to subsequent achievement.

I mention all this not to impress but to show how the road of life sometimes offers opportunities in unexpected and unplanned ways. The key though is keeping your "hands on the wheel and your eyes on the prize"—then unlimited opportunities will come your way. And, hey, who knows—maybe that '66 Corvette could be in my garage yet.

Dedication

With my retirement from active teaching in 2018, I wish to recognize all the students—all 10,000+— that I have taught over the last few decades. You have kept me young, on my toes, and appreciative of your choosing me as your instructor. I honor you all and wish you the best in life and your career.

To the Student—Study and Read Smart

When I was an undergraduate student, I hardly ever read the "To the Student" section (if indeed one existed) in my textbooks because the section rarely contained any information of importance.

This one does, so please read on.

In college, I was a mediocre student until my junior year. Why? Mainly because I did not know how to study properly, and, important here, I did not know how to read a textbook effectively. My textbooks were filled with underlined sentences (highlighters hadn't been invented yet!) without any plan on how I would use this "emphasized" information. In fact, most textbooks *assume* you know how to read a textbook properly. It is not like reading a fictional novel.

Reading a textbook is difficult if you are not properly prepared. So that you can take advantage of what I learned as a student and have learned from instructing thousands of students, I have worked hard to make this text user friendly with a reading style that is not threatening or complicated. Still, there is a substantial amount of information to learn and understand, so having the appropriate reading and comprehension skills is critical. Therefore, I encourage you to spend 20 minutes reading this section, as I am going to give you several tips and suggestions for acquiring those skills. Let me show you how to be an active reader.

Be a Prepared Reader

Before you jump into reading a section of a chapter in this text, prepare yourself by finding the place and time and having the tools for study.

Place. Where are you right now as you read these lines? Are you in a quiet library or at home? If at home, are there any distractions, such as loud music, a blaring television, or screaming kids? Is the lighting adequate to read? Are you sitting at a desk or lounging on the living room sofa? Get where I am going? When you read for an educational purpose—that is, to learn and understand something—you need to maximize the environment for reading. Yes, it should be comfortable but not to the point that you will doze off.

Time. All of us have different times during the day when we perform some skill the best, be it exercising or studying. The last thing you want to do is read when you are tired or simply not "in the zone" for the job that needs to be done. You cannot learn and understand the information if you fall asleep or lack a positive attitude. I have kept the chapters in this text to about the same length, so you can estimate the time necessary for each and plan your reading accordingly. If you have done your preliminary survey of the chapter or chapter section, you can determine about how much time you will need. If 40 minutes is needed to read—and comprehend (see below)—a section of a chapter, find the place and time that will give you 40 minutes of uninterrupted study.

Brain research suggests that most people's brains cannot spend more than 45 minutes in concentrated, technical reading. Therefore, I have avoided lengthy presentations and instead have focused on smaller sections, each with its own heading. These should accommodate shorter reading periods.

Reading Tools. Lastly, as you read this, what study tools do you have at your side? Do you have a highlighter or pen for emphasizing or underlining important words or phrases? Notice, the text has wide margins, which gives you the space to make notes or to indicate something that needs further clarification. Do you have a pencil or pen handy to make these notes? Lastly, some students find having a ruler is useful to prevent your eyes from wandering on the page and to read each line without distraction.

Be an Explorer Before You Read

When you sit down to read a section of a chapter, do some preliminary exploring. Look at the section head and subheadings to get an idea of what is discussed. Preview any diagrams, figures, tables, graphs, or other visuals used. They give you a better idea of what is going to occur. We have used a good deal of space in the text for these features, so use them to your advantage. They will help you learn the written information and comprehend its meaning. Do not try to understand all the visuals, but try to generate a mental "big picture" of what is to come. Familiarize yourself with any symbols or technical jargon that might be used in the visuals.

Be a Detective as You Read

Reading a section of a textbook requires you to discover the important information (the terms and concepts) from the forest of words on the page. So, the first thing to do is read the complete paragraph. When you have determined the main ideas, highlight or underline them. However, I have seen students highlighting the entire paragraph in yellow, including every *a, the,* and *and.* This is an example of highlighting before knowing what is important. So, I have helped you out somewhat. Important terms and concepts in the textbook are in **bold face** followed by the definition (or the definition might be in the page margin). So, in many cases, you should only need to highlight or underline essential ideas and key phrases—not complete sentences. By the way, the important microbiological terms and major concepts also are in the **Glossary** at the back of the text.

What if a paragraph or section has no boldfaced words? How do you find what is important here? From an English course, you may know that often the most important information is mentioned first in the paragraph. If it is followed by one or more examples, then you can backtrack and know what was important in the paragraph.

Say It in Your Own Words

Brain research has shown that everyone can only hold so much information in short-term memory. If you try to hold more, then something else needs to be removed—sort of like having a full computer disk. So that you do not lose any of this important information, you need to transfer it to long-term memory—to the hard drive if you will. In reading and studying, this means retaining the term or concept; so, write it out in your notebook *using your own words.* Memorizing a term does not mean you have learned the term or understood the concept. By actively writing it out in your own words, you are forced to think and actively interact with the information. This repetition reinforces your learning.

Be a Patient Student

In textbooks, you cannot read at the speed that you read your text messages, email, or a magazine story. There are unfamiliar details to be learned and understood in a textbook—and this requires being a patient, slower reader. If you are like me and not a fast reader to begin with, it may be an advantage in your learning process. Identifying the important information from a textbook chapter requires you to *slow down* your reading speed. Speed-reading is of no value here.

Know the What, Why, and How

Have you ever read something only to say, "I have no idea what I read!" As I've already mentioned, reading a microbiology text is not the same as reading *Sports Illustrated* or *People* magazine. In these entertainment magazines, you read passively for leisure or perhaps amusement. In *Microbes and Society, Fifth Edition,* or any other textbook, you must read actively for learning and understanding—that is, for *comprehension.* This can quickly lead to boredom unless you engage your brain as you read—that is, be an active reader. Do this by knowing the what, why, and how of your reading.

- What is the general topic or idea being discussed? This often is easy to determine because the section heading might tell you. If not, then it will appear in the first sentence or beginning part of the paragraph.
- *Why* is this information important? If I have done my job, the text section will tell you why it is important, or the examples provided will drive the importance home. These surrounding clues further explain why the main idea was important.
- *How* do I "mine" the information presented? This was discussed under being a detective.

A Marked-Up Reading Example

So, let's put words into action. A passage from *Microbes and Society, Fifth Edition* is included on the next page. I have marked up (highlighted) the passage as if I were a student reading it for the first time.

It uses many of the hints and suggestions I have provided. Remember, it is important to read the passage slowly, and concentrate on the main idea (concept) and the special terms that apply.

Have a Debriefing Strategy

After reading the material, be ready to debrief. Verbally summarize what you have learned. This will start moving the short-term information into the long-term memory storage—that is, *retention.* Any notes you made concerning confusing material should be discussed as soon as possible with your instructor. For microbiology, allow time to draw out diagrams. Again, repetition makes for easier learning and better retention.

In many professions, such as sports or the theater, the name of the game is practice, practice, practice. The hints and suggestions I have given you form a skill that requires practice to achieve and use efficiently. Change will not happen overnight; perseverance and willingness though will pay off with practice. You might also check with your college or university academic resource center or center for learning. These folks will have more ways to help you to read a textbook better and to study well.

Send Me a Note

In closing, I would like to invite you to email me. Let me know what is good about this textbook so I can build on it and what may need improvement, so I can revise it. Don't be shy; let me know your thoughts. Also, I would be pleased to hear about any news of microbiology in your community, and I'd be happy to help you locate any information not covered in the text.

I wish you great success in your microbiology course. Welcome! Let's now plunge into the wonderful and often awesome world of microorganisms.

—Dr. P

Email: jeffpommervillephd@gmail.com

Flagella

Many bacterial cells contain one or more hair-like projections called **flagella** (sing. **flagellum**) that are anchored in the cell wall and cell membrane. Different bacterial species have either a single flagellum, a tuft of flagella at one end of the cell, or flagella covering the entire cell surface (**FIGURE 4.8**).

The bacterial flagella are long, rigid protein filaments. These filaments bundle together and rotate like a propeller, pushing a bacterial cell forward. In its watery surroundings, the bacterial flagella help the cells swim toward nutrients or away from toxic chemicals. Using flagella, a bacterium such as *Escherichia coli* can travel about 2,000 body lengths in an hour. This is equivalent to 5-foot 10-inch human walking 2.25 miles in an hour.

Escherichia coli: esh-er-EE-key-ah KOH-lee

Pili

Other appendages projecting out from the cell surface are **pili** (sing. **pilus**). Pili are short, rigid protein rods about 1 μm in length and about 7 nm thick (**FIGURE 4.9**). They help bacterial cells attach to tissues or other surfaces. When causing infection, pathogens having pili can attach to tissues, helping start the infection process. **A CLOSER LOOK 4.2** provides an "upsetting" example.

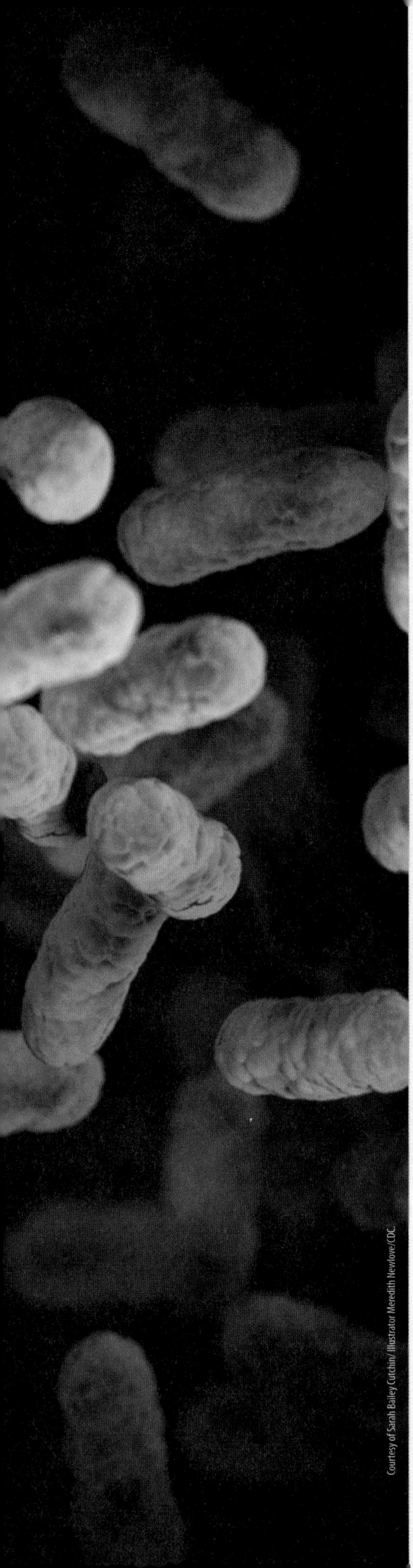

Courtesy of Sarah Bailey Cutchin/ Illustrator Meredith Newlove/CDC.

PART I

The Microbial World

Until the 1980s, some homes in Great Britain were still visited regularly by milkmen who delivered bottles of fresh, pasteurized milk on the doorstep. Unfortunately, magpies, crows, and other birds arrived soon afterward. They used their strong beaks to peck through the foil caps covering the bottles and helped themselves to the cream at the top of the bottles (see photo to left). In doing so, the birds often transmitted bacteria capable of causing intestinal illness, and some unsuspecting families soon suffered the discomfort of diarrhea, cramps, and nausea. Nevertheless, we should not blame all microorganisms for the ills caused by a relatively small number. In fact, most microorganisms are responsible for an unimaginable number of activities essential to life's survival on Earth.

In Part I of *Microbes and Society*, we get to know the microbes close up and personal. In Chapter 1, we become familiar with some of the ways that microbes affect our world and society. We explore in Chapter 2 how their world was discovered and see how the microbes fit into the scheme of the living world. In Chapter 3, we look at the building blocks of life that are used to construct living cells. Chapters 4 through 6 focus on the structure of microorganisms and viruses and survey their remarkable diversity. In the next four chapters, we examine how microorganisms grow (Chapter 7), consider their extraordinary genetics (Chapters 8 and 9), and explore some methods by which they can be controlled (Chapter 10). Chapter 11 brings the microbial world full circle by considering some remarkable social behaviors exhibited by microorganisms. The pages of Part I contain some eye-opening information about the microbial world that is unfamiliar to, and unappreciated by, most of your friends and family—and unfortunately, often by much of society.

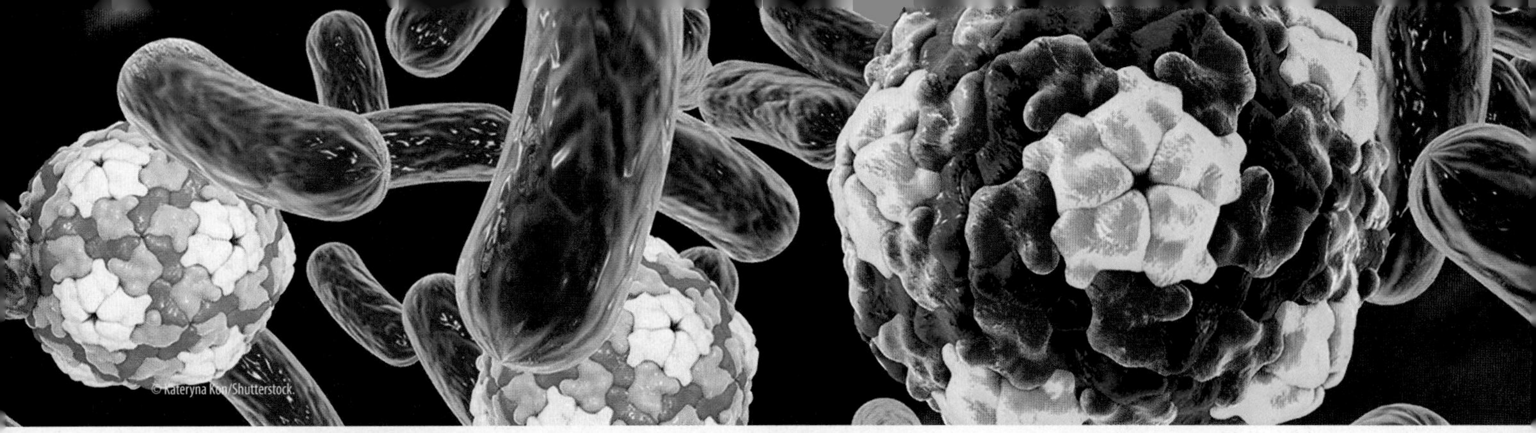

CHAPTER 1

The Microbial World: An Introduction

Good Morning Microbes!

Every day, we have countless contacts with microbes. These interactions begin each morning.

It's 6 A.M. and the wakeup alarm goes off on your cell phone. It's Monday morning and unfortunately, it's time to get up and get ready for class. You shut off the alarm, but while holding the phone, you decide to check what text messages you received overnight.

- **Microbial contact:** Scientists at the University of Arizona report that cell phones carry 10 times more bacteria than most toilet seats! So, unless you regularly clean your cell phone, and as few as 17% do, the phone is crawling with bacteria; as many as 4,000 per cm^2. The good news is that few, if any, of these bacteria are **pathogens**. Most came from your own skin surface or mouth. They shouldn't make you sick.

  Stumbling out of bed, you decide to take a shower before breakfast. You turn on the faucet, warm up the water, and step into the shower stall.

- **Microbial contact:** Human skin is covered with microorganisms. These primarily are bacteria and fungi, many of which expel volatile chemicals responsible for underarm, foot, and other body odors (**FIGURE 1.1**). Showering does remove a large number, so you should shower well. However, don't overly scrub your skin because many of these skin-dwellers help to fend off potential pathogens you might contact during the day.

  Stepping out of the shower, you dry yourself off. You apply an underarm deodorant or antiperspirant to help hide the increasing body odor that can arise as the skin microbes multiply during the day.

Pathogen: A disease-causing agent.

CHAPTER 1 OPENER A false-color illustration of bacteria (purple rods) and viruses (multicolored spheres).

FIGURE 1.1 Foot Odor. Smelly feet are due to the bacteria (gray and red structures) normally found on the feet that produce smelly acid products (blue and orange spheres).

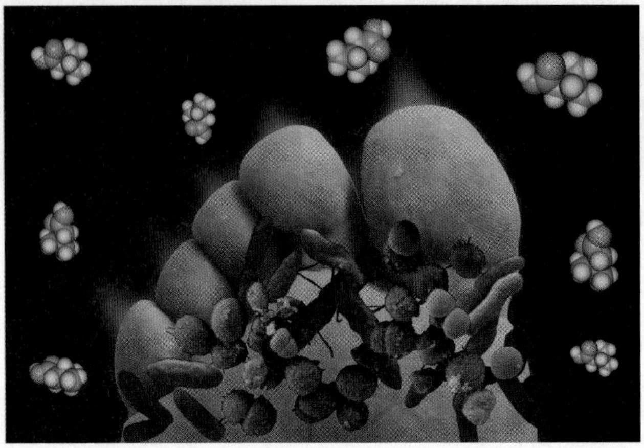

After getting dressed, you take the dog for a walk around the block. When you return, you head to the kitchen to have breakfast. Running late, you decide to have cereal, toast, and coffee.

- **Microbial contact:** Ah, the smell of freshly brewed coffee! Did you know that part of the processing of coffee beans involves placing the beans, fungal yeasts, and bacteria in a **fermentation** tank to degrade some of the exterior layers (pulp) on the beans? During this process, the microbes also produce acids that add flavor, prevent the growth of fungal molds, and keep the beans from sprouting.

Fermentation: In food processing, the process of converting carbohydrates to alcohol or organic acids using microorganisms.

In the refrigerator, you grab the milk carton for your cereal but quickly notice it has a foul smell—it has spoiled!

- **Microbial contact:** Milk inevitably spoils because the dairy production process doesn't eliminate all bacteria in the milk. It's these tough, but not harmful, survivors that eventually produce the acids that give the milk a strong smell and sour taste.

Time is running short for the 9 A.M. class, so you grab a yogurt and put a piece of bread in the toaster.

- **Microbial contact:** Yogurt often contains "active yogurt cultures." These cultures are populations of living bacteria that, through a milk fermentation process, provide texture and a tart flavor to the yogurt (**FIGURE 1.2**). In the breadmaking process, yeasts introduced into the dough undergo another fermentation process that produces carbon dioxide (CO_2) gas. The released gas provides the "push" that causes the bread dough to rise before baking.

Finishing an abbreviated breakfast, you race off to the bathroom to floss and brush your teeth.

- **Microbial contact:** Several oral hygiene products help maintain a healthy **oral cavity** (**FIGURE 1.3A**). Flossing mechanically helps remove much of the food, bacteria, and plaque on and between the teeth. Many toothpastes contain the "silica skeletons" of dead microbes called diatoms (**FIGURE 1.3B**). Brushing with these abrasive skeletons gently removes much of the plaque on the surfaces of the teeth.

Oral cavity: The mouth.

In addition, the plaque itself is a film of different bacterial types, some of which can produce acids that, over many months, slowly dissolve the enamel on the teeth if you do not maintain good daily oral hygiene and

FIGURE 1.2 Part of Breakfast.
(A) Yogurt and some milk products are the result of microbial metabolism.

© Stock-Asso/Shutterstock

(B) Yogurt with "active yogurt cultures" contains millions of bacteria as suggested in this 3D illustration of rod-shaped bacterial cells. (Bar = 1 μm.)

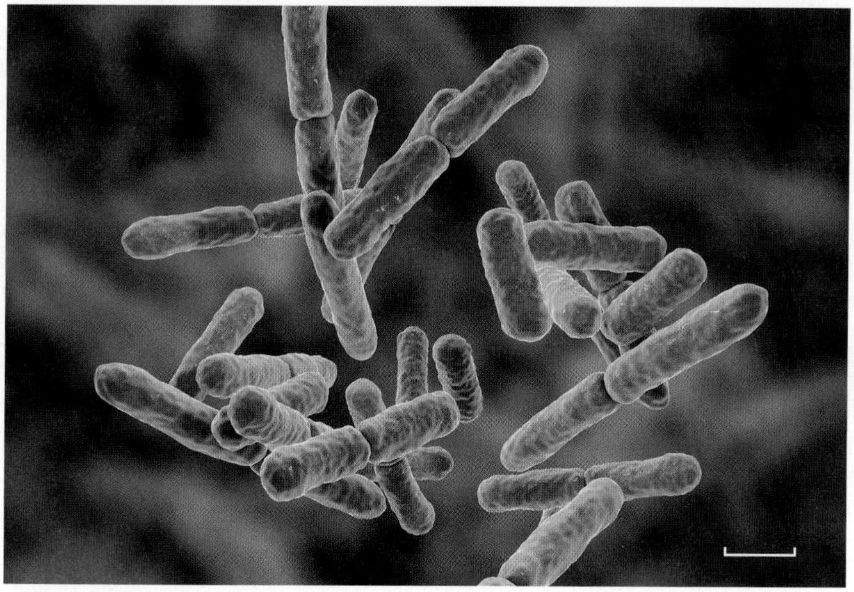

© Kateryna Kon/Shutterstock.

have regular dental checkups. That is why the dentist usually wants to see you every 6 months. Over prolonged periods of time, poor oral hygiene can lead to cavities or even more serious gum diseases.

A quick rinse with a mouthwash (swish, swish, spit!) and you will be off to class. Mouthwash (oral rinses) can reduce more plaque from your teeth and bacteria from your mouth—and give a fresher breath, as those oral bacteria can give off waste products that contribute to bad breath.

Whoops, before you leave, you give your pooch a big hug. In turn, he gives you a big lick on the mouth!

FIGURE 1.3 Products for Good Oral Hygiene.

(A) Everyday products like mouthwash, dental floss, and toothpaste are designed to remove many of the microbes normally found on the teeth and in the mouth.

(B) This light microscope image of diatoms shows the scaly skeletons that act as a mild abrasive to remove dental plaque from tooth surfaces. (Bar = 20 µm.)

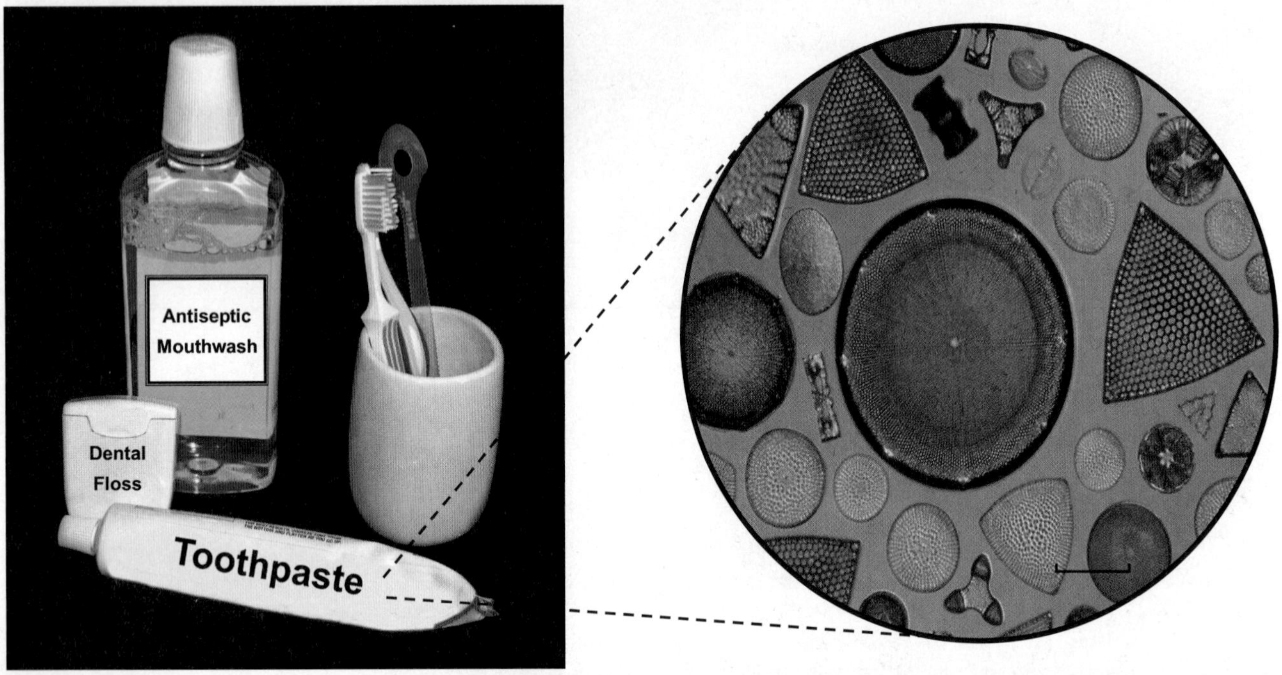

Courtesy of Dr. Jeffrey Pommerville.

© Dr. Norbert Lange/Shutterstock.

- **Microbial contact:** Like humans, dogs have a large variety of microbes cohabitating in their bodies. However, only about 15% of these microbes are common to humans. Consequently, that slobbery lick can deposit untold millions of these foreign bacteria on your skin surface and they can remain there for hours if not washed away.

Your morning routine represents just the beginning of your day's contact and interaction with the microbial world. Incredibly, this is something that occurs each and every day, from birth until after death.

LOOKING AHEAD

After reading and completing this chapter, you will be able to:

1.1 Identify connections between microbiology and society.
1.2 Explain why many microbes are protectors of the planet.
1.3 Summarize the importance of microbes in the human body.
1.4 Describe the role of microbes in infection and disease.
1.5 Provide examples illustrating the influence of microbes to society.
1.6 Discuss some challenges facing microbiology today.

1.1 Microbiology and Society: A Connected World

We share our world with thousands of species of plants and animals we can see with the naked eye. We also occupy this space with many thousands more

FIGURE 1.4 A Microbial Menagerie. This artist's drawing illustrates the types of microbes present in a pinch of rich soil or in a drop of seawater.

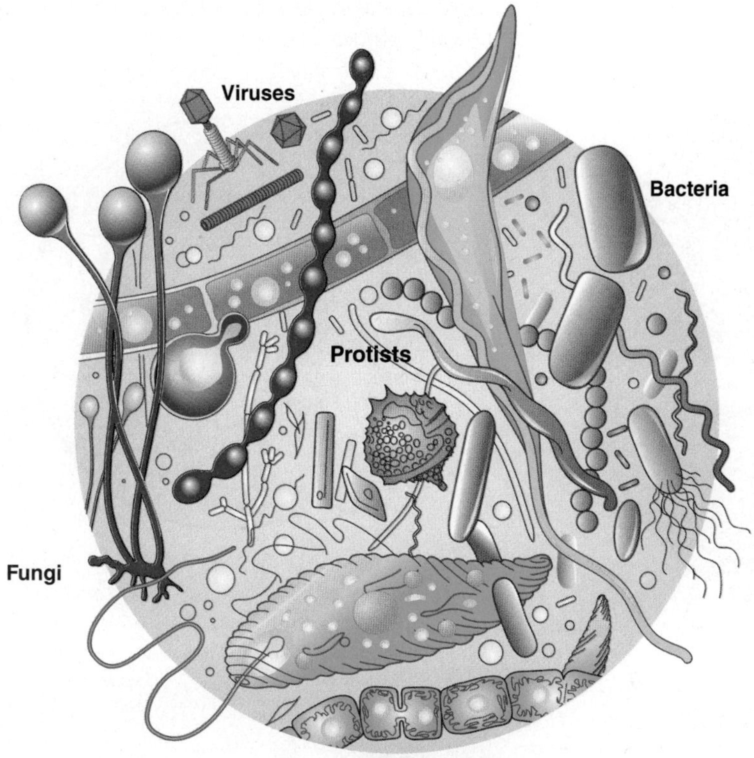

organisms we cannot see. **Microbiology** is the science discipline that studies these invisible agents called **microorganisms** (often simply called **microbes**). Today, it is known that they comprise the bacteria, protozoa, algae, fungi, and viruses (**FIGURE 1.4**). To makes these tiny agents easier to comprehend visually, throughout the text you will see drawings of microbes or view images of them taken with a microscope.

Microbes are Earth's oldest and most successful inhabitants. As a group, they have been on the planet for almost 4 billion years. Microbiologists estimate that there are more than one trillion different types of microbes alive today. In numbers, this amounts to more than 5×10^{30} (5 million trillion trillion) cells, which is more cells than stars in the known universe. During their long evolution, they have had the time and opportunity to adapt and grow in most all earthly environments. Collectively, they have transformed the face of the planet into the form we see today, and they have come to play important roles in the very survival of all life.

Microbes connect the living and nonliving world. To better understand some of those roles, let's uncover why microbes matter.

1.2 Why Microbes Matter: Protecting the Planet

The microbial world is diverse with immense numbers of microbes making their home in the oceans, soils, and atmosphere. Often such microbial communities are referred to as **microbiomes**.

Microbiome: A specific environment characterized by a distinctive microbial community and its collective genetic material.

The Ocean Microbiome

The oceans and seas, which cover more than 70% of the planet, represent the foundation that maintains our planet in a habitable condition. A critical factor

FIGURE 1.5 Marine Bacteria and Viruses.

(A) In this light microscope image, each large dot is one bacterial cell, and each small dot is a virus normally found in one drop of seawater. (Bar = 10 μm.)

(B) This 3D illustration shows what these bacteria (green and blue) and viruses (reddish-brown) might look like at higher magnification. (Bar = 2 μm.)

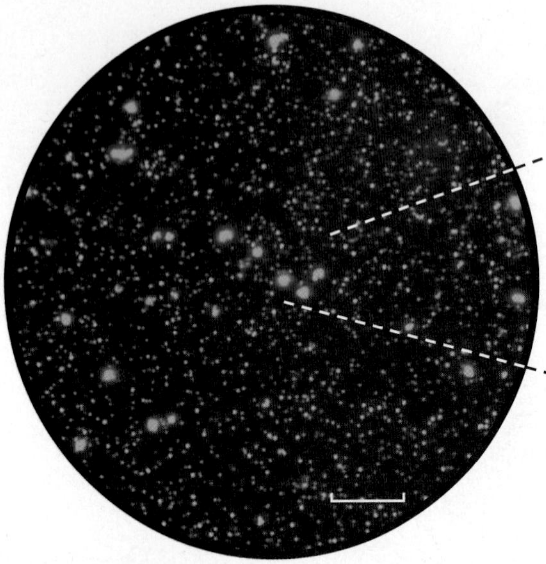

Courtesy of Rachel Parsons, Bermuda Institute of Ocean Science.

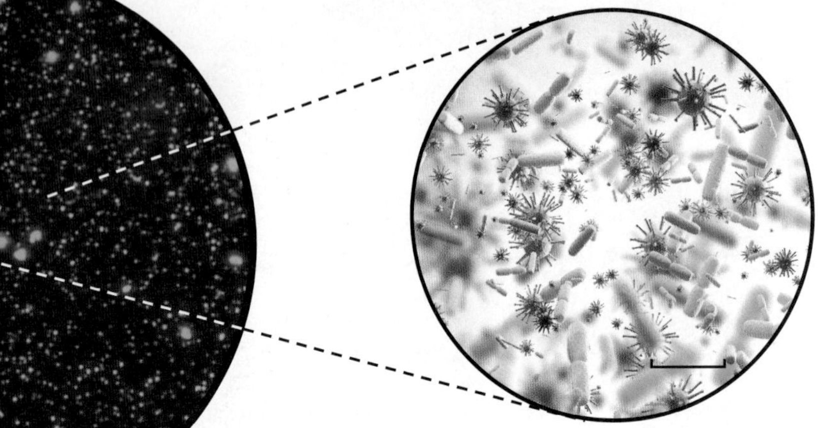

© ImageFlow/Shutterstock.

Ecosystem: A geographic area where plants, animals, and microbes, as well as weather and landscape, work together as a system.

Biomass: Total weight of all organisms in a defined area or environment.

Phytoplankton: The photosynthetic bacteria and algae found in the oceans and fresh water.

Decomposer: A microbe that breaks down dead plant and animal matter.

in safeguarding this **ecosystem** is the marine microbes. They number some 3×10^{30} (3 million trillion trillion) individuals and represent 98% of the ocean's **biomass** (**FIGURE 1.5**). They exist in highly structured and interactive communities that inhabit the ocean surface as well as the extremely cold polar regions and the hot, volcanic thermal vents on the dark seafloor.

Every day, many of these marine dwellers play fundamental roles in natural processes. They:

- Establish the base of the food web (i.e., who eats whom) for all higher-level marine organisms (**FIGURE 1.6A**). The **phytoplankton** are a major food source for all other marine organisms. By floating near the ocean surface, they perform **photosynthesis**, which converts light (solar) energy into chemical energy in the form of sugars. Some members of the food web consume the phytoplankton directly as a food source. Other larger animals (the consumers), including fish and ocean mammals, eat other animals that ate phytoplankton. Therefore, phytoplankton are called the **primary producers** because ultimately, they are the food and energy source for all the food web. Without the primary producers, the food web would collapse.
- Provide up to 50% of the oxygen gas we breath and many other organisms use to stay alive. Again, the photosynthetic process by the phytoplankton produces a large amount of this gas (**FIGURE 1.6B**).
- Control atmospheric events and cloud formation (**FIGURE 1.6C**). In response to heat, phytoplankton release chemical aerosols into the atmosphere. These aerosols then promote cloud formation, which helps cool the planet by reflecting some of the sun's heat. Thus, phytoplankton are key protectors that influence Earth's climate.
- Break down 50% of all dead plant and animal matter. As **decomposers**, marine microbes operate as the engines that recycle essential nutrients,

FIGURE 1.6 Phytoplankton, Trophic Levels, and Cloud Formation. Phytoplankton, which are primary producers, play several roles on a global scale. **(A)** They are the ultimate source of food for all other creatures composing the food web. **(B)** They generate about 50% of the oxygen gas (O_2) in the atmosphere. **(C)** Phytoplankton interactions with the atmosphere produce cloud-forming chemicals (aerosols) that help in cloud formation and climate control.

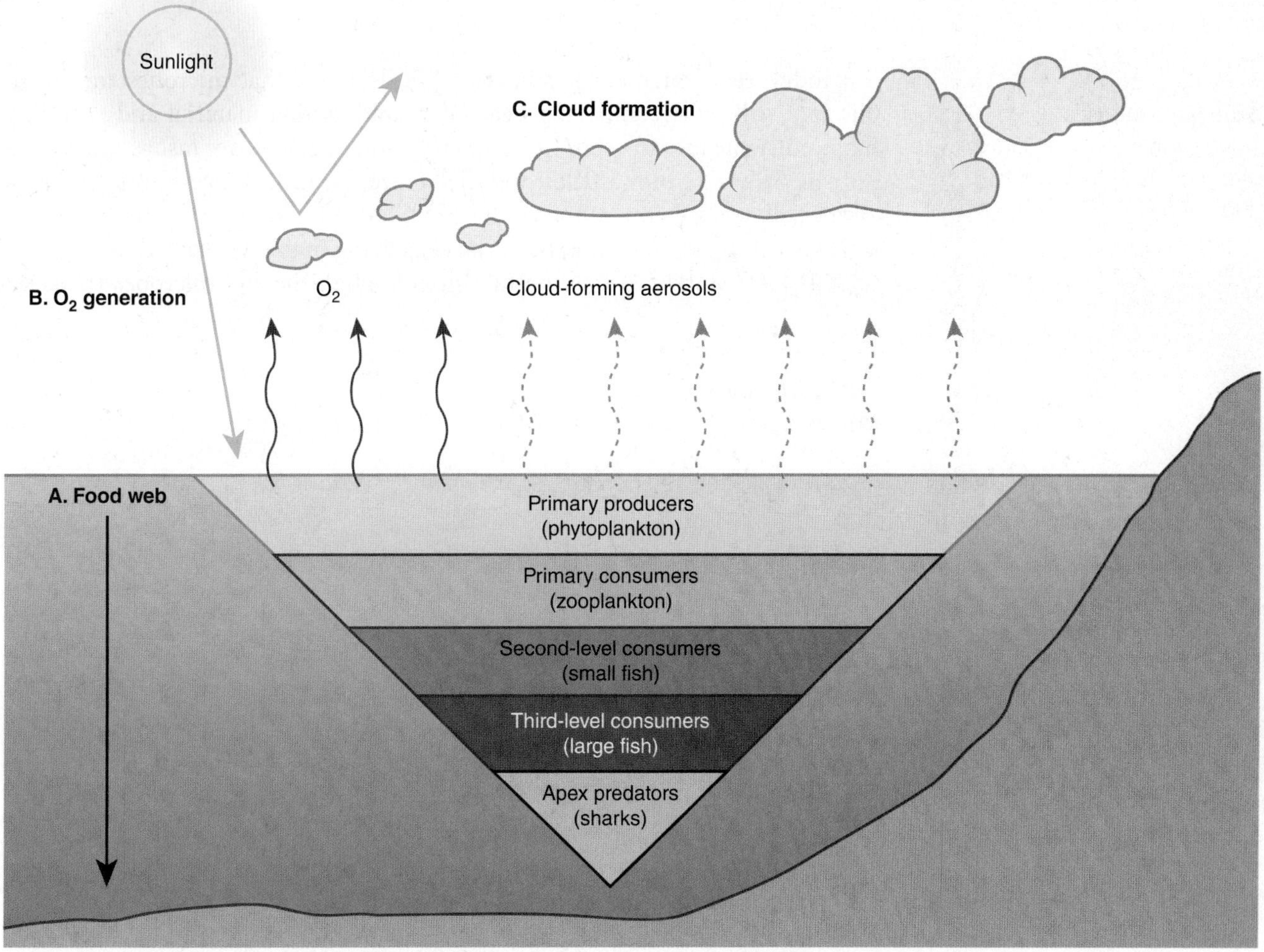

such as carbon, nitrogen, and phosphorus. The decay process carried out by marine microbes underpins the daily activities of marine organisms.

Viruses are the most abundant infectious agents on Earth. In the oceans, they outnumber the bacteria 10 to 1 (see Figure 1.5A). As such, many scientists believe viruses represent the most dominant agents affecting life on Earth. For example, by infecting and destroying members of the phytoplankton, they influence the global nutrient recycling mentioned above. In fact, each year infections by marine viruses are responsible for releasing and recycling some 136 billion kilograms (150 billion tons) of carbon, which then are used by other organisms for growth and reproduction.

Think of it this way—as indicators of ocean change and nutrient recycling, microbes are the protectors of not only the ocean's health but the entire planet's health.

The Soil Microbiome

In the early 1500s, Leonardo Da Vinci remarked:

> *"We know more about the movement of celestial bodies than about the soil underfoot."*

Soil: A mixture of organic matter, minerals, gases, liquids, and organisms that together support life.

Today, that statement is still true. We do know that microbes transform dirt into **soil** by decomposing dead plant and animal material and recycling those nutrients into the soil. Consequently, soil microbes are just as impressive in their daily activities as their marine comrades. In fact, every time you step on the soil, you step on billions of bacteria. Indeed, a kilogram of moist garden soil conceals up to 10 trillion bacterial cells living in the waterfilled pore spaces (**FIGURE 1.7**). Moreover, like the marine microbes, the soil microbes (bacteria

FIGURE 1.7 Soil Microbes.
(A) A sample of good garden soil contains millions of microbes.

(B) If a few grains of soil are placed in a culture dish, aggregations of bacteria and fungi (brownish clusters) grow.

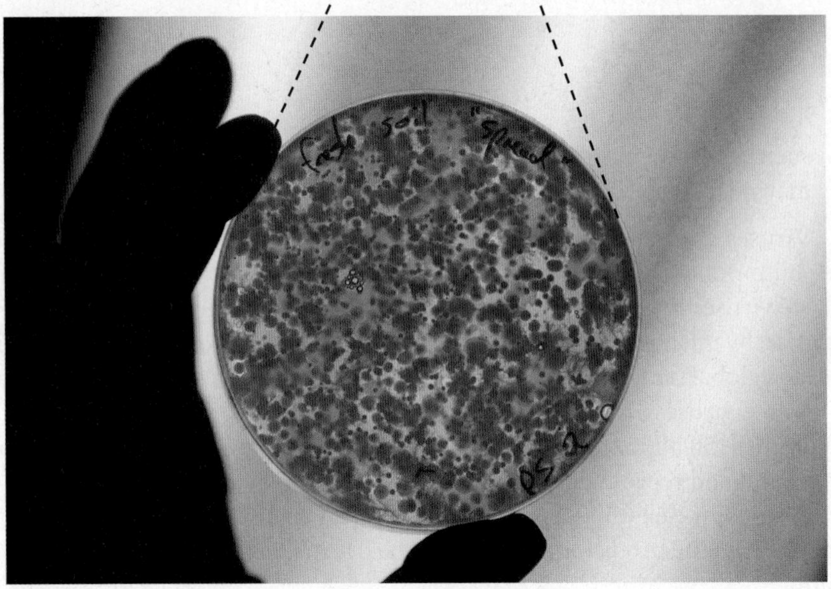

and fungi) can be found in every imaginable place, from the tops of the highest mountains to the deepest caves.

The diverse community of soil microbes also are responsible for many daily activities essential to life. They:

- Produce 70% of today's antibiotics that are used to fight infectious diseases.
- Perform all nutrient recycling occurring in the soil. This recycling by bacteria and fungi maintains soil fertility, which provides the energy and nutrients to directly support crops and to indirectly support livestock.
- Promote crop productivity by directly supplying plants with additional water and nutrients.
- Recycle dead plant and animal material in the soil and release carbon dioxide to the atmosphere.

Think of it this way—microbes are protectors of soil's health, and, as such, they are helping to feed the world.

The Microbiome of the Atmosphere

Many bacteria, mold spores, and viruses also are found in the atmosphere. Most are swept into the air in sea spray and from plants and the soil by winds. However, they do not exist as free units, but rather they are attached to marine materials or soil-dust found in the air near marine and land surfaces. Although many microbes at high altitudes are killed by the ultraviolet radiation from the sun, more than 315 different types of bacteria survive in air masses high above Earth's surface (**FIGURE 1.8**).

The microbes in the atmosphere have been less well studied than those in the oceans and soils. Still, scientists believe the atmospheric communities of microbes also are part of the protectors of the planet. They:

- Help control the weather and influence weather patterns around the world.
- Form water vapor (clouds) to help cool the planet.
- Assist, along with dust particles, in the formation of raindrops and snowflakes.
- Adjust gas concentrations in the atmosphere that influence global temperature balance.

Today, microbiologists believe they have discovered only about 2% of all microbes. Besides the oceans, soil, and air, where else are microbes found?

The Microbiome in Earth's Crust

One surprising location of microbes is kilometers beneath our feet. Recently, scientists have drilled 2.4 kilometers (1.5 miles) into Earth's crust. In the solid, but porous rock, they have discovered extraordinary microbial communities that are sealed off from the rest of the world. These microbes, which seem to survive in these inhospitable conditions, comprise more microbial biomass than that of all the microbes in oceans, soils, and air combined. Scientists believe these "rock eaters" also are involved with the daily recycling of minerals and undoubtedly influence the health of our planet in yet unrecognized ways.

In summary, microbes are the force and protectors that keep our planet operating. Many scientific experts suggest that our future (and our society)

FIGURE 1.8 Microbes Are Found in Clouds.
(A) Researchers have collected air samples from more than 10 kilometers (6.2 miles) over both land and sea.

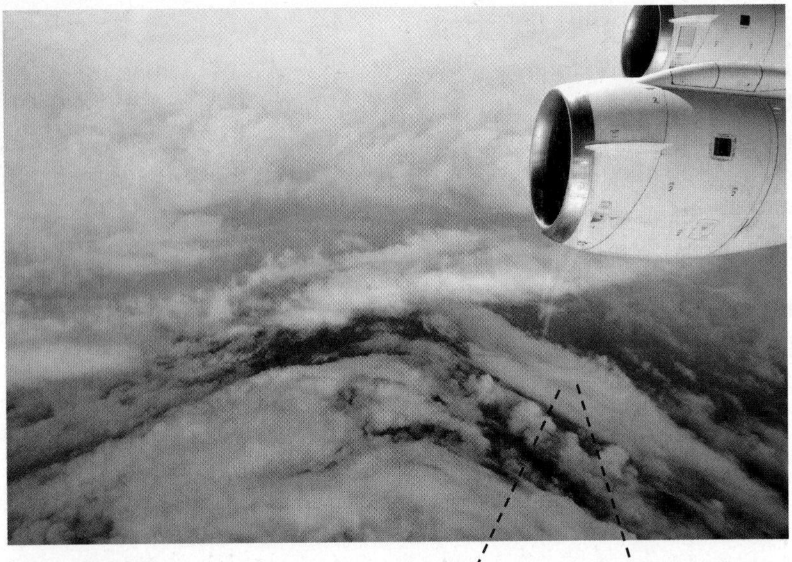

Courtesy of Jane Peterson/NASA.

(B) If an air sample is exposed on a culture dish, many different species of bacteria (small dots) and fungi (larger fuzzy circles) appear.

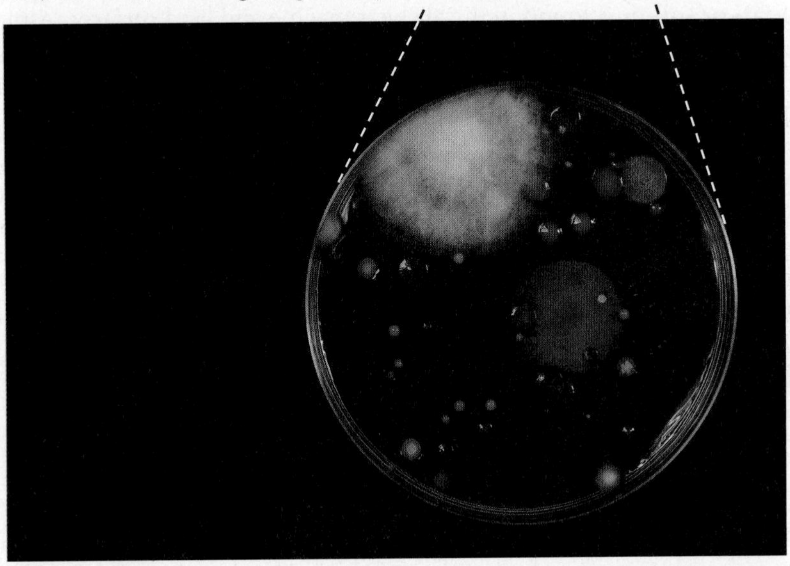

© Khamkhlai Thanet/Shutterstock.

depends on the ability to comprehend the unseen microbial world and understand how it works. As Louis Pasteur, one of the fathers of microbiology who we soon will read about, once stated:

> *"Life [plant and animal] would not long remain possible in the absence of microbes."*

1.3 Why Microbes Matter: Helping Maintain Human Health

The presence and need for microbes can be found much closer to home. Microbiomes cohabit the bodies of all plants and animals. In the case of animals, every animal, from termites to bees, cows, and humans, has a set of microbial communities associated with it. The **human microbiome** consists of an estimated 37 trillion microbes (more than 4,000 different types of bacteria), outnumbering the approximate 30 trillion human cells building the body. All humans do not share the same set of microbes. Rather, each of us has a unique microbiome on our skin, in our mouths, and in our gut (**FIGURE 1.9**). We even have a unique combination of 50 types of bacteria in our belly button! Where did our microbial companions come from, and what good are they?

A healthy human fetus is free of most microbes until birth. Then, during and immediately after birth, body surfaces (e.g., skin, mouth, eyes, nose, gut) on and in the newborn become covered with a large variety of microbes from mom. Soon, this human microbiome expands through contact with family members, other people, family pets, and other sources. The result is the development of unique microbial communities of microbes on the skin, in the mouth and respiratory tract, and especially in the gut.

The human gut microbiome is essential for a healthy life. Many of the microbes existing in the gut spend each day performing an amazing array of essential activities. They carry out an astonishing array of metabolic activities in helping regulate digestion. They help the body resist disease by maintaining a strong immune system. Surprisingly, they even influence our susceptibility to becoming obese as well as to developing asthma and allergies. To be human and healthy, we must share our daily lives and space with these homegrown microbes.

As adults, our various body microbiomes are the product of the environment in which we live (lifestyles) and the foods we eat. If this community of

FIGURE 1.9 The Human Microbiome.

(A) Each human possesses several anatomical areas that each contain a unique microbiome.

(B) The gut microbiome contains hundreds of types of bacteria, as depicted in this artist's illustration. (Bar = 2 μm.)

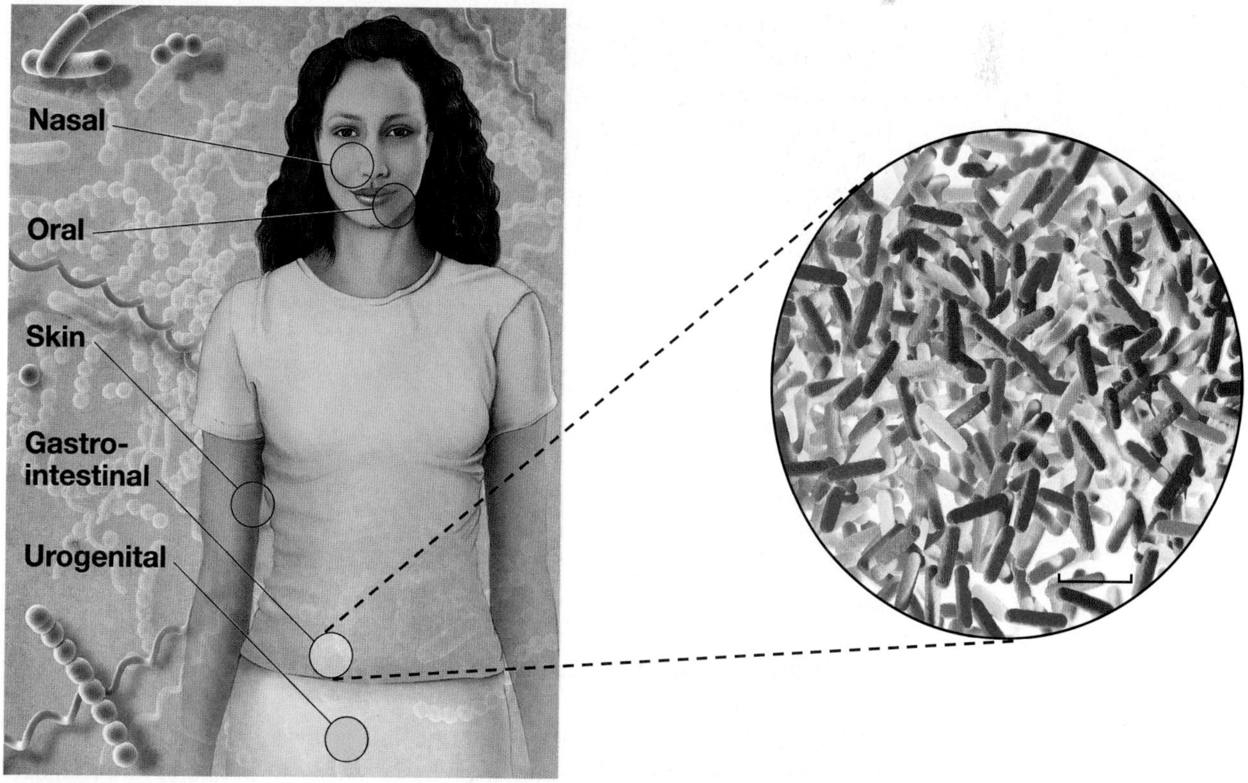

microbes should become unbalanced due to lifestyle or diet changes, we might be more predisposed to developing serious medical conditions. These can range from inflammatory bowel disease (e.g., Crohn's disease and ulcerative colitis) to obesity. Babies born through Cesarean section are not exposed to mother's microbes at birth. This could result in an atypical human gut microbiome, which later in life might lead to increased chances for developing allergies or asthma.

The human microbiome today is one of the most fascinating and hot topics in microbiology. Because of its potential relation to human health, manipulating the human microbiome might resolve some "societal diseases." Such conditions as obesity, diabetes, some cancers, and even depression and autism can be difficult to treat but have been related to a disturbance in the gut microbiome. Therefore, many health experts see the human microbiome as a key component of **personalized medicine**. These experts suggest that a patient's condition might be treated successfully, in part, by altering the individual's "microbiome fingerprint." By correcting the microbial component contributing to the disease, the disease might be resolved. Time will tell.

Personalized medicine: A medical procedure that separates patients into different groups—with medical treatment and/or products targeted to individual patients and their genetic makeup based on their predicted response or risk of disease.

1.4 Why Microbes Matter: Causing Infections and Disease

Most of us are inclined to think of infections and disease when we hear the word "virus" or "bacteria." Often, this negative connection might be justified because some microbes are the agents of illness causing misery and pain. Some pathogens, such as those causing diseases like plague, malaria, and smallpox, throughout history have swept through cities and villages, devastated populations, and killed great leaders and commoners alike. As a result of high death rates, politics, economies, and societies often were transformed on a global scale.

Our perception of microbes as all "bad" also comes from everyday experiences. Public news media today are inclined to report on only those microbes causing human disease. They seldom report on the positive and essential roles microbes play. Consumer advertising emphasizes microbe-destroying products, such as antimicrobial soaps and household cleaning products that can "eliminate 99% of **germs**" (**FIGURE 1.10**). However, as you read through the chapters of this text, remember only a small minority of microbes are dedicated pathogens.

Germ: Another common term for a pathogen.

1.5 Why Microbes Matter: Benefiting Society

In simple terms, **science** is the field of study that builds knowledge based on testing and evidence. It attempts to describe and comprehend the nature of the universe in whole or part, wherever that might lead. By exploring the microbial universe, microbiologists and other scientists are developing and providing information and knowledge to explain phenomena in the natural world. This represents **basic (or pure) science** and embodies the first half of the information in this book.

Microbiology also represents an **applied science**; that is, one that uses the scientific knowledge obtained from basic science to develop more practical

FIGURE 1.10 Antimicrobial Products. Many stores around the world contain aisles of products designed to eliminate or control microbes and germs.

Courtesy of Dr. Jeffrey Pommerville.

applications for, or to help solve, society's needs. Let's briefly look at the applied side of microbiology, which will comprise the second half of this book.

With Most Meals

Besides the morning routine described in the opener to this chapter, rarely does a meal go by when we do not rely on microbes for something on our plate or in our glass. For example, the tangy taste in sausages and the unique tastes of sauerkraut, pickles, and vinegar are due to microbial activity (**FIGURE 1.11**). Microbes are responsible for the final forms of many foods we eat, including dairy products such as yogurt, buttermilk, sour cream, and tofu (fermented bean curd). Virtually all types of bread depend on microbes for their taste and spongy textures. Microbes are the active partners in wine and beer production and often are the flavorful agents in cheeses. All these products, and many others too numerous to mention here, would not exist without microbes. Even the coffee we drink and the chocolate we eat depend, in part, on microbes for their flavor and aroma.

The global food market today is a multibillion-dollar industry. On this industrial scale, microbes are grown in huge batches in enormous tanks, where the microbes produce useful components for many foods and beverages. Citric acid, for example, is produced by some molds and is typically used to enhance flavor in fruit juices, soda drinks, and candies. Lactic acid, produced by several bacteria, is an emulsifier and, like citric acid, also is a food and beverage preservative.

Bacterial proteins produced in industrial quantities are used in baking and to clarify fruit juices. Bacterial and fungal enzymes are employed to tenderize meat and are added to detergents as stain removers. Working like miniature chemical factories, microbes can churn out industrial quantities of vitamins (especially B vitamins). These essential metabolic factors then are added to many of the foods we eat and incorporated into the vitamin supplements many of us take.

FIGURE 1.11 Common Fermented Foods. Many of the foods we eat today are the result of microbial fermentation that produces acids that give the foods distinctive flavors and textures.

©Baloncici/Shutterstock.

Microbes also play an important role in the natural flavoring and aroma of many foods. Flavor enhancers, such as peach, banana, pear, and coconut extracts, are components originating from fungal molds. Monosodium glutamate (MSG), a flavor enhancer often added to classic Chinese meals, canned vegetables, soups, and processed meats, is obtained from the metabolism of specific bacteria. Food thickeners and stabilizers, such as xanthan gum and alginate, are bacterial and algal products. Even the artificial sweetener aspartame (NutraSweet) is derived from the metabolic activity carried out by specific types of bacteria.

In agriculture, humans have put microbes to work on food crops (e.g., corn, wheat, rice, soybeans) as natural insecticides and pesticides. When sprayed on such crops, the poisonous toxins kill the caterpillar (larval) stage of many agricultural pests without any effect on the human consumer. Scientists have even extracted from these bacteria the genes that have the instructions to make these toxins. They then have genetically engineered those genes into crop plants. Remarkably, these genetically modified plants then produce the toxins themselves as they grow in the field. Today, most of the soybean plants currently in American fields contain such bacterial genes.

In the Environment

Besides playing important roles as decomposers in water and soil, microbes can and have been used effectively to "clean up" human-caused pollution. This immensely appealing way of putting microbes to work for society is called **bioremediation**. The process has been used to break down dangerous pesticides and other environmental pollutants and to decontaminate nuclear waste sites containing radioactive materials.

Bioremediation: The use of microbes to degrade toxic compounds in the environment into nontoxic substances.

When an oil spill occurs, technologists might add mineral nutrients to the water to encourage bacterial growth. As their numbers increase, these microbes then gorge themselves on (and help remove) the petroleum. More impressive is the recent action of natural oil-eating microbes. Following the

April 2010 explosion on the *Deepwater Horizon* oil rig in the Gulf of Mexico, tons of oil spilled into the seawater, affecting the shoreline of the Gulf Coast. **A CLOSER LOOK 1.1** summarizes the amazing role natural bacteria played as "bioremediators" in the cleanup.

Microbes also remain the prime factor in sewage treatment. At sewage treatment plants, microbial communities are employed to digest the organic matter and convert the complex mixtures into simple products that can be recycled (**FIGURE 1.12**). Waste treatment plants also rely on microbial chemistry to

A CLOSER LOOK 1.1

Microbes to the Rescue!

On April 20, 2010, the *Deepwater Horizon* drilling rig exploded in the Gulf of Mexico. For 3 months, oil spilled from the blowout well into the Gulf, contaminating nearby shores and wetlands. According to federal government estimates, some 50,000 to 70,000 barrels of oil were spilled, and it was not until September that the well was declared sealed.

Petroleum or crude oil consists of a complex mixture of hydrocarbons of various sizes in liquid, gaseous, and solid forms. Where did all those hydrocarbons go?

In the case of the ruptured Deepwater Horizon well, the entry of oil profoundly altered the natural microbial community. The spill significantly stimulated deep-sea, oil-hungry microbes to consume much of the smaller, dispersed hydrocarbons and methane gas as fuel for their growth and reproduction (see Figure A). Microbial populations exceeded 10^{23} cells and consisted of more than 50 species (some new to science). With these numbers, their efficiency was amazing. Within weeks, some areas of the Gulf were nearly free of suspended oil. Overall, scientists estimate the microbes consumed 200,000 tons (>33,000 barrels) of oil and methane gas. One biogeochemist gave the microbes a "7 out of 10" for their digestive performance.

FIGURE A Sunlight is reflected off the Deepwater Horizon oil spill in the Gulf of Mexico on May 24, 2010. The image was taken by NASA's Terra satellite.

FIGURE 1.12 Sewage Treatment. An aerial view of a wastewater treatment plant that uses sunlight, oxygen, and bacteria to degrade the sewage.

handle the massive amounts of sewage and garbage generated daily. Have you ever passed by a sanitary landfill and smelled the stench emanating from it? Those odors are coming from bacteria and fungi breaking down the garbage.

In the Pharmaceutical/Biotechnology Industries

Biotechnology: The use of microbes and their chemistry to manufacture products that will improve the quality of human life.

Biotechnology is hard at work using bacteria and yeasts to produce effective medicines. Human insulin for diabetics, blood-clotting factors for hemophiliacs, and human growth hormone for children suffering dwarfism are but a few examples of the products being made with the help of microbes. Interestingly, viruses have been genetically altered to provide increased disease resistance in plants and to kill insect pests in the environment. Viruses also have been modified to deliver working genes into human cells to replace the defective ones. Today, biotechnology relies heavily on the metabolic talents of microbes.

1.6 Microbiology Today: Challenges Remain

Global health emergency: A serious public health event that, according to the WHO, endangers international public health and global populations.

Epidemic: Referring to the occurrence of more cases of a disease than expected in individuals within a geographic area.

Outbreak: A more confined and geographically limited epidemic.

Microbiology is one of the most intensively studied fields in the biological sciences, in part from the basic science and applied science topics described above. However, microbiology also faces several challenges, many of which concern infectious diseases.

Microbial pathogens are responsible for some 25% of all human deaths globally (**FIGURE 1.13**). Over the past decade, the World Health Organization (WHO) has declared four **global health emergencies**. Two of them were very recent: the Ebola **epidemic** in West Africa in 2014 to 2016 and the Zika **outbreak** that spread through the Americas in 2015 to 2016. In fact, over the past 30 years, the number of disease outbreaks globally has increased three-fold, from 1,000 per year to more than 3,000 per year (**FIGURE 1.14**). Why the recent increase? Simply stated—it is microbes and society. Here are four important connections.

FIGURE 1.13 Global Mortality from Infectious Diseases. On a global scale, some 15 million people die each year from infectious diseases.

Data from World Health Organization. *World Health Statistics 2011*. Retrieved from: https://www.who.int/whosis/whostat/2011/en/. Accessed February 7, 2019.

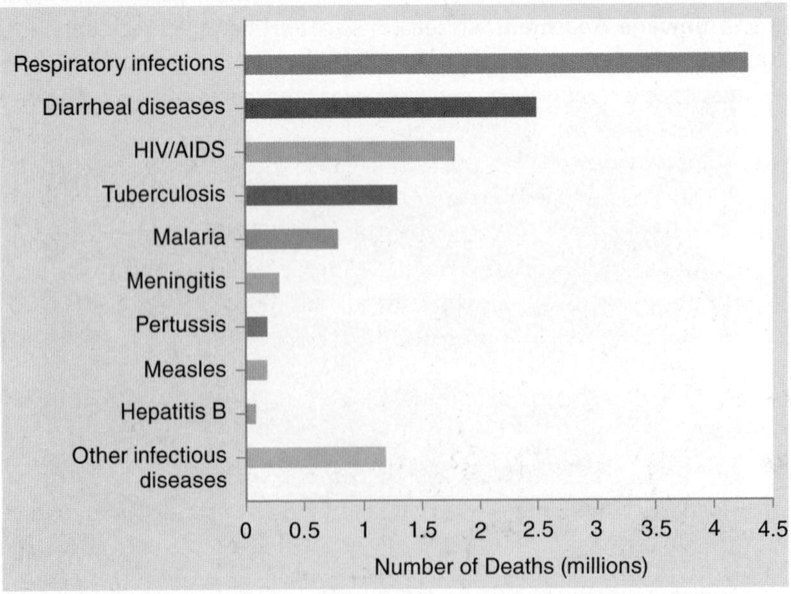

FIGURE 1.14 Global Infectious Disease Outbreaks—May 31, 2018. This HealthMap from Boston Children's Hospital shows a snapshot of current publicly reported, confirmed, and suspected infectious disease outbreaks throughout the world. The large circles indicate country-level alerts, while state, province, and local alerts are indicated by the small circles. Marker color (inset) reflects the level of the outbreak at a location.

Boston Children's Hospital. HealthMap: Flu & Ebola Viruses and Contacts. Retrieved from: http://www.healthmap.org/en/. Accessed May 31, 2018.

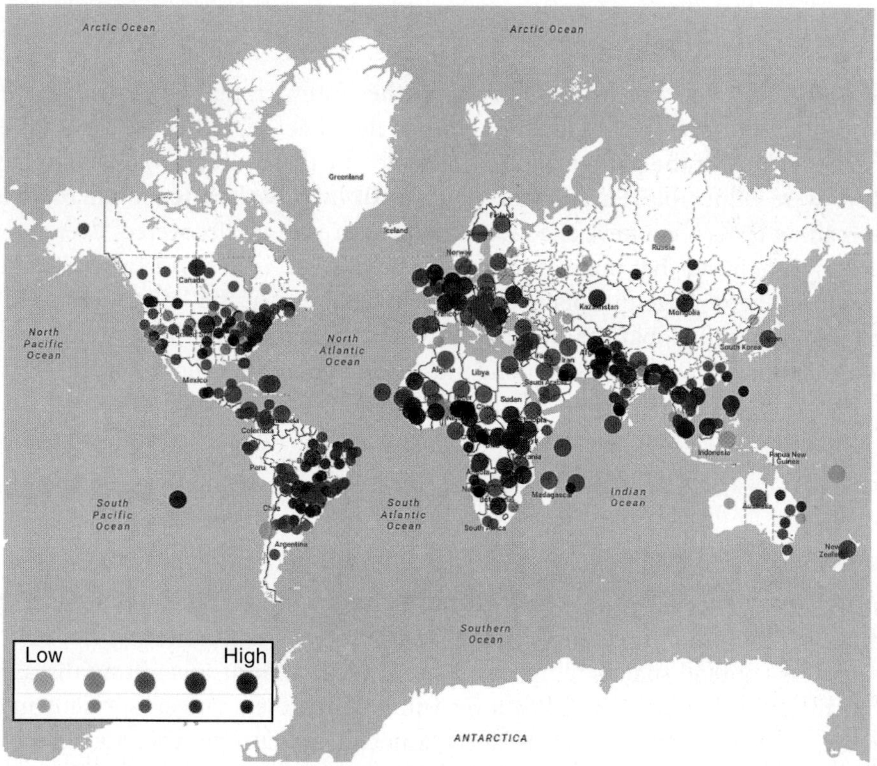

Air Travel

Experts estimate that more than 5 billion people traveled by air in 2018. With so many people traveling globally, a human disease outbreak or epidemic in one part of the world is only a few airline hours away from becoming a potentially dangerous threat in another region of the globe. Moreover, trade in commodities and animals is happening at a faster pace. If these goods or animals carry disease, the pathogens also are transported. Health experts report that almost 75% of newly emerging infectious diseases are spread to humans by animals. But, no matter how a disease is spread, when introduced to a new location, most humans will be susceptible to infection because they lack immunity to the new pathogen. Doctors and health systems also can be caught off guard, as was the case for both the Ebola and Zika virus disease outbreaks/epidemics. So, unlike past generations, today's highly mobile, interdependent, and interconnected world provides numerous opportunities for the rapid spread of infectious diseases. The challenge is to be prepared to recognize and prevent or eliminate these outbreaks.

Urbanization and Poverty

Today, more people live in densely populated urban environments than in past generations. In fact, more than half of the world's population (7.8 billion in mid-2019) now lives in cities, and just about every country around the globe has developed more urbanized societies. That translates to millions, if not billions, of people living in crowded and often unhygienic conditions, where public health infrastructure, and adequate food and water, is lacking. This setting makes for the "perfect storm" for the persistence of pathogens and for infectious disease spread.

Drug-Resistant Pathogens

Many people often turn to antibiotics and other antimicrobial drugs to help fight infections. For example, today the ability to fight bacterial infections is waning because many of the pathogens are unaffected by the antibiotics to which they once were vulnerable. Furthermore, such **antibiotic resistance** in microbes is spreading faster than new, effective antibiotics are being discovered. Ever since it was recognized that many microbes could develop into **superbugs**, a campaign has been waged to make everyone aware of what the WHO sees as the biggest threat to global health today. Physicians need to stop prescribing these antimicrobial drugs, and the public must stop demanding them, for uncalled-for situations. Health experts assert that if actions are not taken to contain and reverse resistance, the world could be faced with previously treatable diseases that have again become untreatable.

Superbug: A bacterial strain that is resistant to multiple antibiotics.

Climate Change

A very controversial issue is how **climate change** might affect the frequency and distribution of infectious diseases and outbreaks around the world. Many scientists contend that as temperatures rise, mosquitoes transmitting diseases like malaria, cholera, and yellow fever will broaden their range, especially into more temperate climates such as North America and Europe. Therefore, microbiologists, health experts, climate scientists, and many others are studying new strategies to limit potential outbreaks, and even **pandemics**, before they can get started, and these scientists need to understand how such climatic changes could affect the health of our society, as well as livestock, plants, and wildlife.

Climate change: A change in the statistical distribution of weather patterns (e.g., temperature, rainfall) when that change lasts for an extended period.

Pandemic: Referring to a disease occurring over a wide geographic area (worldwide) and affecting a substantial proportion of the global population.

A Final Thought

It should be clear from this chapter that microbes, despite their involvement in disease, contribute substantially to the quality of life in our society. To paraphrase what the late industrial microbiologist David Perlman of the University of Wisconsin said about microbes:

1. The microbe is always right, your friend, and a sensitive partner.
2. There are no stupid microbes.
3. Microbes can and will do anything.
4. Microbes are smarter, wiser, and more energetic than chemists, engineers, and others.
5. If you take care of your microbial friends, they will take care of you.

Chapter Discussion Questions

What Was He Thinking?

From your reading of this chapter, identify and discuss five major points about the microbial world that the author was trying to get across to you.

Questions to Consider

1. "Microbes? All they do is make us sick!" From your first "exposure" to microbes in this chapter, how might you counter this statement?
2. If you were to ask someone to describe a microbe, he or she might think of a dot under a microscope. However, the microbial world is quite diverse, and each of its members is unique. Although your experience in microbiology is limited at this moment, the information in this chapter should give you some understanding of the microbial world. How, then, would you now describe a microbe?
3. The poet John Donne once wrote: "No man is an island, entire of itself; every man is a piece of the continent." This statement applies not only to humans, but to all living organisms in the natural world. What are some roles microbes play in their relationships with society?
4. Science, including microbiology, has two arms referred to as basic (pure) science and applied science. How are the two arms similar, and how are they different? Provide a microbiology example for your answer.
5. Our world is somewhat "germ-phobic" (i.e., having an extreme fear of germs and an obsession with cleanliness). The news and social media cover new outbreaks of disease, we eagerly await new antibacterial medicines, and we hear of new ways to "fight germs." That being said, suppose there were no microbes or viruses on Earth. From your reading of this chapter, what do you suppose life would be like?

CHAPTER 2

Microbes in Perspective: The Pioneers, Classifiers, and Observers

What Shall We Call This One?

It's 1730 and Carolus Linnaeus, a Swedish physician and botanist, was in trouble. Sailing ships from around the world were arriving in port, carrying all types of trading goods as well as newly discovered plants and animals. Overwhelmed with so many new organisms, Linnaeus would ask, "What shall we call this one?" "Or this one?" "Or that one?" The museums were full of newly discovered organisms (such as penguins, manatees, kangaroos, tobacco plants, bananas, and potatoes) arriving from distant corners of the globe, and people were bringing new specimens to him almost daily (he soon learned to dread the sight of arriving ships). Linnaeus even added to this frenzy: He inspired an unprecedented worldwide program of specimen hunting by sending his students around the globe in search of new and unknown plants and animals. One of his most famous students, Daniel Solander, accompanied Captain James Cook on his first round-the-world voyage. From this voyage, Solander brought back the first plant collections from Australia and the South Pacific. Other students soon followed, each returning with newly discovered specimens from other parts of the world.

Before Linnaeus' time, the names of organisms varied. Different individuals used arbitrary criteria and long, unwieldy Latin names for different organisms. There was no uniformity in the rules for naming and the need to apply a

CHAPTER 2 OPENER Although life on Earth might have had many fits of starts and failures (represented by the tree stumps), eventually one line succeeded and evolved into the "tree of life" we know today.

FIGURE 2.1 Carolus Linnaeus. Karl von Linné, the Swedish botanist known in scientific history as Carolus Linnaeus, took on the daunting task of classifying the known plants and animals of the biological world and giving them scientific names.

scientific name to all organisms, called "nomenclature," had gotten thoroughly out of hand. To be sure, the biological world needed some order from the chaos of unnamed specimens, but who had appointed Linnaeus king of nomenclature? For Linnaeus, it was exasperating, to say the least!

Linnaeus (also known as Karl von Linné) is just one of the pioneers, classifiers, and observers we will encounter in this chapter (**FIGURE 2.1**). First, we will highlight some of the early pioneers in the study of microbes. Then, we will use Linnaeus' system for naming organisms. Before we are done, we will consider how microbes and all living organisms became organized with respect to one another. Linnaeus could not have anticipated what the future would hold.

LOOKING AHEAD

After reading and completing this chapter, you will be able to:

- **2.1** Compare the contributions to microbiology that were made by each of the four pioneers.
- **2.2** Write correctly an organism's scientific name.
- **2.3** Explain the differences between (a) the light microscope and the electron microscope (EM) and (b) the transmission EM and scanning EM.
- **2.4** Identify the general characteristics of the microbes and viruses.
- **2.5** Describe the tree of life, and assess its significance to microbiology and the evolution of life.

2.1 The Roots of Microbiology: The Pioneers

You already have learned a bit about the microbial world and its essential role. New discoveries make today a remarkable, and sometimes humbling, time to be studying microbes and their relationships with society. However, we must pause before we get too deeply into the study of microbes, for, as an old proverb states:

"To understand where you are going, you must know where you have been."

Therefore, before we launch into our explorations of microbes and society, we must briefly examine microbiology's origins and how microbes and all organisms came to be categorized.

Many individuals mistakenly perceive science as nothing more than a collection of facts and data. They often view scientific theories and principles as something one simply "plucks out of a hat." To the contrary, **science** is a human effort. It is only through careful observations and properly designed experiments that newly found knowledge is built into important and logical frameworks. Here, we see that this "scientific behavior" is exhibited by some very inquisitive individuals who first discovered and studied microorganisms.

Robert Hooke (1635–1703)

The microbial world was virtually unknown until the mid-1600s. At that time, an English scientist named Robert Hooke became fascinated with a newly developed instrument, the microscope that could magnify objects about 20 times (20×). In a book he published, called *Micrographia*, Hooke recorded his observations and descriptions of small compartments in slices of cork (which he named "cells") and drew with amazing accuracy the anatomy of a flea. He also drew the microscopic details of threadlike fungi he found growing on the sheepskin cover of a book. Hooke's stunning illustrations were among the first visual descriptions of living microorganisms.

Meanwhile, across the North Sea in Delft, Holland, more astonishing discoveries of the microbial world were being made.

Antony van Leeuwenhoek (1632–1723)

The individual who first brought the diversity of microbes to the world's attention was a Dutch cloth merchant named Antony van Leeuwenhoek. In the 1670s, Leeuwenhoek developed the skill of grinding glass lenses for magnifying and inspecting the quality of cloth. Knowing of Hooke's *Micrographia*, Leeuwenhoek began using his own lenses to satisfy his curiosity about the invisible world (**FIGURE 2.2A**). With his simple microscope that magnified about 250×, Leeuwenhoek studied the eye of an insect, the scales of a frog's skin, and the intricate details of muscle cells. In 1673, while peering into a drop of pond water with his microscope, he observed tiny microscopic forms of life. Darting back and forth and rolling and tumbling about, he called these creatures organisms **animalcules** (he assumed they were tiny animals). At first, they delighted him and amazed him with their variety of shapes and sizes. Then, they puzzled him as he pondered their meaning.

Leeuwenhoek excitedly communicated his findings to the Royal Society of London, one of the most respected scientific societies of the time. Between 1673 and 1723, he wrote almost 300 letters describing the new microscopic creatures and structures he observed. His letter dated September 17, 1683, is particularly

FIGURE 2.2 Leeuwenhoek's Microscope and Drawings of Bacteria.
(A) To view his animalcules, Leeuwenhoek placed his sample on the tip of the specimen mount that was attached to a screw plate. An elevating screw moved the specimen up and down, while the focusing screw pushed against the metal plate, moving the specimen toward and away from the lens.
(B) With such an instrument, Leeuwenhoek drew the animalcules (bacteria in this drawing) he saw.

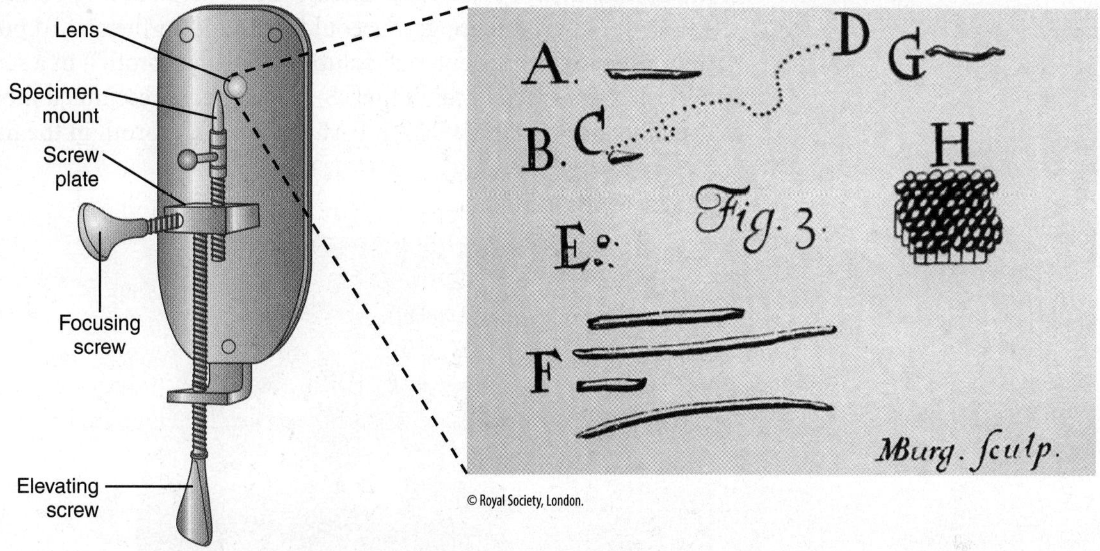

noteworthy. In that letter, Leeuwenhoek, for the first time, describes and draws what are believed to be bacterial cells (**FIGURE 2.2B**).

The observations and descriptions made by Hooke and Leeuwenhoek opened the eyes of the world to the microbial universe.

Louis Pasteur (1822–1895)

After the death of Leeuwenhoek, interest in animalcules gradually waned. Most people assumed the animalcules were simple curiosities of nature and had little or no influence on their lives. However, with improvements in the design and magnification of microscopes, in the 1850s, Louis Pasteur, the renowned French chemist and scientist, called attention to the animalcules (**FIGURE 2.3**). He believed some were the agents of infectious disease.

Pasteur believed the discoveries of science should have practical applications to society, so in 1857 he seized the opportunity to unravel the mystery of why some French wines were turning sour. The prevailing theory held that wine **fermentation** resulted from the purely chemical breakdown of grape juice to alcohol; no living organisms were involved. Yet, with his microscope, Pasteur consistently saw large numbers of tiny yeast cells in the wine. Moreover, he noticed sour wines also contained populations of the barely visible bacteria like the ones described by Leeuwenhoek. In a classic series of experiments, Pasteur boiled several flasks of grape juice and removed all traces of yeast cells from the flasks; he then set the juice aside to see if it would ferment. Nothing happened. Next, he carefully added pure yeast cells back into the flasks, and soon the fermentation was proceeding normally. Moreover, he found that when he used heat to remove all bacteria from the grape juice, the wine would not turn sour; it would not "get sick." Pasteur's recommendation of using heat to control bacterial contamination led to the process that today we call **pasteurization**.

Fermentation: In beverage production, the process of converting carbohydrates to alcohol or organic acids using microorganisms.

Pasteur's work rocked the scientific community. It showed that microscopic yeast cells and bacteria were tiny, living factories that can cause chemical changes. Pasteur then wondered if bacteria also could cause change by making people sick. In 1857, he published a short paper in which he proposed that some microbes might be related to human illness. Thus, his work on fermentation was the foundation for what would become known as the **germ theory of disease**. This fundamental theory held that some microbes (germs) play significant roles in the development of infectious disease. But where did they come from?

Pasteur believed these germs could be found in the air. To prove his idea, Pasteur prepared a nutrient-rich solution (called a "broth") in a series of swan-neck flasks (so named because their S-shaped necks resembled those of swans), as explained in **A CLOSER LOOK 2.1**. Pasteur boiled the broth in the flasks, thereby

FIGURE 2.3 Louis Pasteur.

Courtesy of the National Library of Medicine.

A CLOSER LOOK 2.1

Experimentation and Scientific Inquiry

Science certainly is a body of knowledge, as you can see from the pages in this text! However, science also is a process—a way of learning. Often, we accept and integrate into our understanding new information because it appears consistent with what we believe is true. But, are we confident our beliefs are always in line with what is accurate? To test or challenge current beliefs, scientists must present logical arguments supported by well-designed and carefully performed experiments.

The Components of Scientific Inquiry

There are many ways of finding out the answer to a problem. In science, scientific inquiry—or what has been called the "scientific method"—is a variety of procedures used to investigate a problem. Let's understand how scientific inquiry works by following the logic of the experiments Louis Pasteur published to show that microbes are present in the air.

Pasteur set up an experiment to test the hypothesis.

- Pasteur boiled the broths in swan-necked flasks to kill any microbes that might be present. Microbes can enter the flasks in the air, but the microbes will be trapped at the bottom of the neck.
- After several days, Pasteur either (A) snapped off the neck of some of the flasks or (B) tipped the flasks so the broth could come in contact with the bottom of the neck.

Results

- No growth (microbes) was seen in the intact flasks.
- Microbes (cloudy liquid) were found in the flasks after the neck was snapped off, or the flask tipped so the broth entered the neck.

Pasteur and the Microbes in the Air. First, broth in a swan-necked flask (flask with a long, tapered neck) is boiled. If allowed to cool and remain undisturbed, air can enter the flask, but the curvature of the neck traps dust particles and microorganisms, preventing them from reaching the broth. If a similar flask of sterilized broth has the neck snapped off **(A)**, or the flask is tipped so broth enters the neck **(B)**, organisms will contact the broth and grow.

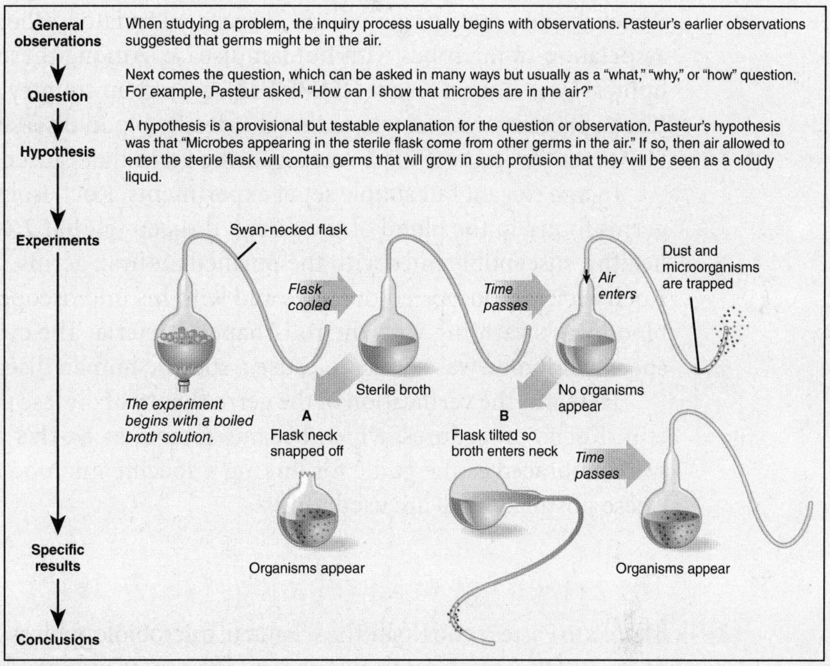

Conclusions

Let's analyze the results. Pasteur had a preconceived notion of what should happen if his hypothesis was correct. In his experiments, only one "variable" (an adjustable condition) changed. In the experiment, the flasks were left intact or they were exposed to the air by breaking or tipping. Pasteur kept all other factors the same; that is, the broth was the same in each experiment, it was heated the same length of time, and similar flasks were used in each experiment.

Thus, the experiment had rigorous "controls" (the comparative condition), which were the intact swan-necked flasks. Pasteur's finding that no microbes appeared in these flasks is interesting but tells us very little by itself. Its significance comes by comparing this to the broken neck (or tipped) flasks where microbes quickly appeared. These experiments validated his hypothesis.

Hypothesis and Theory

When does a hypothesis become a theory? Well, there is no set time or amount of evidence specifying the change from hypothesis to theory. A "theory" is defined as a hypothesis that has been tested and shown to be correct every time by many separate investigators. So, at some point, if validated by others, enough evidence exists to say a hypothesis is now a theory. However, theories are not written in stone. They are open to further experimentation and so can be proven false.

As a side note, today a theory often is used incorrectly in everyday conversation and in the news media. In these cases, a theory is equated incorrectly with a hunch or belief—whether there is evidence to support it or not. In science, a theory is a general set of principles supported by large numbers of experimental studies.

destroying all microbes, then he left the flasks open to the air. The S-shaped neck trapped any microbes in the air and prevented their entry into the broth. Thus, when the flasks were set aside to incubate in a warm environment, no microbes appeared in the broth.

Pasteur then demonstrated that when the neck was broken off a flask, microbes present in the air could contact the broth and grow, as the broth liquid soon became cloudy with microbes. These experiments clearly showed microbes were in the air and probably in other environments as well.

Although Pasteur's work suggested some microbes (germs) could cause disease, he was stymied by his inability to associate a specific germ with a specific disease. Adding to his frustration was the death of three of his five children, two dying from typhoid fever, a germ-caused disease.

Robert Koch (1843–1910)

Pasteur's idea that germs caused disease stimulated others to investigate the association of microbes with human disease. Among these so-called "microbe hunters" was Robert Koch (**FIGURE 2.4A**), a German country doctor. In the 1870s, Koch's primary interest was anthrax, a deadly blood disease in cattle and sheep. He wanted to know if there was a specific germ that caused anthrax.

In one elegant but simple set of experiments, Koch isolated and purified the germs found in the blood of the diseased sheep (**FIGURE 2.4B**). He then injected healthy, susceptible mice with the purified anthrax germs. The mice soon died. Koch excitedly autopsied the mice and with his microscope saw that the mouse blood was swarming with the rod-shaped bacteria. The cycle was complete—a specific microbe was shown to cause a specific human disease.

Here was the verification of the germ theory of disease that had escaped Pasteur. Koch's procedures, which became known as **Koch's postulates**, quickly were embraced as the guide for linking a specific microbe to a specific disease. These postulates still are used today.

The Golden Age of Microbiology (1857–1917)

Thanks to Pasteur and Koch, the science of microbiology blossomed during a period of about 60 years. During this the golden age of microbiology, Pasteur worked

FIGURE 2.4 Koch's Postulates.
(A) Robert Koch.

Courtesy of the National Library of Medicine.

(B) Koch was the first to verify Pasteur's germ theory by using a set of postulates to relate a single microbe to a single disease. The photo (inset) shows the rod shape of the bacterial cells as Koch observed them. Many rods are swollen with spores.

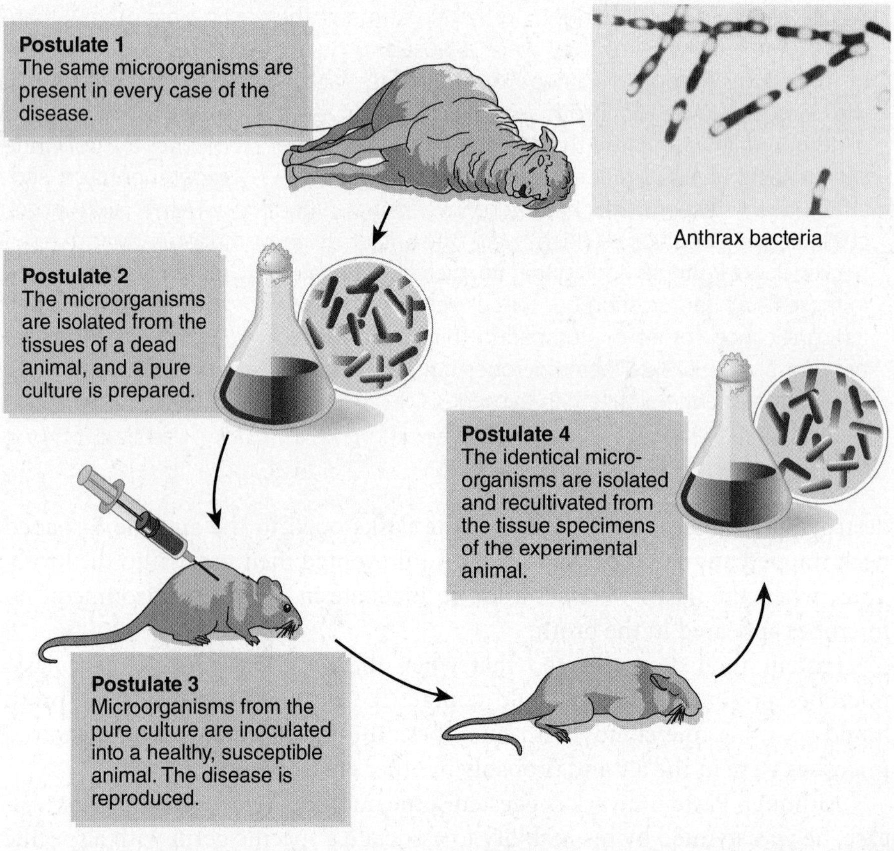

on vaccines for chicken cholera, anthrax, and human rabies. Koch identified the microbes causing tuberculosis and human cholera. Their coworkers and others identified yet more bacteria associated with specific diseases. In fact, the "microbe hunters" had become international in scope. For example:

- David Bruce, a Scottish microbiologist, identified the bacterium causing undulant fever.
- Shibasaburo Kitasato, a Japanese investigator, was one of two investigators to identify the bacterium responsible for bubonic plague.
- Howard Taylor Ricketts, an American microbiologist, discovered the infectious agent of Rocky Mountain spotted fever.
- Walter Reed, an American army physician, isolated the agent causing yellow fever.

With this Golden Age came the realization that microbes have diverse impacts on the world and society far beyond the medical field. Besides Pasteur's discovery of the positive effects of yeasts on wine production, Sergei Winogradsky, a Russian soil microbiologist, discovered how harmless soil bacteria use carbon dioxide to produce sugars. Martinus Beijerinck, a Dutch microbiologist and botanist, isolated soil bacteria that trap nitrogen and convert it into a form usable by plants. An interest in microbial ecology was growing and the realization that microorganisms had diverse impacts on the world and society was developing.

2.2 Naming Microbes: What's in a Name?

Giving all organisms scientific names is important because it eliminates confusion between scientists when talking about a specific organism. Linnaeus came up with a simple system for providing scientific names. In a 1753 book he published titled *Systema Naturae*, he supplied scientific names for more than 6,000 known and newly discovered plants. In the following years, as more and more plant and animal specimens were sent to him, Linnaeus continued to update his book with scientific names for these organisms. Order was being brought to the chaotic mess of organism names.

By the early 1900s, as the microbe hunters discovered and identified more and more microbes, these organisms were given scientific names based on Linnaeus' system. Even today, many microorganisms still are being discovered, and they too are subject to the Linnaean system that now is managed by various internationally agreed codes of rules. So, let's look at this system, primarily using microbes as examples.

Binomial Nomenclature

Linnaeus' system for naming organisms is called **binomial nomenclature**; that is, each organism is assigned a two-word name, the binomial, derived from Latin (or sometimes Greek roots). The first word of the binomial represents the **genus** (pl. **genera**). It is followed by a second word, called the species modifier or the specific epithet. The genus and specific epithet together (the binomial) indicate the **species** (pl. **species**) of organism. For instance, a bacterial species normally found in the human colon is called *Escherichia coli* (**FIGURE 2.5**). The first word of the binomial, *Escherichia*, is the genus name and is derived from (and to honor) the scientist, Theodor Escherich, who first identified the microbe in 1888. The second word of the binomial, *coli*, is the specific epithet derived from the word "colon," which is where Escherich first found the bacterium. [The same rules apply to humans, who scientifically are known as *Homo sapiens* (*Homo* = "man"; *sapiens* = "wise").]

Escherichia coli: esh-er-EE-key-ah KOH-lee

FIGURE 2.5 ***Escherichia coli.*** A false-color electron microscope image of *E. coli* cells. (Bar = 2 μm.)

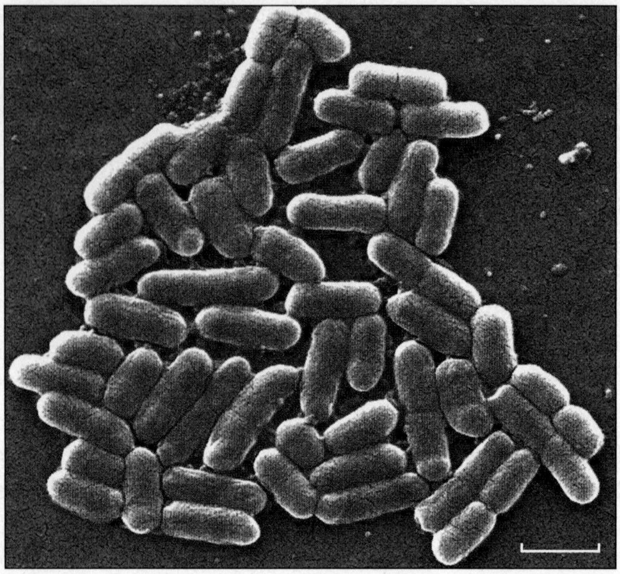

Courtesy of CDC.

When a new species is discovered, inventiveness by the discoverers can go into providing a scientific name. The binomial name might reflect a scientist's name, as in *Escherichia*. The name also might refer to the organism's manner of growth or the location where it was first found. Look at **A CLOSER LOOK 2.2**, which provides several examples of how microbes got their name.

Here are the scientific guidelines for writing a species name.

- Capitalize the first letter of the genus and leave the specific epithet all in lowercase letters (even if it was named after a person).
- *Italicize* (if typed) or underline (if written) the binomial.
- Abbreviate the genus name after it is first spelled out. Once the full species name has been introduced in a piece of writing, the name can be abbreviated by using the first letter of the genus name and the full specific epithet. Thus, *Escherichia coli* is abbreviated *E. coli*.

Exceptions to these guidelines occur. In today's news media and other publications for the public, an organism's binomial often is written in normal text; for example, you might see Escherichia coli or E. coli.

Many microbial species have several genetic variants referred to as strains or subspecies. To name these, additional words are added to the end of the binomial. An example, there is a pathogenic strain of *E. coli* called *E. coli* O157:H7. This strain can cause a severe intestinal illness if ingested in contaminated food. Thus, the designation O157:H7 is useful for identification purposes and distinguishes this strain from other *E. coli* strains, such as the harmless ones normally found in the colon.

2.3 The Microbial World: The Observers

Historically, microbes were discovered and studied by observation through a microscope. Although the microbe hunters of the 19th and 20th centuries discovered many microbes, they would be amazed today to learn that only about 2% of the estimated 1 trillion different microbe species have been observed visually.

A CLOSER LOOK 2.2

"What's in a Name?"

In Shakespeare's *Romeo and Juliet*, Juliet tells Romeo a name is an artificial and meaningless convention. Perhaps from her perspective in convincing Romeo that she is in love with the man called Montague and not the family Montague, there is a love-based reason for "What's in a name." As you read this book, you have and will come across many scientific names for microbes. Not only are many of these names tongue twisting to pronounce (many are listed with their pronunciation in Appendix A), but how in the world did the organisms get those names? Most are derived from Greek or Roman word roots. Here are a few examples.

Species	Meaning of Name
Genera Named After Individuals	
Bordetella pertussis	Named after the Belgian Jules Bordet, who in 1906 identified the small bacterium (*ella* means "small") responsible for pertussis (whooping cough).
Neisseria gonorrhoeae	Named after Albert Neisser, who discovered the bacterial organism in 1879. As the specific epithet points out, the disease it causes is gonorrhea.
Genera Named for a Microbe's Shape	
Vibrio cholerae	*Vibrio* means "comma-shaped," which describes the shape of the bacterial cells causing cholera.
Staphylococcus epidermidis	The stem *staphylo* means "cluster" and *coccus* means "spheres." So, these bacterial cells form clusters of spheres found on the skin surface (epidermis).
Genera Named After an Attribute of the Microbe	
Saccharomyces cerevisiae	In 1837, Theodor Schwann observed yeast cells and called them *Saccharomyces* (*saccharo* = "sugar"; *myce* = "fungus") because the yeast converted grape juice (sugar) into alcohol; *cerevisiae* (Ceres was the Roman goddess of agriculture) refers to the use of yeast since ancient times to make beer.
Myxococcus xanthus	The stem *myxo* means "slime," so these are slime-producing spheres that appear as a yellow (*xantho* = "yellow") growth in culture.

Lastly, in keeping with Shakespeare's poetic style, there is the organism *Thiomargarita namibiensis*. This bacterial species was first isolated in 1997 from sediment samples in the Atlantic Ocean off the coast of Namibia, a country in southwestern Africa (*ensis* = "belonging to"). These spherical-shaped bacterial cells accumulate sulfur (*thio* = "sulfur") so when they are observed with the microscope, the cells appear white and look like a microscopic string of pearls (*margarit* = "pearl"). Thus, we have *Thiomargarita namibiensis*—the "Sulfur Pearl of Namibia." Juliet would be impressed!

So, let's look at this instrument, the microscope, and see how it provided a window by which microbes could be observed and studied in greater detail.

Microbial Measurements and Cell Size

One of the defining features of microbes is their extremely small size. Distinct from small animals like insects that are measured in inches or millimeters (mm), microorganisms are much smaller, and they are measured in **micrometers** (**μm**) (**FIGURE 2.6A**). For example, the length of a typical bacterial cell is about

Micrometer (μm): A measure of length equal to a thousandth of a millimeter.

2 μm, although common species range in size from 0.1 μm to 10 μm. To appreciate that small size, look at **FIGURE 2.6B**, which shows a human cheek cell (60 μm in diameter) from inside the mouth. About 30 bacterial cells, which typically are found attached to cells of the cheek in the mouth, lying end to end would span the diameter of the cheek cell.

Other microorganisms are larger. Yeast cells are approximately 5 μm in diameter, and some protozoan cells can be up to 200 μm long.

Because viruses are much smaller than bacterial cells, viruses are measured in **nanometers (nm)** (see Figure 2.6A). In **FIGURE 2.6C**, about 12 of these bacterial viruses (each 100 nm in length) would span the length of the bacterial cell.

Nanometer (nm): A measure of length equal to a thousandth of a micrometer.

FIGURE 2.7 sums up the spectrum of microbial sizes, from the incredibly tiny viruses to the molds, some of which are visible to the naked eye.

In most cases, two types of microscopes are used to observe the microbial world.

The Light Microscope

Since Leeuwenhoek's time, great strides have been made in the design and construction of the **light microscope** (**FIGURE 2.8A**). Today's instrument routinely

FIGURE 2.6 Measurement of Size.

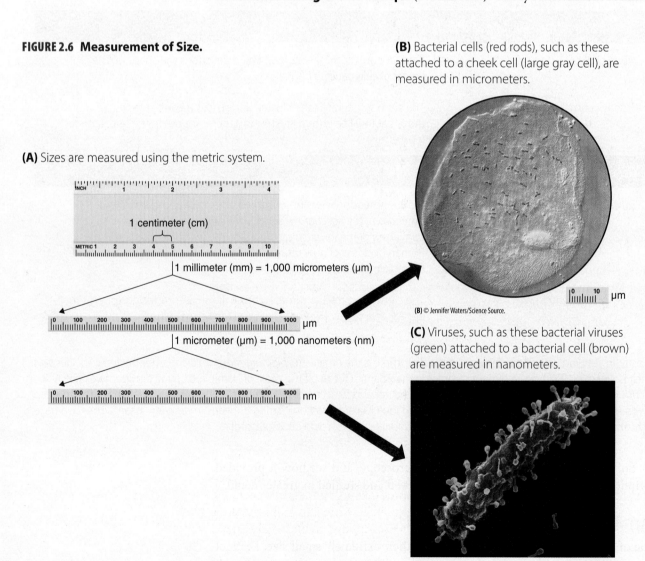

(A) Sizes are measured using the metric system.

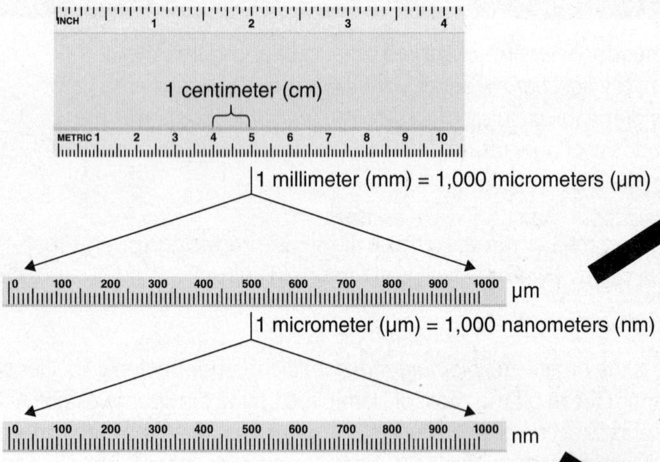

(B) Bacterial cells (red rods), such as these attached to a cheek cell (large gray cell), are measured in micrometers.

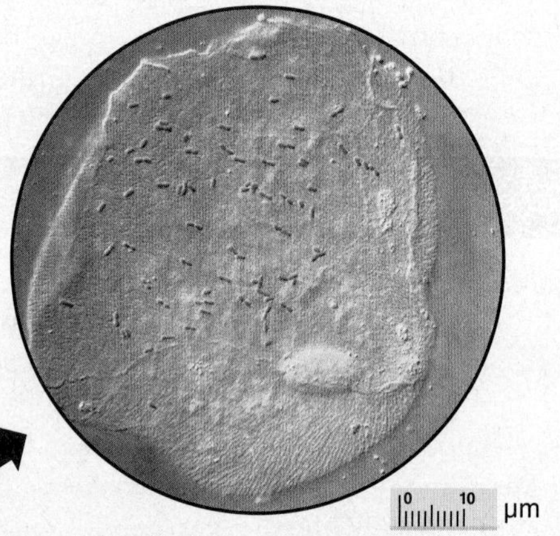

(B) © Jennifer Waters/Science Source.

(C) Viruses, such as these bacterial viruses (green) attached to a bacterial cell (brown) are measured in nanometers.

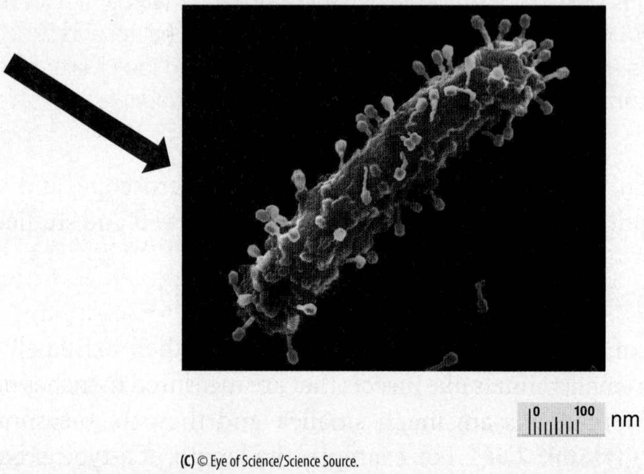

(C) © Eye of Science/Science Source.

FIGURE 2.7 Size Comparisons Among Atoms, Molecules, Viruses, and Microbes (not drawn to scale). Although tapeworms and flukes usually are macroscopic, they often are studied by microbiologists because of their disease potential.

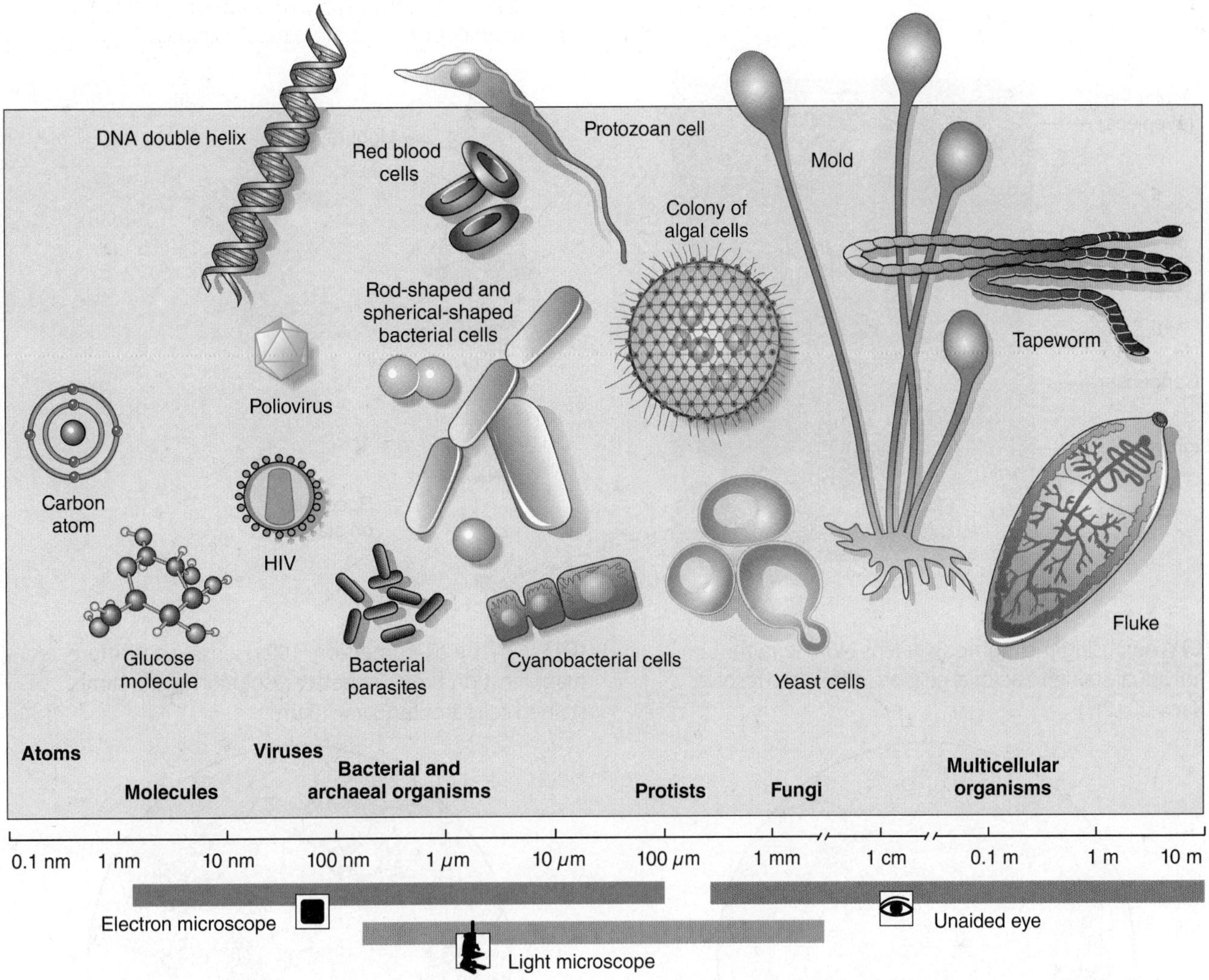

achieves magnifications of 1,000×. The main parts of the light microscope are the ocular lenses (or eyepieces), the objective lenses, the stage (on which lies a glass slide with the specimen), the substage condenser, and the light source.

Most light microscopes have three or more glass objective lenses to magnify the object. These are the low-power lens (10×), the high-power lens (40×), and the oil-immersion lens (100×). Each objective lens creates an image that is further magnified by the eyepiece to form the final image seen by the observer (**FIGURE 2.8B**). The **total magnification** (how many times the object has been enlarged) is calculated by simply multiplying the magnification of the objective lens being used by the magnification of the eyepiece. For example, the high-power objective (40×) when used with a 10× eyepiece yields a total magnification of 400× (**FIGURE 2.8C**).

To focus and study tiny microbes like bacterial cells require maximum magnification from the microscope. This is accomplished using the oil-immersion lens. This lens uses a drop of immersion oil placed between lens and glass slide (specimen) to obtain enough light for viewing. With this lens, the total magnification would be 1,000× (**FIGURE 2.8D**).

Although more details can be seen with the oil-immersion lens, the magnification still is insufficient to see the viruses or the structures inside cells. Even greater magnification is needed to see these very small objects.

FIGURE 2.8 Light Microscopy.

(A) A student light microscope.

(B) Image formation with the light microscope requires the light to pass through the objective lens and an ocular lens before reaching the eye.

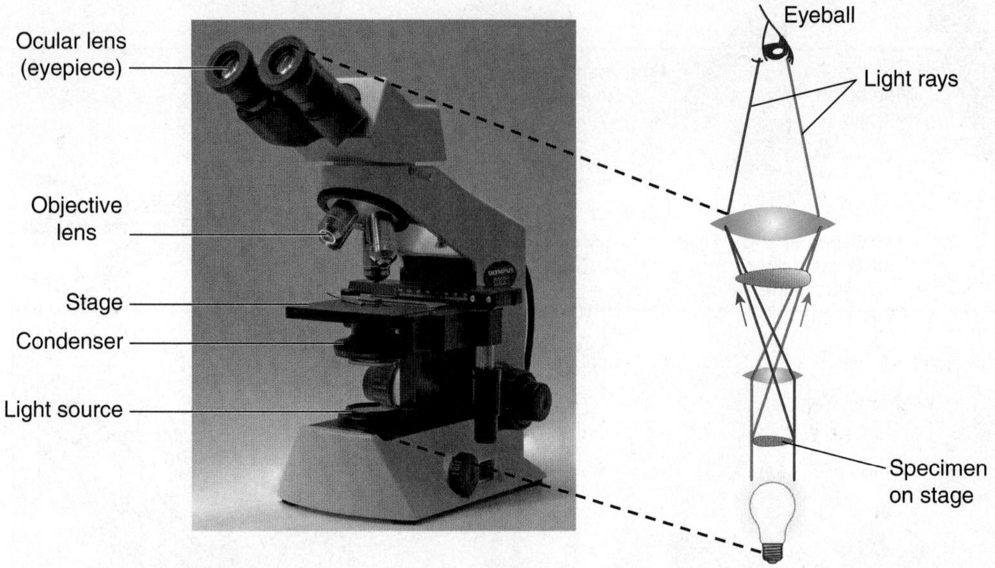

(C) When using the high-power lens (40×) lens, the individual stained bacterial cells are difficult to resolve. (Bar = 20 μm.)

(D) When the oil-immersion (100×) lens is used, more magnification provides better resolution of the same stained cell sample. (Bar = 10 μm.)

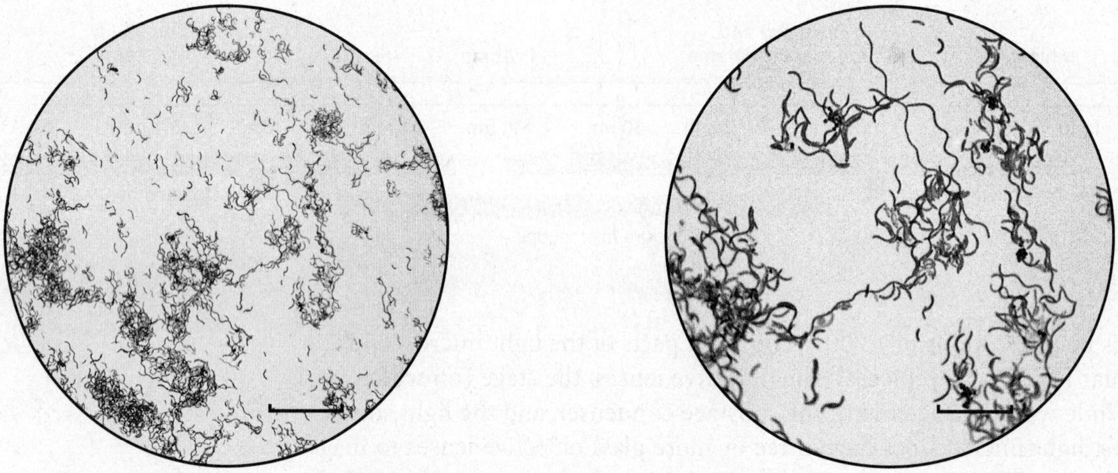

The Electron Microscope

With the development of the **electron microscope** in the 1940s, a whole new world opened to observers. Microscopists now could not only observe cells closer up, they also could see viruses and the small structures inside microbial cells.

An electron microscope uses a beam of electrons (rather than light) passing through a vacuum tube and magnets (rather than glass lenses) to see objects. The electrons bounce off, are absorbed by, or are transmitted through the object to create a final image. The image can be viewed on a microscope screen or monitor, or the image can be captured in digital format for a permanent record.

Images taken with an electron microscope are black and white (gray scale). However, for many publications, these images often are falsely colored to highlight specific features. Most of the electron microscope images you will see in this chapter and throughout the book are such false-color images.

Two types of electron microscopes are in widespread use today.

The Transmission Electron Microscope

The **transmission electron microscope** (**TEM**) is a large instrument (**FIGURE 2.9A**). It produces images of a specimen that previously had been cut into thin slices (100 nm thick) before being placed in the TEM for viewing. The final magnification possible with the TEM is approximately 200,000×, and objects as small as 2 nm can be resolved (**FIGURE 2.9B**). With the TEM, the observer also can see the shapes and patterns of viruses (**FIGURE 2.9C**).

The Scanning Electron Microscope

The **scanning electron microscope** (**SEM**) commonly is used to observe whole cells and viruses. The SEM produces a final magnification of about 20,000×, and the observer can see objects as small as about 7 nm. A key feature of the SEM is that it provides a three-dimensional view of cells and viruses (**FIGURE 2.9D**).

FIGURE 2.9 The Electron Microscope.

(A) A transmission electron microscope (TEM).

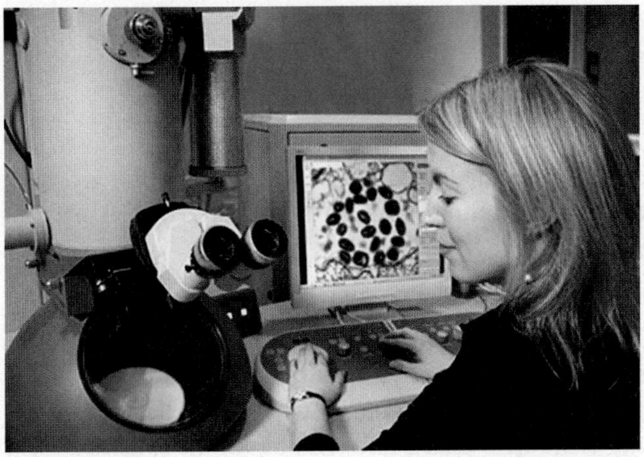

Courtesy of James Gathany/CDC.

(B) A false-colored TEM image of bacterial cells. (Bar = 1 μm.)

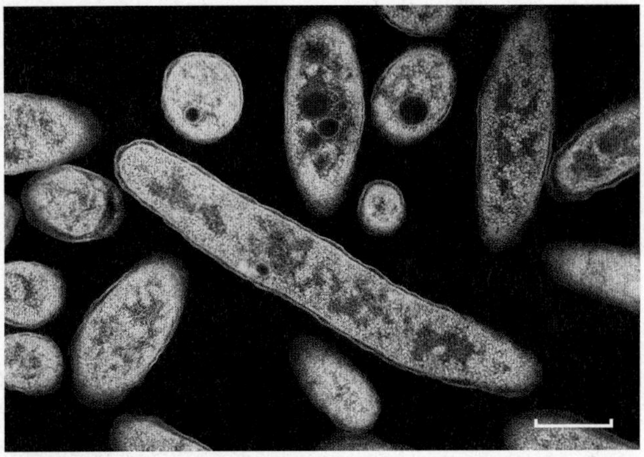

© Phanie/CDC/ Alamy Stock Photo.

(C) False-colored TEM image of viruses. (Bar = 50 nm).

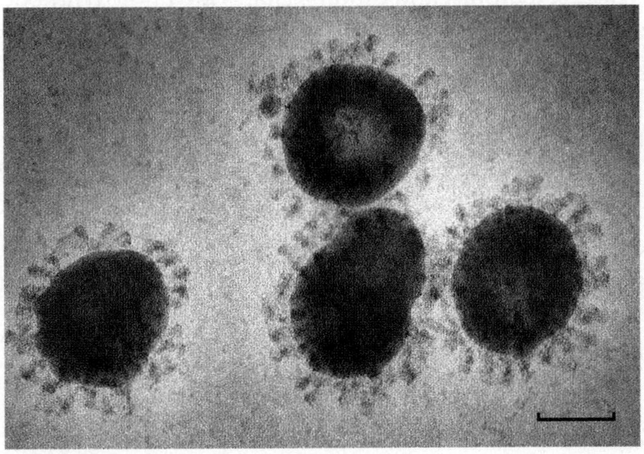

Courtesy of CDC.

(D) A false-colored image of bacterial cells as seen with the scanning electron microscope (SEM). (Bar = 2 μm.)

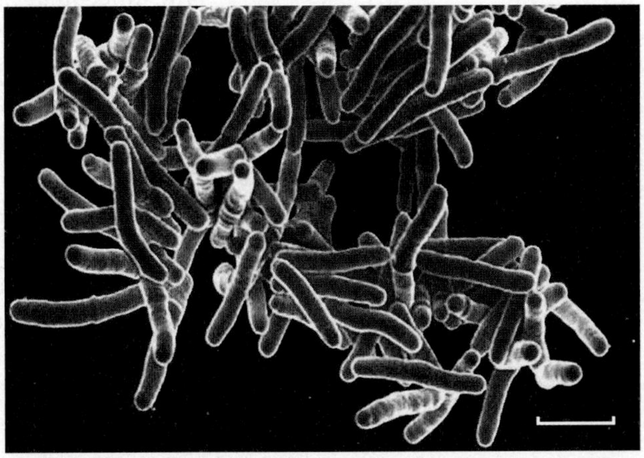

Courtesy of CDC.

2.4 Microbial Groups: Eukaryotes, Prokaryotes, and Viruses

Organism: A living, biological unit made of one or more cells.

So, let's look a little deeper at what can be seen using these microscopes?

The cell is the basic building block of life, and all living **organisms** are built from one or more cells. Most microbes are one-celled (unicellular) creatures, while molds, many algae, as well as plants and animals, are multicellular. An adult human, for example, consists of about 30 trillion cells. When examined with a microscope, all cells are organized in one of two ways.

Fungi, protists, plants and animals are comprised of cells that contain a membrane-enclosed **cell nucleus** that houses the genetic instructions (**deoxyribonucleic acid [DNA]**) for growth and reproduction (**FIGURE 2.10A**). The DNA is wrapped into thread like structures called **chromosomes**. These cells also contain several internal structures called **organelles**. A cell having this type of organization is called a **eukaryotic cell** (*eu* = "true"; *karyon* = "nucleus"), and such organisms are referred to as "eukaryotes."

Bacterial cells lack a cell nucleus. They do contain a bacterial chromosome built from DNA, but it is not surrounded by a membrane (**FIGURE 2.10B**). Few organelles are found in these cells. Therefore, a bacterial cell represents what is called a **prokaryotic cell** (*pro* = "before") and as a group are referred to as "prokaryotes."

Although we will focus on the eukaryotic and prokaryotic organisms more in other chapters, here we will briefly get acquainted with them.

Eukaryotic Microorganisms

The eukaryotic microbes compose the largest part of the eukaryotic group of organisms. These are the protists and fungi, which first appeared on Earth about 2 billion years ago.

Protists

The **protozoa** (sing. **protozoan**) and **algae** (sing. **alga**) compose the **protists**. They are a large and extremely diverse group of mostly single-celled organisms that share certain characteristics with plants and animals. Some species appear to be ancestors of those multicellular life forms.

Many protozoa live freely in the environment (**FIGURE 2.11A**), while other species can be pathogens, such as the ones causing malaria.

The unicellular green algae, diatoms, and dinoflagellates are photosynthetic and, along with the cyanobacteria, are part of the **phytoplankton** inhabiting the ocean surfaces in enormous numbers (**FIGURE 2.11B**). As such, they form the base for all food webs, serving as food sources for other organisms. Algae are commercially important in many parts of the world. Nori, a seaweed used in sushi, is quite common, and some algae are being studied today as a source of biofuels that might serve as a clean energy source in the future.

Algae do not infect humans; however, a few can cause serious poisoning. In addition, so-called "algal blooms," representing enormous growths of an algal population, can cause massive fish kills by using up much of the oxygen dissolved in the water. Such areas often are referred to as "dead zones."

Fungi

The **fungi** (sing. **fungus**) includes the molds, mushrooms, and yeasts. Experts believe there might be up to 5 million different species of fungi. The molds

FIGURE 2.10 Eukaryotic and Prokaryotic Cells Compared.
(A) This photo taken with a light microscope shows several eukaryotic (algal) cells, each containing a cell nucleus and organelles. (Bar = 10 μm.)

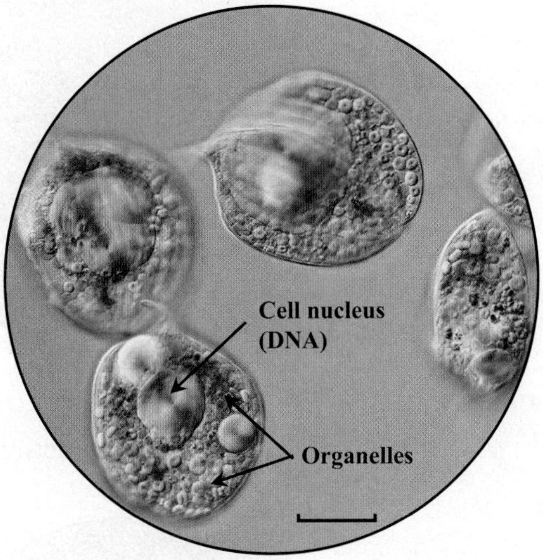

(B) This light microscope image of red-stained bacterial cells lacks the structures typical of a eukaryotic cell. The green ovals are bacterial spores. (Bar = 2 μm.)

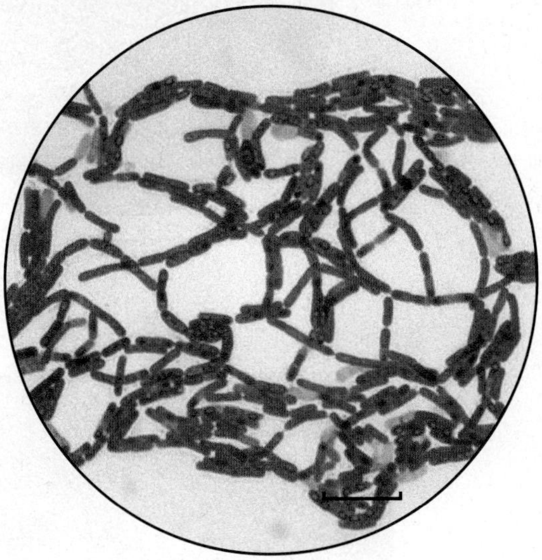

and mushrooms are formed from long, branching cellular filaments that typically grow underground. In contrast, the yeasts often exist as single cells (**FIGURE 2.11C**).

In their feeding patterns, fungi secrete enzymes into the environment that break down the nearby dead plant and animal matter. In fact, many of the molds are among the major decomposers of dead plants and animals. Along with bacterial organisms, they provide essential raw materials for the growth of other organisms.

Many fungi benefit society. For example, several types of mushrooms are edible and quite tasty. Some antibiotics are produced naturally by molds. Other molds are employed in the production of fermented foods, which are foods produced or preserved by the action of microorganisms (e.g., fermented vegetables,

FIGURE 2.11 Gallery of Microbes.

(A) These protozoan cells have engulfed several yeast cells (red structures). (Bar = 10 μm.)

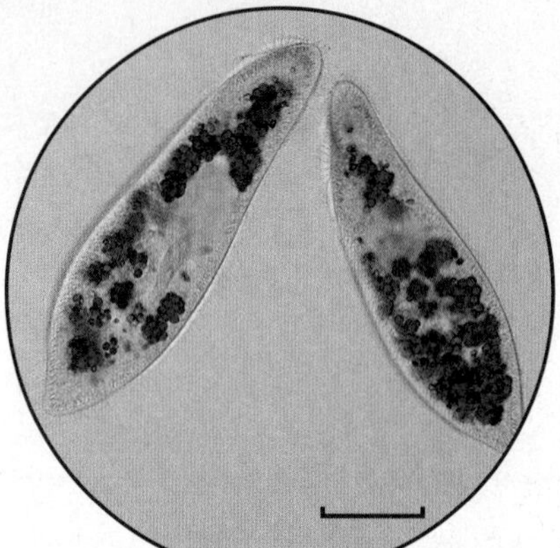

© Lebendkulturen.de/Shutterstock.

(B) A sample of phytoplankton. (Bar = 10 μm.)

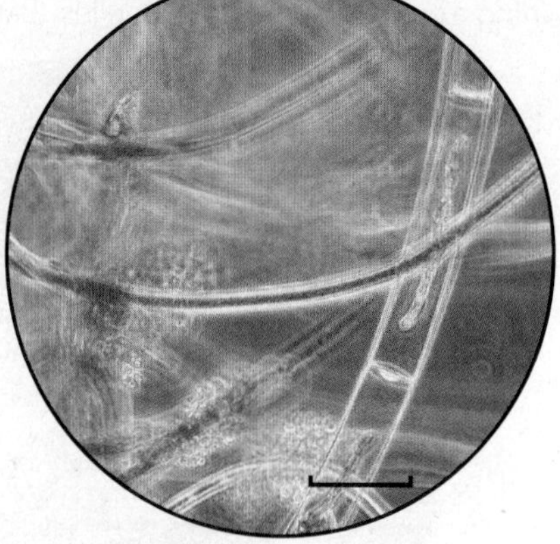

© Rattiya Thongdumhyu/Shutterstock

(C) A light microscope image of oval yeast cells. (Bar = 10 μm.)

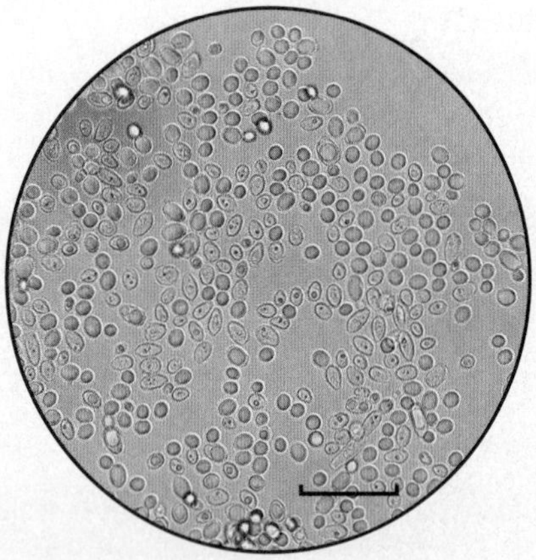

© Rattiya Thongdumhyu/Shutterstock.

(D) Fungal mold spores (small dark dots) and their spore-bearing structures. (Bar = 10 μm.)

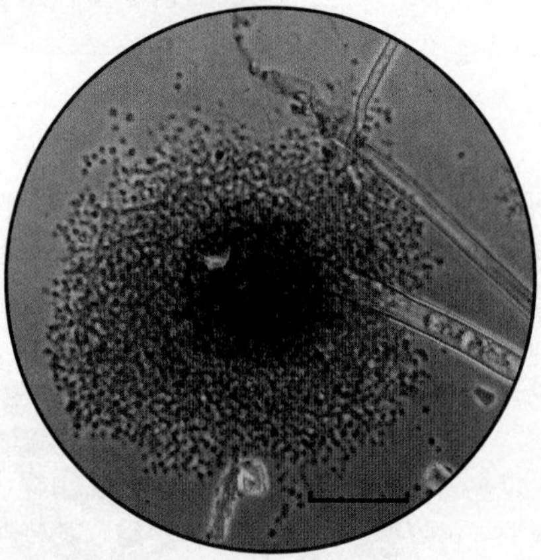

© DE AGOSTINI PICTURE LIBRARY/Getty Images.

(E) A light microscope image showing stained bacterial cells. (Bar = 10 μm.)

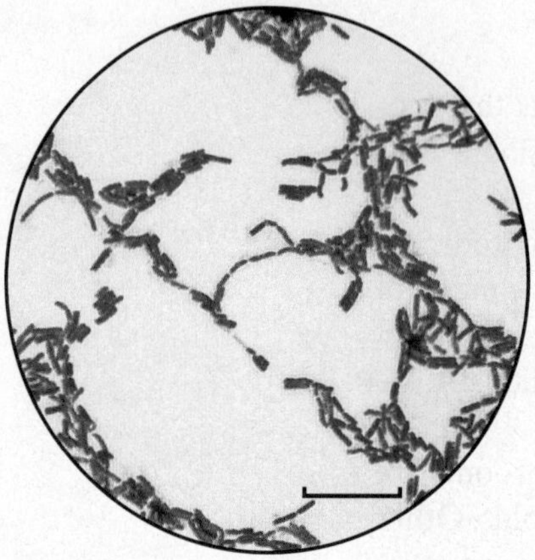

Courtesy of Dr. Jeffrey Pommerville.

(F) False-color TEM image of flu viruses. (Bar = 60 nm.)

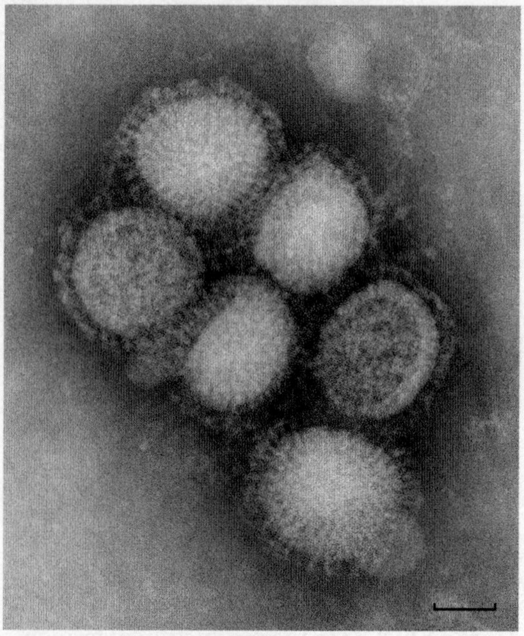

Courtesy of CDC.

yogurt, sour cream). Some types of yeast are involved in wine and beer production, and they are essential in many bread and bakery products.

Unfortunately, some fungal species also cause food spoilage (**FIGURE 2.11D**) while others are serious agricultural pests, causing severe economic damage. Although they are not dominant agents of human disease, some fungal pathogens can cause minor infections (e.g., athlete's foot), while others cause serious infections and sometimes life-threatening diseases of the respiratory tract (e.g., valley fever).

Prokaryotic Organisms

The largest group of organisms on Earth is the prokaryotes. They first appeared approximately 4 billion years ago on an Earth that is very different from today. Over their extraordinarily long period of evolution, they have transformed the planet into what we see today. In doing so, prokaryotes have come to occupy almost every conceivable niche and environment on Earth. There are prokaryotes living in the outer reaches of the atmosphere, at the sunless bottoms of the oceans, in the frigid valleys of the Antarctic continent, and in the guts of all animals, including humans.

Bacteria and Archaea

All prokaryotes once were thought to be closely related because most look alike when observed with the microscope. Today, based on molecular characteristics, scientists have split the prokaryotes into two separate groups.

One group, the **Bacteria**, consists of small, unicellular organisms that lack extensive internal structure (**FIGURE 2.11E**). Among the millions of different species of bacteria, many are master recyclers, decomposing and returning dead plant and animal matter into raw materials used by other organisms.

The second group of unicellular prokaryotes is the **Archaea**. No one is sure how many different species of Archaea exist, but they probably number in the millions. Archaeal cells microscopically look like bacterial cells. However, they are quite different from bacterial cells in having unique chemical features, genetic characteristics, and evolutionary relationships. (More on this in just a few pages.)

Archaea: are-KEY-ah

Many prokaryotes acquire their food by decomposing dead plant and animal matter. Others, such as the **cyanobacteria**, make their own food by photosynthesis in a process very similar to that in the eukaryotic algae and green plants. Although all known archaea and most bacteria do not cause disease, a small number of bacterial species can lead to human infections and sometimes result in serious diseases. Another chapter will study Bacterial Diseases of Humans.

The Viruses

Viruses are the most diverse and numerous entities on Earth; they outnumber the prokaryotes 10 to 1. Viruses are not cellular. Rather, they consist of genetic material surrounded by a coat of protein; in some cases, a membranous envelope encloses the protein shell (**FIGURE 2.11F**). They have no cell nucleus, they do not grow or produce waste products, and they display none of the chemical metabolism associated with prokaryotes and eukaryotes.

Viruses cannot reproduce independently. Rather, they must infect an appropriate **host** cell, and eventually take over the control of that cell. Then, their genetic information hijacks the chemical machinery of the host cell to produce simultaneously hundreds more viruses. These new particles then burst out of the cell, often leaving the host cell damaged or destroyed. As this wave of

Host: A cell or organism in which a microbe or virus can live, feed, and reproduce (replicate).

cell destruction spreads from cell to cell, the infected tissue suffers damage and the symptoms of disease develop.

So, how are all these microbes and viruses cataloged?

2.5 Taxonomy: Cataloging Life

Trying to catalog items always has been part of human nature. Therefore, ever since Linnaeus' time, scientists have tried to figure out how all the forms of life are related. To finish off this chapter, let's examine the two schemes scientists have devised in an attempt to illustrate this relationship between all life, including the microbes.

Cataloging and Organizing Life

In addition to devising binomial nomenclature, Linnaeus also established the ground rules of **taxonomy**, which involves the **classification** (cataloging) of organisms into organized groups. In this tiered system, the least inclusive, most fundamental tier is the species. A group of closely related species is gathered together to form a genus. Likewise, a group of similar genera comprise a **family**.

For instance, let's make a tiered classification for humans (*H. sapiens*), gorillas (*Gorilla gorilla*), domestic dogs (*Canis lupus*), and coyotes (*Canis latrans*) (**FIGURE 2.12**). All four are separate species, but coyotes and dogs are otherwise very alike, so both are in the genus *Canis*. However, humans and gorillas only become similar at the family level, as both organisms have 99% genetic similarity, similar senses, similar hands (with unique fingerprints), and comparable gestation periods.

FIGURE 2.12 A Hierarchical Classification of Humans, Gorillas, Domestic Dogs, and Coyotes.

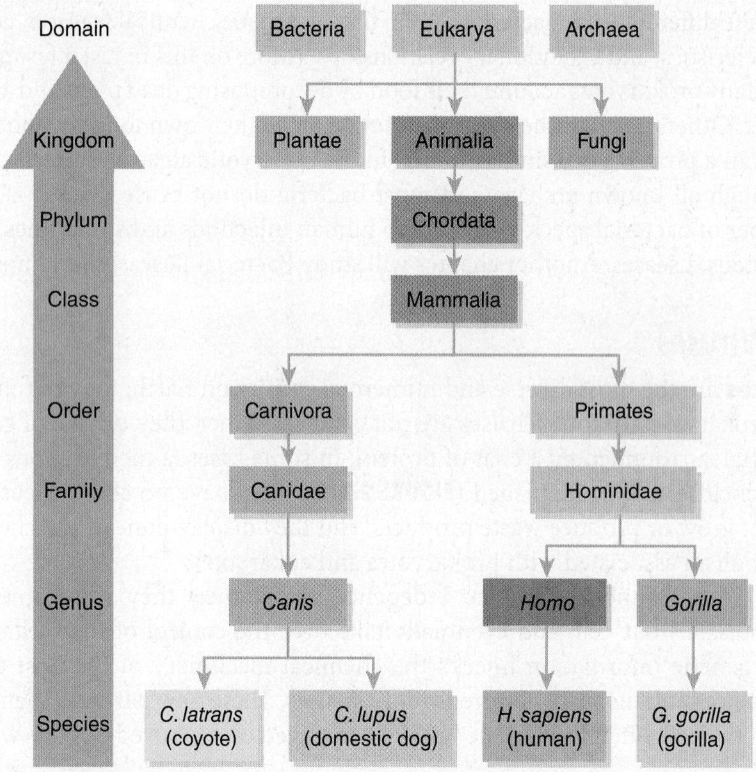

TABLE 2.1 Taxonomic Classification of Humans and Three Microbes[1]

Taxon	Human	Brewer's Yeast	Gut Bacterium	Environmental Microbe
Domain	Eukarya	Eukarya	Bacteria	Archaea
Kingdom	Animalia	Fungi		
Phylum	Chordata	Ascomycota	Proteobacteria	Euryarchaeota
Class	Mammalia	Saccharomycotina	Gammaproteobacteria	Halobacteria
Order	Primates	Saccharomycetales	Enterobacteriales	Halobacteriales
Family	Hominidae	Saccharomycetaceae	Enterobacteriaceae	Halobacteriaceae
Genus	*Homo*	*Saccharomyces*	*Escherichia*	*Halobacterium*
Species	*Homo sapiens*	*Saccharomyces cerevisiae*	*Escherichia coli*	*Halobacterium salinarum*

[1] See **Appendix A** for organism pronunciations.

Taxonomists then use progressively more inclusive tiers of classification. Related families are organized into an **order**, orders are brought together in a **class**, and various classes comprise a **phylum** (pl. **phyla**). In our classification scheme for humans, gorillas, dogs, and coyotes, humans certainly have different characteristics from dogs and coyotes. However, all four animals are similar in producing milk (females), giving birth to live young, and possessing hair or fur. So, all are classified together in the class called Mammalia.

Finally, all phyla of plants, animals, and fungi are grouped together in a **kingdom**. The prokaryotes and the protists currently are not part of any kingdom because of the uncertainty of species relationships. **TABLE 2.1** shows the taxonomic classification for humans, a fungal yeast, a gut bacterium, and an environmental microbe.

The Tree of Life

Besides organizing life into a hierarchical system, taxonomists since Darwin have worked to arrange all life into a single scheme that shows the evolutionary relationships among and between all organisms.

To illustrate how organisms are related to one another and how all organisms originated, more and more organism characteristics (genetic, molecular, biochemical) have been used. DNA sequencing techniques and information technology especially have been useful to build what is referred to as the **tree of life** (**TOL**) (**FIGURE 2.13**). As shown, all life has been assigned to one of three groups called **domains**. Consequently, a **three-domain system** comprises the TOL, which has been simplified for our purposes. There are many more branches and "twigs" on the full version that taxonomists work with today.

One of the modern taxonomy pioneers was Carl Woese. Through DNA sequencing, Woese and his coworkers uncovered two very different evolutionary histories for the prokaryotes. Therefore, they proposed that the prokaryotes be split into two domains, the **domain Bacteria** and the **domain Archaea**. The third domain, which includes all the eukaryotic organisms (protists, fungi,

FIGURE 2.13 The Tree of Life. All organisms are assigned to one of three domains. Microorganisms (red branches) make up the clear majority of the tree. The branches and "twigs" are only a small representation of all the known groups.

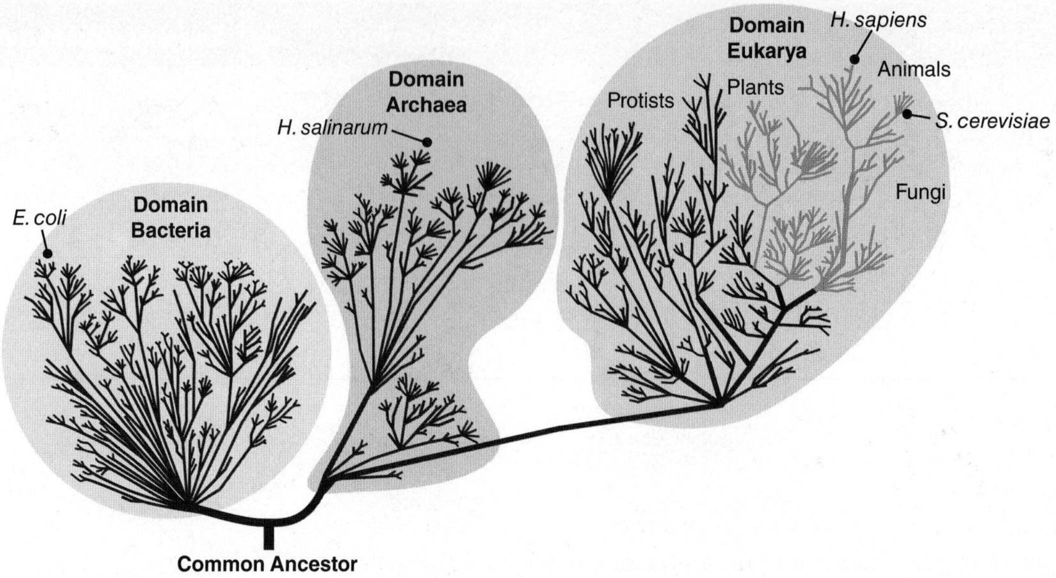

Eukarya: YOU-care-ee-ah

plants, animals—and humans), had a third unique set of characteristics based on DNA sequencing. These organisms were assigned to the **domain Eukarya**.

So, what does the TOL reveal ?

About 4 billion years ago, life took hold after many unsuccessful attempts (see chapter opening image). At that time, an ancestral prokaryotic microbe (common ancestor) gave rise to two lineages. One would evolve into organisms that today are in the domain Bacteria. The other lineage evolved into organisms in the domain Achaea. As Darwin put forward in his **Theory of Natural Selection** (commonly call "survival of the fittest"), as organisms evolved and diverged (drawn as separate branches and twigs on the tree), some were better fit and survived to reproduce. Less fit ones were unable to adapt and slowly died off and became extinct. So, the branches and "twigs" on the TOL represent how today's organisms (at the tips of the branches) came to be.

Some 2 to 3 billion years ago, taxonomists believe a particularly "well-fit" branch of the domain Archaea "bloomed" and gave rise to protists and to the ancestors of the domain Eukarya. With the evolution of multicellularity, plants, fungi, and animals appeared, forming considerably more branches and twigs on the TOL. Note that in the TOL:

- The microbes make up almost 70% of the organisms. They represent all organisms in the domains Bacteria and Archaea, and a large part of those in the domain Eukarya (protists and many fungi).
- All the currently living organisms are the product of almost 4 billion years of evolution. In other words, all current life is connected through a common ancestry by the branches to distant relatives and extinct ancestors.
- The viruses are not included. This omission is intentional because the viruses are not organisms. As mentioned earlier, viruses are not made of cells, which is characteristic of all prokaryotes and eukaryotes.

Today, the TOL continues to expand as taxonomists continue to discover and study new organisms.

The Tree of Life and Society

You might be thinking that all this work about the TOL is interesting scientifically. But what about its relationship to society? In fact, taxonomists and other biologists hope that a comprehensive understanding of the TOL will provide enormous benefits to society. Here are but a few examples.

- Millions of people around the world are discovering their genetic heritage, ethnicity, and family history by having their DNA sequenced. The technology that is used to identify one's heritage has its origins in the tools used to build the TOL.
- Hundreds of new life forms are discovered each year. Many could have useful benefits for human health. Today, there have been many patent filings in industry for medical diagnostics whose products were identified through TOL studies.
- Knowledge of organism relationships among the world's most important crop plants (cereal grains) and domestic animals is crucial for their continued genetic improvement. TOL studies are helping to point out the relationships to other plants and animals.
- Billions of dollars are spent worldwide each year controlling invasive species (microbial, plant, and animal). TOL studies can help identify potential invasive species before they do economic and ecological damage in the environment.
- Around the world, 60,000 people die each year from snakebites! Developing effective medications to counteract snake venom can come from TOL research because these evolutionary studies help correlate venom properties to other organisms that naturally produce antivenom compounds.

A Final Thought

The diversity of life has astounded scientists and explorers for centuries. The challenge of naming and categorizing this vast array of organisms is nowhere more complicated than it is with the microbial world. With so many new species being discovered every year, scientists today are dependent on different forms of microscopy and various biochemical, genetic, and molecular tools to set organisms apart from one another. The fruits of this labor have resulted in the three-domain system and the TOL, which visibly describe the various forms of microscopic life. The history of taxonomy, from Linnaeus to Woese, is a magnificent study of the evolution of life and one that continues to challenge the thinking about the basic nature of life itself.

Chapter Discussion Questions

What Was He Thinking?

From your reading of this chapter, identify and discuss five major points about microbes that the author was trying to get across to you.

Questions to Consider

1. Louis Pasteur made several contributions to the understanding of microbes and the field of microbiology. In this figure, identify the four contributions he made, as indicated by the four panels.

© Bilha golan/Shutterstock.

2. An online news site contains an article about "the famous fungal genus ecoli." How many errors can you find in this phrase? Rewrite the phrase with the mistakes corrected.
3. In a respected scientific journal, an author wrote, "Linnaeus gave each life form two Latin names, the first denoting its genus and the second its species." A few lines later, the author wrote, "Man was given his own genus and species *Homo sapiens*." What is conceptually and technically wrong with both statements?
4. Which person (Hooke, Leeuwenhoek, Linnaeus, Pasteur, Koch, or Woese) discussed in this chapter do you think had the most impact on the science of his day [Hooke, Leuwenhoek (late 1600s), Linnaeus (early 1700s), Pasteur, Koch (late 1800s), and Woese (late 1900s)]?
5. What infectious disease breakthrough made by Robert Koch is being celebrated on this stamp from the African nation of Burkina Faso (formerly called Upper Volta)? Hint: the image at the left of the stamp is an X-ray image of the lungs.

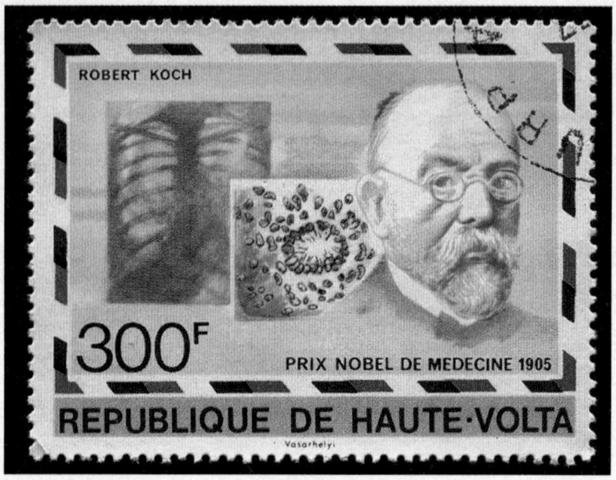

© rook76/Shutterstock.

6. Explain why viruses are not included in the tree of life.
7. Maria missed class last week and is copying your notes from that class. While copying, Maria comes across the term "organism" many times, but there are no notes as to its meaning. She asks you "What is an organism?" How would you reply to her question?
8. Every state has an official animal, flower, and/or tree, but one state has an official bacterial species named in its honor: *Methanohalophilus oregonense*. What's the state? Decipher the meaning of the species name. (Note: *methano* = methane gas-producing; *halophilus* = salt loving; *ense* = "belonging to.")
9. If Leeuwenhoek, Linnaeus, and Darwin each got in a time machine and came to today's world, what do you think each would make of microbial diversity and the evolution of life as it is known today?
10. Fill in the following chart with the relevant information, as provided in this chapter.

Group	Domain	Cell Organization Type	Presence of Cell Nucleus	Presence of Organelles	Causes Infectious Diseases in Humans
Bacteria					
Archaea					
Protists					
Fungi					
Viruses					

CHAPTER 3

Molecules of the Cell: The Building Blocks of Life

The Spark

Stanley Miller was intrigued. As a young graduate student at the University of Chicago, he became fascinated with the ideas presented in a lecture by Professor Harold Urey. Urey told the audience he believed the atmosphere of primitive Earth was very different from today's atmosphere. He proposed that billions of years ago, the air consisted primarily of methane, ammonia, hydrogen sulfide, and hydrogen gases. Urey further suggested that using this atmosphere, it might be possible to generate some of the chemical building blocks needed for the emergence of life.

Life on Earth began around 4 billion years ago, evolving from some form of "**primordial soup**" into the incredible diversity of microbes and life we see today. So, could the very first molecules of life on Earth arise as Urey proposed?

Primordial soup: A pond or body of water rich in substances that could provide favorable conditions for the emergence of life.

In 1952, Miller took a research position in Urey's lab and asked to test Urey's hypothesis. After much argument and Miller's persistence, Urey gave his approval to the experiment. After 3 months of planning, Miller's experiment attempted to simulate the primitive Earth atmosphere. Miller put water vapor along with methane, ammonia, and hydrogen gases in a closed reaction vessel (**FIGURE 3.1**). But how would he get these gases to interact with one another? There needed to be a spark.

Because lightning storms would have been very common in Earth's early atmosphere, Miller had the brilliant idea of putting an electric charge through the gas mixture to simulate lightning going through the atmosphere. So, using electrical sparks in the reaction vessel, Miller allowed the reaction to continue for

CHAPTER 3 OPENER Many ideas have been put forward to explain how life began on planet Earth. Here Dr. Stanley Miller is shown recreating his famous experiment performed in 1953. A spark was sent through a flask containing gases believed present in the atmosphere of the Earth billions of years. The experiment produced several amino acids, one of the building blocks of life.

FIGURE 3.1 The Miller-Urey Experiment. The experiment consisted of a reaction vessel containing Earth's primordial gases in which an electrical charge provided the spark for the chemical reactions.

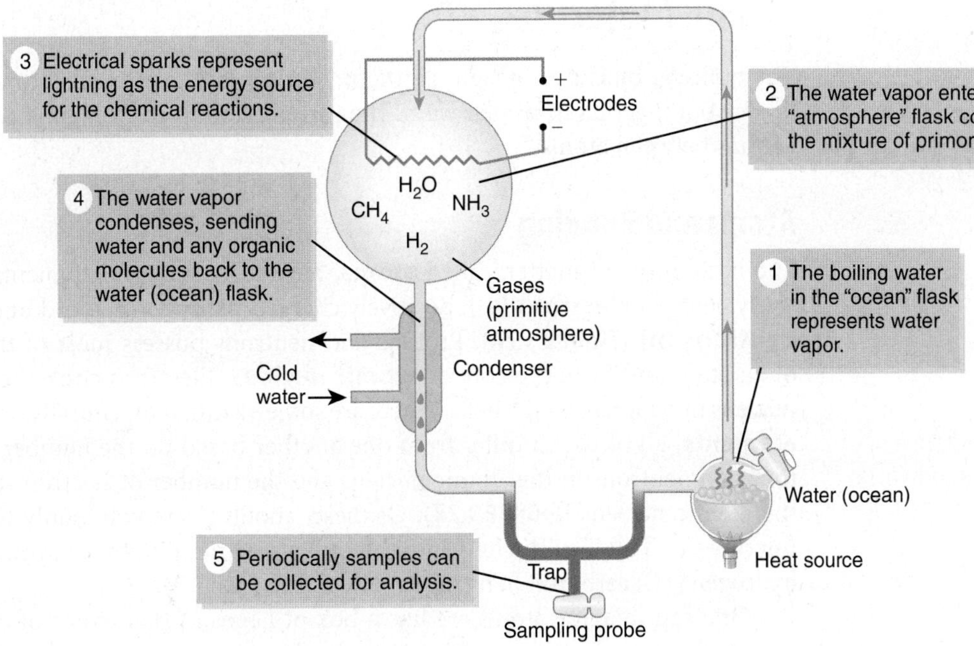

a week. As the days passed, he noticed the accumulation of a brown slime on the reaction vessel walls and a yellow-brown color in the water. When analyzed, Miller detected five amino acids, which are some of the building blocks of proteins.

The Miller-Urey experiment provided evidence that some of the building blocks of life, like amino acids, could be made under conditions simulating early Earth. Today, we know Earth's primitive atmosphere did not have the exact composition Urey proposed. Still, Miller's work represents a landmark experiment. In fact, in 2008, a year after Miller died, more modern techniques of analysis turned up more amino acids and other substances of interest from the original reaction vessels Miller used in 1952.

Today, we still are not certain how life began, but we can study the building blocks and other large molecules that microbes, viruses, and all life need to survive. These are the carbohydrates, lipids, proteins, and nucleic acids—and they are the topic in the coming pages.

LOOKING AHEAD

After reading and completing this chapter, you will be able to:

3.1 Draw the structure of an atom.
3.2 Describe the various simple sugars and polysaccharides.
3.3 Compare fats, phospholipids, and sterols.
3.4 Draw an amino acid, and contrast the four levels of protein structure.
3.5 Identify the parts of a nucleotide and distinguish DNA from RNA.

3.1 Chemistry Basics: Atoms, Bonds, and Molecules

Organic: In chemistry, refers to any carbon-containing substance except carbon monoxide and carbon dioxide.

The building blocks of life and all the **organic** substances needed to generate life are put together in similar ways. They are all built from atoms and the interactions between atoms.

Atoms and Bonding

The basic units of matter, called **atoms**, have three basic components: negatively charged **electrons (e^-)**, positively charged **protons (p^+)**, and uncharged **neutrons (n)** (**FIGURE 3.2A**). Protons and neutrons possess most of the mass of an atom and form the core or **atomic nucleus**. Electrons circle the atomic nucleus in regions called "shells." There are some 92 different naturally occurring **elements**, all of which differ from one another based on the number of protons and neutrons in the atomic nucleus and the number of electrons orbiting the atomic nucleus (**FIGURE 3.2B**). Of these, about 25 are commonly found in microbes and all life. The four most abundant (with their atomic symbols) are hydrogen (H), carbon (C), nitrogen (N), and oxygen (O).

Element: A pure substance that cannot be broken down into simpler substances by ordinary chemical means.

One can imagine atoms are like a box of Legos in that atoms of different sizes and configurations can "snap together" to form different chemical structures. When this happens, the electrons of the interacting atoms form links called **chemical bonds** that hold the atoms together. So, a **molecule** is two or more atoms bonded together by their electrons. For example, the gases used in the Miller-Urey experiment included methane and ammonia. Methane consists of one carbon atom bonded to four hydrogen atoms and is written as CH_4 (Note: if there is just one atom in a molecule, there is no subscript number used); ammonia is written as NH_3.

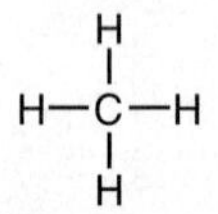

Methane

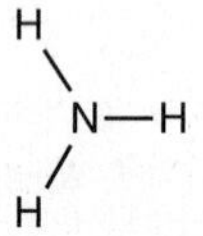

Ammonia

Water

All life occurs in watery surroundings and approximately 70% of the mass of a cell is water. Therefore, another essential molecule for life is water, which consists of two hydrogen atoms bonded to one oxygen atom (H_2O).

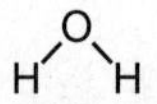

Water

FIGURE 3.2 The Structure of Atoms.

(A) An atom consists of protons and neutrons in the atomic nucleus and surrounding shells of electrons.

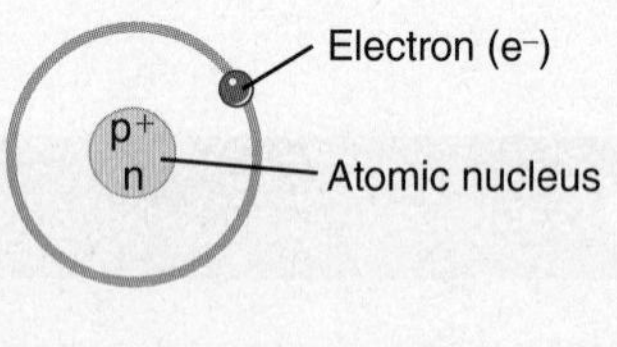

(B) Different atoms have different numbers of protons, neutrons, and electrons.

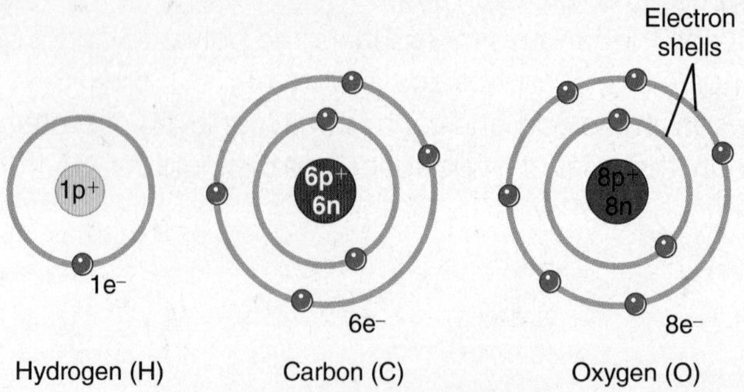

All the chemical reactions occurring in cells take place in water because water acts as the universal **solvent** in cells. Take for example what happens when you put a **solute** like salt or sugar in water. The salt or sugar dissolves, forming an **aqueous solution**; that is, one or more solutes dissolved in water. Water molecules also are part of many chemical reactions, as we will see later in this chapter.

Solvent: The substance doing the dissolving in a solution.

Solute: The substance dissolved in a solution.

Besides its solvent properties, water also has other characteristics that make it an ideal molecule for life, as pointed out in **A CLOSER LOOK 3.1**.

A CLOSER LOOK 3.1

Water and Life

Life on Earth certainly evolved in a watery primordial soup some 4 billion years ago and remained in a water environment for some 3 billion years before spreading to land where survival still depended on water. Without water, the human body would be unable to survive for more than 1 week. In fact, water is such a necessity for life as we know it, that when scientists send spacecraft and rovers in search for traces of microbial life on other worlds, such as Mars, water is one of the most important molecules they look for (see Figure A).

Besides its property as the solvent for life, what are water's other life-supporting properties? We can identify three.

Cohesion

Water molecules will stick together due to weak bonding. The sticking together is called cohesion. For example, it is the cohesion between water molecules that allows water to be transported up trees from the roots to the leaves.

Also, by sticking together, water has a high surface tension; it is hard to break water molecules apart. That means small insects can "walk on water" as if there was a film on the water surface.

FIGURE A The white bits in this photo are Martian water ice.

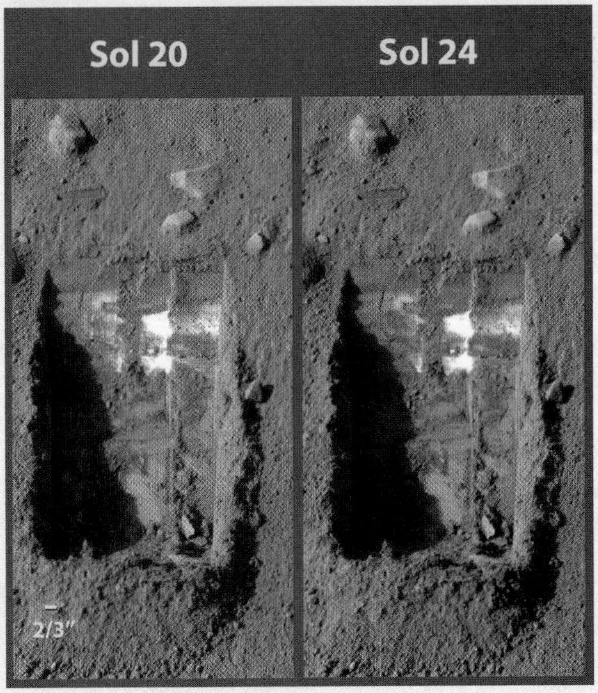

Courtesy of JPL-Caltech/University of Arizona/Texas A&M University/NASA.

Temperature Moderation

It takes a lot of energy (heat) to raise the temperature of water. For instance, if you have ever accidently touched a metal pot full of water being heated, the metal heats up much faster than the water. The bonding between water molecules gives water a stronger resistance to temperature changes than occurs with most other substances. Likewise, coastal areas, bordering an ocean, such as San Diego, in the summertime tend to have more moderate air temperatures than deserts, such as around Phoenix, because of the higher humidity (water content) in the atmosphere. There is more water to absorb the heat along the ocean coast.

In the human body, water also moderates body temperature by evaporative cooling. When water evaporates, the water left behind cools down because the water molecules with the greatest energy (the "hottest" ones) vaporize first. That's why sweating is important—it helps prevent an individual from overheating.

Insulation

As you know, frozen water (ice) floats, unlike most substances that will sink on freezing. These substances sink because the atoms move closer together and become denser than the surrounding liquid when frozen. Due to bonding between water molecules, these molecules move farther apart, making the ice less dense than the surrounding liquid water and the ice therefore floats.

By ice floating, it acts as insulation. If frozen water was denser than liquid water, it would sink, and in the winter all ponds and lakes would eventually freeze solid, freezing any living creatures (including microbes). However, because ice floats, it forms an insulating "blanket" over the body of water in winter, allowing the water underneath to remain liquid—and life to survive.

So, drink up and stay hydrated. And that goes for microbes too!!

The Molecules of Microbes

Most of the molecules in a cell are significantly larger than simple molecules like methane and water. Rather, the molecules of life often are hundreds to billions of atoms in number. For this reason, they are called "macromolecules" (*macro* = "large") and form the carbohydrates, lipids, proteins, and nucleic acids. Their structural differences are a consequence of the way the atoms in each macromolecule interact with one another.

3.2 Carbohydrates: Simple Sugars and Polysaccharides

Carbohydrates are organic molecules containing carbon, hydrogen, and oxygen, generally in a ratio of 1:2:1. Thus, the basic formula unit for a carbohydrate is CH_2O. Carbohydrates vary from relatively small, simple sugars to extremely large, complex macromolecules called polysaccharides.

Simple Sugars

Monosaccharides can contain three to seven carbon atoms. Among the most common simple sugars are the five-carbon sugars (pentoses), such as ribose and deoxyribose. Also common are six-carbon sugars (hexoses), including glucose, fructose, and galactose. These hexoses have the same numbers of carbon, hydrogen, and oxygen atoms ($C_6H_{12}O_6$). However, their atoms are bonded in a different order, which gives rise to their unique structures. Monosaccharides, especially glucose, have two roles. They serve as the essential building blocks for polysaccharides and as sources of energy for cellular activities.

The monosaccharides are bonded together in a chemical reaction that involves the removal of H_2O from the sugars. When water is lost during the formation (synthesis) of a molecule, it is known as a **dehydration synthesis reaction** (**FIGURE 3.3A**). These reactions involve the action of cellular **enzymes**, protein molecules capable of rearranging the atoms of organic molecules while themselves remaining unchanged.

Disaccharides are formed by linking together two monosaccharides through a dehydration synthesis reaction (see Figure 3.3A). Among the common disaccharides are:

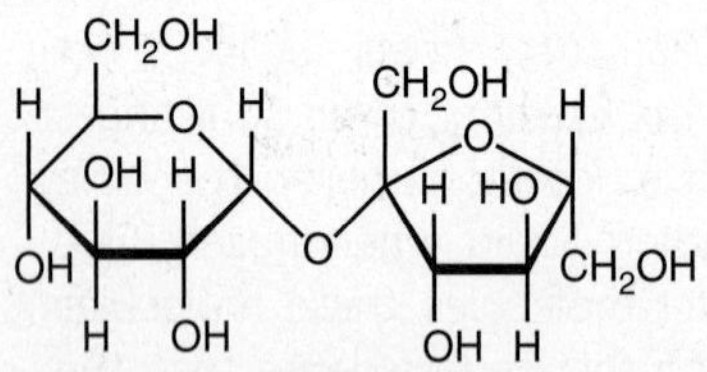

Lactose

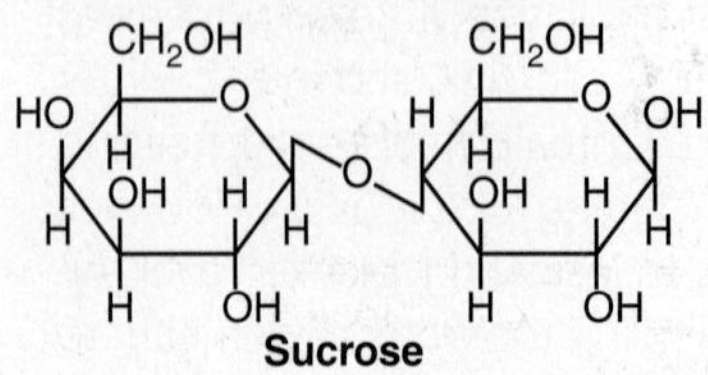

Sucrose

- **Maltose**, also known as malt sugar. This disaccharide is found in cereal grains such as barley, where the sugar is fermented by yeast cells to produce the alcohol in beer.
- **Lactose**, the principal sugar in milk. This disaccharide is a combination of a glucose molecule and another monosaccharide called galactose. Lactose can be chemically changed to lactic acid by certain species of bacteria. The acid causes milk to become sour. However, the reaction can be controlled and used in the dairy industry to produce yogurt, buttermilk, and sour cream.
- **Sucrose**, commonly known as table sugar. Sucrose is a combination of a glucose molecule and the monosaccharide fructose. Sucrose is a source of energy for cells and, as a natural sugar, is extracted from sugarcane and sugar beets.

FIGURE 3.3 Carbohydrates Consist of the Monosaccharides, Disaccharides, and Polysaccharides.

(A) Glucose is a monosaccharide that can bond with another glucose in a dehydration synthesis reaction to form maltose, a disaccharide.

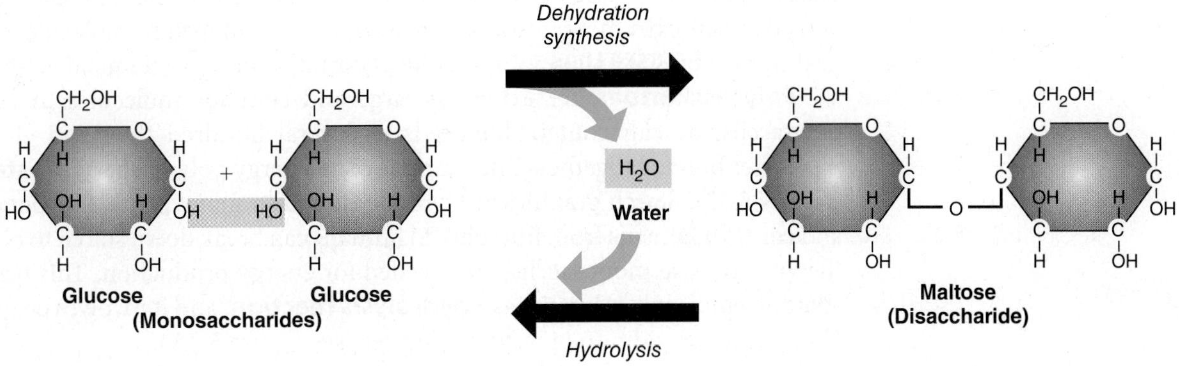

(B) The bonding of additional glucose molecules leads to the formation of a polysaccharide, such as starch. Note that each hexagon represents a glucose.

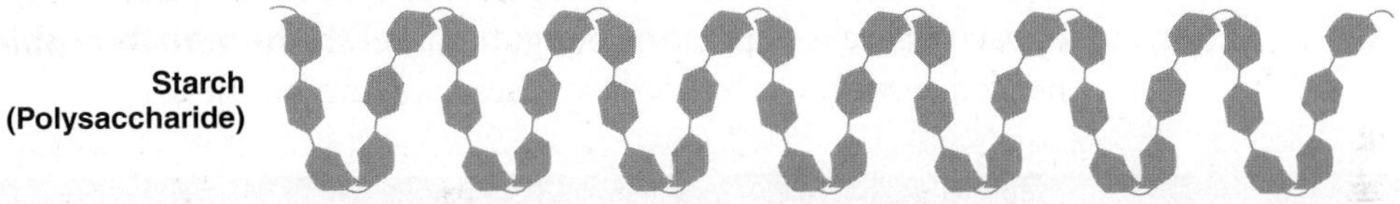

(C) *N*-acetylmuramic acid (NAM) and *N*-acetylglucosamine (NAG) are modified simple sugars that can bond together to form the bacterial cell wall peptidoglycan.

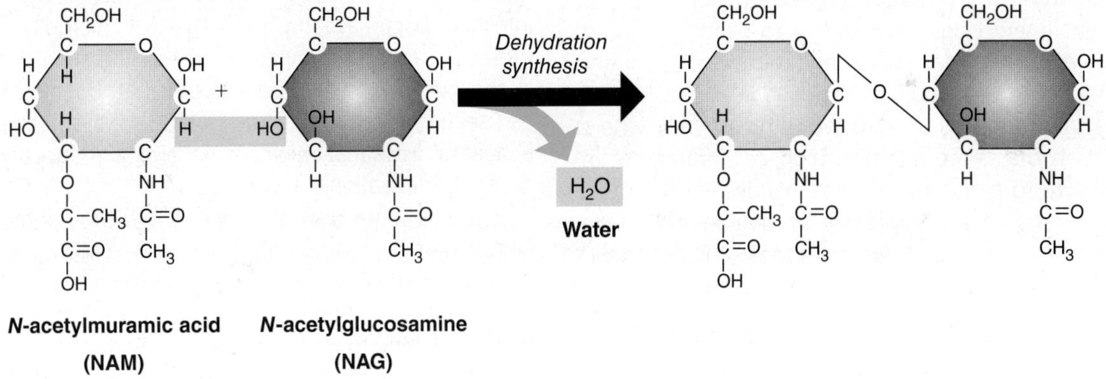

(D) Peptidoglycans are held together by side chains of amino acids.

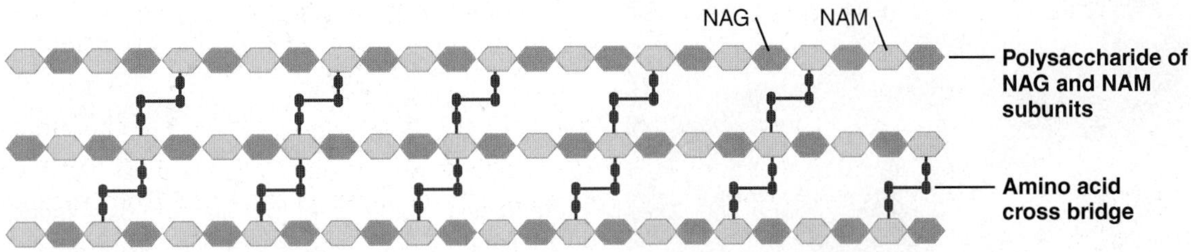

Oligosaccharides and Polysaccharides

Oligosaccharides are carbohydrates that contain 3 to 10 simple sugars bonded together. These carbohydrates are less familiar to most people. However, they are extremely important at birth and as important molecules for a healthy gut. **A CLOSER LOOK 3.2** spotlights their importance to human health.

Polysaccharides are extremely large and complex molecules. A single polysaccharide can contain hundreds to several hundred thousand glucose molecules bonded together. One example of an energy polysaccharide is **starch** (**FIGURE 3.3B**). Starch granules are typically found in many plants like potatoes and corn. Organisms from microbes to humans can break down starch to obtain the vital glucose monosaccharides needed for energy production. This tearing apart of a molecule is known as a **hydrolysis reaction**, and it involves enzymes along with the addition of water molecules (see Figure 3.3A).

Other large carbohydrates in cells are structural polysaccharides. For example, **cellulose**, a major part of plant cell walls, is composed of long chains made of glucose. However, the glucose molecules in cellulose are bonded together differently than in starch. Interestingly, humans lack the necessary enzyme to break the bonds between glucose molecules in cellulose, so we cannot digest the cellulose in the plant foods we eat. However, many of the trillions of bacterial cells populating our gut (part of the **human microbiome**) have the needed enzymes to break down these polysaccharides.

A CLOSER LOOK 3.2

How Sweet It Is

An anonymous adage says, "Breastfeeding is nature's health plan." As we learn more about ourselves and the microbial world, this saying becomes even more true.

Recent interest has been drawn to oligosaccharides because of an important discovery: these carbohydrates promote and protect human health. One natural source of oligosaccharides is in a mother's breast milk where the carbohydrates are found in high concentrations. In fact, after lactose and fat, oligosaccharides are the most abundant chemical component in human milk. One oligosaccharide, called 2′-fucosyllactose (2′FL) is found in the breast milk of 80% of all mothers. 2′FL has been shown to support a healthy immune system in infants. However, newborns (and even adults, for that matter) lack the enzymes to break down these slightly sweet oligosaccharides. So, why are they present in mom's milk?

© Goodluz/Shutterstock.

2′FL and other human milk oligosaccharides (HMOs) have a prebiotic effect that helps support the growth of intestinal bacteria composing the gut microbiome. The presence and dominance of these intestinal bacteria in the gut reduce the chances that pathogenic bacteria can secure a foothold in the gut. So, HMOs help ensure a healthy intestinal microbiome and a reduced risk of dangerous intestinal infections.

Here is one way that some HMOs might work. Researchers have discovered that some HMOs mimic the sites on intestinal cells to which bacterial pathogens would attach. Therefore, because of the prevalence of HMOs, the pathogens bind to the HMOs rather than the intestinal cells.

Importantly, HMOs might reduce the risk of premature infants becoming infected with a potentially life-threatening disease, such as necrotizing enterocolitis (NEC). NEC is a devastating disease that affects mostly the intestine of premature infants. The wall of the intestine is invaded by bacteria, which results in an infection that can ultimately perforate and destroy the wall of the intestine.

Some HMO metabolites also might affect the nervous system or the brain, influencing the long-term development and behavior of children.

Today, HMOs are appearing as supplements (prebiotic) in some infant formulas and baby foods. These supplements would ensure that babies who are not being breastfed still get the essential oligosaccharides they need for good health.

The bacterial cell wall represents another structural polysaccharide. It is constructed from two sugars that have been modified to contain nitrogen (**FIGURE 3.3C**). Long linear chains of these two sugars form fibers called **peptidoglycan**. In many bacteria, multiple layers of peptidoglycan build the cell wall. The adjacent chains are stabilized and held together by cross bridges made of five amino acids. The chapter Exploring the Prokaryotic World will describe the cell wall in more detail.

3.3 Lipids: Fats, Phospholipids, and Sterols

We are all familiar with some types of **lipids**, such as the animal fats and plant oils in our diet. Like the carbohydrates, these lipids are used by some microbes and many other organisms for energy and energy storage. Lipids are built from carbon, hydrogen, and oxygen atoms, but there are more carbon–hydrogen bonds in a lipid than in a similar sized carbohydrate.

If you shake a mixture of water and oil, the two liquids will separate from one another because lipids do not dissolve in water. Therefore, lipids are **hydrophobic** (*hydro* = "water;" *phobic* = "fearing"), whereas simple sugars like glucose and sucrose are **hydrophilic** (*philic* = "loving") because they dissolve in water. Based on their chemical composition, lipids are split into three different groups: fats and oils, phospholipids, and sterols.

Fats and Oils

Fats and **oils** are composed of a three-carbon molecule called glycerol and three long chains of carbon atoms called **fatty acids** (**FIGURE 3.4**). Like carbohydrates, lipids are the result of enzyme-regulated dehydration synthesis reactions in cells. The key difference between fats and oils is the fatty acid chains. A fatty acid is **saturated** if it contains the maximum number of hydrogen atoms bonded to the carbon backbone. A fatty acid is **unsaturated** if it contains less than the maximum number of bonded hydrogen atoms. Notice in Figure 3.4 that two of the chains are saturated and straight. The third is unsaturated and bent. Importantly, saturated fatty acids tend to make the lipid solid at room temperature (e.g., butter and animal fat), while unsaturated fatty acids tend to make the lipid liquid at room temperature (e.g., vegetable oils and fish oils).

Phospholipids

Phospholipids have a phosphate group (PO_4) in place of one fatty acid chain (**FIGURE 3.5A, B**). The phosphate group forms the "head" end of the phospholipid and is hydrophilic. The fatty acid tails are hydrophobic. This mix of hydrophilic and hydrophobic properties is key to membrane formation. Remember, cells live in a watery environment, and water and oils do not mix. By organizing as two parallel layers (or bilayer), phospholipid heads of one layer point toward the watery environment surrounding the cell. The hydrophilic heads of the other layer point toward the watery interior of the cell (**FIGURE 3.5C**). Such an arrangement of phospholipid layers allows the hydrophobic fatty acid tails to associate toward one another and not be exposed to water.

Sterols

Other types of lipids include the **sterols**, which are very different from the other groups of lipids. They are included with lipids because they too are hydrophobic molecules. Sterols play structural roles in microbes by stabilizing cell

FIGURE 3.4 Glycerol and Fatty Acids Combine to Form a Fat or Oil. Glycerol is a three-carbon molecule, and fatty acids are long carbon-hydrogen chains that can be saturated or unsaturated. Three fatty acid chains can combine with each glycerol in dehydration synthesis reactions to form a fat or oil.

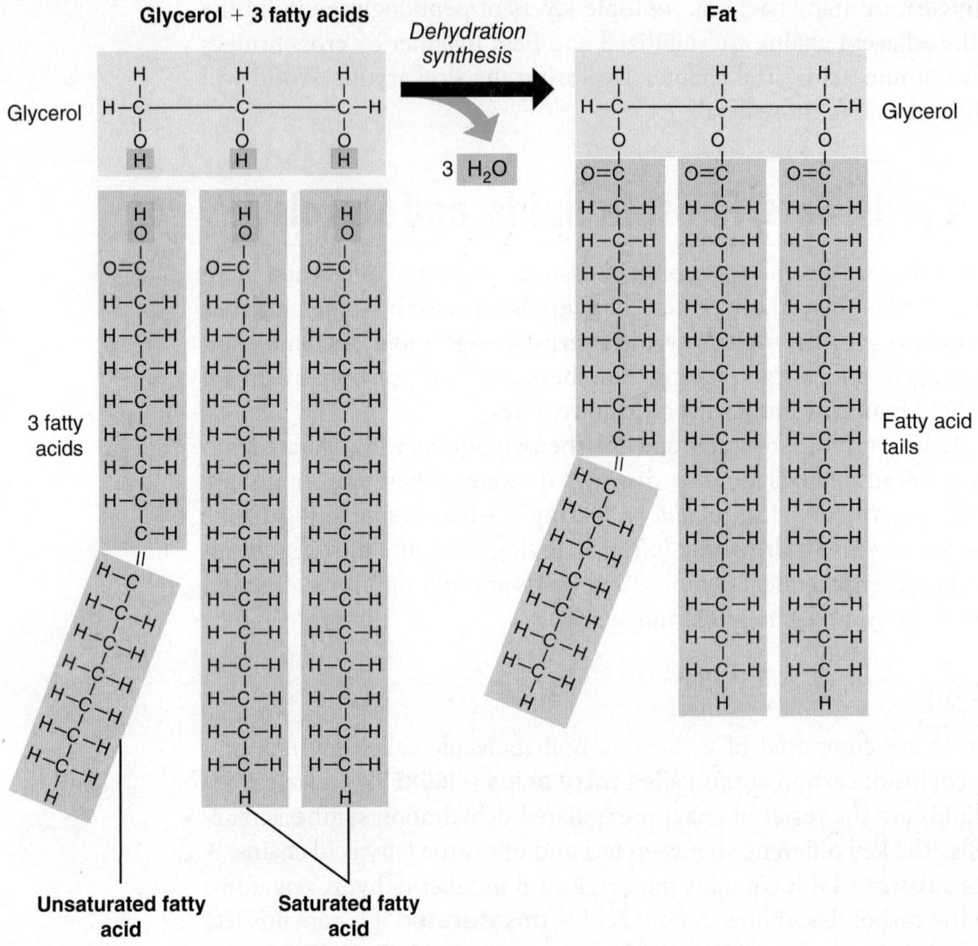

membranes of a few bacterial species as well as the membranes of protists and fungi. The sterol cholesterol stabilizes the plasma membranes of human cells.

3.4 Proteins: Amino Acids and Polypeptides

Dry weight: The weight of the materials in a cell after all the water is removed.

Proteins are the most abundant large molecules in all living organisms. Composed of carbon, hydrogen, oxygen, nitrogen, and, usually, sulfur atoms, proteins make up about 60% of a microbial cell's **dry weight**. This high percentage suggests that proteins have essential and diverse roles in cells. Many proteins function as structural components of cells and cell walls, as transport agents in membranes, and as enzymes that control all the chemical reactions in cells. In addition, viruses are composed of genetic information enclosed in a protein shell.

Amino Acids

All proteins are built from subunits called **amino acids** (**FIGURE 3.6A**). This structure includes one amino group (NH_2) and one carboxyl group (CO_2H) bonded together through a central carbon atom.

The amino acids vary from one another based on the atoms attached as a side group to the central carbon. This side group of atoms, known as the "R-group," can be as simple as a hydrogen in the case of glycine, or involve

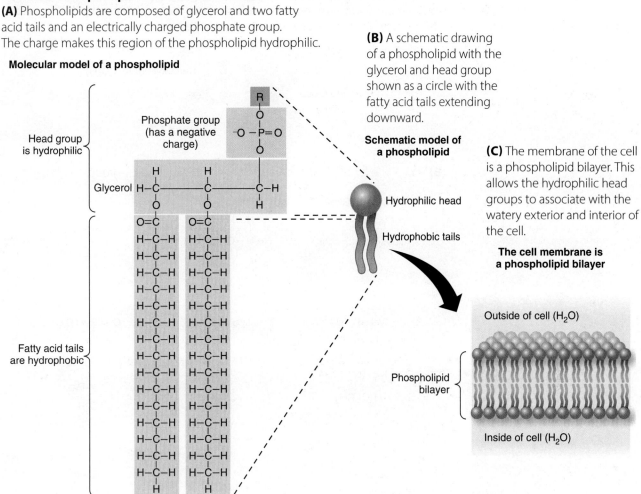

FIGURE 3.5 Phospholipids and Cell Membranes.
(A) Phospholipids are composed of glycerol and two fatty acid tails and an electrically charged phosphate group. The charge makes this region of the phospholipid hydrophilic.

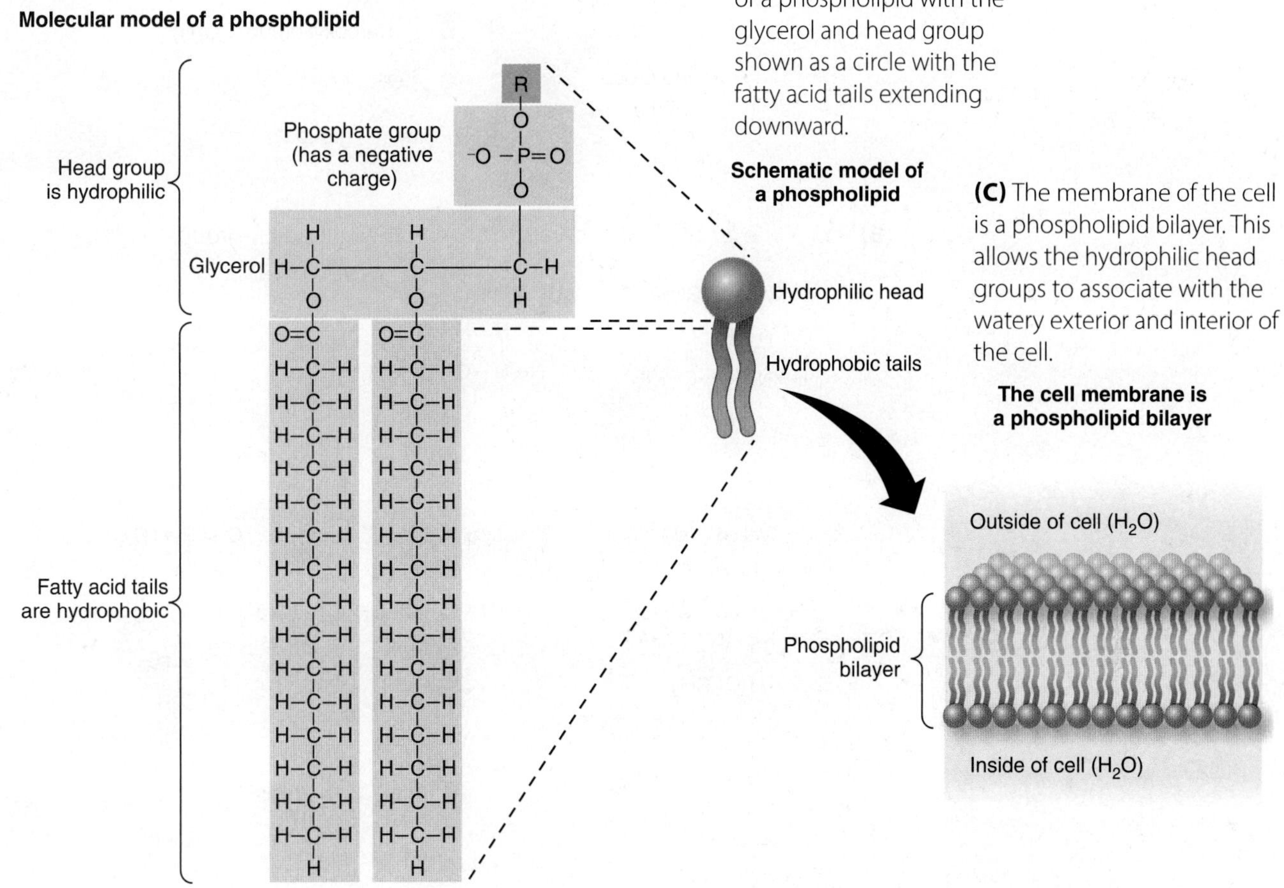

(B) A schematic drawing of a phospholipid with the glycerol and head group shown as a circle with the fatty acid tails extending downward.

(C) The membrane of the cell is a phospholipid bilayer. This allows the hydrophilic head groups to associate with the watery exterior and interior of the cell.

other combinations of atoms in other amino acids (**FIGURE 3.6B**). In all, there are 20 different R-groups, which means there are 20 different amino acids available to build proteins.

Like the polysaccharides and lipids, proteins are built from amino acids by means of bonding through dehydration synthesis reactions (**FIGURE 3.7**). The bond linking two amino acids is referred to as a **peptide bond**. By forming successive peptide bonds, more and more amino acids can be attached to the growing chain. The final number of amino acids in a chain can vary from a very few to thousands making a **polypeptide**. An extraordinary variety of proteins can be formed from the 20 available amino acids in one or more polypeptides.

Polypeptides and Protein Shape

Proteins must have tremendous differences in shape and structure to perform their diverse roles in cells.

Primary Structure

The specific sequence of amino acids in a polypeptide is referred to as the **primary structure** (**FIGURE 3.8A**). This sequence is unique to each polypeptide. However, a long chain of amino acids does not form the final shape and structure of the protein.

FIGURE 3.6 The Structure of Amino Acids.

(A) All amino acids have the same basic structure, but each varies by the R-group, which is a set of atoms attached to the central carbon.

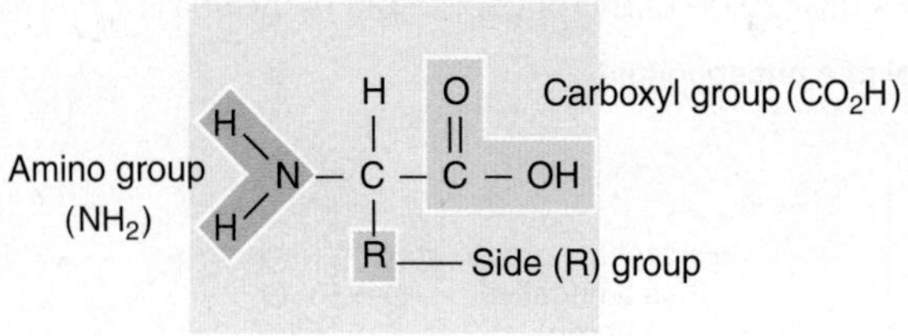

(B) Five of the 20 amino acids are shown with their unique R-group.

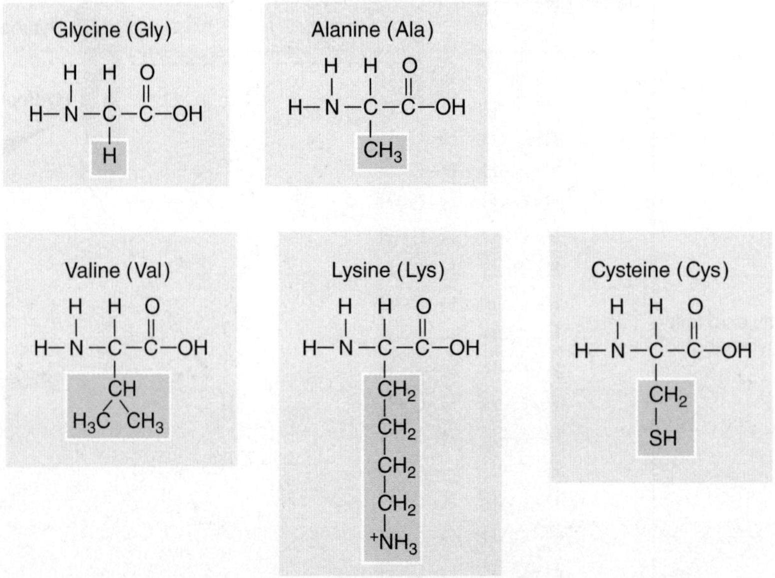

FIGURE 3.7 Formation of a Dipeptide. The amino acids alanine and valine are shown. The OH group from the carboxyl group of alanine combines with the H from the amino group of valine to form water. The carbon atom of alanine and the nitrogen atom of valine then link together, forming a peptide bond. Continued dehydration synthesis reactions that add additional amino acids will form a polypeptide.

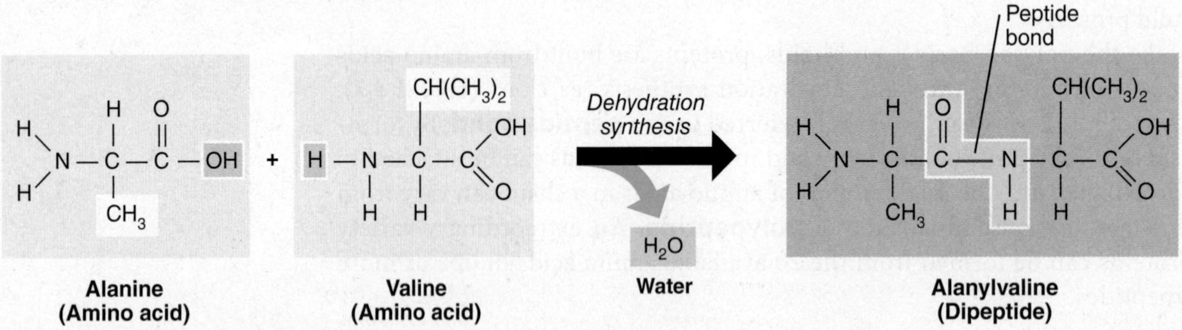

Secondary Structure

The amino acids in the primary sequence interact with one another, forming the polypeptide's **secondary structure** (**FIGURE 3.8B**). Often this secondary structure takes on the form of a helix (coil) or a folded sheet-like structure. Yet, this too does not determine the final shape and structure of the protein.

Tertiary and Quaternary Structures

Most polypeptides further fold to form the **tertiary structure** (**FIGURE 3.8C**). This shape depends on the interactions between R-groups of various amino

FIGURE 3.8 The Four Levels of Protein Structure.
(A) Primary structure refers to the sequence of amino acids. Each type of amino acid is represented here by a different geometrical shape.

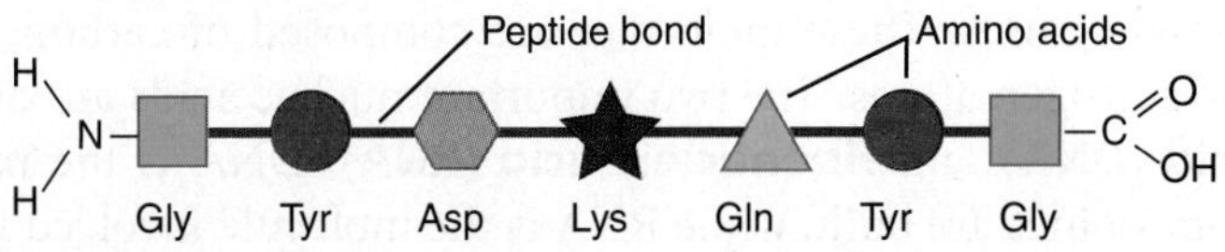

(B) Secondary structure involves interactions between amino acids, causing changes in shape of the. The resulting shape is usually a helix or sheet-like structure.

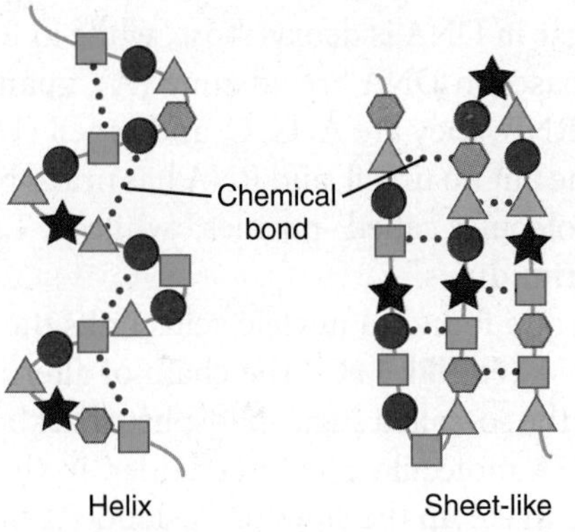

(C) Tertiary structure is formed when the polypeptide folds back on itself through interactions between R-groups. In addition, some proteins require more than one polypeptide to function, and the polypeptide configuration of these proteins is known as the **quaternary structure**.

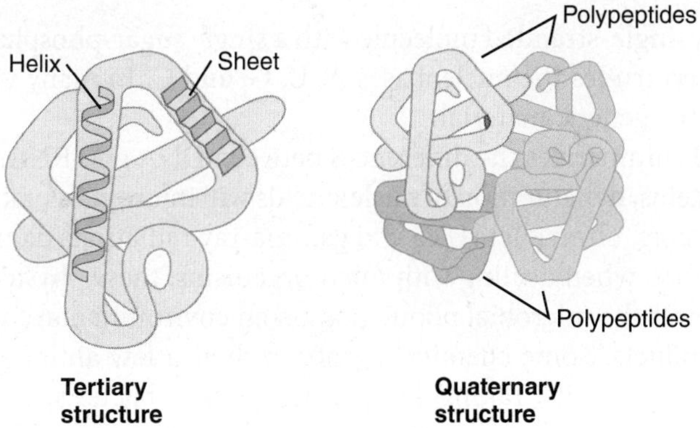

acids of the polypeptide. The resulting bonds between interacting R-groups give the polypeptide a globular or fibrous structure.

In many proteins, two or more polypeptides must bond together to form the final functional protein. This represents the **quaternary structure**. Examples include the human red blood cell protein hemoglobin and antibodies, both of which consist of four polypeptides.

In tertiary structures, the bonds between R-groups are relatively weak and can be easily disrupted. The unraveling and loss of protein shape is called **denaturation**. The resulting loss of protein function might kill microbes or inhibit their growth. For microbes, denaturation can occur as a result of exposure to physical agents like heat. It also can occur with the application of chemical agents, such as antiseptics and disinfectants. The chapter on Controlling Microbes will address the control of microbes through physical and chemical agents.

3.5 Nucleic Acids: DNA and RNA

Nucleic acids are the fourth major group of macromolecules found in all organisms and viruses. These molecules are composed of carbon, hydrogen, oxygen, and nitrogen atoms. The two important nucleic acids are **deoxyribonucleic acid** (**DNA**) and **ribonucleic acid** (**RNA**). DNA is the molecule of which chromosomes are built, while RNA is the molecule involved in converting the information in the DNA into a polypeptide.

Nucleic acids are built from subunits called **nucleotides** (**FIGURE 3.9A**). Each nucleotide is composed of three parts: a five-carbon sugar, a phosphate group (PO_4), and a nitrogen-containing molecule called a nitrogenous base, or simply, a **nucleobase**. The sugar in DNA is deoxyribose, while in RNA it is ribose.

The four nucleobases in DNA are adenine (A), guanine (G), cytosine (C), and thymine (T); in RNA, they are A, G, C, and uracil (U) (**FIGURE 3.9B**). (Note that DNA has thymine but no uracil, and RNA has uracil but no thymine.) A and G are double-ring molecules called "purines," while C, T, and U are single-ring molecules called "pyrimidines."

The phosphate group found in nucleic acids links the sugars to one another in both DNA and RNA (**FIGURE 3.9C**). The chain of alternating sugar and phosphate subunits forms the so-called sugar-phosphate backbone of the nucleic acid.

To visualize a DNA molecule, picture a ladder. In the molecule, two sugar-phosphate backbones make up the sides of the ladder, and the rungs (the steps) of the ladder are composed of the nucleobases (see Figure 3.9C). On one side of each rung is a purine molecule, and on the other side is a pyrimidine molecule. Thus, in DNA, an A always bonds with a T (and vice versa) and a G with a C (and vice versa). The ladder then is twisted into a spiral staircase-like structure called the **DNA double helix**. Microbes and all other living organisms have their DNA in this form.

RNA is a single-stranded molecule with a single sugar-phosphate backbone from which protrudes the nucleobases A, U, G, and C. In many viruses, RNA, not DNA, is the genetic material.

TABLE 3.1 summarizes the differences between DNA and RNA.

Like proteins, denaturation of nucleic acids will injure cells or kill the organism. For example, ultraviolet (UV) and gamma-ray radiations damage or break DNA. Therefore, when dealing with microorganisms, these physical agents can be used to lower the microbial population on an environmental surface or sterilize food products. Some chemical agents, such as a few antibiotics, interfere with nucleic acid activity. Taking these medicines for a bacterial infection might slow the growth or kill the bacterial cells. The chapter on Controlling Microbes will discuss antimicrobial drugs and their effect on microbes.

A Final Thought

We could discuss microbes without talking about their chemistry. However, it would be like trying to describe a Big Mac without knowing what is in the hamburger. To some people, the word "chemistry" is equivalent to "root canal," but it is important to remember that chemical molecules are the nuts and bolts of all living organisms and viruses. In other chapters, we discuss milk products containing carbohydrates, membranes composed of lipids, antibodies consisting of proteins, genes composed of nucleic acids, and a host of other concepts that include a little chemistry. To understand how yeasts cause bread to rise, you must understand the chemistry of the process. To explain how viruses make more of themselves and cause disease, some chemistry is needed. In addition,

FIGURE 3.9 The Molecular Structure of DNA.

(A) The sugars in the nucleotides are ribose and deoxyribose, which are identical except for one additional oxygen atom in ribose.

(B) The nucleobases include the purines adenine (A) and guanine (G) and the pyrimidines thymine (T), cytosine (C), and uracil (U).

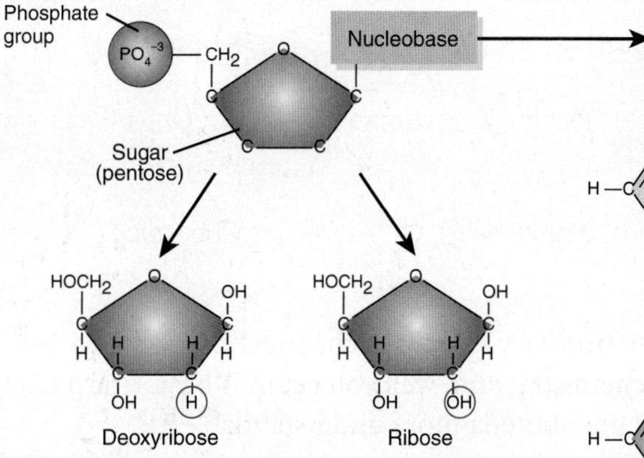

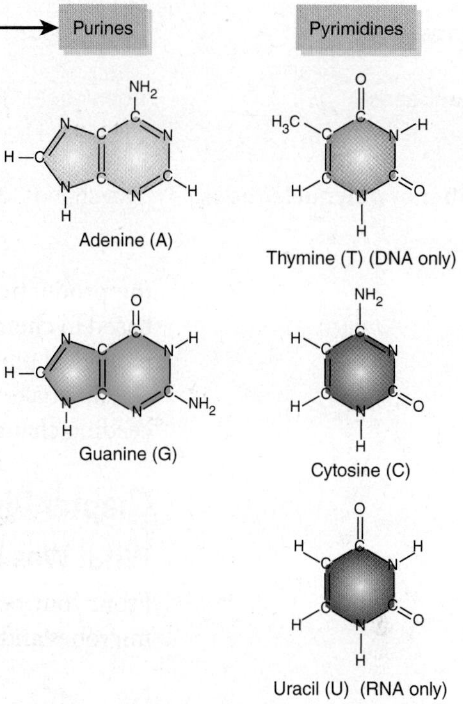

(C) Nucleotides are bonded together by dehydration synthesis reactions. The two polynucleotides of DNA are held together by chemical bonds between A and T and between G and C to form the DNA double helix.

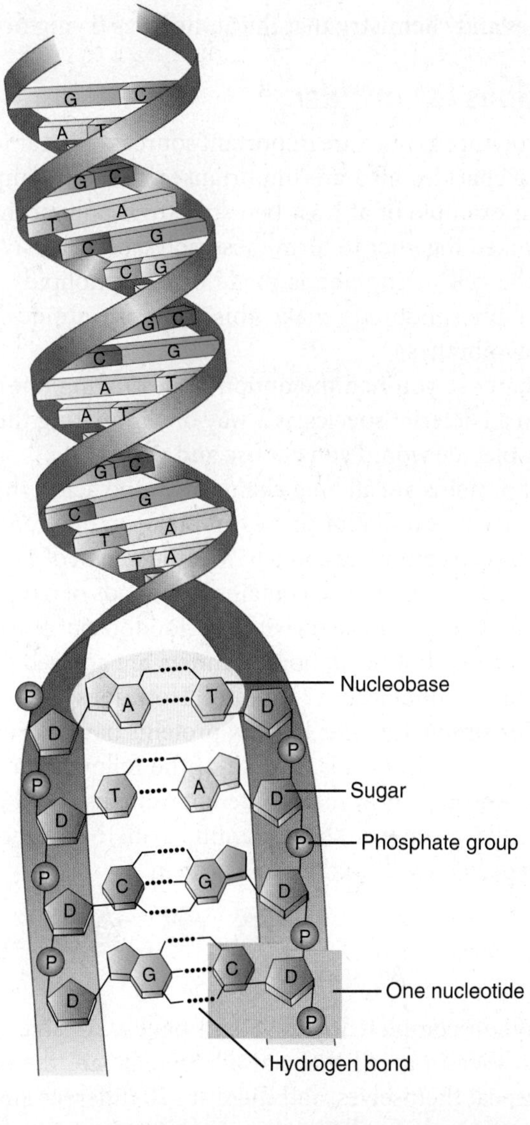

TABLE 3.1 A Comparison of DNA and RNA

Property	DNA	RNA
Pentose sugar	Deoxyribose	Ribose
Nucleobases	Adenine (A), guanine (G), cytosine (C), thymine (T)	Adenine (A), guanine (G), cytosine (C), uracil (U)
Number of polynucleotides	Two strands (double helix)	One strand

the production of yogurt is a chemical process, the process of disinfection is based in chemistry, and, well you get it. A basic grasp of chemistry makes other biological phenomena more understandable.

So, make sure you give this chapter on chemistry a careful reading. In succeeding chapters, you will find that your investment of time was worthwhile.

Chapter Discussion Questions

What Was He Thinking?

From your reading of this chapter, identify and discuss five major points about microbes and chemistry that the author was trying to get across to you.

Questions to Consider

1. Polysaccharides are important sources of sugar for energy; however, polysaccharides also are important structural components of the cell. Give an example of at least two structural polysaccharides. Which sugars are linked together to form these polysaccharides?
2. The cell membrane is made of phospholipids. What unique properties of phospholipids make this molecule uniquely adept at forming cell membranes?
3. Suppose you had the option of destroying one type of organic molecule in a bacterial species as a way of eliminating the microbe. Which type of molecule would you choose and why?
4. If proteins are all long chains of amino acids, then how can different proteins have different shapes and take on different functions?
5. Oxygen comprises about 65% of the weight of a living organism. This means a 120-pound person contains 78 pounds of oxygen. How can this be?
6. The toxin associated with the foodborne disease botulism is a protein. To avoid botulism, home canners are advised to heat preserved foods to boiling for at least 12 minutes. How does the heat help?
7. For simplicity sake, let's say proteins have a primary sequence consisting of five amino acids. If each of the following icons represents a different amino acid, how many different proteins could you form from a sequence of five geometric shapes (amino acids)? Assume no shape (amino acid) is repeated twice in the same protein.

Amino acid 1 2 3 4 5

When completed, you should have a sizeable set of possible sequences. Realize in a cell that not all proteins are the same length, amino acids repeat themselves, and there are 20 different amino acids.

CHAPTER 4

Exploring the Prokaryotic World: The Bacteria and Archaea Domains

Hail Caesar!

It's a popular salad. Some people have it Hawaiian-style with pineapple and ham, some have it Italian-style with pasta and tomatoes, and some toss it with grilled chicken, flank steak, or grilled salmon. Regardless of the addition, the salad remains the same—it is a Caesar salad.

After a day of classes, you meet a classmate for dinner at the local deli. You decide to have a Caesar salad with grilled shrimp. The warm and smoky char of the grilled shrimp give this meal a delightful taste. Better than the meatball sandwich that your friend ordered.

Many believe the first Caesar salad was created on July 4, 1924, in the mind of Caesar Cardini. Cardini, the owner of Caesar's Place in Tijuana, Mexico, was desperate for a fill-in dish during a very busy Fourth of July. So, he threw together some Romaine lettuce, Parmesan cheese, lemon, garlic, oil, and raw eggs. With a dramatic flair of the tableside tossing "by the chef," his customers were entertained and delighted.

You too enjoy your salad, then say good night to your friend, and head to your apartment. During the night you wake up several times with an unsettled stomach. By morning, you have a very upset stomach (**FIGURE 4.1A**). Your symptoms of diarrhea, abdominal cramps, and some vomiting get worse as the morning progresses and you decide to stay home. You take your temperature and find that you have a low-grade fever.

CHAPTER 4 OPENER This is a three-dimensional, computer-generated image of two *Salmonella enterica* bacterial cells. If this organism contaminates food products, the person consuming the product might suffer a foodborne illness.

You wonder what caused your ailment. Maybe it is food poisoning. Could it be the Caesar salad you had for dinner? You call your friend who you had dinner with, and he says he is feeling fine. "I'll bet it was that Caesar salad I ate," you declare. You do an Internet search and discover that sometimes the raw eggs used to make a Caesar salad can be a problem, especially if the raw eggs are contaminated with the bacterial species *Salmonella enterica* (**FIGURE 4.1B**). Your search finds that in such cases a foodborne infection called **salmonellosis** can occur. Luckily, it appears you have a mild case. You bear through the day and by the next morning you are starting to feel better.

Salmonella enterica: sal-mon-EL-lah en-TAIR-eh-kah

FIGURE 4.1 ***Salmonella*** **Can Contaminate Eggs.**
(A) An upset stomach might be one of the symptoms from consuming a food product contaminated with *Salmonella.*

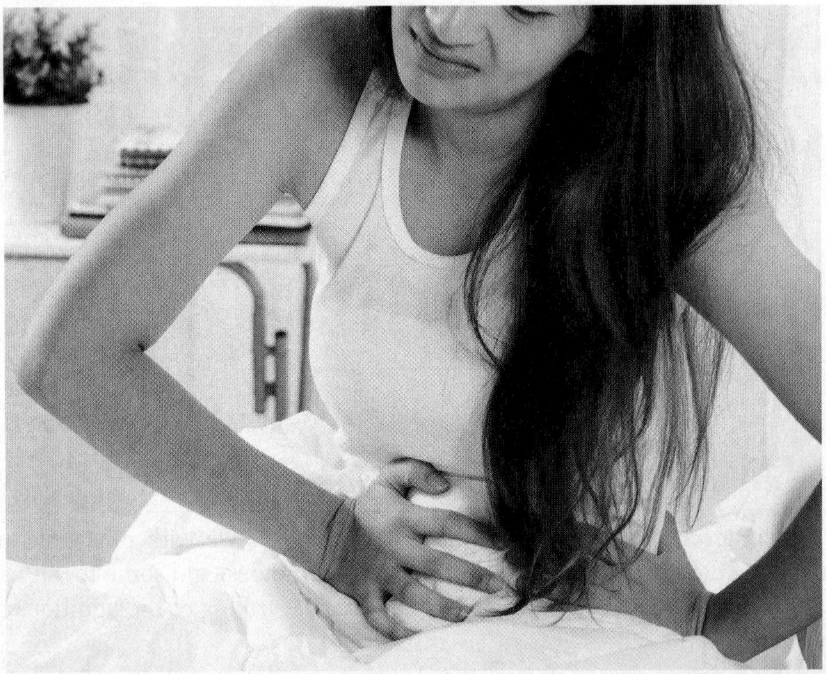

© SizeSquare's/Shutterstock.

(B) Artist's illustration of *Salmonella* cells that can contaminate eggs. (Bar = 2 μm.)

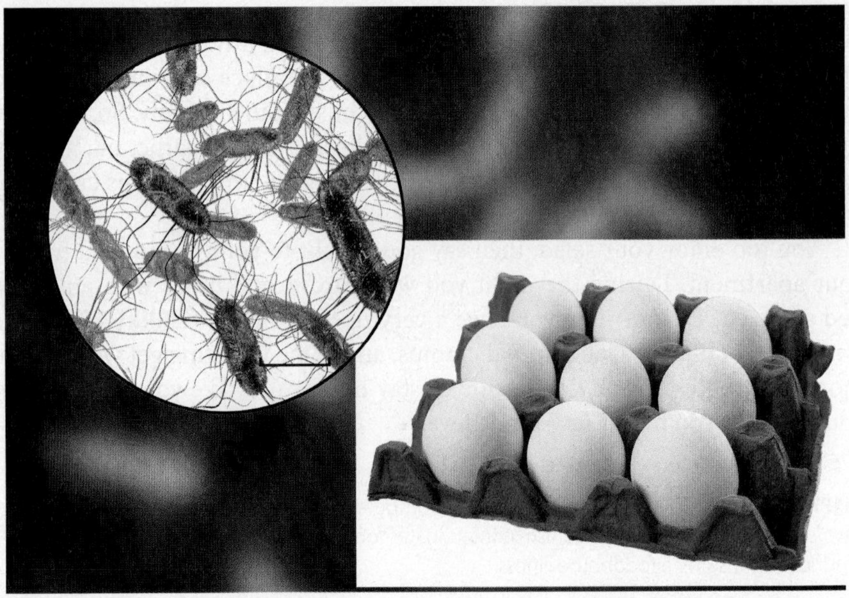

© Kateryna Kon/Shutterstock.

So, how could the bacteria grow so fast that they caused the gastrointestinal disease? Then again, why did the symptoms disappear just as quickly?

Answering these questions requires a basic understanding of prokaryotic growth, which is one of the main topics in this chapter. However, before discussing their growth, we need to examine prokaryotic cell structure (cell anatomy, if you will). As a preface to this chapter about prokaryotes, especially bacteria, read **A CLOSER LOOK 4.1**, which provides some interesting details about the bacteria.

A CLOSER LOOK 4.1

Bacteria in Eight Easy Lessons[1]

Mélanie Hamon is a member of the Cell Biology and Infection Department at the Institut Pasteur in Paris. Several years ago, when she was an assistante de recherché (research assistant) at the Institut, she introduced herself as a bacteriologist. She often was asked, "Just what does that mean?" To help explain her discipline, she gives us, in eight letters (BACTERIA), what she calls "some demystifying facts about bacteria."

Basic principles: Their average size is 1/25,000th of an inch. In other words, hundreds of thousands of bacteria fit into the period at the end of this sentence. In comparison, human cells are 10 to 100 times larger with a more complex inner structure. While human cells have copious amounts of membrane-contained subcompartments, bacteria more closely resemble pocketless sacs. Despite their simplicity, they are self-contained living beings, unlike viruses, which depend on a host cell to carry out their life cycle.

Astonishing: Bacteria are the root of the evolutionary tree of life, the source of all living organisms. Quite successful evolutionarily speaking, they are ubiquitously distributed in soil, water, and extreme environments such as ice, acidic hot springs, or radioactive waste. In the human body, bacteria account for 10% of dry weight, populating mucosal surfaces of the oral cavity, gastrointestinal tract, urogenital tract, and surface of the skin. In fact, bacteria are so numerous on earth that scientists estimate their biomass to far surpass that of the rest of all life combined.

Crucial: It is a little known fact that most bacteria in our bodies are harmless and even essential for our survival. Inoffensive skin settlers form a protective barrier against any troublesome invader while approximately 1,000 species of gut colonizers work for our benefit, synthesizing vitamins, breaking down complex nutrients, and contributing to gut immunity. Unfortunately for babies (and parents!), we are born with a sterile gut and "colic" our way through bacterial colonization.

Tools: Besides the profitable relationship they maintain with us, bacteria have many other practical and exploitable properties, most notably, perhaps, in the production of cream, yogurt, and cheese. Less widely known are their industrial applications as antibiotic factories, insecticides, sewage processors, oil spill degraders, and so forth.

Evil: Unfortunately, not all bacteria are "good," and those that cause disease give them all an often undeserved and unpleasant reputation. If we consider the multitude of mechanisms these "bad" bacteria—pathogens—use to assail their host, it is no wonder that they get a lot of bad press. Indeed, millions of years of coevolution have shaped bacteria into organisms that "know" and "predict" their hosts' responses. Therefore, not only do bacterial toxins know their target, which is never missed, but bacteria can predict their host's immune response and often avoid it.

Resistant: Even more worrisome than their effectiveness at targeting their host is their faculty to withstand antibiotic therapy. For close to 50 years, antibiotics have revolutionized public health in their ability to treat bacterial infections. Unfortunately, overuse and misuse of antibiotics have led to the alarming fact of resistance, which promises to be disastrous for the treatment of such diseases.

Ingenious: The appearance of antibiotic-resistant bacteria is a reflection of how adaptable they are. Thanks to their large populations they are able to mutate their genetic makeup, or even exchange it, to find the appropriate combination that will provide them with resistance. Additionally, bacteria are able to form "biofilms," which are cellular aggregates covered in slime that allow them to tolerate antimicrobial applications that normally eradicate free-floating individual cells.

A long tradition: Although "little animalcules" were first observed in the 17th century, it was not until the 1850s that Louis Pasteur fathered modern microbiology. From this point forward, research on bacteria has developed into the flourishing field it is today. For many years to come, researchers will continue to delve into this intricate world, trying to understand how the good ones can help and how to protect ourselves from the bad ones. It is a great honor to be part of this tradition, working in the very place where it was born.

[1] Reprinted with permission of the author, the Institut Pasteur, and the Pasteur Foundation. The original article appeared in *Pasteur Perspectives* Issue 20 (spring 2007), the newsletter of the Pasteur Foundation, which may be found at www.pasteurfoundation.org

LOOKING AHEAD

After reading and completing this chapter, you will be able to:

4.1 List the three major shapes of prokaryotic cells and their possible arrangements.

4.2 Identify and explain the function of (a) bacterial surface structures and (b) cytoplasmic structures.

4.3 Describe binary fission and explain the four phases of a bacterial growth curve.

4.4 Contrast broth and agar culture media and compare selective and enriched media.

4.5 Give examples illustrating the diversity within the domain Bacteria and domain Archaea.

4.1 Cell Structure: Shapes and Arrangements

In the chapter on Microbes in Perspective, the prokaryotes were separated from the eukaryotes by their structural differences. A prokaryotic organism, such as *S. enterica*, is much smaller than most eukaryotic organisms. They also lack a cell nucleus that is a hallmark of eukaryotic cells.

Prokaryotic Cell Shape and Arrangement

Most prokaryotic cells have one of three common shapes (**FIGURE 4.2**). Rod-shaped cells are known as **bacilli** (sing. **bacillus**). They vary in size, with the shortest ones measuring about 0.5 μm in length and the longer ones about 20 μm. Other prokaryotic cells are spherical and are called **cocci** (sing. **coccus**). They measure roughly 1 μm in diameter. The third common shape for prokaryotic cells is in the form of a long, spiral that can be bent (called **vibrios**), flexible and wavy [called **spirilla** (sing. **spirillum**)], or rigid and "cork screw-" shaped (called **spirochete**).

Although prokaryotic cells are unicellular, in some species the individual cells remain attached by their adjacent cell walls. In these cases, the groups of cells exhibit different arrangements (see Figure 4.2).

Bacilli can be single cells, arranged as pairs called **diplobacilli** (*diplo* = "two"), or in short or long chains called **streptobacilli** (*strepto* = "chain"). The cocci can be arranged in pairs (**diplococci**), packets of four cells called **tetrads** (*tetra* = "four"), in clusters called **staphylococci** (*staphylo* = "cluster"), or in chains (**streptococci**).

In medical labs, technicians can use a microscope to distinguish these arrangements and assist physicians in their diagnoses. For example, besides *Salmonella*, other forms of bacterial food poisoning can be caused by *Bacillus cereus*, which is rod-shaped and *Staphylococcus aureus*, which is sphere-shaped. Note that sometimes, as in these two examples, the genus name might also identify the cell's shape or arrangement.

Bacillus cereus: bah-SIL-lus SEH-ree-us

Staphylococcus aureus: staff-ih-loh-KOK-us OH-ree-us

Other shapes and arrangements also exist. For example, the cells of different species of cyanobacteria can be arranged in filaments or exist as chains of rods or spheres (**FIGURE 4.3**). The cells of archaeal organisms, besides having the typical rod and sphere shapes, also can be square, triangular, or star-shaped.

Most prokaryotic cells lack color, so it can be very difficult to see the individual cells and their arrangements with the light microscope. To solve this

FIGURE 4.2 Bacterial Cell Shapes and Arrangements. (All bars = 10 μm.)

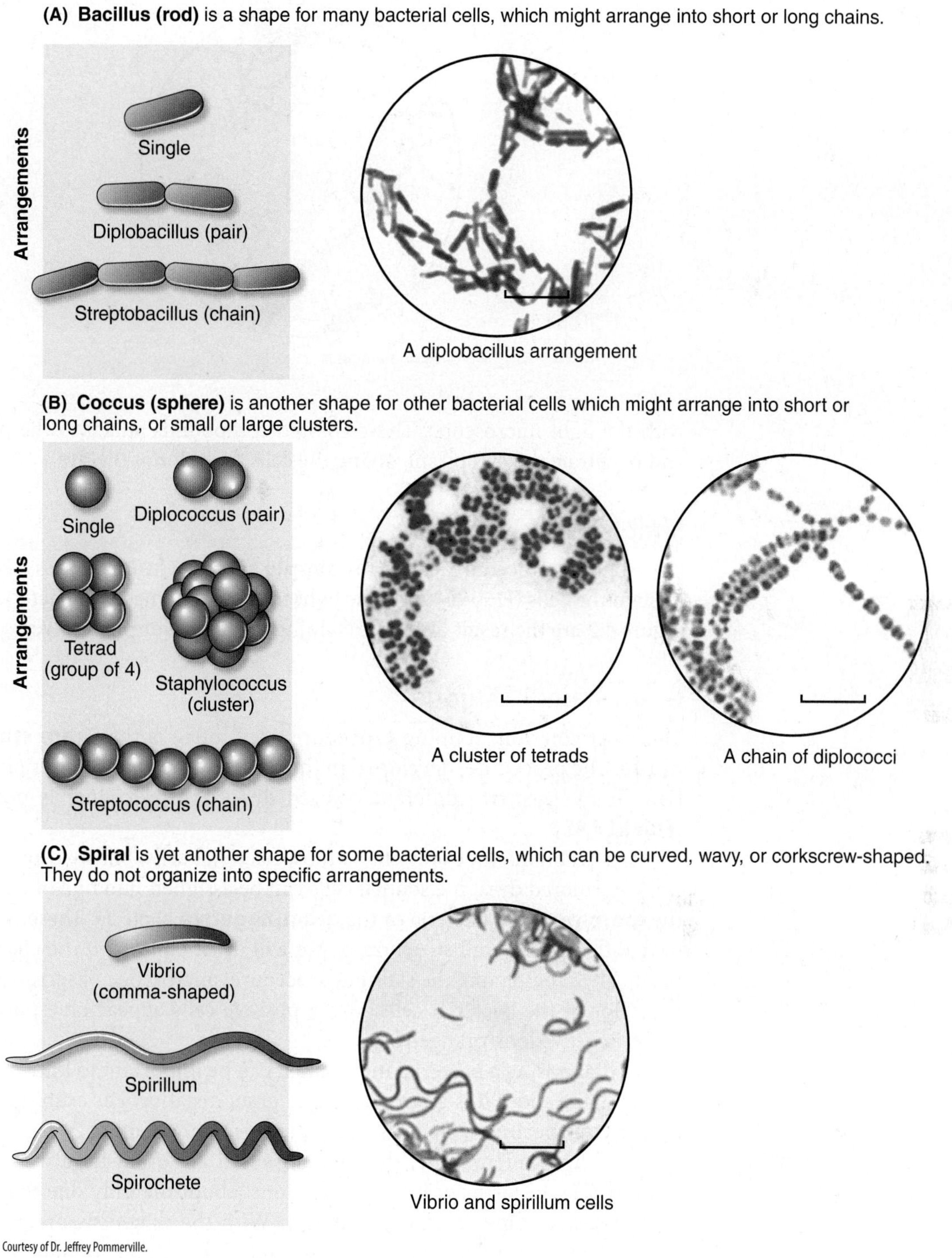
(A) Bacillus (rod) is a shape for many bacterial cells, which might arrange into short or long chains.
Arrangements
Single
Diplobacillus (pair)
Streptobacillus (chain)
A diplobacillus arrangement
(B) Coccus (sphere) is another shape for other bacterial cells which might arrange into short or long chains, or small or large clusters.
Arrangements
Single
Diplococcus (pair)
Tetrad (group of 4)
Staphylococcus (cluster)
Streptococcus (chain)
A cluster of tetrads
A chain of diplococci
(C) Spiral is yet another shape for some bacterial cells, which can be curved, wavy, or corkscrew-shaped. They do not organize into specific arrangements.
Vibrio (comma-shaped)
Spirillum
Spirochete
Vibrio and spirillum cells

Courtesy of Dr. Jeffrey Pommerville.

problem, the cells are heat killed and then stained with a dye to give them the contrast needed for microscope observations.

Staining Techniques

In the late 1800s and early 1900s, scientists such as Robert Koch and Paul Ehrlich started using colored dyes to improve their ability to see bacterial cells

FIGURE 4.3 Cyanobacteria. A light microscope image of a cyanobacterial species that is arranged into long chains or filaments. (Bar = 50 μm.)

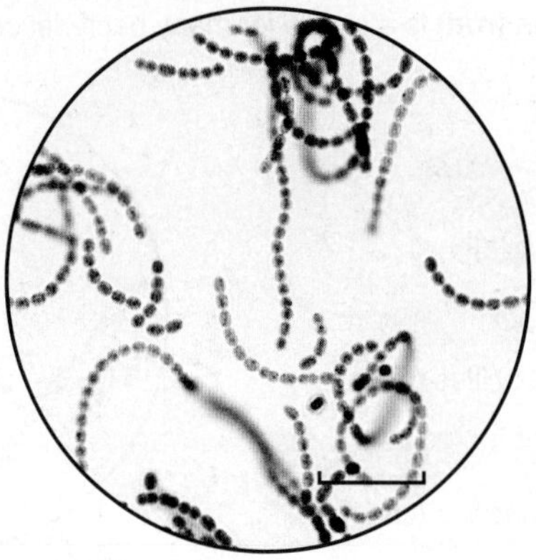

Courtesy of Dr. Jeffrey Pommerville.

with the light microscope. These colored compounds attach to the dead cells and penetrate the cytoplasm, giving the cells the color of the dye.

Simple Stain Technique

One staining procedure, called the **simple stain technique**, uses a single dye to stain the cells (**FIGURE 4.4A**). The light microscope images of bacterial cells in Figure 4.2 are the result of simple staining using blue or purple dyes.

Gram Stain Technique

The most common staining procedure used today is the **Gram stain technique**. The procedure, developed in the 1880s by the Danish physician Christian Gram, uses two different colored dyes. The procedure is outlined in (**FIGURE 4.4B**).

This technique not only stains most bacterial cells, but by using two contrasting colored dyes, the stained cells can be assigned into one of two groups: the **gram-positive** bacteria or the **gram-negative** bacteria. These two groups have different cell wall structure, as we will describe later in the chapter. As a result, after completing the staining procedure, microscope observation reveals the color of the bacterial cells. Gram-positive cells appear blue-purple while gram-negative look orange-red.

Urethral: Referring to the tube that carries urine (and semen in males) out of the body.

In diagnosing a bacterial infection, it can be important to know the bacterium's shape and if it is gram positive or gram negative. For example, suppose a **urethral** specimen taken from a patient is Gram stained. When examined with the light microscope, the cells observed are gram-negative diplococci (**FIGURE 4.5**). In terms of human infections, about the only one that involves gram-negative diplococci is gonorrhea. With the diagnosis made, the physician now must select an appropriate antibiotic. Because many antibiotics are effective only against gram-positive species, the physician would need to select an antibiotic for this patient that works well against gram-negative bacteria. So, the gram-staining results helped guide in the diagnosis and treatment therapy.

Let's now turn to an examination of prokaryotic cell structure. We will focus on bacterial cells because they have been studied in more detail than most archaeal cells.

FIGURE 4.4 Stain Reactions in Microbiology.
(A) In the simple stain technique, the dye stains the bacterial cells and makes them visible in the light microscope.

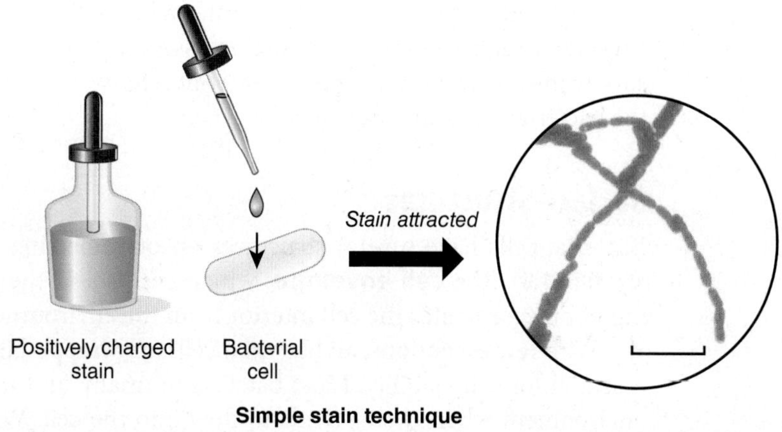

(B) In the Gram stain technique, the dead bacterial cells first are stained with a crystal violet dye (Step 1) and then with Gram's iodine solution (Step 2). At this point, all the bacterial cells are blue-purple due to the crystal violet. Now, the stained cells are washed with ethyl alcohol (Step 3), which removes the dye from the gram-negative cells only; the gram-positive cells remain blue-purple. Because the gram-negative cells are now colorless and would be difficult to see in the light microscope, the final step of the technique involves the addition of a red dye called safranin (Step 4). This dye is picked up by the gram-negative cells, causing them to appear orange-red in color; the gram-positive cells remain blue-purple. In the stained image, gram-positive and gram-negative cells are stained. (Bar = 10 μm.)

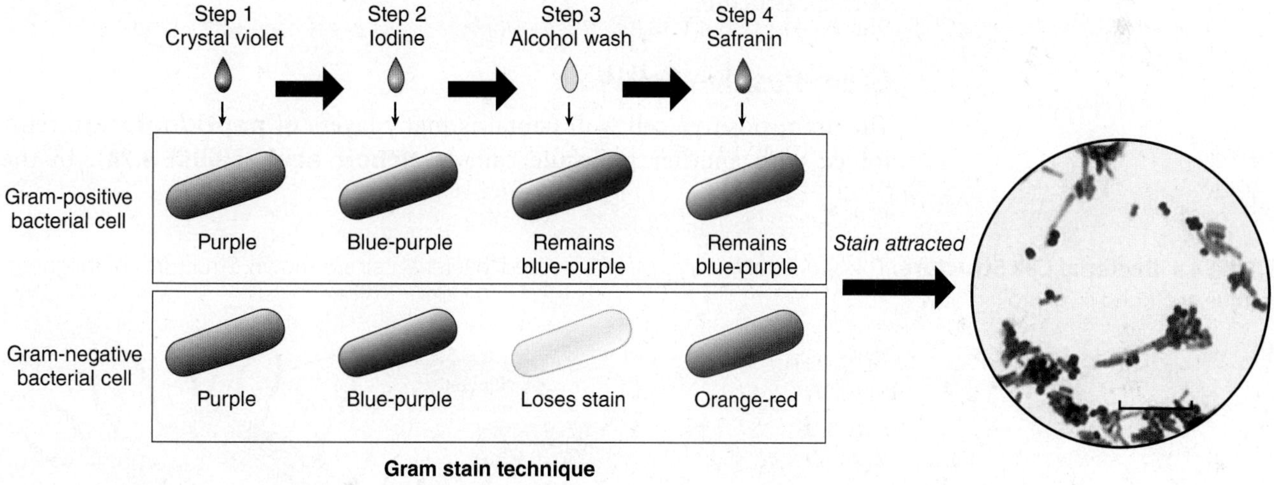

FIGURE 4.5 Disease Diagnosis. Light microscope image of stained gram-negative diplococci (small red dots) among many stained white blood cells (large red cells). Such cell arrangements and Gram staining are diagnostic for gonorrhea. (Bar = 10 μm.)

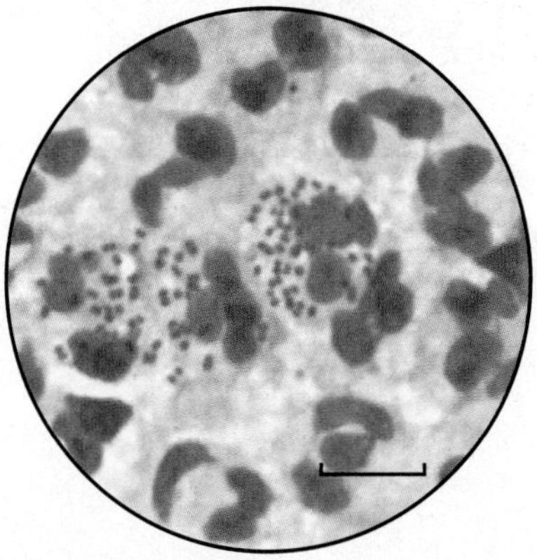

Courtesy of CDC.

4.2 Cell Structure: Anatomy of a Cell

Although prokaryotic cells lack the internal membrane structures (organelles) common to eukaryotic cells, they still possess some distinctive structures. As we survey the structural makeup of bacterial cells, watch for the various functions the bacterial structures perform.

Surface Structures

Bacterial cells have several structures on or projecting from the cell surface (**FIGURE 4.6**). The **cell envelope**, which consists of the cell wall and the cell membrane, separates the cell interior from the environment.

With few exceptions, all bacterial cells are wrapped in a **cell wall**, which is essential for cell viability. Most bacteria normally find themselves in a watery environment where water tends to flow into the cell. Without a cell wall, the increasing water pressure in the cell soon would burst the cell, like blowing too much air into a balloon. However, the cell wall counteracts the internal water pressure, and the cell remains undamaged. The cell wall also helps determine cell shape.

As mentioned earlier in this chapter, bacteria are divided into two groups based on Gram staining (see Figure 4.4). Gram-positive cells stain blue-purple, while gram-negative cells stain red-orange. This staining difference is a result of cell wall structure.

Gram-Positive Cell Wall

The gram-positive cell wall contains many layers of **peptidoglycan**, reinforced with another molecule called **teichoic acid** (**FIGURE 4.7A**). In the

FIGURE 4.6 Bacterial Cell Structure. The structural features of an "idealized" bacterial cell are shown. Structures highlighted in blue are found in all prokaryotes.

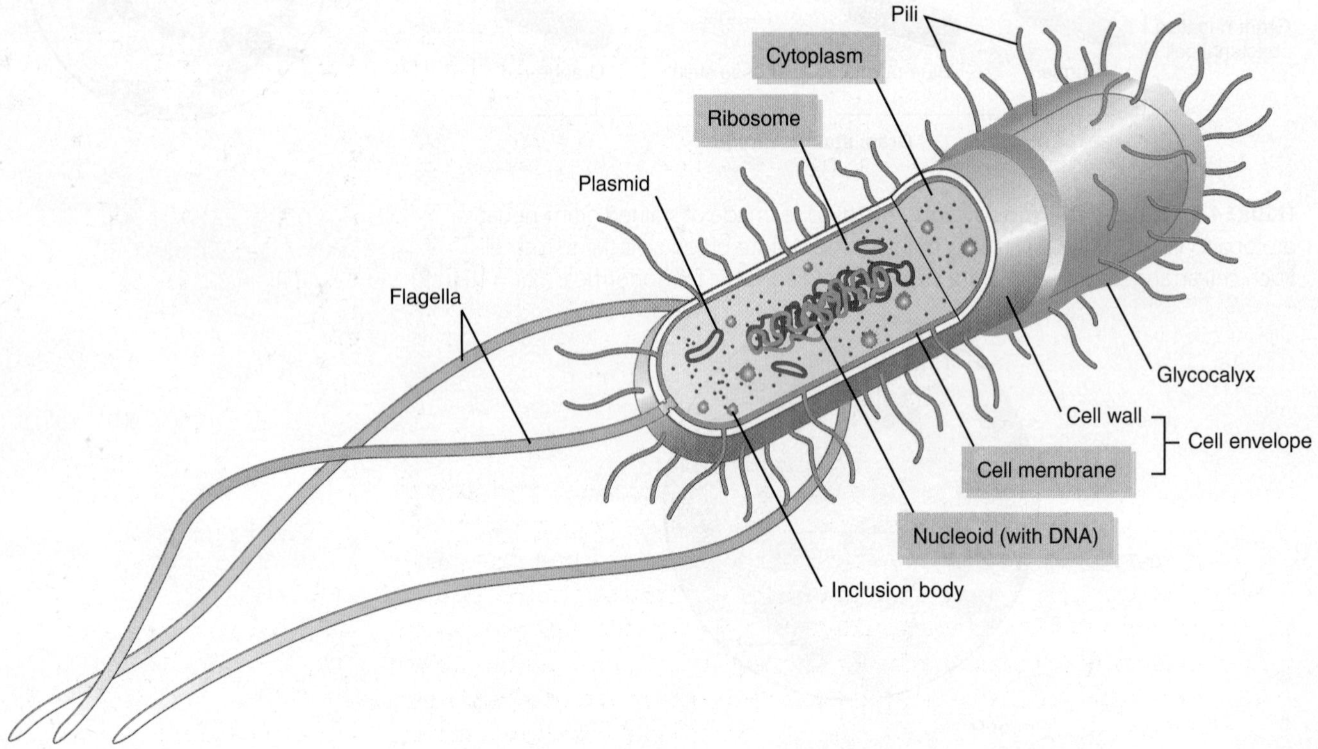

chapter on the Molecules of Cells, peptidoglycan was described as tough fibers of carbohydrate cross-linked by a short chain of amino acids. So, think of the cell wall like a block wall. The rows of blocks would represent the polysaccharide, the proteins would be the mortar holding the blocks in place, and the teichoic acids are the rebar (steel rods) providing reinforcement. Unlike solid blocks, water and nutrients can percolate through the gram-positive cell wall.

Some antibiotics affect the assembly of the peptidoglycan chains. For example, penicillin and its relatives (e.g., amoxicillin and methicillin) disrupt the assembly of peptidoglycan chains. Without a strong cell wall, the internal water pressure bursts the cell. All these drugs will be discussed in the chapter on Controlling Microbes.

Gram-Negative Cell Wall

The walls on gram-negative cells are organized differently than those of the gram-positive bacteria (**FIGURE 4.7B**). First, there is only one or two chains of peptidoglycan and no teichoic acid. Unique to the gram-negative bacteria is an **outer membrane** that covers the peptidoglycan. The outer half of this membrane is a lipid and polysaccharide layer called the **lipopolysaccharide** (**LPS**). The food poisoning described in the chapter introduction was caused by *S. enterica*, a gram-negative species.

During an infection by *S. enterica*, the lipid portion of the LPS is released in the body. Its presence triggers an immune reaction resulting in the symptoms of salmonellosis. Antibiotics like penicillin would not be prescribed because the antibiotic cannot cross the outer membrane to interfere with peptidoglycan assembly. However, other antibiotics can penetrate the wall and inhibit internal metabolic reactions.

Cell Membrane

All cells are surround by a **cell** (**plasma**) **membrane**. As described in the chapter on Molecules and Cells, the cell membrane is a double layer (a bilayer) of phospholipids. Inserted in the bilayer are numerous types of proteins, many spanning the bilayer (see Figure 4.7). Some of these proteins function as enzymes for cell metabolism, while others transport nutrients across the membrane. Because of its ever-changing nature, the membrane is called a "fluid mosaic" because the proteins often change position (the mosaic) within the fluid phospholipid bilayer.

Notice in Figure 4.7B that gram-negative cells have a gap between the outer membrane and the cell membrane. This area, called the **periplasmic space**, is an active and important processing center where nutrients too large to pass through the cell membrane are first broken down. The peptidoglycan also is found here.

Glycocalyx

Some bacterial species have a carbohydrate-rich coat called the **glycocalyx** that surrounds the cell envelope. The glycocalyx provides protection for the cell, shielding it from drying (desiccation), chemicals, and environmental stresses. The glycocalyx also can attach the bacterial cell to a surface. For example, the glycocalyx of *Streptococcus mutans* gives the cells the ability to attach to tooth enamel. In the pockets between the teeth and gums the attached cells break down sugars and other carbohydrates we consume and produce large amounts of acid. If the teeth are not regularly cleaned and flossed to remove

Streptococcus mutans: strep-toe-KOK-us MEW-tanz

FIGURE 4.7 A Comparison of the Cell Walls of Gram-Positive and Gram-Negative Bacterial Cells.
(A) The cell wall of a gram-positive bacterial cell is composed of multiple peptidoglycan chains intermixed with teichoic acid molecules. The protein cross-bridges are not shown. Below the wall is the cell membrane that consists of a phospholipid bilayer in which are embedded transport proteins.

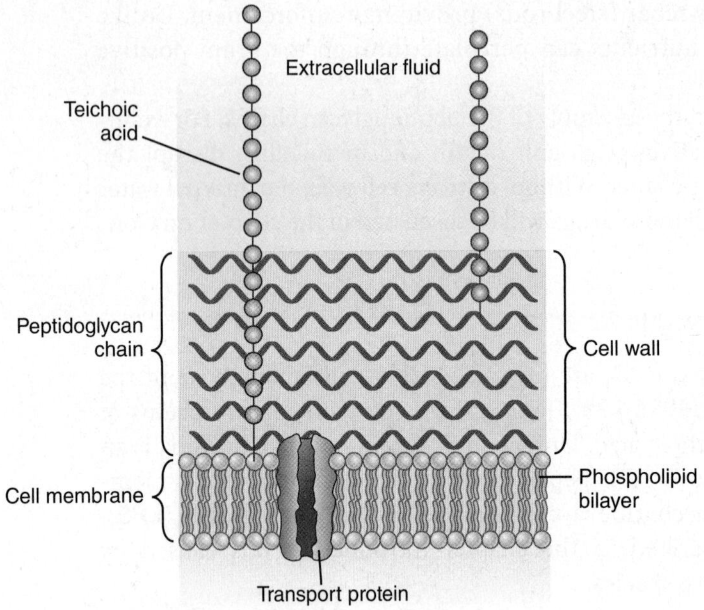

Gram-positive cell wall

(B) In the gram-negative cell wall, the peptidoglycan chains are few and there are no teichoic acid molecules. Moreover, an outer membrane overlies the peptidoglycan layer in the periplasmic space. Note that the outer membrane contains special transport proteins, and the outer half is unique in containing lipopolysaccharide (LPS) molecules.

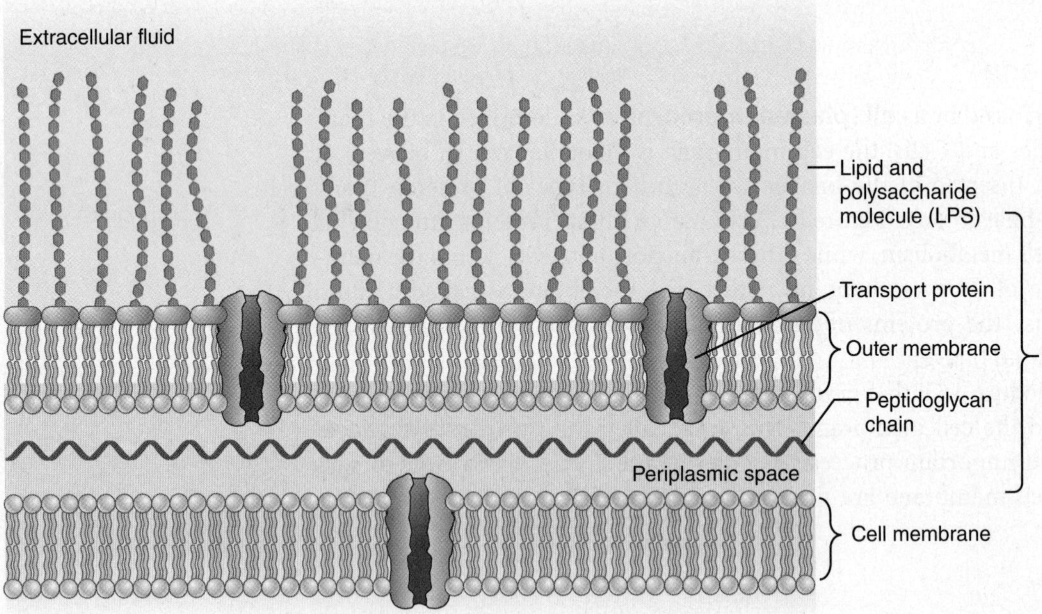

Gram-negative cell wall

most of the bacteria, the acid gradually will eat away at the tooth enamel, forming a depression, or cavity.

Surface Projections

Many bacterial species also have protein projections that extend from the cell surface. These are the flagella and pili.

Flagella

Many bacterial cells contain one or more hair-like projections called **flagella** (sing. **flagellum**) that are anchored in the cell wall and cell membrane. Different bacterial species have either a single flagellum, a tuft of flagella at one end of the cell, or flagella covering the entire cell surface (**FIGURE 4.8**).

The bacterial flagella are long, rigid protein filaments. These filaments bundle together and rotate like a propeller, pushing a bacterial cell forward. In its watery surroundings, the bacterial flagella help the cells swim toward nutrients or away from toxic chemicals. Using flagella, a bacterium such as *Escherichia coli* can travel about 2,000 body lengths in an hour. This is equivalent to 5-foot 10-inch human walking 2.25 miles in an hour.

Escherichia coli: esh-er-EE-key-ah KOH-lee

Pili

Other appendages projecting out from the cell surface are **pili** (sing. **pilus**). Pili are short, rigid protein rods about 1 μm in length and about 7 nm thick (**FIGURE 4.9**). They help bacterial cells attach to tissues or other surfaces. When causing infection, pathogens having pili can attach to tissues, helping start the infection process. **A CLOSER LOOK 4.2** provides an "upsetting" example.

FIGURE 4.8 Bacterial Flagella. A light microscope image showing stained bacterial flagella (gray structures) that extend from the cell surface. Note that the flagella have a characteristic wavy arrangement. (Bar = 10 μm.)

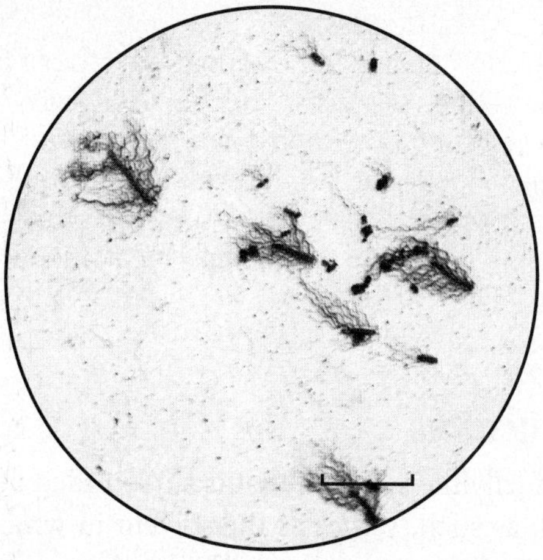

FIGURE 4.9 Bacterial Pili. A false-color transmission electron microscope image of an *Escherichia coli* cell (oval) with many pili (red). (Bar = 1 μm.)

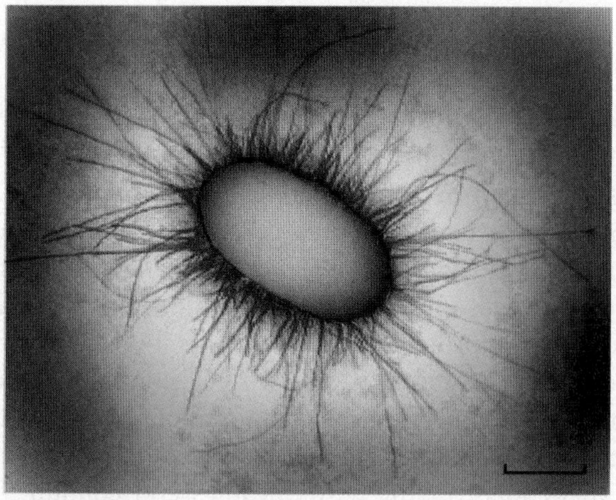

A CLOSER LOOK 4.2

Diarrhea Doozies

They gathered at the clinical research center at Stanford University to do their part for the advancement of science (and earn a few dollars as well). They were the "sensational 60"—60 young men and women who would spend 3 days and nights and earn $300 to help determine whether hair-like structures called pili have a significant place in disease.

Several nurses and doctors were on hand to help them through their ordeal. The students would drink a fruit-flavored cocktail containing a special diarrhea-causing strain of *Escherichia coli* (see **FIGURE**). Thirty cocktails had *E. coli* with normal pili, and 30 had *E. coli* with pili mutated beyond repair. The hypothesis was that the bacterial cells with the normal pili should latch onto intestinal tissue and cause diarrhea, while those with mutated pili should be swept away by the rush of intestinal movements and not cause intestinal distress—at least that's what the sensational 60 were out to verify or prove false.

On the fateful day, the experiment began. Neither the students nor the health professionals knew who was drinking the diarrhea cocktail and who was getting the "free pass"; it was a double-blind experiment. Then came the waiting. Some volunteers experienced no symptoms, but others felt the bacterial onslaught and clutched at their last remaining vestiges of dignity. For some, it was 3 days of hell, with nausea, abdominal cramps, and numerous bathroom trips; for others, luck was on their side (investing in a lottery ticket seemed like a good idea at the time).

When it was all over, the numbers appeared to bear out the theory: the great majority of volunteers who drank the cocktail with mutated bacterial pili experienced no diarrhea, while most of those who drank the cocktail with normal bacterial pili had attacks of diarrhea, in some cases, real doozies.

In the end, all appeared to profit from the experience: the scientists had some real-life evidence that pili contribute to infection, the students made their sacrifice to science and pocketed $300 each, and the local supermarket had a surge of profits from sales of toilet paper and antidiarrheal medications.

Courtesy of James Archer/ Illustrated by Alissa Eckert and Jennifer Oosthuizen/CDC.

Cytoplasmic Structures

The **cytoplasm** is a jellylike solution that is surrounded by the cell membrane. It is 70% water and, as such, serves as the solvent in which many small molecules and other substances are dissolved. These substances include amino acids and proteins, simple sugars, and nucleotides, as well as numerous minerals and growth factors needed for cell growth and reproduction. In addition, there are a few larger subcompartments and structures (see Figure 4.6).

Nucleoid and Plasmids

The genetic information in prokaryotic cells exists in the form of chromosomes and plasmids. The full complement of DNA in a cell (or virus) is called the **genome**.

The **bacterial chromosome** is a long, closed loop of DNA. In the cytoplasm, the chromosome forms the area called the **nucleoid**, which lacks the surrounding membranes typical of the eukaryotic cell nucleus (**FIGURE 4.10**). Organized on the chromosome are thousands of **genes** that contain the essential information to produce proteins needed for metabolism, growth, and reproduction. For example, a typical *S. enterica* cell contains about 3,800 genes. By comparison, the human genome is composed of about 22,000 genes.

Many bacterial species also have **plasmids**. These small, closed loops of DNA are independent of the chromosomes and contain 5 to 100 genes. These genes encode the information to make proteins that are not essential for

FIGURE 4.10 The Bacterial Nucleoid. In this false-color transmission electron microscope image, the yellow-colored nucleoid of each cell is shown. The brown granular area represents the cytoplasm, with the small dots being the ribosomes. (Bar = 0.5 μm.)

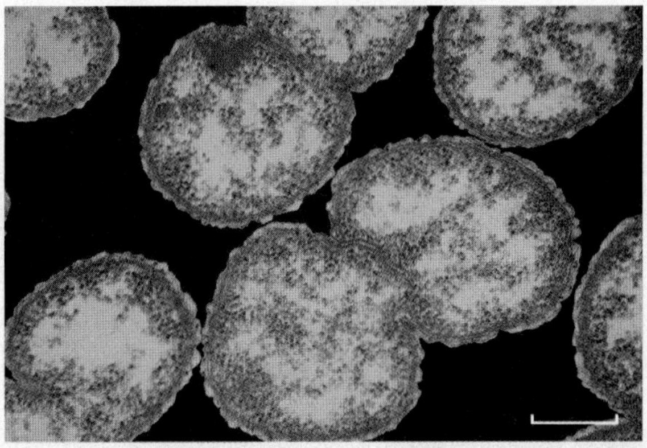

everyday survival. Rather, the information might be for resistance to antibiotics, toxins, or other potentilly toxic environmental chemicals. DNA and genes will be studied in more detail in the chapter on The DNA Story.

Ribosomes

All prokaryotes have thousands of **ribosomes** spread throughout the cytoplasm. They consist of 60% RNA and 40% protein. Although ribosomes are smaller and less dense than their eukaryotic counterparts, they all function to decode the information in genes (DNA) into protein molecules (amino acid chains). Many antibiotics work by inhibiting ribosome function, which either slows bacterial growth or, by disrupting metabolism, kills the bacterial cell.

Inclusion Bodies

Some prokaryotic cells contain **inclusion bodies** in the cytoplasm. In response to the environment, these structures might store carbohydrates or lipids as future energy sources. Other inclusion bodies might store phosphate. Phosphate is an essential component to make phospholipids and the phosphate-sugar backbone of DNA and RNA.

4.3 Prokaryotic Reproduction and Survival: Binary Fission and Endospores

One of the characteristics of life is the ability to reproduce; that is, to make more copies of itself and, as a result, build a larger population of such organisms. Let's look at the reproduction process in the prokaryotes, again emphasizing bacteria like *S. enterica*.

Prokaryotic Reproduction

Most prokaryotic species reproduce by a cell division process called **binary fission**. This process ensures that both new cells (often referred to as daughter cells) contain a complete genome and other materials to survive as independent cells. Several events occur prior to the actual binary fission event.

Prior to cell division, the cell lengthens and increases in volume (**FIGURE 4.11**). In addition, all the materials needed for division, such as wall

FIGURE 4.11 Binary Fission. After a bacterial cell has replicated its DNA (chromosome), the fission process (steps A–C) pinches the cell in two as the cell wall and cell membrane converge.

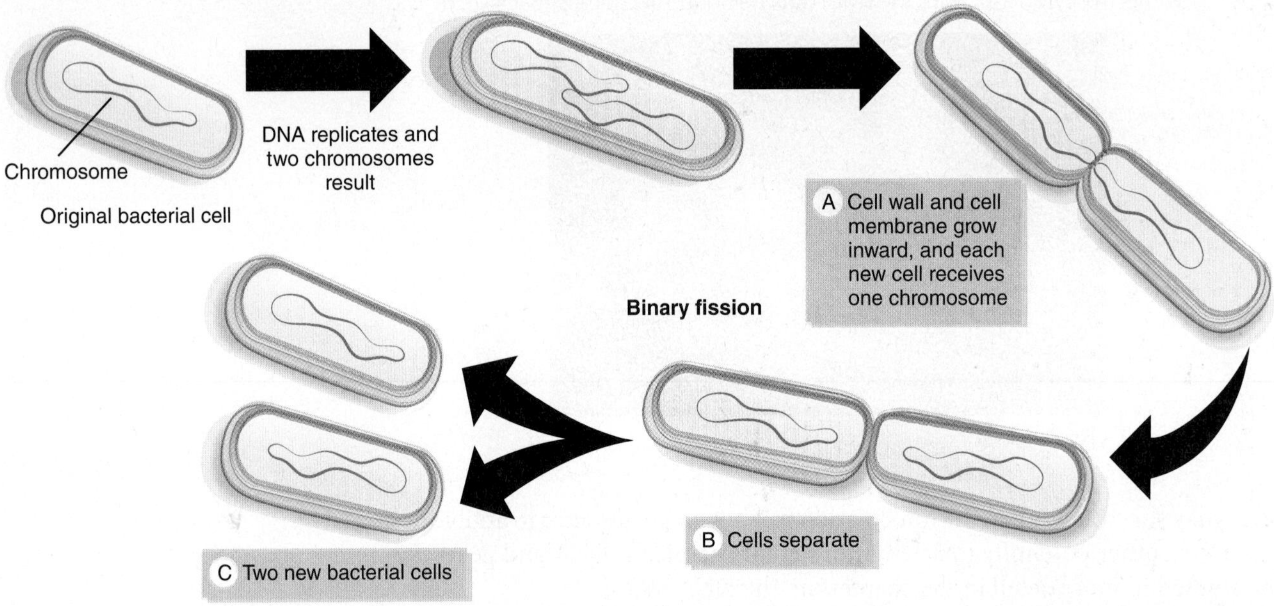

materials, phospholipids, and proteins, are made ready. However, before cell division can occur, the chromosome must be copied, so each new cell after division will have a complete set of genetic instructions. To accomplish this, cytoplasmic enzymes are used to make a copy of the bacterial chromosome. Once the replication process is completed, the two chromosomes separate from one another in the common cytoplasm.

Binary fission then starts. This involves the extension of a new cell wall and cell membrane inward from the middle of the cell (see Figure 4.11). As the wall and membrane come together, the cell is pinched in two; that is, the original cell now exists as two daughter cells, each cell having one chromosome. The effect is somewhat like forming two sausages by pinching the original sausage in two.

The speed of bacterial cell division is quite extraordinary. Under ideal growth conditions, an *S. enterica* cell can undergo binary fission and produce two new daughter cells about every 30 minutes. By comparison, a human liver cell takes about 20 hours to produce two new daughter cells. Consequently, within 1 hour, a single *S. enterica* cell has divided into four cells (two binary fissions); in 2 hours, there will be 16 cells (four binary fissions); and in only 7 hours, there will be a population of more than 16,000 cells (**FIGURE 4.12**).

Now you should start to understand why you would not feel well the next morning after eating the *Salmonella*-contaminated Caesar salad. If it had been 12 hours since you ate the salad, there would be almost 17 million cells, assuming you ingested but one cell. More than likely, a contaminated salad would already have millions of *Salmonella* cells. So, after 12 hours, your gut would contain billions of cells as the result of binary fission. No wonder you did not feel well!

Interestingly, one enterprising mathematician has calculated that a single *E. coli* cell reproducing every 20 minutes could yield in 36 hours (108 binary fission events) enough bacterial cells to cover the surface of the Earth a foot thick!

Measuring Prokaryotic Growth

If these calculations for prokaryotic growth of bacteria like *E. coli* and *S. enterica* are correct, why are we not smothered in a sea of bacterial cells? The answer is that all prokaryotic cells are susceptible to the same dynamics as any population

FIGURE 4.12 A Skyrocketing Bacterial Population. The number of *Salmonella enterica* cells progresses from one cell to more than 16,000 cells in 7 hours.

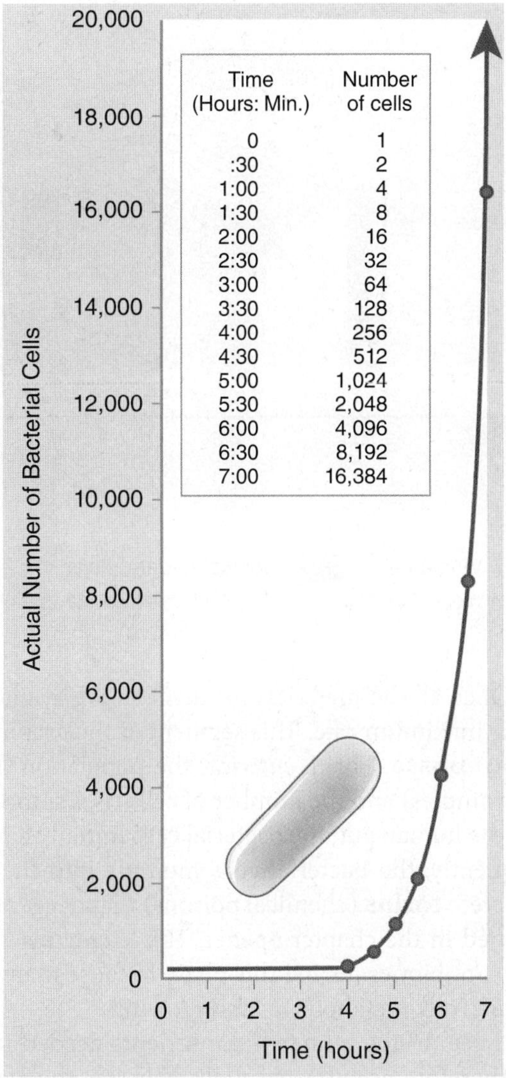

of plants, animals, or humans. Eventually, nutrients become scarce, waste products accumulate, water is in short supply, or the environmental temperature changes. From the organism's point of view, life gets stressful. Fewer cells reproduce, more cells die, and the population stabilizes or declines in number.

Such a rise and potential fall of microbial cells in a population can be presented graphically as a **growth curve**. As an example, let's continue to use *S. enterica* example from the chapter opener.

As described above, a population of bacterial cells will increase in number as the cells undergo binary fission. In the growth curve shown in **FIGURE 4.13**, the growth in numbers is plotted with logarithmic numbers (powers of 10) versus time. Various phases in the curve are identified, and these phases show the dynamics of the population as time passes. A typical growth curve can be separated into four phases.

- **Lag Phase.** The first segment in the graph is called the **lag phase**. During this preparation phase, no increase in cell number is observed. Rather, the *S. enterica* cells are synthesizing cell parts and enzymes for digesting the nutrients in the gut. The actual length of the lag phase depends on the metabolic activity of the microbial population, including the preparations for binary fission.

FIGURE 4.13 The Growth Curve for a Bacterial Population. This growth curve illustrates the four phases typically occurring during the growth of a bacterial cell population.

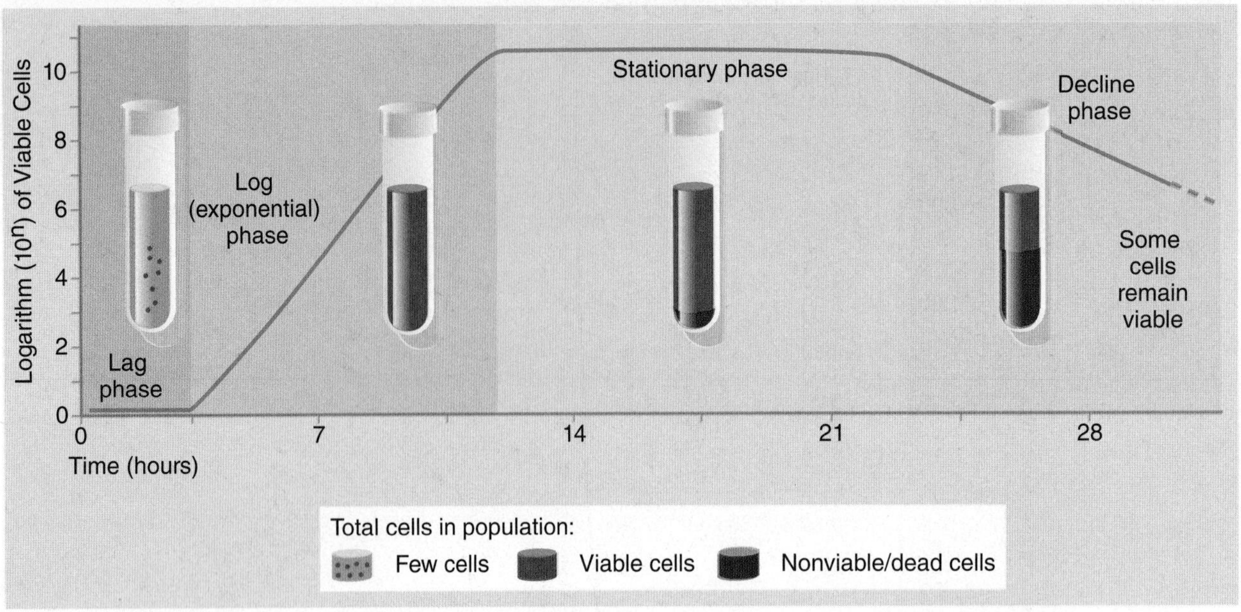

- **Log Phase.** Once all the materials for division are available, the microbes divide at their maximum rate. This segment of the graph is called the **logarithmic (log) phase**. For *S. enterica*, the population doubles at regular intervals (30 minutes) and the number of cells rises smoothly in an upward direction. In the human gut, the bacterial cells initially have plenty of nutrients. Consequently, the bacterial cells multiply into the billions. As they grow, they secrete **toxins** (chemical poisons) that trigger the disease symptoms mentioned in the chapter opener. The *S. enterica* cells also are most susceptible to antibiotics during this phase because many antibiotics affect cells that are actively metabolizing and growing.
- **Stationary Phase.** In the human gut, nutrients needed for growth are not unlimitless. Therefore, as nutrient supply dwindles, the *S. enterica* cells compete for limited nutrients with other bacteria that are part of the human gut microbiome. The vigor of the population changes, and the *S. enterica* cells enter a plateau, called the **stationary phase**. Although most cells remain alive, they stop dividing and population growth is arrested.
- **Death Phase.** If nutrients in the gut are not replenished (and they probably would not since you are very sick), metabolism will stop and eventually the nongrowing cell population enters a **decline (death) phase**. Some *S. enterica* cells can survive for a short time using the nutrients that are released from the dead and dying cells. Thus, the population enters a state of balanced cell death, some cells surviving for a time as other cells die.

So, now you can understand how the foodborne infection gradually disappears. With a lack of nutrients, competition for nutrients between pathogen and your gut microbiome is fierce. In a human infection like salmonellosis, there are two additional factors in pathogen elimination. One involves vomiting and diarrhea. These responses help eliminate bacterial cells and the toxins produced by the pathogens. The second factor is your immune system. White blood cells and immune defenses have evolved to recognize the infection and help eliminate the infecting cells and the toxins. We will have much more to say about immunity in the chapter on Disease and Resistance.

Endospores

Although most prokaryotic species will die when nutrients are depleted, a few bacterial species might survive. Many species of *Bacillus* and *Clostridium* produce an extraordinarily resistant structure called the **endospore** (**FIGURE 4.14**). Its sole function is to survive in the environment that is unfavorable for growth. The spore is formed during a complex process during which the mother cell produces an internal spore (i.e., endospore) that contains a complete set of genetic information. The spore is surrounded by a very thick set of peptidoglycan-containing walls. Once formed and mature, the dormant spore is released with the death of the mother cell.

Clostridium: kla-STRIH-dee-um

After release, the free spore can remain in a dormant state for decades, centuries, or even thousands of years. These dormant structures probably are the most resistant biological structures known. By containing only 10% water, desiccation has little effect on the spore. They are heat resistant and impervious to most chemicals. Endospores can remain viable in boiling water (100°C/212°F) for 2 hours. When placed in 70% ethyl alcohol, endospores have survived for 20 years. Humans can barely withstand 500 **rems** of radiation, but endospores can survive 1 million rems. Yet, when suitable nutrients return, the endospores quickly germinate and produce an actively metabolizing cell that can grow and reproduce.

rem (roentgen equivalent in man): A measure of radiation dose related to biological effect.

These extreme properties make endospores difficult to eliminate from contaminated medical materials and food products. In fact, a few serious human diseases are the result of endospore contamination. The most newsworthy has been *Bacillus anthracis,* the agent of the 2001 anthrax bioterror attack. Spores were sent through the mail, and, when inhaled, the endospores germinated in the recipients' lungs. The growing cells then secrete two deadly toxins.

Bacillus anthracis: bah-SIL-lus an-THRAY-sis

Botulism: A rare but serious bacterial disease caused by a toxin that attacks the body's nerves, weakening the muscles involved in breathing, which can lead to difficulty breathing and even death.

Likewise, **botulism** and **tetanus** are diseases caused by *C. botulinum* and *C. tetani*, respectively. Clostridial endospores often are found in soil, so a puncture wound contaminated with *C. tetani* endospores provides an environment where the endospores germinate. As they grow, the cells produce tetanus toxin, which causes a painful tightening of the muscles. Similarly, in a sealed can of food inadvertently contaminated with *C. botulinum* endospores, the spores will germinate, and the growing cells produce botulism toxin, which causes muscle weakness.

Tetanus: A bacterial disease marked by rigidity and spasms of the voluntary muscles that can lead to an inability to open the mouth (called lockjaw) and difficulty swallowing and breathing.

Clostridium botulinum: kla-STRIH-dee-um bot-you-LIE-num

FIGURE 4.14 Bacterial Endospores. This false-color scanning electron microscope image shows the germination of an endospore. The green cells are the growing cells. (Bar = 1 μm.)

4.4 Prokaryotic Growth: Culturing Bacteria

With this understanding of bacterial population growth, let's now look at the methods used to grow prokaryotes in the lab. In some cases, trying to grow them is quite challenging. Again, we primarily will examine bacterial species.

Growth Methods

Bacterial organisms are grown (cultivated) on or in **culture media** (sing. **medium**), which contain various nutrients that promote cell growth. Culture media are used as liquids or in a gel-like form. In liquid form, the culture medium is called a **broth**. A typical example is nutrient broth, a "beef/vegetable soup" containing nutrients called extracts or digests that come from various animal and/or plant sources.

The gel, or semisolid, form of a culture medium is used to culture microbes in a culture (Petri) dish. This type of medium consists of a broth solidified with **agar**, a complex carbohydrate extracted from red algae. The agar in this **nutrient agar** plate is not used as a nutrient by most bacterial species; rather, it is simply a solidifying substance that remains solid at temperatures as high as 80°C/176°F (human body temperature, for comparison, is 37°C/98.6°). When a sample of *S. enterica* cells are placed onto the surface of a nutrient agar plate, each cell will divide repeatedly and produce large numbers of cells, forming a visible **colony** (**FIGURE 4.15**).

There are a tremendous variety of culture media used to grow microbes. Many of these media are general purpose, meaning the nutrients in the medium will support the growth of many different microbial species. There also are many types of specialized media, called **selective media**. These will promote (select for) the growth of desired species while inhibiting the growth of all other species. There even is a selective medium for the isolation of *Salmonella* species from clinical samples and from food. With our chapter opening example, this

FIGURE 4.15 Bacterial Colonies. Colonies of *Salmonella enterica* growing on blood agar in a culture dish. Blood agar is a mixture of nutrient agar and sheep red blood cells. It often is used for growing many different bacterial species.

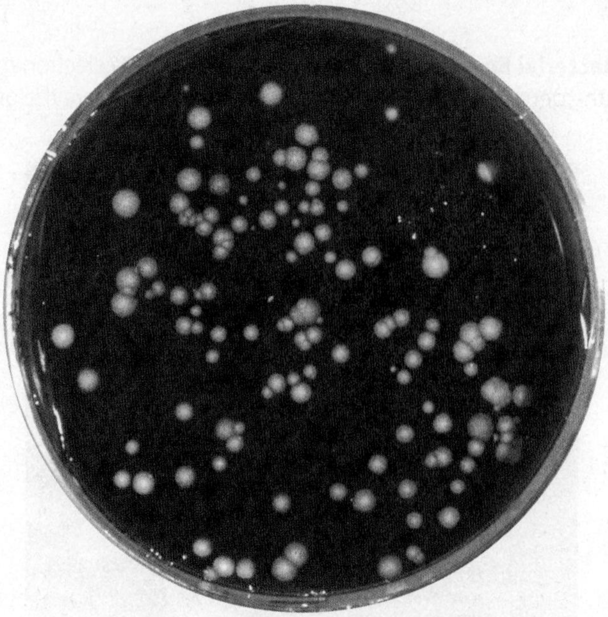

Courtesy of CDC.

FIGURE 4.16 Selecting for *Salmonella enterica*. This culture plate contains a selective medium that only allows *S. enterica* cells to grow.

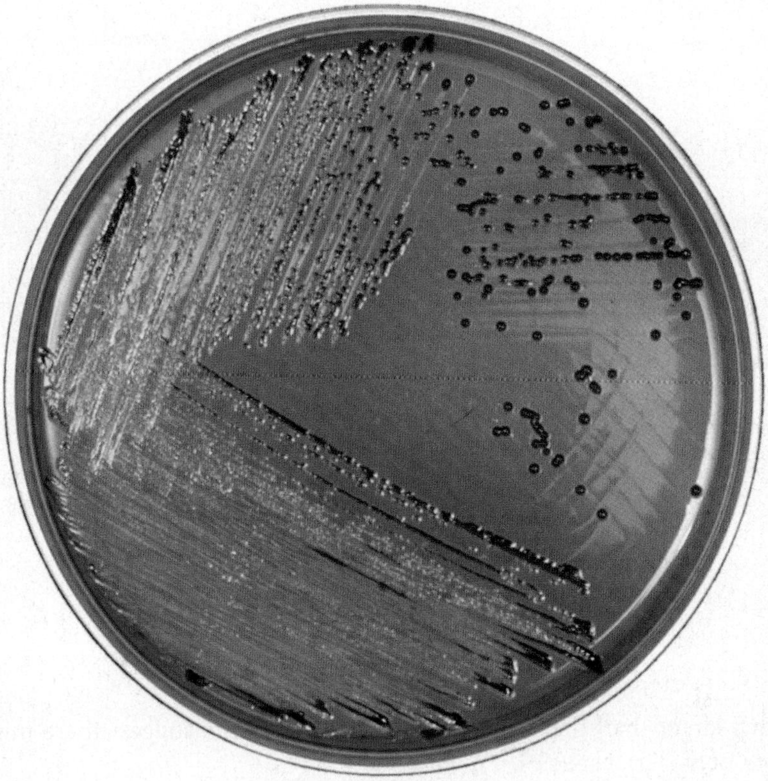

Courtesy of CDC.

selective medium could have been used by a health agency's microbiology lab to identify *S. enterica* from a gut sample and to identify the organism in the contaminated Caesar salad (**FIGURE 4.16**).

Sometimes microbes will not grow on any known standard culture medium because the medium lacks one or more essential nutrients. To stimulate growth of these species, microbiologists might add unusual nutrients to a standard medium. Such an **enriched medium** attempts to grow these uncultured species.

Today, only about 2% of all known prokaryotic species can be cultured. This means 98% of all prokaryotic species cannot be grown in any culture medium yet devised. The cells of these species often can be observed when viewed with the microscope, but when placed on an agar medium, no colonies grow. Such organisms are called **viable but noncultured** (**VBNC**), meaning the cells are alive but will not reproduce in the unfamiliar culture environment. Presumably, some unique growth requirement (nutrient) is missing, and no known enriched medium has the needed factor(s). For example, a gram of ordinary garden soil contains thousands of bacterial species, only a few of which can be cultivated and studied. This again highlights the tremendous diversity of prokaryotic species that exist, which is the topic for the last section of this chapter.

4.5 Meet the Prokaryotes: The Domains Bacteria and Archaea

Currently, there are more than 13,000 known prokaryotic species. More species are being discovered each year, so the actual number of species certainly

FIGURE 4.17 The Tree of Life for the Bacteria and Archaea. The tree shows several of the bacterial and archaeal lineages discussed in this chapter.

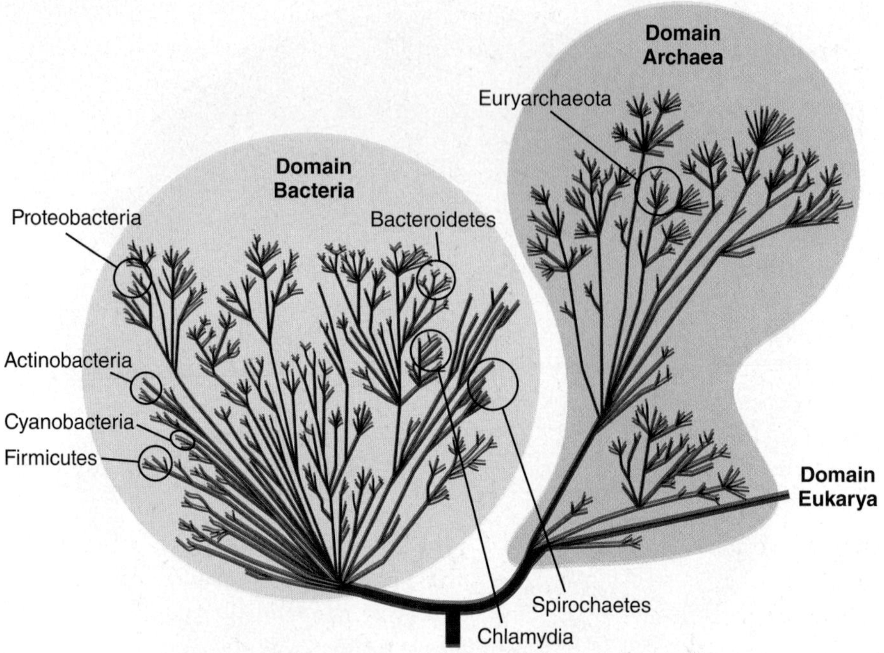

is much larger than the 13,000. Some microbiologists suggest there might be billions of bacterial species.

Archaea: are-KEY-ah

In this section, we will highlight a few common phyla and groups using the "tree of life," which, as described in the chapter on Microbes in Perspective, organizes all living organisms into one of three domains: Bacteria, **Archaea**, and Eukarya (**FIGURE 4.17**). The prokaryotes compose the domain Bacteria and domain Archaea.

Domain Bacteria

Members of the **domain Bacteria** have adapted to the diverse environments on Earth. They inhabit the water, soil, and air. They can be isolated from arctic ice, thermal hot springs, the fringes of space, and the tissues of plants and animals. Bacterial species, along with their archaeal relatives, have so completely colonized every part of Earth that their mass is estimated to outweigh the mass of all plants and animals combined. Let's look briefly at some of the more common phyla and other groups, which give us a sense of the bewildering diversity that exists.

Photosynthetic Bacteria

Bloom: A marked visible discoloration of water caused by a sudden increase in the number of cells of an organism in an environment.

The members of the **Cyanobacteria** represent one of the largest and oldest phyla on Earth. They thrive in freshwater ponds, lakes, and oceans in such numbers that they change the color of the water. The periodic pea-soup look of Lake Erie is due to **blooms** of cyanobacterial species (**FIGURE 4.18**) that exist as unicellular, filamentous, and colonial forms. They get their energy from photosynthesis, producing much of the oxygen gas we breath today. In fact, ancestors to the present-day cyanobacteria were the first oxygen gas-producing organisms on Earth, helping transform life on the young planet into what it is today. Today's cyanobacterial species occupy a key position in the nutritional patterns of nature by providing organic matter at the base of the food web. Undoubtedly, they are one of the most important groups of bacteria on Earth.

FIGURE 4.18 Cyanobacteria in Lake Erie. This satellite image shows a massive bloom of cyanobacteria (swirls of light green colors) extending many miles across Lake Erie.

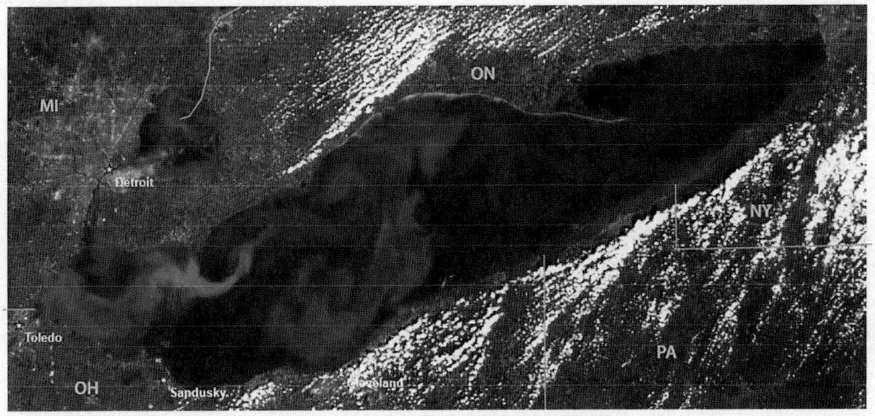

Courtesy of NOAA.

Gram-Negative Bacteria

Most bacterial species do not carry out photosynthesis, but rather obtain their food and energy from sources other than the sun. Many do this by playing key roles as **decomposers**, organisms that break down chemical compounds in the environment. They then can recycle nutrients back into the water or soil for use by other organisms. Although only a small fraction of the thousands of species of known bacteria on the planet Earth cause disease in humans, a few of the more prominent pathogens are gram-negative species.

- **Proteobacteria.** The **Proteobacteria** (*proteo* = "first") contains the largest and most diverse group of bacterial species, many of which can be cultured. The phylum includes familiar genera, such as *Escherichia* and some of the most recognized human pathogens, including genera responsible for food poisoning (*Shigella, Salmonella*), the plague (*Yersinia*), cholera (*Vibrio*), and the sexually transmitted disease gonorrhea (*Neisseria*).

  Among the other members of the Proteobacteria are the rickettsiae (sing. **rickettsia**). These organisms are called **obligate intracellular parasites** because they need the nutrients inside host cells to survive and reproduce. These tiny bacterial cells are transmitted among humans primarily by ticks, fleas, and mosquitoes. Rocky Mountain spotted fever is transmitted by ticks infected with the pathogen.

  Many other species are of medical, industrial, or environmental importance. Among the more interesting are members of the genus *Pseudomonas*. One species, *P. aeruginosa*, is responsible for "hot tub rash" where the cells in contaminated water infect the skin. In the soil, other species of the genus produce a large variety of enzymes that contribute to the breakdown of pesticides and similar waste chemicals.

  Finally, it is worth mentioning one more species, *Serratia marcescens*, a rod-shaped species distinguished by the blood-red pigment it produces when it forms colonies. Its "blood" has historical significance, as **A CLOSER LOOK 4.3** details.

- **Bacteroidetes.** Another phylum of gram-negative bacteria is the Bacteroidetes. These rod-shaped cells live in water and soil that is free of oxygen gas. They also live in the oxygen-free conditions of the gut. In fact, the human gut microbiome is dominated by several Bacteroides species that are very beneficial to our digestive health. Numerous research studies suggest that normal-weight and lean adults have a high level of Bacteroides species and

Shigella: shih-GEL-lah

Yersinia: yer-SIN-ee-ah

Vibrio: VIB-ree-oh

rickettsiae: rih-KET-sea-eye

Pseudomonas aeruginosa: sue-doh-MOH-nahs ah-rue-gih-NO-sah

Serratia marcescens: ser-RAH-tee-ah mar-SES-senz

Bacteroidetes: BAK-teh-roy-deh-teez

Bacteroides: BAK-teh-roy-deez

A CLOSER LOOK 4.3

The Blood of History

The cells of the bacterium *Serratia marcescens* contain a red pigment, which when observed as colonies on agar appear blood red (see Figure A). Because of its characteristic blood-red pigment and propensity for contaminating bread, *S. marcescens* has had a notable place in history. For example, the dark, damp environments of medieval churches provided optimal conditions for bacterial growth on sacramental wafers used in Holy Communion. Often such occurrences were viewed as miracles.

One such event happened in 1263 in the church of Santa Cristina in Bolsena, Italy, when bread of the eucharist began to "bleed" and then dripped onto the tablecloth in the shape of a cross. The event was later commemorated by Raphael in his fresco *The Mass of Bolsena* (1512–1514) that is in the Vatican's Apostolic Palace. Such miracles also were recorded earlier in history and had great historical significance.

In 332 BCE, Alexander the Great and his army of Macedonians laid siege to the city of Tyre in what is now Lebanon. The siege was not going well. Then, one morning blood-red spots appeared on several pieces of bread. At first it was thought to be an evil omen, but Alexander's astrologer named Aristander indicated the "blood" was coming from within the bread. He predicted that blood would be spilled within Tyre and the city would fall to Alexander. Alexander's troops were buoyed by this interpretation and with renewed confidence they charged headlong into battle and captured the city. The victory opened the Middle East to the Macedonians, and their march did not stop until they reached India.

In 1819, Bartholemeo Bizio, an Italian pharmacist, demonstrated that these and other "bloody" miracles were caused by a living organism. Bizio named it *Serratia* after Serafino Serrati, a countryman whom he considered the inventor of the steamboat. The name *marcescens* came from the Latin word for "decaying," a reference to the decaying of bread that led Alexander the Great to victory.

FIGURE A Colonies of *Serratia marcescens* growing on a nutrient agar plate.

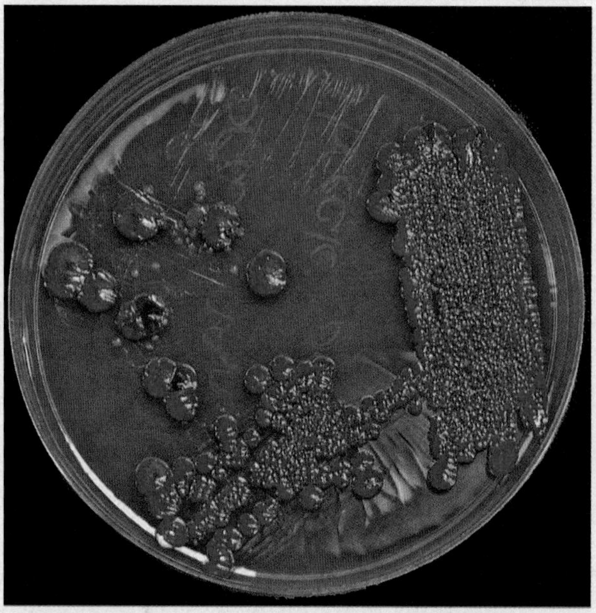

a low number of Firmicutes (see below) composing their gut microbiome. Obese adults have just the opposite; that is, they have a higher abundance of Firmicutes and lower amount of Bacteroidetes. The conclusion has been made that the dynamics of these two groups of bacteria in the gut can influence weight gain. The exact interaction between these different bacteria requires more investigation.

Chlamydiae: kla-MIH-dee-eye

- **Chlamydiae.** Members of the phylum Chlamydiae also are obligate, intracellular parasites. Most species are pathogens, and one species is the causative agent of the sexually transmitted infection (STI) called chlamydia. In 2016, more than 1.6 million cases of chlamydia were reported to the United States Centers for Disease Control and Prevention (CDC). This is the largest number of cases ever reported for the infection.

Spirochaetes: spy-row-KEY-teez

- **Spirochaetes.** The species in the phylum Spirochaetes consist of gram-negative cells possessing a unique cell body that is coiled like a corkscrew (see Figure 4.2). The spirochetes include free-living species found in mud and sediments. Other species inhabit termite guts, where they assist in the insect's digestion of wood. Still others are pathogens in the urogenital tracts of vertebrates. Among the human pathogens is *Treponema pallidum,* the causative agent of syphilis, one of the most common STIs. A species of *Borrelia* is transmitted by ticks and is responsible for Lyme disease. Recent estimates from the CDC suggest that as many as 300,000 people might get Lyme disease each year in the United States.

Treponema pallidum: treh-poh-NEE-mah PAL-eh-dum

Borrelia: bore-RELL-ee-ah

Gram-Positive Bacteria

The gram-positive bacterial groups rival the Proteobacteria in diversity and are divided into two phyla.

- **Firmicutes.** The **Firmicutes** (*firm* = "strong"; *cuti* = "skin") contain some soil genera, such as *Bacillus* and *Clostridium* species already mentioned, that are endospore formers. Other species in the Firmicutes do not form endospores. Species within the genera *Streptococcus* and *Staphylococcus* are responsible for several mild to life-threatening human illnesses. For example, about 30% of humans carry *Staphylococcus aureus* in their noses where it usually does not cause any harm. When uncontrolled, an *S. aureus* infection can be serious or fatal by infecting the blood, heart, lungs, or bone.

  Another member of the Firmicutes is the genus *Lactobacillus.* Some species live in the female genital tract and help guard against infection by other microbes. Other species are used in the large-scale manufacturing of cheese, sour cream, yogurt, and other fermented milk products.
- **Actinobacteria.** Another phylum consisting of gram-positive species is the **Actinobacteria**. Unique from most bacteria, these soil organisms form a system of branched filaments and are responsible for the characteristic mustiness of garden soil. The genus *Streptomyces* is the source for more than 500 antibiotics, some of which are used to treat human bacterial diseases. Another medically important genus is *Mycobacterium,* one species of which is responsible for tuberculosis.

Streptococcus: strep-toe-KOK-us

Lactobacillus: lack-toe-bah-SIL-lus

Streptomyces: strep-toe-MY-seas

Mycobacterium: my-koh-back-TIER-ee-um

Domain Archaea

We close our discussion of prokaryotic diversity by surveying the members of the domain Archaea. Although archaeal cells often resemble bacterial cells when observed with the light microscope, they have an unusual wall and membrane structure, and unique physiology and biochemistry. The recent discovery of many new species (most VBNCs) has expanded the "limbs" and "twigs" on the domain.

Many archaeal organisms are referred to as **extremophiles** because they grow best at physical or geochemical extremes. Some grow well at very high temperatures, high salt concentrations, or even in highly acidic volcanic lakes (**FIGURE 4.19**). Many more species exist in very cold environments. Although there also are archaeal genera that thrive under more mild conditions, there are no known species causing disease in any plants or animals. The archaeal genera can be placed in one of several "supergroups."

- **Euryarchaeota.** The **Euryarchaeota** contain organisms with varying physiologies. Some species are called **methanogens** (*methano* = "methane"; *gen* = "produce") because they produce methane (natural) gas. In fact, these archaeal species release more than 2 billion tons of methane gas into the atmosphere every year. About one-third comes from the archaeal species living in the stomach (rumen) of cows and is a contributing factor to climate change.

Euryarcheota: ur-ee-are-kee-OH-tah

FIGURE 4.19 The Grand Prismatic Spring. In and around the Grand Prismatic Spring in Wyoming's Yellowstone National Park, bands of green, yellow, and orange surrounding the spring represent microbial mats containing extremely heat-tolerant archaeal species.

Also, within this supergroup is the **extreme halophiles** (*halo* = "salt"; *phil* = "loving"). These species require high concentrations of salt (up to 30% NaCl) to grow and reproduce. They can be found in such places as Utah's Great Salt Lake.

Other archaeal organisms within the Euryarchaeota compose the **hyperthermophiles** (*hyper* = "high;" *thermo* = "heat") that grow optimally at temperatures above 80°C/176°F (remember water boils at 100°C/212°F). They typically are found in volcanic terrestrial environments and deep-sea hydrothermal vents where the temperature can be as high as 113°C/235°F; indeed, they will not grow if the temperature drops below 90°C/194°F because it's just too cold! Many archaeal species also grow in very acidic environments at a pH of 1.0 (more acidic than the acid in a car battery).

- **The Other Supergroups.** Among the other supergroups, the **Crenarchaeota** is the best studied. The phylum consists mostly of hyperthermophiles, typically growing in hot springs and marine hydrothermal vents. Other species are dispersed in open oceans, often inhabiting the cold ocean waters (–3°C/27°F) of the deep-sea environments and polar seas.

Crenarchaeota:
cren-are-key-OH-tah

TABLE 4.1 summarizes some of the characteristics shared or are unique among the domain Bacteria and Archaea.

A Final Thought

When you pick up this book for the first time, you might experience a moment anticipating your first look at microbes. Perhaps you leaf through the pages, turn back to see a photograph a second time, then pause and think: "Is that all there is? Little rods and spheres?"

One's first encounter with microbes often is a disappointing (and perhaps exasperating) experience. Since early childhood, we have been taught to loathe

TABLE 4.1 Some Similarities and Major Differences Between the Prokaryotic Domains

Characteristic	Bacteria	Archaea
Cell size	1 µm	1 µm
Cell nucleus	No	No
Chromosome form	Single, circular	Single, circular
Plasmids	Yes	Yes
Chlorophyll-based photosynthesis	Yes (cyanobacteria)	No
Peptidoglycan cell wall	Yes	No
Cell (plasma) membrane	Yes	Yes (different lipid structure)
Membrane-bound organelles	No	No
Ribosomes	Yes	Yes
Growth above 80°C	Some	Some
Growth above 100°C	No	Some
Pathogens	Some	No

and despise bacteria, and we have been schooled in all the dastardly deeds they do. "Wash your hands" we are admonished; "Don't eat it if it falls on the ground." And on and on. We expect bacterial life to be filled with grotesque and fearsome monsters. But they turn out to be little sticks and rods, not very dangerous-looking at all.

So, perhaps we need to wipe away any preconceived notions of "bad bacteria" and start rebuilding our views. It's going to take some time to absorb the importance of the prokaryotic and eukaryotic microbes to our lives, but for now try to appreciate the prokaryotes for more than the tiny rods you see on these pages or under the microscope. The electron microscope has yielded a wealth of information about their structure (as this chapter demonstrates) and studying prokaryotic structure gives us a clue to what they do. This "what they do" part is key because it explains their importance to us and society.

Chapter Discussion Questions

What Was He Thinking?

From your reading of this chapter, identify and discuss five major points about the bacterial and archaeal organisms that the author was trying to get across to you.

Questions to Consider

1. Today, many microbiologists along with other biologists, biochemists, and engineers are trying to build cells from scratch; that is, build a new

type of cell from the ground up. If you were designing a bacterial cell, what parts must your cell have to be alive, grow, and reproduce?

2. Extremophiles are of interest to industrial corporations, who see these prokaryotic species as important sources of enzymes that function at temperatures of 100°C and in extremely acidic conditions (the enzymes have been dubbed "extremozymes"). What practical uses can you foresee for these enzymes?
3. Several years ago, public health officials found the water in a midwestern town was contaminated with sewage bacteria. The officials suggested homeowners boil their water for 10 minutes before drinking it. Would this treatment remove all traces of bacterial cells from the water? Why?
4. About 100 to 250 grams (3 to 8 ounces) of feces are excreted by a human adult each day. Thirty percent of the solid matter consists of bacterial cells that were present in the colon. If that is the case, about how much bacterial mass do we "produce" in a week? In a year? What topic in this chapter allows us to produce this mass of bacterial cells?
5. "Bacteria are all the same. Once you've seen one, you've "seen 'em all!" What evidence could you present to counter this statement?
6. There are thousands, if not millions, of bacterial species, yet most have variations of only three shapes: the rod, the sphere, and the spiral. Do you find this strange? Why or why not?
7. A bacterial species has been identified as a gram-positive rod. What cell structures would it most likely have? Explain your reasons for each choice.
8. Some evolutionary biologists argue that bacterial pathogens have acted as great "slate wipers" of history (e.g., cholera, plague, typhoid fever, and syphilis) and are the key agents of natural selection for the human species; that is, in a sense, pathogens help improve our species by selecting the fitter individuals through the distasteful task of infectious disease. Do you agree or disagree with this statement? [Note: you might want to read Dan Brown's mystery thriller, *Inferno*, published by Doubleday (2013).]
9. Using the tree of life below, label the phylum location in the tree from the phylum descriptions given below.

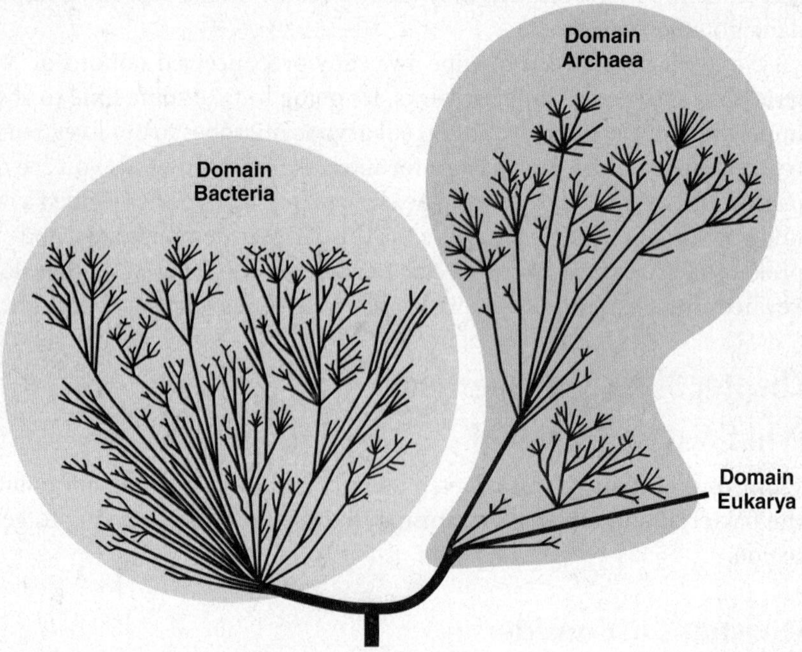

Phylum Description

a. Contains the largest and most diverse group of gram-negative bacterial species.
b. Consist of gram-negative cells possessing a unique cell body coiled into a corkscrew-like shape.
c. Some of these gram-positive species produce endospores.
d. Members of this phylum produce oxygen gas through photosynthesis.
e. Contains the methanogens, extreme halophiles, and many hyperthermophiles.

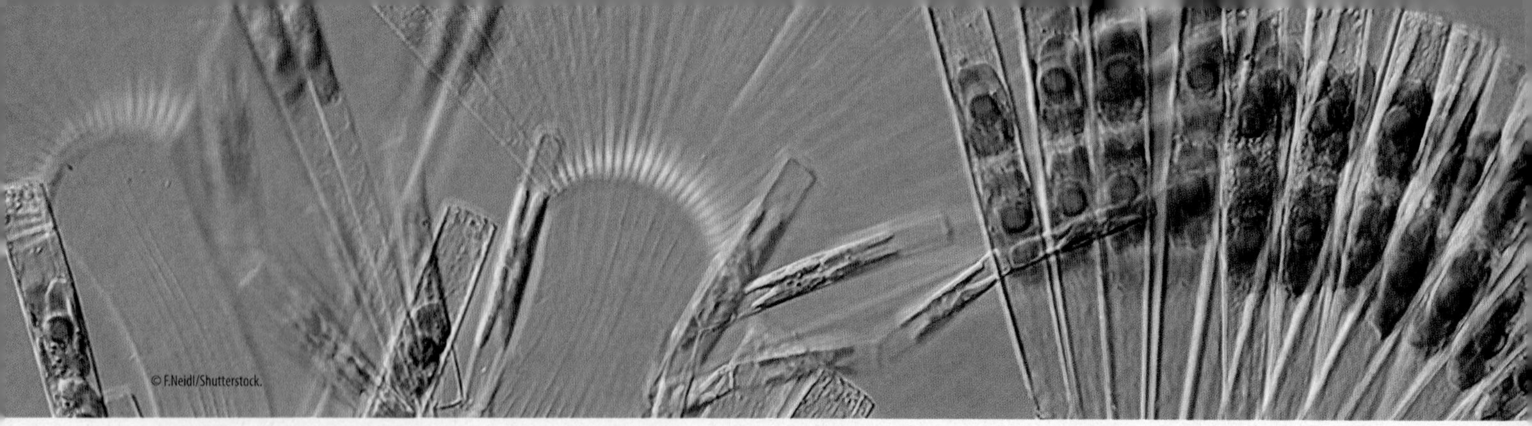

© F.Neidl/Shutterstock.

CHAPTER 5

The Eukaryotic World: Protists and Fungi

The Big Ditch

Heralded at the time as The Eighth Wonder of the World, the construction of the Panama Canal would be a spectacular engineering achievement. With its completion global trade would be stimulated through much shorter transit times for goods and products shipped between the Atlantic and Pacific coasts (**FIGURE 5.1A**). However, construction was being hampered by malaria, a tropical disease that doomed France's attempt at canal construction in 1883. When the United States took over construction in 1904, malaria was still a major problem. By 1906, over 21,000 of the 26,000 employees working on the canal had been hospitalized at some point with malaria. Successful completion of the "Big Ditch" would have to control this disease.

In 1878, Alphonse Laveran, a French military doctor posted in Algeria, started studying malaria, which was a serious problem in that country. Working from his anatomical observations and blood samples from patients with malaria, Laveran concluded that a protozoan parasite was the agent causing malaria. But how was the parasite transmitted to people?

Ronald Ross was a British medical officer stationed in Calcutta, India where malaria also was killing thousands of people (**FIGURE 5.1B**). In 1892 he began the hunt to discover how the malarial parasite was transmitted. The breakthrough came in 1897 when Ross observed infectious protozoal parasites in mosquitoes. He proposed that when taking blood meals from individuals, the infected mosquitoes transmitted the parasite to humans.

Ross' discovery had a great impact on building the Panama Canal. In 1904, Colonel W. C. Gorgas of the U.S. Army Medical Corps headed the Isthmian

CHAPTER 5 OPENER A light microscope image of a marine golden alga isolated from the Mediterranean Sea. These fan-shaped colonies are just one example of the diverse types of eukaryotic protists.

FIGURE 5.1 Malaria and the Panama Canal.
(A) The Panama Canal, which connects the Atlantic and Pacific Oceans, opened on August 15, 1914. The control of malaria was vital for the construction of the Canal.

(B) The discovery by Major Ronald Ross that malaria was transmitted by mosquitoes had tremendous impact on development programs in the tropics.

DR. RONALD ROSS, C.B., THE HERO OF THE MOSQUITO THEORY OF MALARIA.

Canal Commission whose task was to implement a mosquito control program. Through the introduction of insecticide spraying, the number of malaria hospitalizations and deaths quickly dropped. Work on the canal now could continue at a fast pace up to its completion in 1913. The canal officially opened to ship commerce in 1914.

Society owes a great debt of thanks to Laveran and Ross. Had they not been intrigued about the cause of malaria and the source of its transmission, it is possible that American construction of the canal would have taken much longer. Global ocean commerce between the Atlantic and Pacific Oceans would have remained lengthy and economic development would have been delayed.

Today, we know malaria is caused by one of four species in the genus *Plasmodium*. It is one of the protists that, along with the fungi, we will meet as we continue to explore the world of microbes—the eukaryotic members—and continue to witness how they affect society.

Plasmodium: plaz-MOH-dee-um

LOOKING AHEAD

After reading and completing this chapter, you will be able to:

5.1 List and summarize the functions for the major eukaryotic cell organelles.
5.2 Describe the theory proposed for the origins of the mitochondria and chloroplasts.
5.3 Identify the characteristics of the protists and distinguish between the animal-like, plant-like, and fungal-like protists.
5.4 Recognize the structure of fungi and give examples of organisms in the ascomycete and basidiomycete groups, including those with symbiotic relationships.

5.1 Eukaryotic Cells: Their Structure

Historically, prokaryotic and eukaryotic cells were distinguished by their structural differences. With the light microscope, eukaryotic cells were found to contain a cell nucleus and cytoplasmic structures that were absent from prokaryotic cells. However, prokaryotic cells and all living cells contain deoxyribonucleic acid (DNA) as their genetic information. In addition, a cell (plasma) membrane to separate the cytoplasm from the environment and ribosomes to manufacture proteins are universal structures.

Let's examine the major eukaryotic **organelles**, which are labeled in (**FIGURE 5.2**). Although we will focus on eukaryotic microbes, the principle organelles mentioned also are common to most plants and animals.

Endomembrane and Cytoskeletal Systems

Eukaryotic microbes have a group of membrane-enclosed organelles that compose the cell's **endomembrane system**. This system manufactures and transports proteins and lipids through and out of the cell (**FIGURE 5.3**). The transport is controlled by a cytoskeletal network.

Endomembrane System

- **The Endoplasmic Reticulum** (ER). The **endoplasmic reticulum** (**ER**) is a membrane network consisting of flat membranes on which **ribosomes** are attached (called the "rough endoplasmic reticulum"). Most of the proteins produced by these ribosomes are exported from the cell. A second,

FIGURE 5.2 A Stylized Drawing of a Eukaryotic Cell. Eukaryotic cells contain a variety of internal structures absent from prokaryotic cells. Structures that are common to eukaryotic and prokaryotic cells are indicated in red.

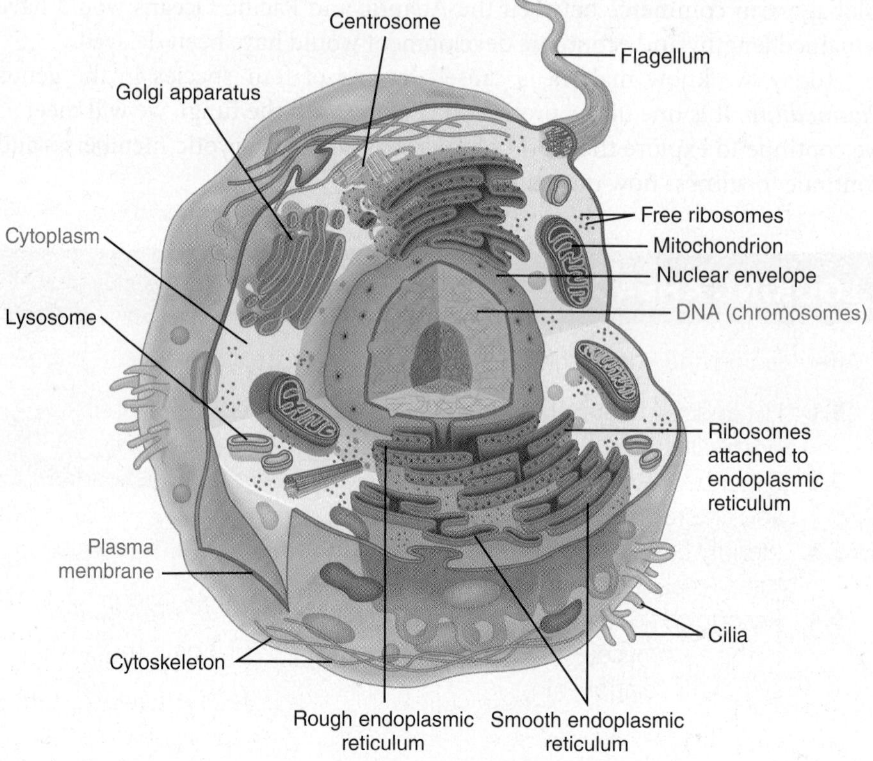

FIGURE 5.3 The Endomembrane System in Eukaryotic Cells. The rough endoplasmic reticulum (RER), Golgi apparatus, and vesicles work together to modify, package, and transport lipids and proteins, the latter made by the ribosomes attached to the RER.

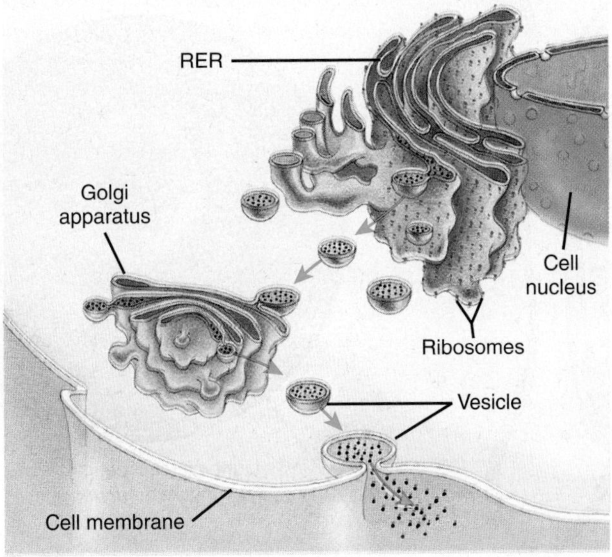

tube-like form of ER is free of ribosomes (called the "smooth endoplasmic reticulum"). It is involved in lipid synthesis and transport.

- **The Golgi Apparatus.** The **Golgi apparatus** consists of a group of independent stacks of flattened membranes and **vesicles** (see Figure 5.3). Scattered throughout the cell, each Golgi apparatus modifies, sorts, and packages proteins and lipids received from the ER.
- **Lysosomes. Lysosomes** are membrane-enclosed structures containing digestive (hydrolytic) enzymes. In humans, the white blood cells of the immune system use the lysosomes to destroy bacterial and viral pathogens that have infected the body.

Vesicle: A small structure within a cell, consisting of fluid enclosed by a lipid bilayer.

Even though prokaryotic cells lack an endomembrane system, they can manufacture and modify molecules just as their eukaryotic relatives do.

Cytoskeletal System

Eukaryotic microbes contain an interconnected network of cytoplasmic fibers and threads called the **cytoskeleton**. This protein lattice extends throughout the cytoplasm. It gives shape to the cell and, acting like railroad tracks, it transports proteins and lipids from the endomembrane system to their destination in or outside the cell.

Cell Motility

Many organisms, especially microorganisms, live in watery or damp environments. To move from place to place, they have structures that provide for **cell motility**. The motility structures include:

- **Flagella.** Some protists and a few fungi have long, thin protein projections called **flagella** (sing. **flagellum**). These structures are covered by the plasma membrane and they project outward from the cell surface (**FIGURE 5.4**). Composed of protein tubules, these appendages beat back and forth in a wave-like motion, which provides the mechanical force to push the cell forward.

FIGURE 5.4 Flagella. Some eukaryotic cells, such as the photosynthetic green alga in this drawing, are motile by means of flagella. Several other eukaryotic organelles are labeled. (Bar = 5 μm.)

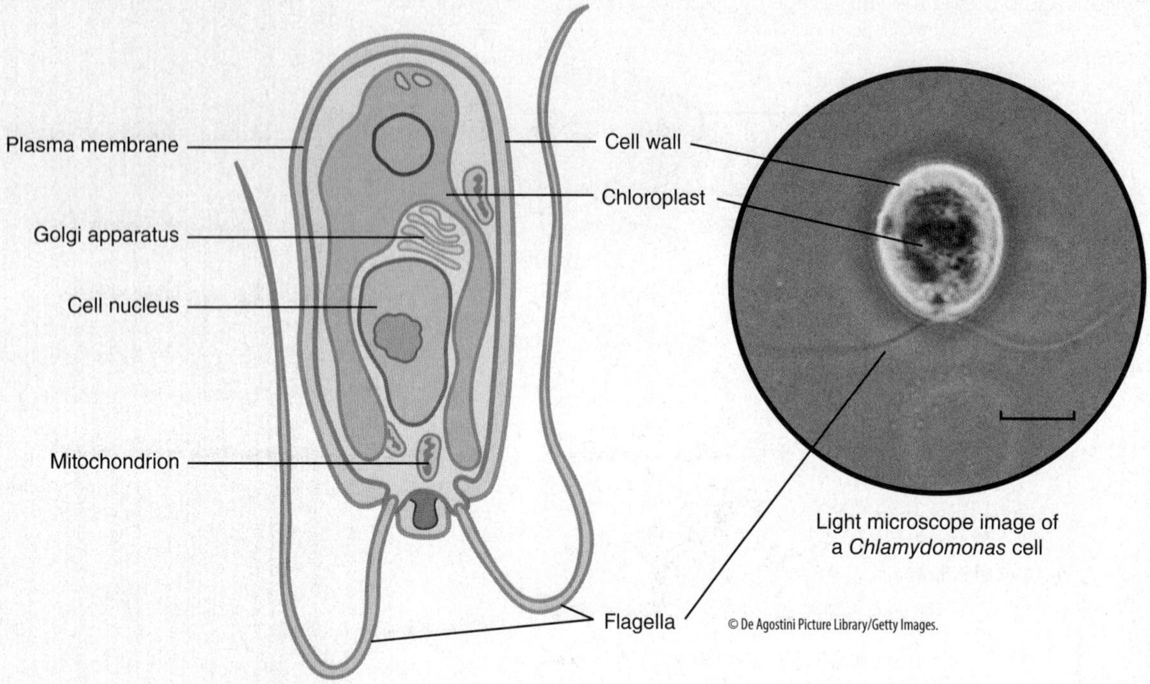

As mentioned in the chapter on Exploring the Prokaryotic World, many prokaryotic cells also have flagella for motility. However, bacterial flagella are quite different in structure from the eukaryotic flagella.

- **Cilia.** Some protists have similar membrane-enclosed protein tubules called **cilia** (sing. **cilium**) that are more numerous, but shorter in length than flagella. Extending from the cell surface, all the cilia on a cell generate a wave-like synchrony. Like oars on an ancient sailing ship, the cilia propel the cell forward. No prokaryotic cells have similar structures solely dedicated to motility.

Energy Organelles

Eukaryotic microbial cells have one or two types of organelles to carry out important energy transformations.

- **Mitochondria.** Most protists and fungi contain membrane-enclosed organelles called **mitochondria** (sing. mitochondrion) (see Figure 5.4). These organelles are responsible for the production of **adenosine triphosphate** (**ATP**), the universal energy currency of all cells. ATP is an organic molecule that provides the energy needed to drive many processes in living cells. This process is described in the chapter on Growth and Metabolism.
- **Chloroplasts.** Among the eukaryotic microbes, the algae are the only group that contain **chloroplasts** (see Figure 5.4). These membrane-enclosed structures carry out **photosynthesis**, the process that converts light energy (sunlight) into chemical energy in the form of sugar molecules. Some bacterial groups, such as the Cyanobacteria, also carry out an almost identical set of photosynthetic reactions. However, these prokaryotes do not require chloroplasts to complete the process.

Although, prokaryotes lack these energy organelles, they still carry out the same cellular processes as found in eukaryotes. The origin of these energy organelles is described below.

TABLE 5.1 Comparison of Eukaryotic and Prokaryotic Cell Structures/Processes

Cell Structure/Process	Function	Prokaryotes	Eukaryotes
Cell/plasma membrane	Semipermeable barrier separating environment from cell cytoplasm	Yes	Yes
Cell nucleus	Houses the genetic information needed for growth and metabolism	No structure	Yes
	■ Presence of DNA	Yes, as a single circular chromosome in cytoplasm	Yes, as multiple, linear chromosomes surrounded by a double membrane
Cytoplasm	Fluid and contents that fill the cell and in which most metabolism occurs	Yes	Yes
■ Endomembrane system	Membranes that divide the cell into functional and structural compartments and regulate protein traffic	No	Yes
■ Cytoskeleton	Cytoplasmic cellular scaffolding for transport and cell division	Yes (organization unique from eukaryotes)	Yes
■ Ribosomes	Site of protein manufacture	Yes	Yes
■ Mitochondria	Conversion of chemical energy to cellular energy (ATP)	No structure	Most
	■ Make ATP	Yes (on cell membrane)	Yes
■ Chloroplasts	Conversion of light energy to chemical energy (photosynthesis)	No structure	Algae and plants
	■ Carry out photosynthesis	Some (on cell membrane)	Algae and plants
Exterior structures			
■ Flagella	Cell movement (motility)	Some (structurally unique from eukaryotes)	Some
■ Cilia	Cell movement (motility)	No	Some protists and animals
■ Cell walls	Cell structure and water balance	Most	Algae, fungi, and plants

Cell Walls

Like most prokaryotic cells, fungi and algae contain a **cell wall** exterior to the cell (plasma) membrane (see Figure 5.4). Although the structure and organization of these cell walls are unique to each group, all cell walls provide support for the cells, give the cells their shape, and help the cells resist cell rupture.

A summary of the prokaryotic and eukaryotic processes and their associated structures is presented in **TABLE 5.1**.

5.2 Eukaryotic Cells: Their Origin

Eukaryotic cells appear to be much more complex in structure than prokaryotic cells. Scientists believe the first eukaryotes appeared about 2.7 billion years ago. Before that time, the world consisted only of prokaryotic organisms. How the first ancestral eukaryote came into being is a puzzle that has received much thought and study in contemporary biology.

FIGURE 5.5 Internal Membranes and the Endosymbiont Theory. Two major events have been proposed to explain the evolution of the first eukaryotic cell.

(**A**) Internal membranes might have evolved when an ancestral prokaryote developed the ability to fold its plasma membrane inward.

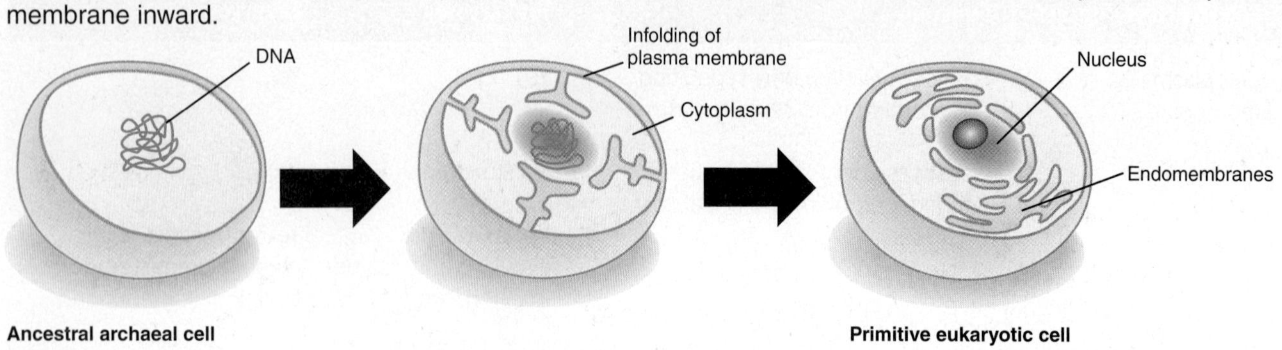

(**B**) The mitochondrion and choroplast might have originated from an independent ATP-manufacturing bacterial cell and a photosynthetic cyanobacterium that were taken into primitive eukaryotic cells through endosymbiosis that evolved into the organisms alive today.

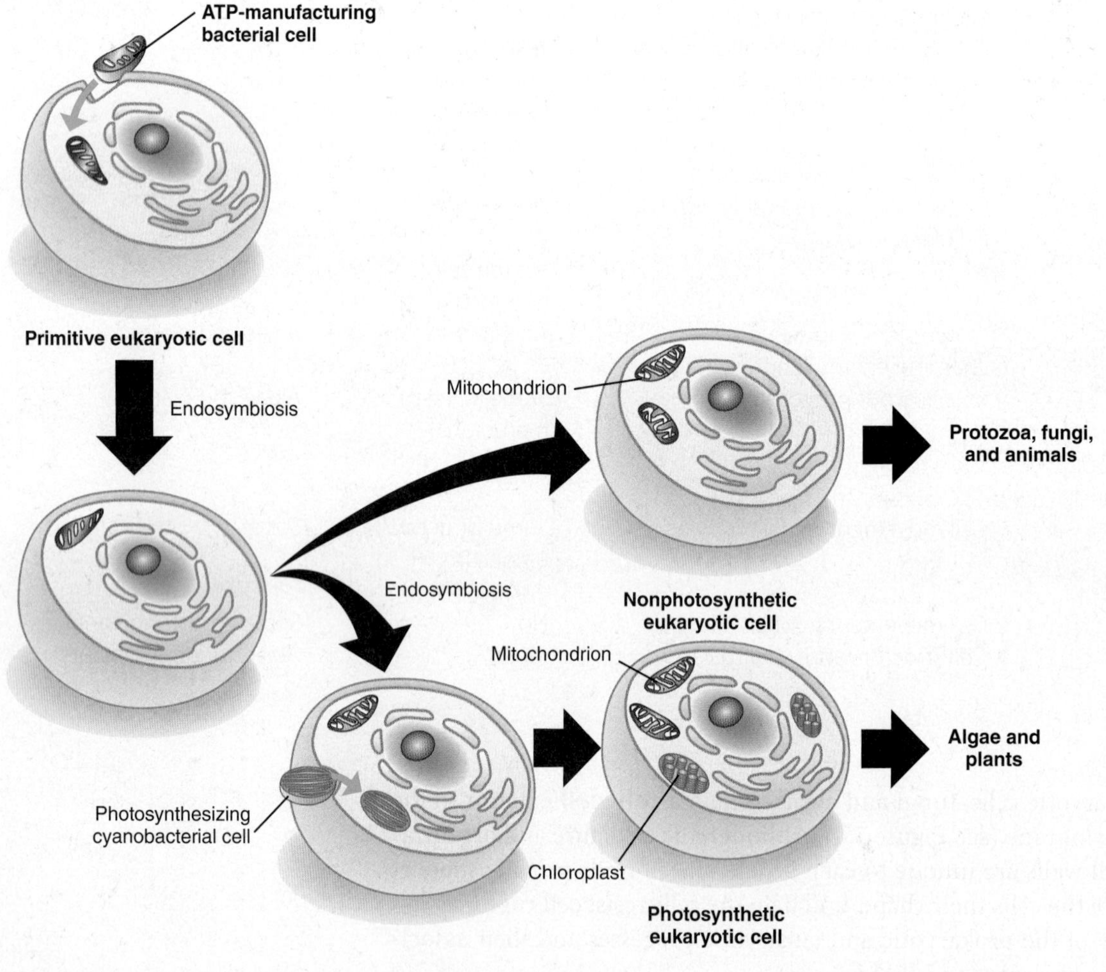

The Evolution of the Eukaryotic Cell

One current hypothesis suggests that eukaryotic cells evolved through a series of structural changes to an ancestral cell of the domain Archaea. In this hypothesis, two major events are proposed.

One event involved an ancestral archaeal cell developing the ability to fold parts of its plasma membrane inward (**FIGURE 5.5A**). By forming these folds into the cytoplasm, separate compartments could be constructed. Some folds might have surrounded the DNA and eventually evolved into the cell nucleus. Other enclosing membranes might have evolved into the endomembrane system present in all eukaryotic cells today. The advantage to the ancestral cell was to divide metabolic processes into separate compartments so their complex chemistries would not compete and interfere with one another. In addition, the additional membrane provided more surface or "workbench space" on which metabolic reactions could occur.

The Endosymbiont Theory

The other event in the evolution of an ancestral eukaryotic cell concerns the origin of the mitochondria and chloroplasts. The late Lynn Margulis and her colleagues supported the **endosymbiont theory**. A **symbiosis** is a living together of two or more organisms, one inside the other. The theory therefore suggests that one organism, a primitive eukaryotic cell, "swallowed up" an ATP-generating bacterial cell (the **endosymbiont**) (**FIGURE 5.5B**). Rather than destroy the symbiont, the host cell maintained it within (endo) the cytoplasm. Eventually, the host cell relinquished the responsibility for energy metabolism (ATP manufacture) to the symbiont. As the two partners became more dependent on one another for energy and survival, they developed as a single eukaryotic cell with the bacterial partner eventually evolving into the present-day mitochondrion.

Carrying the endosymbiont theory further, a similar scenario can be drawn for the evolution of the chloroplast in algae and plants (see Figure 5.5B). In this case, uptake of a photosynthetic bacterium, perhaps a member of the Cyanobacteria, occurred. Again, tolerance of the photosynthetic symbiont gave the host cell a way to manufacture its own food (organic molecules). The result would be the evolution of the endosymbiont into the present-day chloroplasts.

Of course, this is all speculation. Unless we had a time machine to go back some 3 billion years, we will never know for sure how the eukaryotic cell evolved. However, the endosymbiont model is considered a theory because there is substantial evidence supporting the idea. Listed in **TABLE 5.2** are several biochemical and physiological similarities between mitochondria, chloroplasts, and bacterial cells. All contain DNA and RNA. In addition, ribosomes of mitochondria and chloroplasts are very similar to those in bacterial species. Also, the energy organelles and bacterial cells reproduce through binary fission. The similarities are too striking to ignore.

With the evolution of an endomembrane system and energy organelles, eukaryotic cells could become more complex structurally and evolve into a diverse domain of organisms—the domain Eukarya.

TABLE 5.2 Similarities Between Mitochondria, Chloroplasts, and Bacteria

Characteristic	Mitochondria	Chloroplasts	Bacteria
Average size	1–5 µm	1–5 µm	1–5 µm
Nuclear envelope present	No	No	No
DNA molecule shape	Circular	Circular	Circular
Ribosomes	Yes; bacterial-like	Yes; bacterial-like	Yes
Protein synthesis	Make some of their proteins	Make some of their proteins	Make all of their proteins
Reproduction	Binary fission	Binary fission	Binary fission

5.3 The Protists: A Microbial Grab Bag

The protists are a mixed group of microbes. Although they are quite distinct from prokaryotes, many protists more closely resemble various members of the animals, plants, or fungi than they do to other members of the protists. Here, we will simply examine the characteristics of the protists and then meet several representative species.

Protist Characteristics

Protists compose much of the diversity within the domain Eukarya (**FIGURE 5.6**). Over 200,000 species of protists are currently known, and they are exceeded only by the prokaryotes in the number of environments to which they have adapted. Although all protists are unicellular organisms, many exist in multicellular arrangements that form colonies or filaments of cells. Some are animal-like protists and fungal-like protists, capturing their food by engulfing other organisms, or by absorbing simple organic nutrients across their plasma membranes.

Parasite: An organism that lives on or in a host and gets its nutrients from or at the expense of the host.

Some protists are **parasites** that cause human and animal diseases. Other protists, such as the algae, have chloroplasts to trap the Sun's energy and, through photosynthesis, form carbohydrates such as glucose. There are even some protists that can do both; that is, they carry out photosynthesis at one time and then change their nutrition and capture food at another time. All in all, the protists are a real "mixed bag" of organisms.

Sexual reproduction: The process of producing new living organisms by combining genetic information from two individuals of different types (sexes).

Asexual reproduction is the most common means by which protist cells increase in number. **Sexual reproduction** also occurs in most eukaryotic microbes—it is among the evolutionary traits first appearing in the protists.

Finally, it is even hard to characterize protist habitats because they too are very diverse. Most of these environments are moist and many species reside in ponds, lakes, and oceans. Others do fine in moist soils and leaf litter. And as mentioned earlier, there are pathogens and parasites, such as the malarial parasite, that take up temporary or permanent residence in a host organism.

To illustrate their diversity, let's look at a few examples of protists representing different protist groups.

FIGURE 5.6 The Tree of Life. The prokaryotic domains Bacteria and Archaea share the tree of life with the eukaryotes. Note that most of the domain Eukarya consists of an enormous diversity of protists. Endosymbiosis gave rise to present-day mitochondria and chloroplasts.

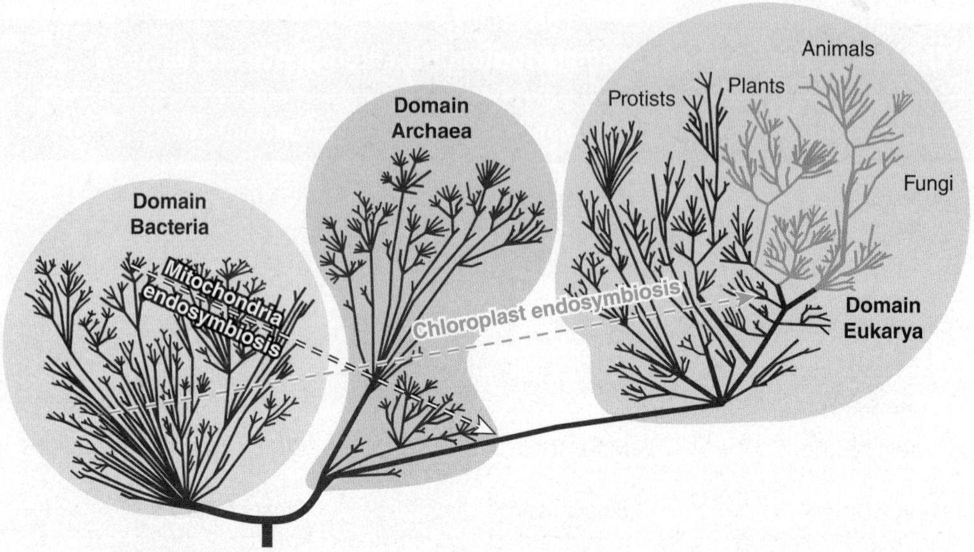

Meeting the Protists

The **protozoa** are an old lineage of microbes and are so named because they were once believed to be the first animals (*proto* = "very first"; *zoa* = "animal"). Approximately 8,000 species of protozoa have been named and another 35,000 are predicted to exist; most are found in aquatic environments, in moist soil, or as parasites. Many are essential participants to nutrient recycling in the soil, where they decompose the remains of dead animals and plants and recycle the nutrients.

Many protozoa are part of the **zooplankton**, the animal-like component of aquatic food webs. By feeding on microscopic cyanobacteria and unicellular algae that compose the phytoplankton (see below), the zooplankton become food for other "consumer" organisms, such as sponges, jellyfish, worms, and tiny marine invertebrates (animals without backbones). On land, protozoa perform the same nutrient-releasing functions in the digestive tracts of cattle, goats, and other so-called ruminant animals.

Because the diversity of protists has made their classification difficult and one needing more study, we will separate them informally as animal-like, plant-like, and fungus-like protists.

The Animal-Like Protists

Four groups of protists compose what has been referred to as the "protozoa" (**FIGURE 5.7**).

- **Pseudopod Protists.** Some of the nonphotosynthetic protists have no definite form or shape. Rather, protists like the **amoebae** (sing. **amoeba**) constantly change shape by sending out cellular extensions called **pseudopods** (see Figure 5.7A). The pseudopods allow the cells to creep over their environment such as the bottom of a pond. Pseudopods also are used to capture prey.

  A serious human pathogen is *Entamoeba histolytica*, which is the causative agent of amoebic **dysentery**. This form of **gastroenteritis** is more common in tropical areas where poor sanitary conditions exist. Consequently, transmission typically comes from consuming contaminated food and water. Symptoms of amoebic dysentery include intestinal ulcers and sharp appendicitis-like pain. According to the World Health Organization (WHO), amoebic dysentery is responsible for up to 100,000 deaths globally every year.

  Another health problem is caused by a species of *Acanthamoeba*. This protozoan can cause **keratitis** in people who wear contact lenses. This rare disease usually results from the contact lenses becoming contaminated after cleaning the contacts with solutions containing the protist. Other amoebas are found in some home humidifiers. When aerosolized, the cells might be inhaled into the respiratory tract and cause an allergic reaction called "humidifier fever."

  The **foraminiferans** (**forams**) are amoeboid-like in shape. However, they form hardened, shell-like casings (see Figure 5.7A). Over millions of years, the fossil remains of the forams built up as dense deposits on the ocean floor. Geologic upthrusting then raised these deposits to the surface. The White Cliffs of Dover in England is a well-known example (**FIGURE 5.8**). Foram remains also are associated with ancient oil deposits. Foram shells therefore serve as depth markers for geologists drilling for oil.
- **Flagellated Protists.** Some protozoa have one or more flagella. These **flagellates** form an impressive array of protists living within animals. In the gut of a wood-eating termite, for example, protistan flagellates, along with bacteria, break down the cellulose in wood and release the glucose for use by the termite. This symbiotic relationship benefits both the termite and

Entamoeba histolytica: en-tah-MEE-bah hiss-toe-LIH-tih-kah

Dysentery: A disease of the lower intestine (colon) that is characterized by severe diarrhea, inflammation, and the passage of blood and mucus.

Gastroenteritis: An inflammation of the stomach and the intestines, causing vomiting and diarrhea.

Acanthamoeba: a-kan-thah-ME-bah

Keratitis: An infection of the cornea of the eye.

FIGURE 5.7 The Four Major Groups of Animal-Like Protists. The organelles of motion (pseudopods, flagella, and cilia) are shown for members of three groups. The nonmotile protists lack such structures.

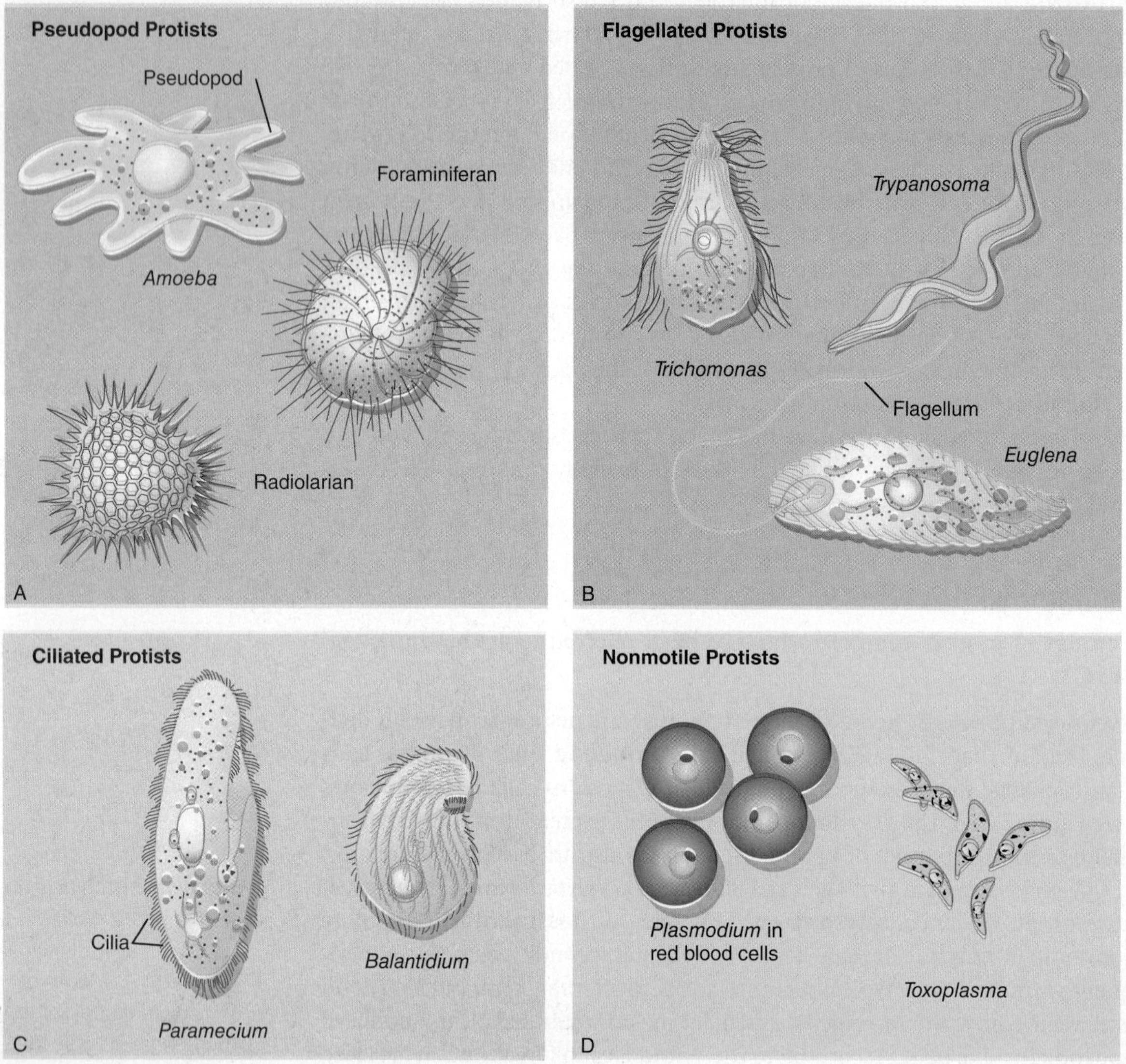

protozoan (but is of little interest to the homeowner who must repair the structural damage caused by termites).

Trypanosoma brucei: trih-PAH-no-soe-mah BREW-sea-ee

Trypanosoma cruzi: trih-PAH-no-soe-mah CREWZ-ee

Among the flagellates, a few species are human pathogens. Two species, *Trypanosoma brucei* and *T. cruzi* are the causes of African and American trypanosomiasis, respectively (see Figure 5.7B). *T. brucei* is transmitted by tsetse flies while *T. cruzi* is transmitted by cockroach-like triatomid bugs. When introduced into the bloodstream of a human, the flagellates move through the blood and invade the spinal cord and brain, where they cause a coma-like condition. American trypanosomiasis, also called Chagas disease, can cause cardiac and intestinal complications. Today, 8 to 11 million cases of Chagas disease are reported annually in Mexico, Central America, and South America. Each year, there are some 300,000 cases reported in the United States, primarily in people who have emigrated from Latin America.

Giardia intestinalis: gee-ARE-dee-ah in-tes-TIN-al-iss

Giardia intestinalis, the curse of campers, hikers, and backpackers, is contracted from contaminated water, especially in mountain streams and lakes. The disease, called giardiasis, causes nausea, gastric cramps, and a foul-smelling watery diarrhea. The protist infects wild animals, which can be the source of water contamination. Giardiasis is the most common protistan-caused

FIGURE 5.8 The White Cliffs of Dover. Situated on the English coastline facing the Strait of Dover and France are gigantic cliffs 350 meters (110 feet) high composed of the remains of foraminiferans that were thrust up from the ocean floor millions of years ago.

disease in the United States. More than 14,000 cases are reported annually to the United States Centers for Disease Control and Prevention (CDC).

Trichomonas vaginalis is the protozoan species that causes trichomoniasis ("trich") (see Figure 5.7B). It is a sexually transmitted infection that affects almost 4 million Americans annually. Patients suffer intense itching, burning pain, and a frothy discharge from the reproductive tract. The disease is more common in women than men, and it can be easily treated and cured with antiprotistan drugs.

Trichomonas vaginalis: trick-o-MOAN-as vah-gin-AL-iss

- **Ciliated Protists.** The **ciliates** are an extremely diverse group that typically is found in watery environments. One of the most studied is *Paramecium*. This slipper-shaped, rigid cell is covered with cilia that beat in synchronized waves to propel the cell through the water (see Figure 5.7C).

Paramecium: pair-ah-ME-sea-um

- **Nonmotile Protists.** The fourth and final group of protozoa lacks any form of motility. Nearly all are parasitic, and many of them cause serious diseases in humans and other animals.

The most life-threatening of the nonmotile protists (and probably among all microbes) is the genus *Plasmodium* that, as the chapter opener describes, is the infectious agent of malaria. Malaria has been infecting humans for more than 5,000 years. Today, between 300 and 500 million of the world's population suffer from malaria, which exacts its greatest toll in Africa. According to the 2018 World Malaria Report from the WHO, more than 435,000 people died from malaria. Children are the most susceptible because their immune system has not had time to develop partial immunity from mild malaria infections. In 2018, one child died every 60 seconds from malaria! No infectious disease of contemporary times can claim such a dubious distinction.

Several species of *Plasmodium* can cause malaria, and mosquitoes transmit the parasite (see chapter opener). The cycle involves two hosts, mosquitoes and humans (**FIGURE 5.9**). An infected mosquito bites (takes a blood meal from) an unsuspecting individual, transmitting the parasites to the human host. In that person, the parasites first attack the liver, where they reproduce asexually before moving on to infect and destroy red blood cells. The destruction of

FIGURE 5.9 The Life Cycle of *Plasmodium*. The malarial parasite *Plasmodium* requires two different hosts (mosquito and human) in which specific stages of the life cycle occur.

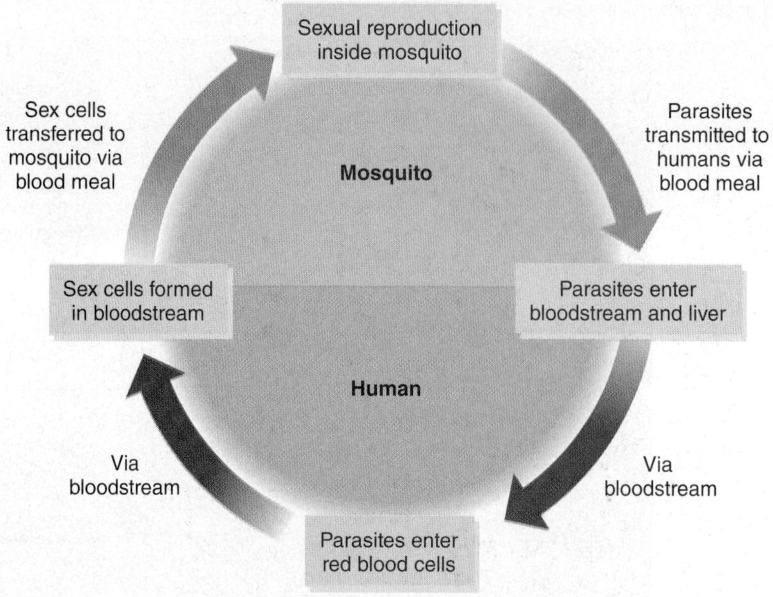

red blood cells causes a wave of intense chills followed by intense fever, the so-called "malaria attack." Extensive anemia develops, and the hemoglobin from ruptured red blood cells makes the urine dark (giving rise to the name "blackwater fever"). Red blood cell fragments and clusters accumulate in the small vessels of the brain, kidneys, heart, liver, and other vital organs and cause clots to form. Heart attacks, cerebral hemorrhages, and kidney failure follow.

If an uninfected mosquito now takes a blood meal from the infected human individual, some parasite cells are transferred to the mosquito host where sexual reproduction occurs. In the mosquito, new parasites mature into a form capable of being transmitted through the insect's bite to another unsuspecting person, starting another cycle of infection.

Toxoplasma gondii: toxs-oh-PLAZ-mah GONE-dee-ee

One other nonmotile protist of societal concern is *Toxoplasma gondii*, the agent of toxoplasmosis ("toxo"). Up to 50% of the world's population carries the parasite. This includes 50 million Americans, making toxo one of the most common human parasitic infections.

Toxo also is a blood disease often acquired through contact with parasite-containing cat feces, such as when changing the litter box or from contaminated soil while working in the garden. In addition, ingestion of the parasite can occur by consuming improperly cooked beef infected with the parasite. Healthy adults experience a mild flu-like illness, but a pregnant woman can transmit *T. gondii* cells across the placenta to the fetus. The result might be a child born with serious nerve damage. In persons with weakened immune systems, such as patients with acquired immunodeficiency syndrome (AIDS), a toxo infection can cause seizures and brain lesions.

The Plant-Like Protists

The **algae** (sing. **alga**) often are considered plant-like protists. The importance of these unicellular, photosynthetic organisms to society cannot be overstated. As a major type of life in the seas, the **phytoplankton** (*phyto* = "plant"; *planktos* = "wandering"), along with the cyanobacteria, comprise drifting communities of microbes that live near the ocean surface. They use the sun's energy during photosynthesis to generate much of the molecular oxygen available in the atmosphere. The phytoplankton is key to running the oceanic food webs because virtually every animal in the sea depends directly or indirectly on phytoplankton for food. In fact, about half the world's organic matter is produced by phytoplankton, as **A CLOSER LOOK 5.1** describes.

Let's examine a few of the important types of algae.

- **Dinoflagellates.** A key member of the phytoplankton is the **dinoflagellates**. These algae have cells encased in rigid walls composed of cellulose coated with silicon. They move by using their two flagella to generate a spinning form of motility (*dino* = "whirling").

  Some species are **bioluminescent**; that is, chemical reactions occurring in their cytoplasm yield energy in the form of light, allowing the organisms

A CLOSER LOOK 5.1

Tuna Sandwiches

Yum! A tuna fish sandwich for lunch. Take a can of tuna, add some mayonnaise and celery, and you have what has been called "the mainstay of almost everyone's American childhood." In fact, in the United States, 52% of canned tuna is used for sandwiches. But let's examine the microbial contribution to that tuna sandwich.

"What eats what" in the world often is shown as a trophic pyramid where organisms are grouped by the role they play in the food web (see **FIGURE A**). Keeping in mind that trophic levels are rarely this simple because organisms often feed at more than one trophic level), we can say:

- The first level forms the base of the pyramid and is made up of producers, in this case the phytoplankton, the most abundant and widespread producers in the marine environment.
- The second level is made up of consumers. In our figure, the zooplankton in the marine environment consumes phytoplankton.
- The higher trophic levels include carnivorous consumers such as small and midsize fish, along with other consumers like squid and octopus. In simple terms, the zooplankton is eaten by smaller fish, smaller fish are eaten by midsize fish, and midsize fish are eaten by tuna, sharks, and other consumers.
- In our food web diagram, the top-level consumers are humans.

On average, only 10% of the energy from a trophic level is transferred to its consumer (the rest is lost as waste and heat energy). Therefore, that tuna sandwich you ate came from: 1 kilogram (kg) (2.2 pounds) of tuna, which had to eat 10 kg (22 lbs.) of midsized fish; each midsized fish had to eat 100 kg (220 lbs.) of small fish; each small fish had to eat 1,000 kg (2,200 lbs.) of zooplankton, which in turn had to eat 10,000 kg (22,000 lbs.) of phytoplankton. In other words, the energy you obtained from your tuna fish sandwich can be traced back to 10,000 kg of phytoplankton.

Whew! Those calculations made me hungry. I better finish my tuna fish sandwich and get more energy—and yes, thank you phytoplankton!

FIGURE A **Tuna Sandwiches.** A simple trophic pyramid and energy.

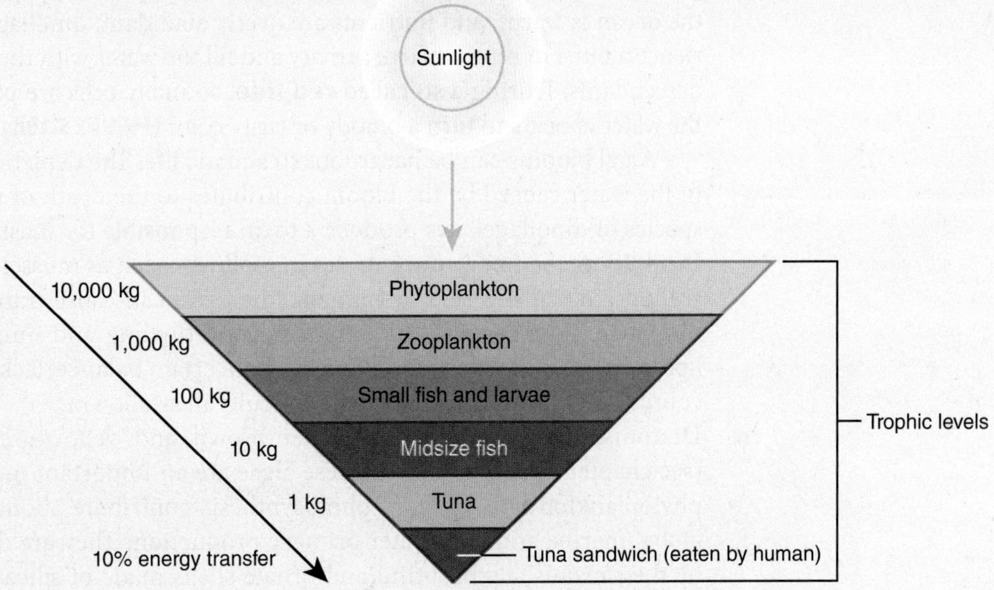

FIGURE 5.10 Phytoplankton.

(A) Some types of phytoplankton can bioluminesce, in this case producing an eerie blue glow visible at night.

(B) The sudden growth of some dinoflagellates is responsible for algal blooms, such as this red tide.

to give off a greenish or bluish glow (**FIGURE 5.10A**). Unfortunately, dinoflagellates sometimes undergo an "algal bloom," or population explosion. When the ocean is warm, and nutrients are overly abundant, dinoflagellates experience a burst of reproductive activity and fill the water with their trillions of descendants. During a so-called **red tide**, so many cells are produced that the water appears to turn a bloody or rusty color (**FIGURE 5.10B**).

Algal blooms can be hazardous to aquatic life. The depletion of oxygen in the water caused by the bloom contributes to the death of plants. Some species of dinoflagellates produce a toxin responsible for massive fish kills. In addition, the toxin concentrates in mollusks such as mussels, clams, and scallops. When ingested by humans, the toxin can cause transient neuromuscular disturbances. This might include tingling and numbing of the lips, tongue, and fingertips, followed by uncertain balance, lack of muscular coordination, slurred speech, and difficulty in swallowing.

- **Diatoms.** The **diatoms** are golden-brown and yellow-green in color (see chapter opening photo). These algae are an important member of the phytoplankton and through photosynthesis contribute about 20% of the global marine and freshwater primary production. They are distinguished by their exquisitely beautiful and ornate shells made of silica that overlap much like the two halves of a shoe box (**FIGURE 5.11A**).

Diatoms have economic importance because they are used as filtering, polishing, and insulating materials. For example, many people with swimming pools or aquaria use **diatomaceous earth** to filter out contaminants in the water. Diatoms also are used as a mild abrasive in toothpaste, a pesticide for bed bugs, a thermal insulator for fire-resistant safes, and an anti-caking agent for animal feed.

- **Green Algae.** Perhaps the most well-known group of unicellular photosynthetic protists is the **green algae**. A well-studied member of the green algae is the unicellular, flagellated genus *Chlamydomonas* (see Figure 5.4). Another member is the genus *Volvox* that is organized into colonies of cells (**FIGURE 5.11B**). Each cell forming the surface of the hollow ball resembles a *Chlamydomonas* cell. The green balls inside the sphere

Chlamydomonas: klam-ih-do-MO-nahs

Volvox: VOLE-vox

FIGURE 5.11 Single-Celled Algae.

(A) Diatoms, like the other phytoplankton, trap the Sun's energy through photosynthesis. Note the striking geometric shapes of the algal cells when viewed with special optics on a light microscope. (Bar = 100 μm.)

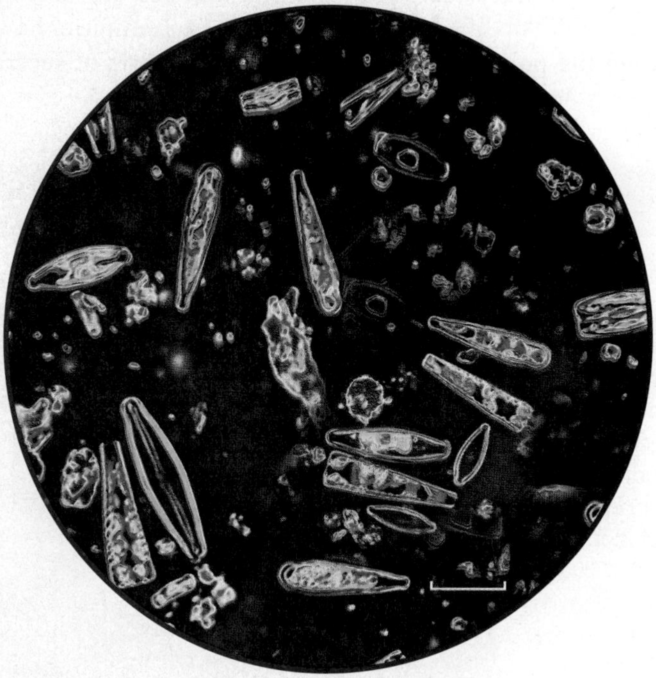

(B) With the light microscope, *Volvox* appears as colonial green algae consisting of *Chlamydomonas*-like somatic cells and internal daughter colonies. (Bar = 300 μm.)

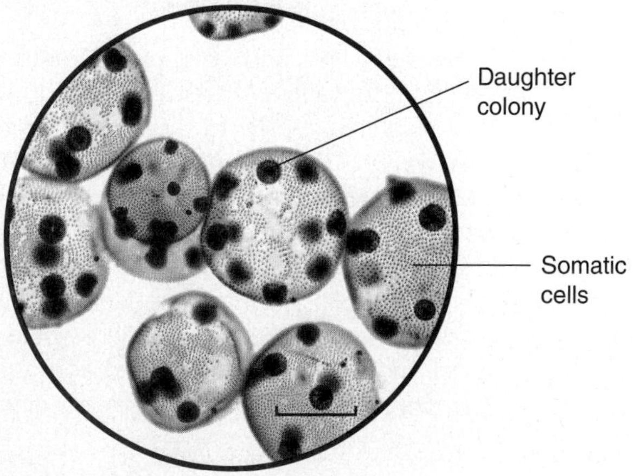

represent daughter colonies that are released when the parent colony ruptures. Although not directly related to any multicellular organism, *Volvox* suggests how multicellularity might have evolved. It is conceivable that some 400 million years ago, when multicellularity evolved, one or more kinds of green algae were among the first to try this experiment in community living.

The Fungus-Like Protists

Phytophthora infestans: fi-TOF-tho-rah in-FES-tanz

Some protists have a growth form that resembles that of fungal molds. These so-called **water molds** live in water and damp soil, and they cause the unsightly and furry parasitic growths that plague fish in home and commercial aquaria. Several species of water molds are of significant economic importance as plant pathogens. One species infects grapes, producing yellow circular spots on the leaves and another has destroyed hundreds of thousands of oak trees in California. Still another species, *Phytophthora infestans*, is responsible for late blight of potatoes, a disease that destroyed the Irish potato crop during the extremely damp years of the mid-1800s. This infection, which is described in **A CLOSER LOOK 5.2**, is perhaps the best example of how microbes have impacted the political, economic, and social fabric of several nations, especially North America.

A CLOSER LOOK 5.2

"Black '47"

Early Spanish explorers to the New World discovered the potato in Central and South America and brought it back to Europe. Because most of the plant is poisonous, except the tuber, it wasn't until about 1800 that Europeans grew the potato and its tuber as a food crop. Especially in Ireland, the potato grew well in the moist, cool climate. Although Ireland of the 1840s was an impoverished country of tenant farmers, the farming of potatoes brought a population explosion and the country went from about 4.5 million people in 1800 to more than 8 million by 1845, the population depending in great part on the potato season after season.

Early in the 1840s, heavy rains and dampness portended calamity. Then, on August 23, 1845, *The Gardener's Chronicle and Agricultural Gazette* reported: "A fatal malady has broken out amongst the potato crop. On all sides we hear of the destruction. In Belgium, the fields are said to have been completely desolated." Beginning as black spots, the potato disease, called late blight, decayed the leaves and stems, and left the potatoes a rotten, mushy mass with a peculiar and offensive odor (see **FIGURE A**). Even the harvested potatoes rotted. In the damp climate, a fungus, the water mold *Phytophthora infestans*, was running rampant through the potato fields.

The winter of 1845 to 1846 was a disaster for Ireland. Farmers left the rotten potatoes in the fields, allowing the disease to spread. For food, farmers first ate the animal feed and then the farm animals. They also devoured the seed potatoes, leaving nothing for spring planting. After 2 years, the late blight seemed to slacken, but in 1847 ("Black '47"), an unusually cool and wet year, it returned with a vengeance and once again destroyed the Irish potato crop within days.

FIGURE A Late Blight. Potatoes destroyed by *Phytophthora*.

Between 1845 and 1860, over 1 million Irish people died from starvation. At least 1.5 million Irish left the land and emigrated, mostly to the eastern United States. As great waves of Irish immigrants came to the United States, their Irish American descendants in the next century would influence American culture and politics. From Boston policemen, to workers on the transcontinental railroad, to political leaders (President John Kennedy was an Irish American), American society would never be the same. And to think—it all resulted from the water mold *Phytophthora infestans,* which remains a difficult organism to control even today.

5.4 Fungi: Yeasts and Molds

The **fungi** (sing. **fungus**) usually are an unheralded and often overlooked group of microbes. In actual fact, they are a critical link in the web of life on Earth. They serve as important decomposers of organic matter. They make incalculable contributions to ecosystems by liberating nutrients from organic materials and making those nutrients available to insects, worms, and countless other organisms in the environment. Without fungi (and bacteria), the nutrients in organic materials would be locked up, the cycles of elements would grind to a halt, the fertility of soil would decline precipitously, and ecosystems would collapse.

So, exactly what are fungi?

Characteristics of the Fungi

The fungi are a diverse group of eukaryotic organisms consisting of some 75,000 identified and named species. Scientists believe there might be another 1.5 million species still waiting to be discovered. Until the mid-1900s, those wishing to study the fungi (called mycologists) typically would enroll in a botany course because fungi were thought to be simple plants. Since then, fungi have been found to be more closely related to animals, and they have been placed in their own kingdom Fungi, within the domain Eukarya in the "tree of life" (see Figure 5.6). The study of fungi is called **mycology** (*myco* = "fungus"; *ology* = "the science of").

Fungal Structure

The fungi have the characteristic set of eukaryotic organelles, including a cell nucleus, mitochondria, and ribosomes. The fungi consist of the microscopic molds and yeasts and the clearly visible mushrooms. The **yeasts** usually grow as single cells and are several times larger than most bacterial cells (**FIGURE 5.12A**). The **molds**, on the other hand, usually are composed of threadlike filaments called **hyphae** (sing. **hypha**). The walls of the hyphae are composed of **chitin**, a polysaccharide not found in plants or prokaryotes. As a hypha lengthens, it branches and forms an interwoven mass of hyphae called a **mycelium** (pl. **mycelia**) (**FIGURE 5.12B**). Often this mycelium escapes our attention because it usually is underground where it can maximize its contact with moisture and organic matter.

Growth and Reproduction

Being single cells, yeasts grow in cell number through an asexual reproduction process called **budding**. For the filamentous molds in the soil, hyphal growth extends from the tip where continued mycelial expansion brings the organism in contact with new food sources.

Yeasts and molds digest organic matter by secreting enzymes into the environment. They then absorb the small organic products of digestion (e.g., sugars, amino acids) across the cell wall and plasma membrane. For example, some fungal species produce the enzyme **cellulase** and use it to decompose cellulose (the principal polysaccharide in wood). When cellulose is digested, it yields glucose molecules, which are extremely useful nutrients and sources of energy. Some species of fungi also produce other enzymes to break down additional plant wall fibers. That is why in a forest one typically finds molds growing on leaf litter or a rotting log (**FIGURE 5.13A**). Many of the materials broken down by these **decomposers** generate vast quantities of organic matter that can be recycled and used by other organisms.

Decomposer: An organism that breaks down dead or decaying matter.

FIGURE 5.12 The Fungal Form.
(A) A light microscope image of oval yeast cells. (Bar = 2 μm.)

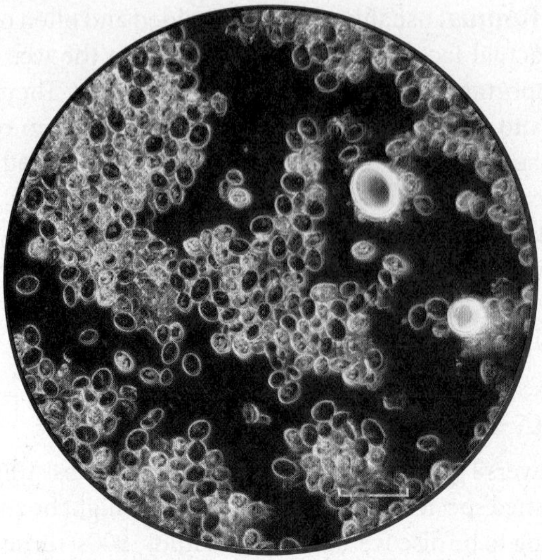

© Bill McVety Photography/Moment Open/Getty Images.

(B) A filamentous mold (white filaments) growing on a damaged tomato.

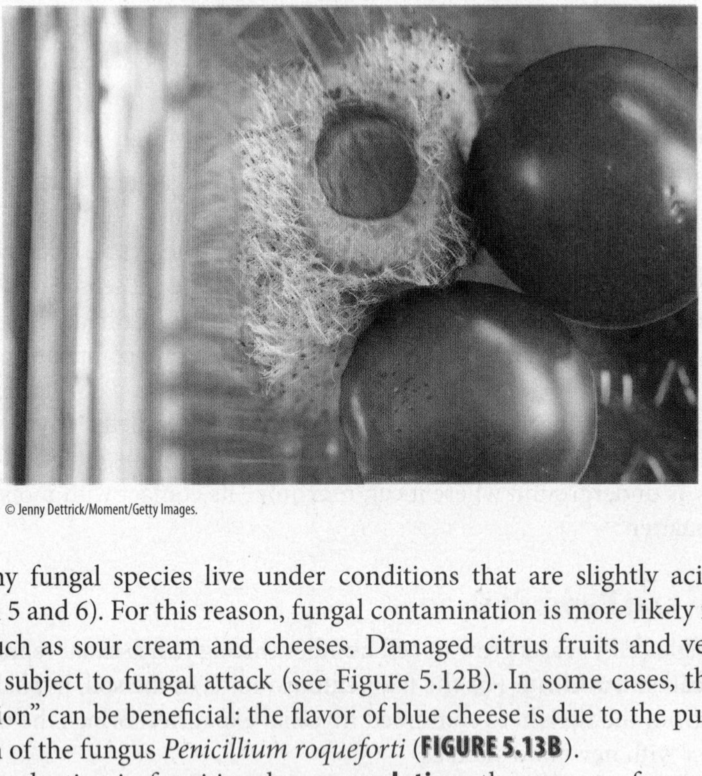

© Jenny Dettrick/Moment/Getty Images.

Many fungal species live under conditions that are slightly acidic (pH between 5 and 6). For this reason, fungal contamination is more likely in acidic foods such as sour cream and cheeses. Damaged citrus fruits and vegetables also are subject to fungal attack (see Figure 5.12B). In some cases, this "contamination" can be beneficial: the flavor of blue cheese is due to the purposeful addition of the fungus *Penicillium roqueforti* (**FIGURE 5.13B**).

Penicillium roqueforti: pen-ih-SIL-lee-um row-ko-FOR-tee

Reproduction in fungi involves **sporulation**, the process of spore formation. In asexual reproduction, thousands of spores are produced, usually on the ends of specialized hyphae (**FIGURE 5.14**). Such asexual spores are extremely light and are disbursed in huge numbers by wind currents. If fungal spores land in an appropriate environment having moisture and nutrients, they will germinate into a new hypha that will elongate and form a new mycelium.

Asexual reproduction is advantageous to the spread of a fungus because it provides enormous numbers of spores, each of which can become a new mycelium. However, the spores are genetically identical. Therefore, sexual reproduction provides the means to increase genetic diversity by bringing genetically different individuals together.

FIGURE 5.13 The Fungal Mycelium.
(A) The fungal mycelium on this rotting log is an example of a decomposer. The fungus produces enzymes to digest the cellulose and other wall polysaccharides.

(B) Fungi are purposely introduced into some cheeses, such as blue cheese. The mycelium grows through the cheese, often imparting flavor and color.

FIGURE 5.14 Asexual Reproduction. False-color scanning electron microscope image of a mold producing thousands of spores at the tips of the hyphae. (Bar = 20 μm.)

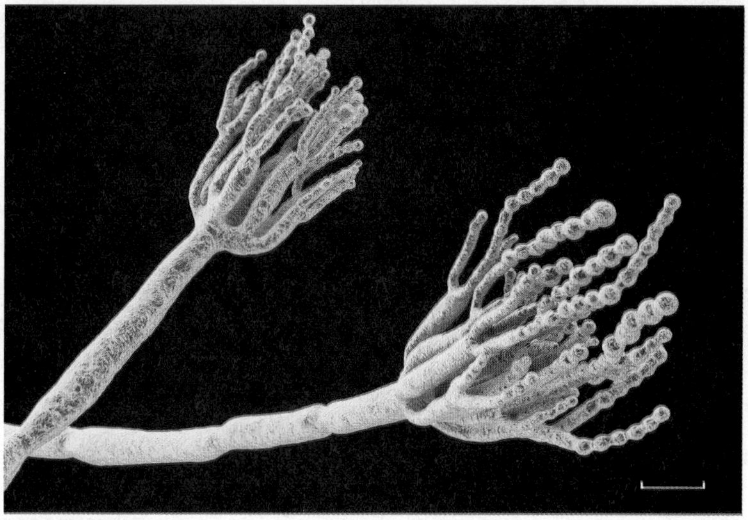

Through sexual reproduction, many fungi also produce spores, which are often contained within a visible reproductive structure that facilitates spore dispersal. Perhaps the most recognized is the mushroom, such as the store-bought white mushroom (**FIGURE 5.15**). Here, hyphae that represent opposite (+ and –) "mating types" (like opposite sexes in animals) come together and fuse. From the "mating," a mushroom will eventually arise, and spores will be produced on the gills forming the underside of the mushroom cap. These spores are released and dispersed by water or wind. When the spores land in a nutritious environment, they germinate and a mycelium will eventually form.

Because the cell nuclei were genetically different in each mating type, the spores produced will have a new genetic composition. Unlike asexual spores,

FIGURE 5.15 A Life Cycle of a Typical Mushroom. Spores are produced from the underside of the mushroom cap. The spores germinate and produce new mycelia of different mating types.

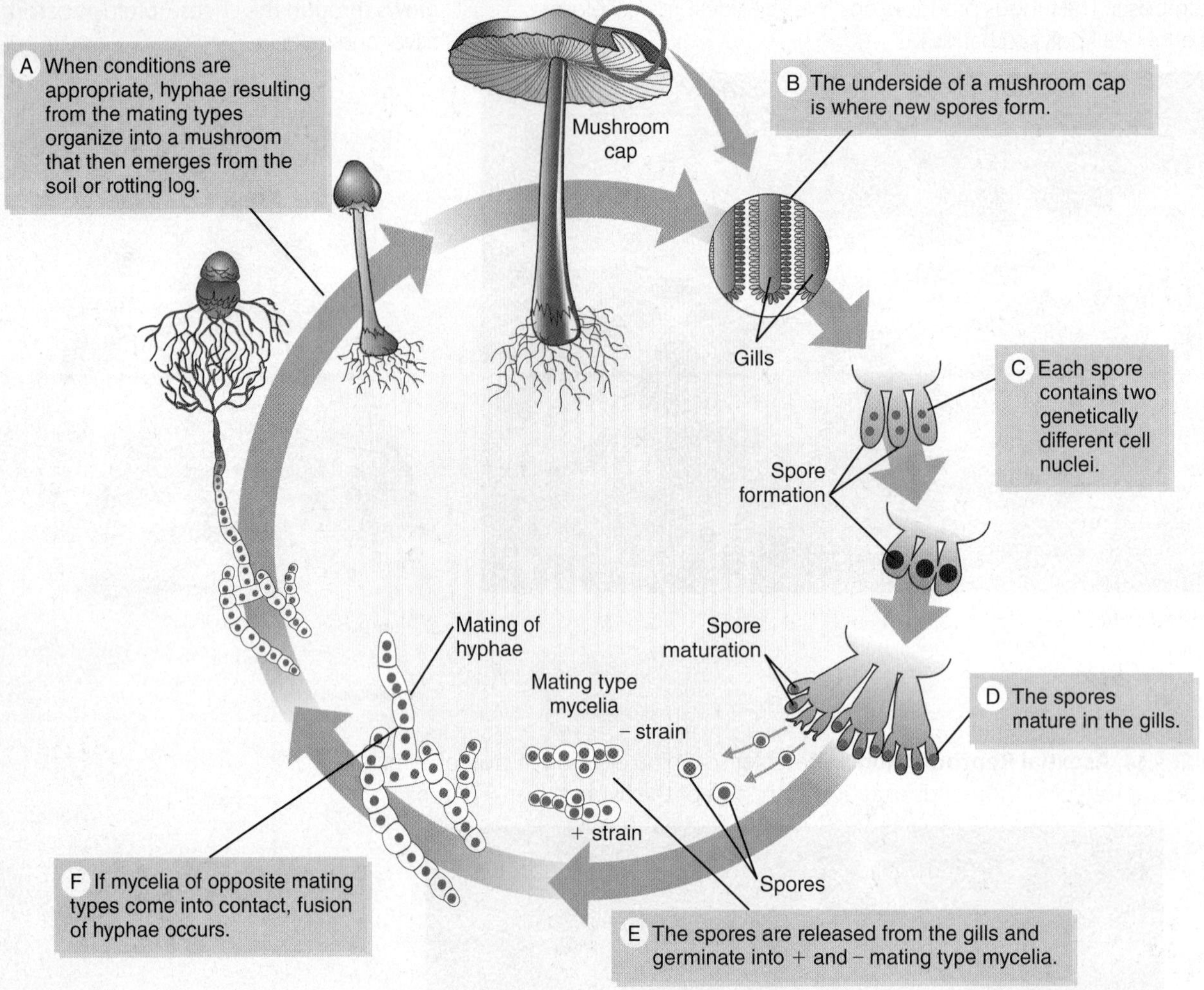

this genetic variability might permit a spore to germinate and survive in a formerly unfavorable environment.

Meeting the Fungi

The fungi are cataloged into several different groups (phyla) based on genetic relatedness and lifestyles. Most are nonpathogens to humans and animals, although a few can cause human diseases referred to as **mycoses** (sing. **mycosis**). In this section, we will highlight a few examples from these groups.

The Chytrids

The oldest known fungi are members of the so-called **chytrids.** These fungi are predominantly aquatic and produce flagellated reproductive cells. Until recently, their impact on the environment appeared to be of little significance. **A CLOSER LOOK 5.3** provides a different assessment.

The Ascomycetes

The so-called **ascomycetes** represent the largest phylum with 65,000 described species. Most are decomposers and are composed of hyphae that

A CLOSER LOOK 5.3

The Day the Frogs Died

Amphibians (frogs, toads, salamanders) are the oldest class of land-dwelling vertebrates, having survived the effects that brought the extinction of the dinosaurs. Now amphibians are facing their own potential extinction from a different, infectious effect.

In the early 1990s, researchers in Australia and Panama reported massive declines in the number of amphibians in ecologically pristine areas. As the decade progressed, massive die-offs occurred in dozens of frog species, and a few species even became extinct. Once filled with frog song, the forests were quiet. "They're just gone," said one researcher. By 2012, scientists estimated that more than 300 frog species had gone nearly extinct worldwide, and many others were facing the same outcome. Why were the frogs dying?

One of the reasons for the decline was due to an infectious disease. Many amphibian species on four continents were being infected with a chytrid called *Batrachochytrium dendrobatidis*. This fungus accumulates in the frog's skin and secretes a chemical that cripples the frog's immune system. In addition, the fungus infection causes the frog to overstimulate production of keratin. This overproduction alters the frog's skin permeability, which leads to a fatal water imbalance and heart failure. Roughly one-third of the world's 6,000 amphibian species are now considered under threat of extinction due to the disease chytridiomycosis (see **FIGURE A**).

It now appears that crayfish can harbor the fungus and are spreading the disease on to the amphibians. But to make matters worse, in 2014 researchers identified two viruses infecting toads and newts and another species of *Batrachochytrium* affecting European newts and salamanders. Conservation biologists now refer to this decline in amphibians as the Global Amphibian Crisis.

Where the chytrid came from remains a mystery. The global amphibian trade might be one origin. Are these amphibian deaths a sign of a yet unseen shift in the ecosystem, much like a canary in a coal mine? Some believe this type of "pathogen pollution" might be as serious as chemical pollution. It represents an alarming example of how emerging pathogens (fungal and viral) could potentially wipe out an entire species and, eventually, an entire ecosystem.

FIGURE A A species of harlequin frog, one of many species being wiped out by a chytrid infection.

form a mycelium. Several species of ascomycetes are of great economic value. Among "foodies," the morels and truffles, are prized edible fungi. Among the ascomycetes are several species of *Penicillium*, some of which produce the antibiotic penicillin. Other species of *Penicillium* are used to ripen and flavor blue cheese (see Figure 5.13B), Roquefort cheese, and Camembert cheese.

Batrachochytrium dendrobatidis: bah-trah-koh-KIH-tree-um den-dro-bah-TIE-diss

Included in the phylum are several plant pathogens that have had devastating economic and social impact in the eastern United States.

- **Chestnut Blight.** The ascomycete that causes chestnut blight has wiped out more than 4 billion chestnut trees in the eastern United States since the early 1900s. Chestnuts are a major food source for many animals, such that tree loss resulted in a drastic decrease of squirrel, deer, and hawk populations and to the extinction of seven native moth species. The blight also resulted in a $2 billion dollar loss to the timber industry.
- **Dutch Elm Disease.** Dutch elm disease has devastated native populations of elms. The fungal pathogen is spread by elm bark beetles (**FIGURE 5.16A**). The disease has had serious implications for forest ecosystems, and efforts to eradicate the disease in the United States have had little success.

A few ascomycetes also are associated with animal and human illnesses.

- **White Nose Syndrome.** An ascomycete pathogen is associated with an infectious disease in bats called white nose syndrome. Since emerging in 2006, the disease has killed more than 7 million hibernating bats (affecting 15 species) in the United States and Canada (**FIGURE 5.16B**).

Claviceps purpurea: KLA-vi-seps pur-POO-ree-ah

- **Ergot.** One particularly interesting ascomycete, *Claviceps purpurea*, has been associated with human behavioral disorders. The fungus can grow on the ears of grains, such as rye. When humans consume the contaminated grain or bread baked from the grain, the fungal chemical deposited in the rye causes a nervous system disorder called ergot disease. The affected individuals initially develop nausea, vomiting, and muscle pain and weakness. There is an intriguing body of evidence, related in **A CLOSER LOOK 5.4**, indicating ergot disease might have been of historical significance in early America. Ironically, *C. purpurea* is now used for medical purposes where, in small quantities, its chemical products can be used to treat migraine headaches or to induce childbirth.

To finish off the ascomycetes, we need to consider the most well-known ascomycetes, the yeasts. The most familiar of the yeasts is the genus

FIGURE 5.16 Ascomycete-Produced Diseases.

(A) This Dutch elm tree has been killed by an ascomycete pathogen.

Winnipeg Free Press "City Approves Additional Funds to Battle Dutch Elm Disease".

(B) A scientist is checking this bat for white-nose syndrome.

© Jay Ondreicka/Shutterstock.

A CLOSER LOOK 5.4

"The Work of the Devil"

As an undergraduate, Linda Caporael was missing a critical history course for graduation. Little did she know that through this class she was about to provide a possible answer for one of the biggest mysteries of early American history: the cause of the Salem Witch Trials. These trials in 1692 led to the execution of 20 people who had been accused of being witches in Salem, Massachusetts (see **FIGURE A**).

Caporael, now a behavioral psychologist at New York's Rensselaer Polytechnic Institute, in preparation of a paper for her history course had read a book where the author could not explain the hallucinations among the people in Salem during the witchcraft trials. Caporael made a connection between the "Salem witches" and a French story of ergot poisoning in 1951. In Pont-Saint-Esprit, a small village in the south of France, more than 50 villagers went insane for a month after eating ergotized rye flour. Some people had fits of insomnia, others had hallucinogenic visions, and still others were reported to have fits of hysterical laughing or crying. In the end, three people died.

Caporael noticed a link between these bizarre symptoms, those of Salem witches, and the hallucinogenic effects of drugs like LSD, which is a derivative of ergot. Could ergot possibly have been the perpetrator in Salem too?

During the Dark Ages, Europe's poor lived almost entirely on rye bread. Between 1250 and 1750, ergotism, then called "St. Anthony's fire," led to miscarriages, chronic illnesses in people who survived, and mental illness. Importantly, hallucinations were considered "the work of the devil."

Toxicologists know eating ergotized food can cause violent muscle spasms, delusions, hallucinations, crawling sensations on the skin, and a host of other symptoms—all of which, Linda Caporael found in the records of the Salem witchcraft trials. Ergot thrives in warm, damp, rainy springs and summers, which were the exact conditions Caporael says existed in Salem in 1691. In addition, parts of Salem village consisted of swampy meadows that would be the perfect environment for fungal growth. And because rye was the staple grain of Salem, it is not a stretch to suggest that the rye crop consumed in the winter of 1691–1692 could have been contaminated by large quantities of ergot.

Caporael concedes that ergot poisoning can't explain all the events at Salem. Some of the behaviors exhibited by the villagers probably represent instances of mass hysteria. Still, as people reexamine events of history, it seems just maybe ergot poisoning did play some role—and, hey, not bad for an undergraduate history paper!

FIGURE A A witch trial in Massachusetts, 1692.

Saccharomyces (*saccharo* = "sugar"). *S. ellipsoideus* is used for alcohol production, and *S. cerevisiae* (brewer's yeast) is used for bread baking.

The *Saccharomyces* yeasts are single-celled, oval microbes about 8 μm in length and 5 μm in diameter (see Figure 5.12A). They usually reproduce by budding. The cytoplasm of *Saccharomyces* is rich in B vitamins, a factor making yeast tablets valuable nutritional supplements (ironized yeast) for people with iron-poor blood. The baking industry relies heavily upon *S. cerevisiae* to supply the texture in breads. Yeast fermentation breaks down glucose and other carbohydrates, producing carbon dioxide gas. The carbon dioxide expands the dough, causing it to rise (**FIGURE 5.17A**).

Wild yeasts are plentiful on the skins of many orchard fruits and grapes (**FIGURE 5.17B**). In natural alcohol fermentations, wild yeasts of various *Saccharomyces* species are crushed with the fruit; in controlled fermentations, *S. ellipsoideus* is added to the prepared fruit juice. The fruit juice bubbles profusely as carbon dioxide is produced. As the oxygen is depleted, the yeast metabolism shifts to fermentation and the production of ethyl alcohol begins. The large share of the economy in many nations taken up by the wine and spirits industries is testament to the significance of the fermentation yeasts.

Saccharomyces ellipsoideus:
sack-ah-roe-MY-seas
ee-lip-SOY-dee-us

Saccharomyces cerevisiae:
sack-ah-roe-MY-seas
seh-rih-VIS-ee-eye

FIGURE 5.17 Yeasts.
(A) A baker is testing rising dough, which is caused by carbon dioxide gas released through yeast fermentation in the dough.

(B) In nature, wild yeasts are commonly found on fruit skins, such as these wine grapes.

S. cerevisiae probably is the most understood eukaryotic organism at the molecular and cellular levels. Its complete genome was sequenced in 1997, the first eukaryote to be completely sequenced. *S. cerevisiae* contains about 6,000 genes. It might appear initially that *S. cerevisiae* would have little in common with human beings. However, both are eukaryotic organisms with a cell nucleus, chromosomes, and a similar mechanism for cell division. *S. cerevisiae,* therefore, has been used to better understand cell function in animals. Yeast cells also have been key to the development of some human vaccines, such as the hepatitis B vaccine.

The Basidiomycetes

The so-called **basidiomycetes** consist of some 32,000 described species. This includes puffballs, shelf fungi, and mushrooms. **Mushroom** is the common name for the spore-producing body, which is composed of densely packed hyphae. As detailed earlier in Figure 5.7, the mycelium forms underground, and the mushrooms arise from the mycelium.

The basidiomycetes also include a few plant parasites that cause so-called rust and smut diseases. Rust diseases are so named because of the distinct orange-red color of the fungus (**FIGURE 5.18A**). Wheat, oat, and rye plants, as well as lumber trees, such as white pines, are susceptible. Smut diseases get their name from the fungus's black sooty appearance on infected plants such as blackberry, corn, and various other grains. These diseases bring about untold millions of dollars of crop damage and loss each year. On the other hand, in parts of Mexico and the American Southwest, when the smut fungus develops on ears of corn, it can be a delicacy in cooking. Called **huitlacoche** (corn smut or Mexican truffle), the fungus is eaten, usually as a filling in tortilla-based foods (**FIGURE 5.18B**).

Huitlacoche:
wheat-la-KOH-cheh

FIGURE 5.18 Rusts and Smuts.
(A) Close-up of bright orange patches of pear rust growing on pear leaves. The orange area represents the fungal mycelium.

(B) Corn smut (gray structures) protruding from an ear of corn.

In ancient Rome, mushrooms were the food of the gods, and only the emperors were permitted to partake of their pleasures. Today, exotic mushrooms enjoy an equally high reputation among the world's gourmets. Some experts know how to spot them in the wild, but for amateurs, the key word is "caution"—in mushroom hunting, ignorance is disaster. In fact, every year in the United States such mushroom hunting results in some 9,000 cases of mushroom poisoning being reported; children under 10 years of age account for most cases. For those who insist on eating wild mushrooms, mycologists recommend joining a local mycological society, reading extensively, and treading lightly into this hobby. According to the Minnesota Mycological Society:

> *"There are old mushroom hunters, and there are bold mushroom hunters, but there are no old, bold mushroom hunters."*

Symbiotic Relationships

As described earlier in this chapter, a symbiosis is a living together or close association between two unrelated organisms. Often, such close associations are beneficial for both partners. There are two interesting and extremely important symbiotic roles that fungi play with algae and plants.

- Lichens. **Lichens** are organisms that represent an association between a fungus and a photosynthetic partner called the **photobiont**. The photobiont usually is a cyanobacterium or a unicellular green alga. Of the approximately 15,000 species of lichens, most of the fungal partners belong to the ascomycetes.

  About 90% of a lichen's mass is fungal, and the specialized fungal hyphae either penetrate or encase the cells of the photobiont (**FIGURE 5.19A**). Lichens can survive with only 2% water by weight (compared to about 90% by weight for other organisms), which means they are extremely resistant to harsh environments. They grow in such diverse settings as arid desert regions and Arctic zones, as well as on bare soil and tree trunks. Lichens often are the first organisms to occur in rocky areas (**FIGURE 5.19B**), and their biochemical activities begin the process of rock breakdown and soil formation. This produces an environment in which mosses and other plants can gain a foothold. To establish a new population, a lichen breaks into fragments of fungus and photobiont, which are blown or carried away. Landing in a new area, they give rise to new populations.

Mycorrhiza: my-core-RYE-zah

- Mycorrhizae. **Mycorrhizae** (sing. **mycorrhiza**) are symbiotic associations between some species of soil fungi and vascular plants (e.g., trees, flowers, vegetables). In fact, scientists believe that over 90% of all green plants have a fungal symbiont. In this relationship, the fungus attaches to the surface or penetrates the roots of the plant (*mycorrhizae* = "fungus roots"), supplying the plant with more nutrients (especially phosphorus and water) than it could absorb through its roots alone. Meanwhile, the plant supplies the fungus with products of photosynthesis that provide some of the raw materials for its metabolism (**FIGURE 5.20A**). As a result of the symbiosis, plants tend to grow larger and more vigorously than plants lacking a fungus, especially in soils of poor nutrient quality (**FIGURE 5.20B**). Populations and societies depend directly or indirectly on crop plants (e.g., wheat, oat, rice) for most everything we eat. A scary fact is that without mycorrhizal fungi, today's important crop plants might not exist.

FIGURE 5.19 Lichen.
(A) A cross section of a lichen, showing the upper and lower surfaces where tightly coiled fungal hyphae enclose photosynthetic algal cells. On the upper surface, airborne clumps (fragments) of algae (photobiont) and fungus are dispersed to propagate the lichen.

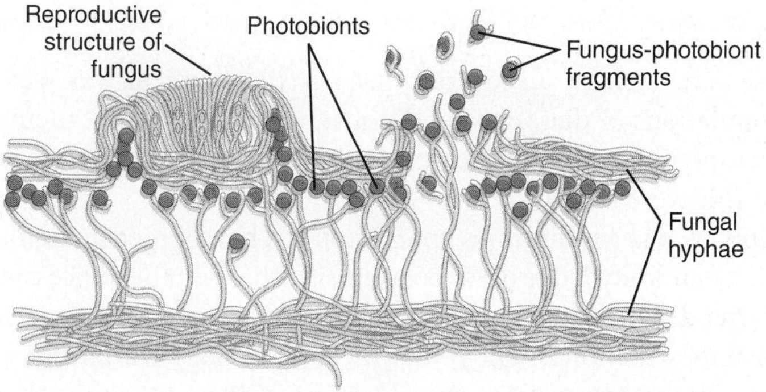

(B) A typical lichen growing on the surface of a rock.

Courtesy of Dr. Jeffrey Pommerville.

FIGURE 5.20 Mycorrhizae and Their Effect on Plant Growth.
(A) The hyphae of the mycorrhiza surround the plant root in a mesh-like network.

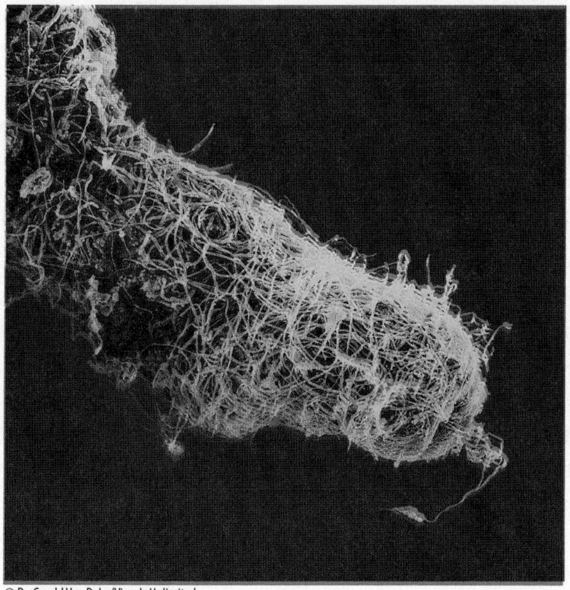

© Dr. Gerald Van Dyke/Visuals Unlimited.

(B) A demonstration showing the mycorrhiza effects on plant growth. CK = Control (without mycorrhizae); GM (with mycorrhizae).

© Science VU/R. Roncadori/Visuals Unlimited.

A Final Thought

This chapter on the eukaryotic microbes illustrates their diversity and roles in the environment, and to health and society. Although their diversity does not reach the level to that in the prokaryotes, it is important to remember that microbiologists and biologists still know little about the diversity in the microbial world. As a chemist once remarked: "If you dip a tennis ball in the ocean, the water dripping from the tennis ball represents all that is known; the ocean represents all that is waiting to be learned."

Chapter Discussion Questions

What Was He Thinking?

From your reading of this chapter, identify and discuss five major points about the protists and fungi that the author was trying to get across to you.

Questions to Consider

1. Suppose a research scientist discovered that a toxic chemical was wiping out the populations of diatoms and dinoflagellates (significant members of the phytoplankton) in the oceans of the world. What horror story might occur if this were true?
2. You and a friend, who is 3 months pregnant, stop at a hamburger stand for lunch. Based on your knowledge of toxoplasmosis, what helpful advice can you give your friend? On returning home, you notice that she has two cats. What additional information might you be inclined to share with her?
3. In an article in *Discover* magazine several years ago, the editor gave this most interesting description for a protist. See if you can identify the organism and disease described.

   "It races through the bloodstream, hunkers down in the liver, then rampages through red blood cells before being sucked up by its flying, buzzing host to mate, mature, and ready itself for another wild ride through a two-legged motel."
4. Suppose you have a protist that is a protozoan. What one characteristic could you use to identify to which one of the four groups it belongs? How did that characteristic permit the identification?
5. Provide a definition for a "eukaryotic microbe."
6. You decide to make bread. You let the dough rise overnight in a warm corner of the room. The next morning you notice a distinct beer-like aroma in the air. What are you smelling, and where did the aroma come from?
7. Fungi are extremely prevalent in the soil, yet we rarely contract fungal disease by consuming fruits and vegetables. Why do you think this is so?
8. A student of microbiology proposes a scheme to develop a strain of bacteria that could be used as a fungicide. Her idea is to collect the chitin-containing shells of lobsters and shrimp, grind them up, and add them to the soil. This, she suggests, will build up the level of chitin-digesting bacteria. The bacteria would then be isolated and used to kill fungi by digesting the chitin in fungal cell walls. Do you think her scheme will work? Why?
9. Why are the symbiotic associations seen with the lichens and with mycorrhizae considered to be "win-win" relationships?

CHAPTER 6

Viruses: At the Threshold of Life

The Tulip Bubble Bursts

Ah. Everybody loves tulips. Their beautiful colors delight the masses and those tulips with variegated (striped) flower petals are especially spectacular (see chapter opening image).

Tulips were first brought to Europe in the late 1500s by traders from Asia Minor (present-day Turkey). Some varieties produced incredibly exotic and beautiful variegated flowers. In Holland, tulips became so popular (a fad, if you will) that people would trade almost anything for just a few tulip bulbs, especially the variegated "Semper Augustus" tulip (**FIGURE 6.1**). "Tulipomania" swept the nation, and tulip speculators traded the flower's bulbs for extraordinary sums of money or possessions. Some people would trade their entire home for 10 tulip bulbs. Others spent a year's salary for a few bulbs. Tulips even began to be used as a form of money. The "Semper Augustus" tulip was, to some people, worth just about anything in trade. Tulipomania represents perhaps the first economic price bubble.

Then the tulip bubble burst. Prices plummeted on February 6 and 7, 1637. Many people were lucky because they had bought tulip bulbs but had not yet paid for them. So, most people did not go bankrupt, but lots of speculators were left holding the "tulip" bag.

So, what does this have to do with microbiology and viruses? During the 19th century, it was discovered that the appearance of variegated tulip flowers was the result of a virus infection, which was transmitted by aphids as they fed on the plants. This virus, called the tulip breaking virus, causes a change (break) in pigmentation that then resulted in the variegated colors of the petals.

Today, botanists have selectively bred variegated tulips to mimic the swirled colors of breaking—all without the virus. Importantly, in this story, here again is an example where a microbe, a virus in this case, changed people's (a whole population's) behavior and affected the economies of nations.

Every science has its borderland where the known and visible merge with the unknown and invisible. Startling discoveries often emerge from this

CHAPTER 6 OPENER A beautiful photograph of tulips, showing the broken petal colors that impart a variegated appearance to the flowers. Could the variegation be caused by a virus?

FIGURE 6.1 "Still Life" by Hans Bollongier. In 1639, Bollongier, a student of Rembrandt, painted this elegant white flamed Semper Augustus tulip, the bulbs of which often brought outrageous prices during tulipomania.

uncharted realm, and certain objects manage to loom large. In the borderland of microbiology, viruses exist at the threshold of life. Yet, they outnumber bacterial cells 10 to 1. In fact, probably every cell (organism), including tulips, has one or more viruses that can infect the organism. In the microbial world, viruses stand out for their simplicity, extraordinarily small size, distinctive method of replication, and disease potential. We highlight these characteristics in this chapter.

LOOKING AHEAD

On completing this chapter, you will be able to:

6.1 Discuss the early discovery of viruses.
6.2 Draw the structure and label the components of enveloped and nonenveloped viruses.
6.3 Describe the five stages of virus replication.
6.4 Explain the two ways by which an emerging virus might arise.
6.5 Describe how viruses can cause tumors.
6.6 Contrast viroids and prions.

6.1 Viruses: Their First Sightings

During the Golden Age of microbiology, Pasteur, Koch, and others pinpointed the bacterial organisms causing tuberculosis, typhoid fever, syphilis, and many other infectious diseases. People began to limit disease spread by improving sanitation methods and food preservation practices. Microbial ecologists, like Winogradsky and Beijerinck, pointed out the importance of microbes in the environment. The viruses also would be recognized as potential disease-causing agents.

A Contagious Living Fluid

In the 1890s, the Russian botanist Dmitri **Ivanowsky** and the Dutch investigator Martinus **Beijerinck** independently were studying tobacco mosaic disease. This disease causes tobacco leaves to shrivel and die (**FIGURE 6.2A**). Believing it must be due to a bacterial infection, they crushed diseased tobacco leaves and filtered the solution. They expected the bacterial cells would be trapped on the filter and not pass through with the liquid. To their surprise, the clear liquid coming through the filter contained the infectious agent. Ivanowsky suggested a "filterable virus" caused tobacco mosaic disease and later Beijerinck often referred to this contagious living fluid as a "virus" (*virus* = "poison").

Ivanowsky: eye-van-OFF-ski

Beijerinck: BY-yer-ink

The ability to see viruses came in the 1930s when scientists discovered they could form crystals of tobacco mosaic virus. Because living organisms cannot be crystallized, the opinion was that viruses were some sort of chemical molecule. Then, in the 1950s, scientists used the electron microscope to see individual tubes that were the tobacco mosaic viruses (**FIGURE 6.2B**).

Today, microbiologists consider viruses to be Earth's most abundant biological units. They make up the so-called **virosphere** that exceeds 10^{31} (that's 1 followed by 31 zeros) viruses. Furthermore, scientists and microbiologists are finding them faster than they can make sense of them, as **A CLOSER LOOK 6.1** illustrates.

Virosphere: The locations where viruses are found and interact with their hosts.

FIGURE 6.2 Tobacco Mosaic Disease and the Virus.
(A) An infected tobacco leaf exhibiting the mottled or mosaic appearance caused by the disease.

(B) This false-color transmission electron microscope image of the tobacco mosaic virus shows the cylindrical (helical) rod-shaped structure of the virus particles. (Bar = 80 nm.)

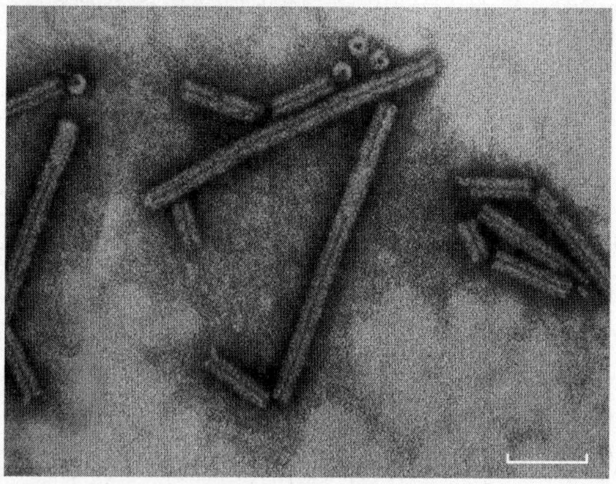

A CLOSER LOOK 6.1

Into the Virosphere

Under the Naica hills in northern Mexico are some amazing caves. These gigantic caverns, some the length of a football field and two stories high, contain 50-ton crystals of gypsum that were formed 500,000 years ago (see **FIGURE A**). Called the Cave of Crystals, the caverns are buried 300 meters (984 feet) below the desert surface, in sweltering temperatures of 45°C to 50°C (113°F to 122°F) with saturating humidity.

When first explored, these caves appeared devoid of life, especially microbial life. However, in 2009, Dr. Curtis Suttle and his group from the University of British Columbia set out to discover if any microorganisms could survive in the hellish conditions of the caves. Water samples were collected from the cave pools, taken back to the lab in British Columbia, and prepared for microscopy. What the samples contained stunned the researchers.

In the water samples that had been cut off from the outside world for perhaps millions of years, bacterial cells were identified. More startlingly, in those samples were tiny, geometric forms—viruses! And there were lots of them, perhaps up to 200 million viruses per drop of water.

Almost halfway around the world, Monika Häring and other microbiologists from several European universities were studying acidic hot springs (85°C to 95°C/185°F to 203°F) in Pozzuoli, Italy. When cultures were prepared from hot spring samples, several archaeal organisms were identified along with virus particles. Meanwhile, much further south in the Antarctic's Lake Limnopolar, a freshwater lake that is frozen 9 months of the year, Spanish scientists found bacterial cells, protists, and—yes—viruses. Again, huge numbers of viruses were detected, representing perhaps 10,000 different types, making the lake one of the most diverse virus communities in the world. Even the world's oceans are filled with viruses. Rachel Parsons from the Bermuda Institute of Ocean Sciences and her collaborators have detected enormous virus populations in the western Atlantic (see **FIGURE B**).

Around the world, the number of viruses exceeds 10^{31}. Most of these viruses infect microbes but there are plenty that infect plants and animals as well. This world of viruses, the so-called virosphere, exists, as Suttle's lab explains, "wherever life is found; [they are] a major cause of [illness and] mortality, a driver of global geochemical cycles, and a reservoir of the greatest unexplored genetic diversity on Earth." And viruses are a major driving force for evolution because each virus infection has the potential to introduce new genetic information into just about any organism in the domains Bacteria, Archaea, and Eukarya. The virosphere is huge, incredibly diverse, and has a tremendous impact beyond infection and disease.

FIGURE A Viruses can be found wherever life exists. This includes isolated environments deep in the earth such as the Cave of Crystals.

FIGURE B Viruses populate the world's oceans. The larger, fluorescent dots are bacterial cells, and the smaller spots are viruses. (Bar = 10 μm.)

6.2 Virus Structure: Geometric Perfection

If you were to stop 100 people on the street and ask if they recognize the word "virus," all 100 would probably nod their heads knowingly and say yes. They might tell you about the flu they recently had or mention the "computer virus" that crashed their system! Were you then to ask the same individuals to describe a flu virus, they might answer it's "a tiny thing," or "something you need a microscope to see," or "a germ." After a couple of moments, they would probably scratch their heads and admit they are familiar with the word virus, but they don't really know what a virus is (unless, of course, they had recently read this chapter). So, what is a virus?

A **virus** is a small, obligate intracellular parasite; that is, a tiny agent that must infect a **host** to produce more viruses. Viruses must act in this way because they lack the cellular machinery for generating energy and building the component parts of a virus.

Host: A cell or organism in which a microbe or virus can live, feed, and reproduce (replicate).

Viruses can vary in size **FIGURE 6.3**. Among the smallest viruses are the polio viruses that are 28 nanometers (nm) in diameter. The larger ones, such as smallpox virus, measure more than 200 nm in diameter. Some more recently discovered viruses, such as the Pandoraviruses, are giants in the virus world, as they can be more complex in structure than other viruses and they match the size of many small bacterial cells.

The Component Parts of Viruses

Viruses, unlike cells, are without much internal structure. Rather, they consist simply of a core of nucleic acid (**genome**) enclosed by some type of covering layer or layers (**FIGURE 6.4**).

Genome: The complete set of genetic information in an organism or virus.

The Viral Genome

Like all cellular organisms, viruses need to carry the information to make more of themselves. Some viruses are classified as **DNA viruses** because their genetic information is in the form of deoxyribonucleic acid (DNA). Many other viruses are unique **RNA viruses**. They have a genome made of ribonucleic acid (RNA).

FIGURE 6.3 Size Relationships Among Cells and Viruses. The sizes (not drawn to scale) of various viruses relative to a eukaryotic cell, a cell nucleus, and the bacterium *Escherichia coli*.

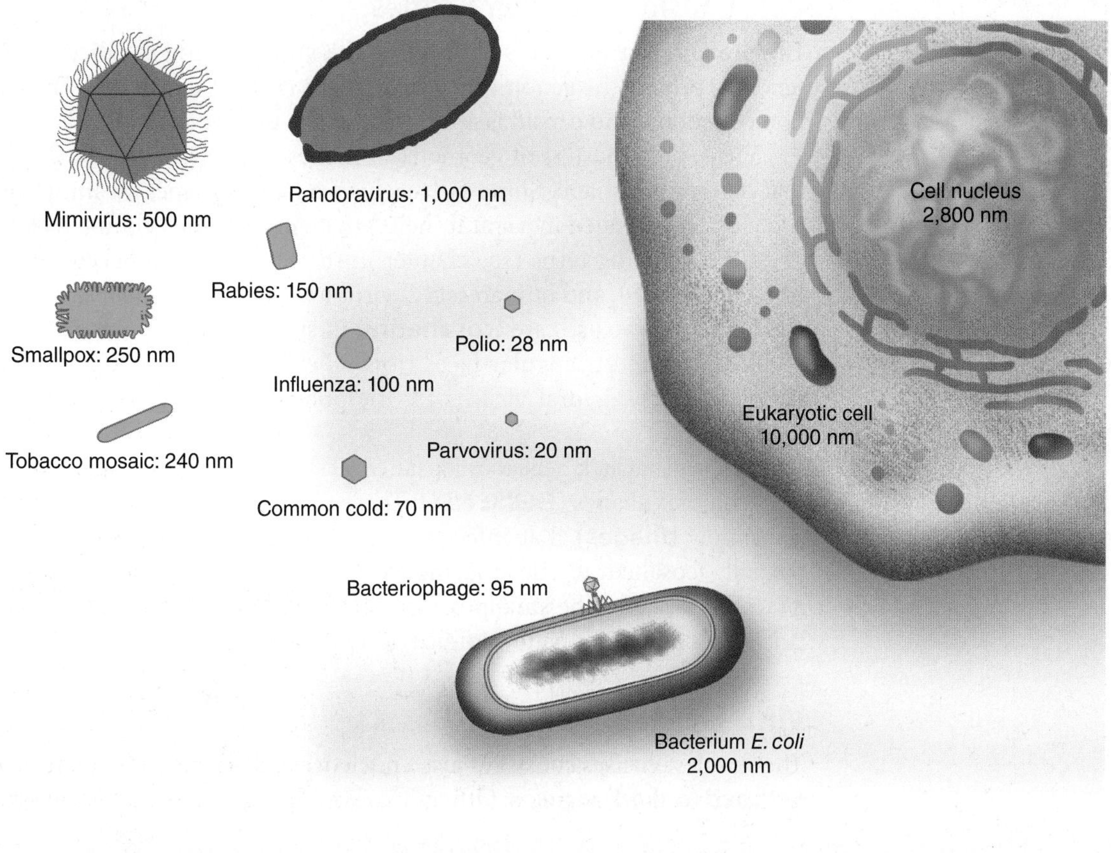

FIGURE 6.4 Virus Anatomy.
(A) Nonenveloped viruses consist of a nucleic acid genome (either DNA or RNA) and a protein capsid. Spikes extend from the capsid. **(B)** Enveloped viruses have an envelope that surrounds the nucleocapsid. Again, spikes usually are present on the envelope.

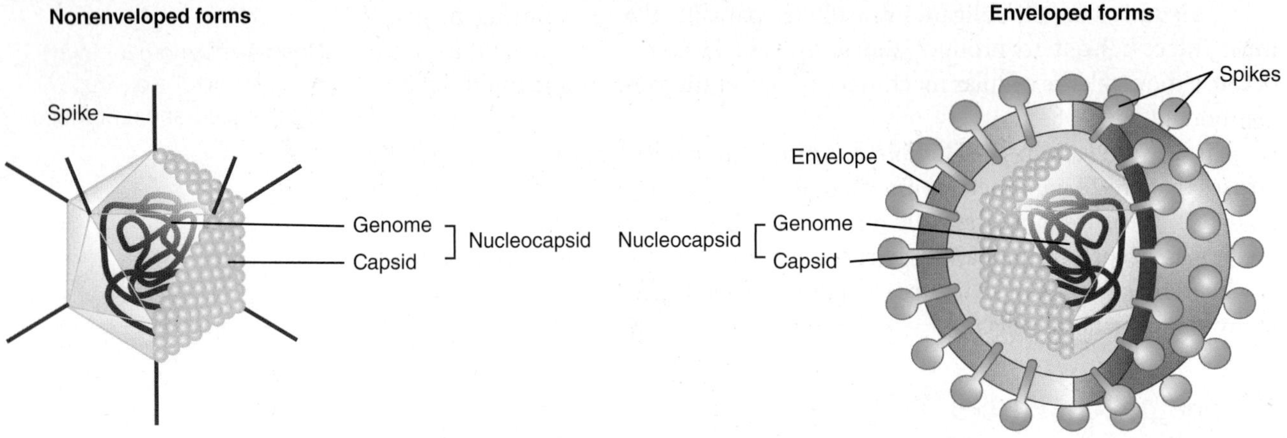

However, in either case, the viral genome contains all the information to build new viruses.

The size of the viral genome also varies from one virus to another. Some viruses only have a genome large enough to produce 2–3 proteins. Other, larger viruses might have a genome capable of producing 100 proteins.

For classification, DNA viruses and RNA viruses are split into separate families. This cataloging often is based on whether the genome is a double-stranded or single-stranded nucleic acid and on the presence or absence of an envelope (see below). Usually the nucleic acid is a single molecule, but occasionally it exists in several segments. The genome of the influenza viruses, for example, consists of eight segments of RNA.

Viral Capsids and Their Shapes

The protein coat or shell surrounding the viral genome is called a **capsid**. It serves to protect the genome and facilitate its entry to a host cell. The combination of genome and capsid is referred to as the **nucleocapsid**.

Viruses are masters of geometrical shapes (symmetry), and it is the capsid that determines shape. Some viruses have a helical shape, meaning the capsid proteins are arranged in a spiral (helix) that encloses the genome (**FIGURE 6.5A**). The tulip breaking virus (see chapter introduction), the tobacco mosaic virus (see Figure 6.2B), and human rabies virus are examples of helical viruses.

Numerous viruses are **icosahedral** in shape, meaning the capsid consists of 20 equal-sized triangular faces (*icos* = "20"; *edros* = "sided") (**FIGURE 6.5B**). Among the icosahedral viruses are the polio viruses and the herpes simplex viruses (HSV).

Viruses that lack helical or icosahedral shape are lumped together as viruses with complex shapes (**FIGURE 6.5C**). These viruses include the **bacteriophages** (or simply **phages**) that infect bacterial cells. They are partially helical and partially icosahedral. The amazing world of the bacteriophages is discussed in **A CLOSER LOOK 6.2**. The smallpox virus also has a complex shape, looking like a brick with a swirling pattern of fibers on its surface.

Virus Envelopes

The simplest viruses consist of just a nucleocapsid and are referred to as **nonenveloped (naked) viruses**. Other viruses have a lipid layer, called an **envelope**,

FIGURE 6.5 Virus Shapes and Symmetry.

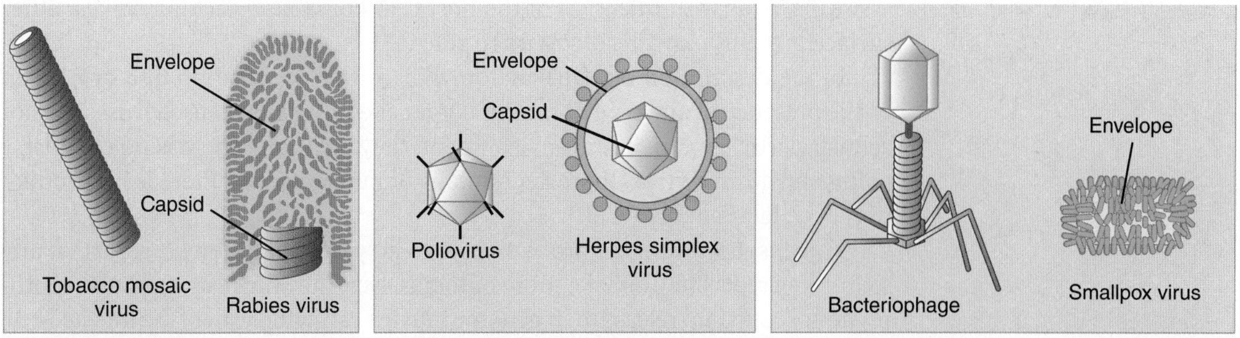

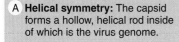

A **Helical symmetry:** The capsid forms a hollow, helical rod inside of which is the virus genome.

B **Icosahedral symmetry:** The capsid forms a 20-sided structure inside of which is the virus genome.

C **Complex symmetry:** The capsid forms a complex structure inside of which is the virus genome.

A CLOSER LOOK 6.2

The Amazing Bacteriophages

Bacteriophages are viruses that target bacteria. These so-called phages that look like lunar lander spacecraft (see **FIGURE A**) are everywhere—in soil, oceans, hot springs, Arctic ice, and human intestines. They are the most diverse and numerous biological agents on the planet, outnumbering the bacteria 10:1. That said, scientists don't have a clue what they're doing. But they have some ideas of their impact—and potential uses. Here are just two examples.

- **In the Environment.** Phages cause a trillion trillion successful bacterial infections per second in the world's oceans. That means phages destroy up to 40% of all oceanic bacterial cells every day. Following their deaths at the hands of phages, those carbon-containing bacteria sink down into the marine sediment, effectively providing nutrients to organisms in the marine sediment.

  Anything that bacteria do as decomposers, from breaking down the carcasses of dead animals to converting atmospheric nitrogen into plant food, is at the mercy of the phages that potentially can infect and kill them. In fact, phages can influence the ocean's food supply by limiting phytoplankton populations that are essential in the food chain.
- **In the Human Body.** There are enormous numbers of phages present in the human digestive tract. Although phages do not infect human cells, they still can have an enormous effect on the microbiome in our gut. Scientists are wondering whether this sea of phages might influence our physiology, perhaps even helping to regulate our immune systems. In humans, phages are more than 4 times as abundant in the mucus layers of the gut as they are in the adjacent environment. By sticking to mucus, phages encounter more of their bacterial prey, suggesting that phages might protect the underlying cells from potential bacterial pathogens.

  In addition, reports have been published that the cells lining the gut can transport phages (30 billion per day!) into the blood and organs of the body. What they are doing there is completely unknown.

As you can see, there is much to be studied and discovered. When scientists better understand the role of phages in the human body, the researchers could start looking at phages to manipulate the bacterial microbiome and even control our own cells. See A Closer Look 6.3 for even more uses for phages in disease therapy.

FIGURE A An illustration depicting several bacteriophages infecting a bacterial cell. (Bar = 100 nm.)

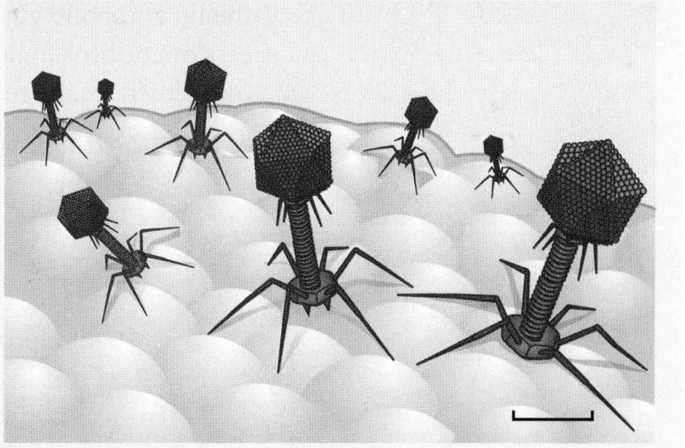

external to the nucleocapsid (see Figure 6.4B). These are referred to as **enveloped viruses**.

In many viruses, the naked nucleocapsid or envelope has protein projections called **spikes** (see Figure 6.4). Sticking out like spines on a cactus, the spikes on the capsid or envelope help the virus contact its host cell, and then assist the virus with infection of the host cell. For viruses lacking spikes, other surface proteins play this recognition role.

In the description of virus anatomy, there was no mention of a cytoplasm or the presence of any type of internal organelles. This is because viruses are not cellular particles, and, in the environment, they are inert. There is no chemistry going on within a virus, there is no intake of nutrients, and there is no production of cellular energy (ATP).

There is, however, one process viruses carry out, and they do it particularly well—they replicate. To make more of themselves, they must attach to the appropriate host cells and release their genome into the cell cytoplasm of the infected cell. Once in the cell, the infecting viruses use proteins encoded by the viral genome to biochemically reprogram the host cell to simultaneously produce hundreds, sometimes thousands of new viruses. So, unlike organism reproduction where one cell gives rise to only two cells, virus replication assembles hundreds of new viruses simultaneously. In so doing, viruses often destroy the cell that was infected.

Let's now look at how a virus can "hack" a host cell and replicate.

6.3 Virus Replication: A Massive Production Factory

Outside a host cell, a virus is an inert particle; that is, it cannot replicate. However, when it encounters and infects the appropriate host cell, the viral genome transforms the host cell into a highly efficient factory to produce and assemble simultaneously hundreds of viruses.

Virus replication covers the entire process by which the viral genome enters a host cell and controls the production of more viruses. Inside the host cell, the proteins encoded by the viral genome take over the cell's metabolic machinery for the sole purpose of producing more viruses. This process can be immediate or delayed, as we will witness.

When Viruses Replicate Without Delay

Many viruses have but one purpose for infection: Get in the host cell, produce more viruses, and get the new viruses out of the cell. Such viruses, which includes the flu and polio viruses, are said to undergo a **productive infection**. This process can be broken down into five steps (**FIGURE 6.6**). Use this figure and **FIGURE 6.7** that illustrates the replication process for HSV, as we go through the five steps.

1. **Attachment**. The first step of virus replication involves attachment of the virus particle to an appropriate host cell. For example, hepatitis viruses only attach to liver cells, and flu viruses attach only to cells of the respiratory tract. This specificity of attachment derives partly from the presence of spikes or other proteins on the surface of the virus that recognize protein "receptors" in the cell membrane of the host cells. In other words, in the host cells, the receptors represent the chemical "locks" to which spike proteins (the key) must fit to "open the doors" to the host cell.
2. **Penetration**. Depending on the type of virus, once attached, the entire virus, the nucleocapsid, or just the viral genome enters the host cell. But no matter how penetration occurs, once in the cytoplasm, a free viral genome is present.
3. **Biosynthesis**. With the viral genome now in the cytoplasm of the host cell, the production (biosynthesis) of new viral parts begins.

FIGURE 6.6 The Steps of Virus Replication. These five steps are common to all viruses undergoing a productive infection. A latency step is found with those viruses that can remain dormant for some period.

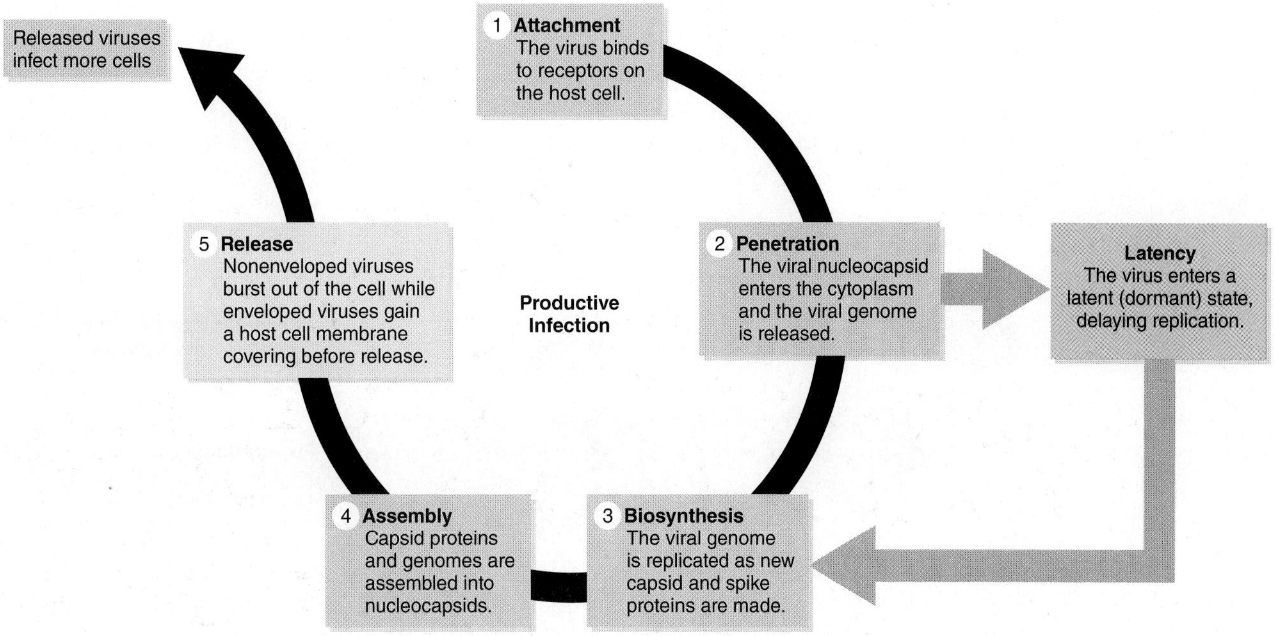

Biosynthesis uses the host cell's metabolic machinery to make new copies of the viral nucleic acid and viral proteins (capsids and spikes).

4. **Assembly**. Like any factory assembly line, once the parts are made, they are assembled into the final product. In the case of viruses, the nucleic acids and viral proteins are assembled into hundreds of new virus particles simultaneously. Depending on the virus, assembly might occur in the cell nucleus or the cytoplasm.
5. **Release**. In the final step of virus replication, the viruses are released from the host cell. The nucleocapsids of enveloped viruses become surrounded by a host cell membrane containing the spike proteins, which represents the new viral envelope. For nonenveloped viruses, often their large number bursts the host cell and the new viruses are released. The host cell usually is killed in this bursting process.

When Viruses Lay Dormant

Some viruses might not immediately undergo a productive infection when they enter a host cell (see Figure 6.6). This delay in replication is called **latency**. For example, HSV and the virus causing chickenpox (varicella-zoster virus [VZV]) initially cause a productive infection leading to cold sores on the lips (HSV) or chickenpox (VZV). Then, these DNA viruses go latent and their genome hides out, but does not replicate, in various neural **ganglia** (sing. **ganglion**) in the body. Months, years, or decades later, the genome might become reactivated and initiate another crop of cold sores or, in the case of VZV, shingles.

Ganglion: A cluster of nerve cell bodies outside the brain and spinal cord.

Another particularly well-studied form of latency occurs in the human immunodeficiency virus (HIV) that causes **acquired immunodeficiency syndrome (AIDS)**. After HIV has entered its host cell (a type of immune cell called a helper T cell), the capsid disassembles, and the RNA is released (**FIGURE 6.8**).

FIGURE 6.7 Replication of a DNA Virus. The replication process illustrated here is for a herpes simplex virus (such as one causing cold sores or fever blisters on the lips). Genome replication and assembly occur in the cell nucleus and biosynthesis in the cytoplasm.

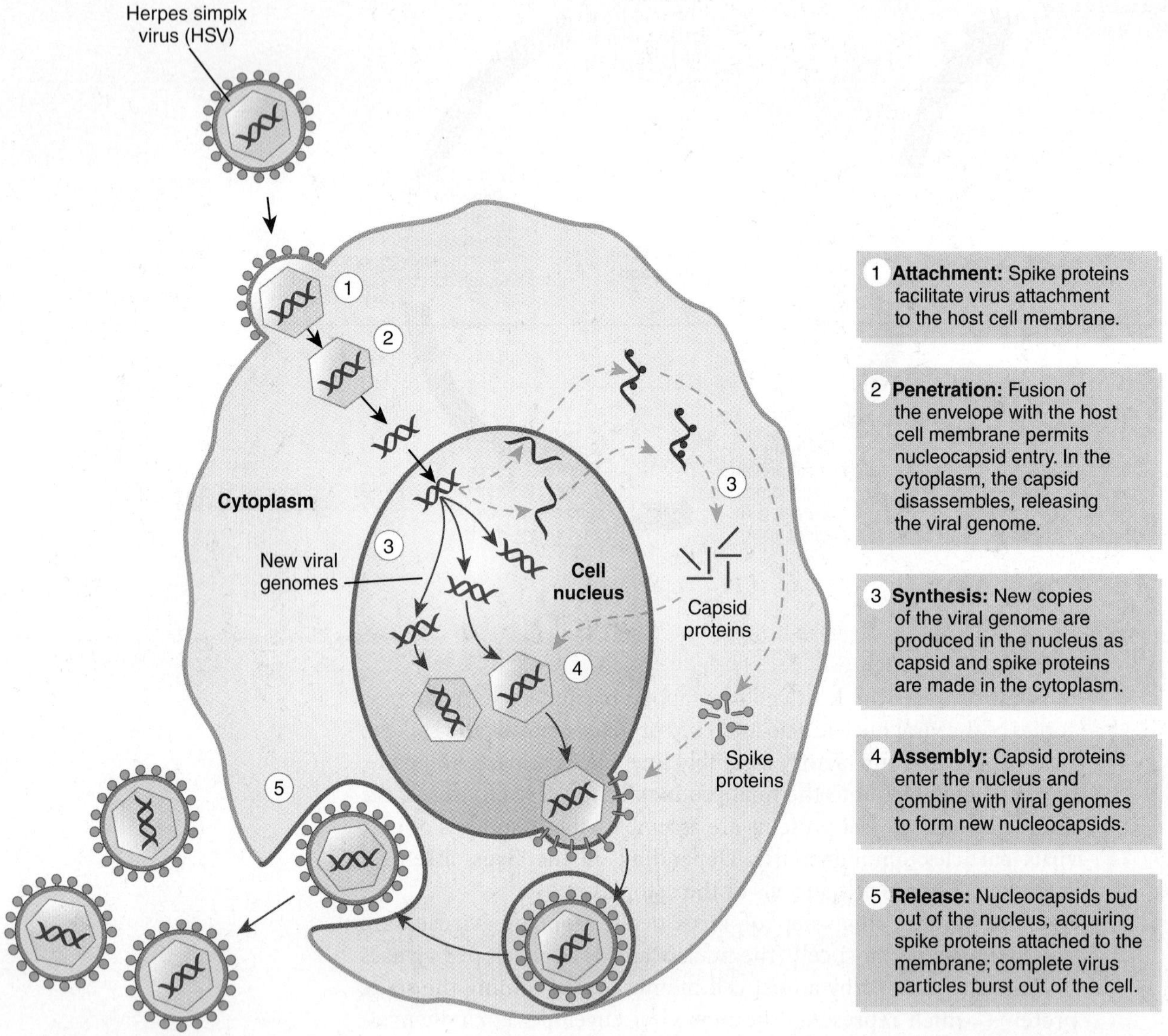
Herpes simplx virus (HSV)
Cytoplasm
New viral genomes
Cell nucleus
Capsid proteins
Spike proteins
1
2
3
3
4
5
1 **Attachment:** Spike proteins facilitate virus attachment to the host cell membrane.
2 **Penetration:** Fusion of the envelope with the host cell membrane permits nucleocapsid entry. In the cytoplasm, the capsid disassembles, releasing the viral genome.
3 **Synthesis:** New copies of the viral genome are produced in the nucleus as capsid and spike proteins are made in the cytoplasm.
4 **Assembly:** Capsid proteins enter the nucleus and combine with viral genomes to form new nucleocapsids.
5 **Release:** Nucleocapsids bud out of the nucleus, acquiring spike proteins attached to the membrane; complete virus particles burst out of the cell.

HIV and a few similar viruses are unique in carrying within the nucleocapsid a special enzyme called **reverse transcriptase**. With the RNA in the cytoplasm, the reverse transcriptase copies the RNA genome into a double-stranded DNA molecule. The DNA then is transported to the cell nucleus, where it integrates into the host cell's DNA; that is, into a chromosome. Such a fragment of viral-derived DNA incorporated into a chromosome is called a **provirus**. From its site within the nucleus, the provirus can remain dormant, causing no illness.

Under circumstances that are not entirely clear, when the provirus eventually reactivates, a new productive infection begins, and new viruses are produced by the provirus genome. These new viruses are released from the cell, and they are capable of infecting more similar type immune cells. As time passes, the number of these immune cells is reduced severely due to infection and there comes a point at which the body cannot mount an effective immune response against other pathogens. The person now is said to have AIDS.

FIGURE 6.8 Latency and the Formation of a Provirus. Shown here is a human immunodeficiency virus (HIV) infection. Following attachment and penetration, the single-stranded (ss) RNA genome is converted into double-stranded (ds) DNA by a reverse transcriptase enzyme. The DNA then enters the cell nucleus and integrates into a chromosome as a provirus.

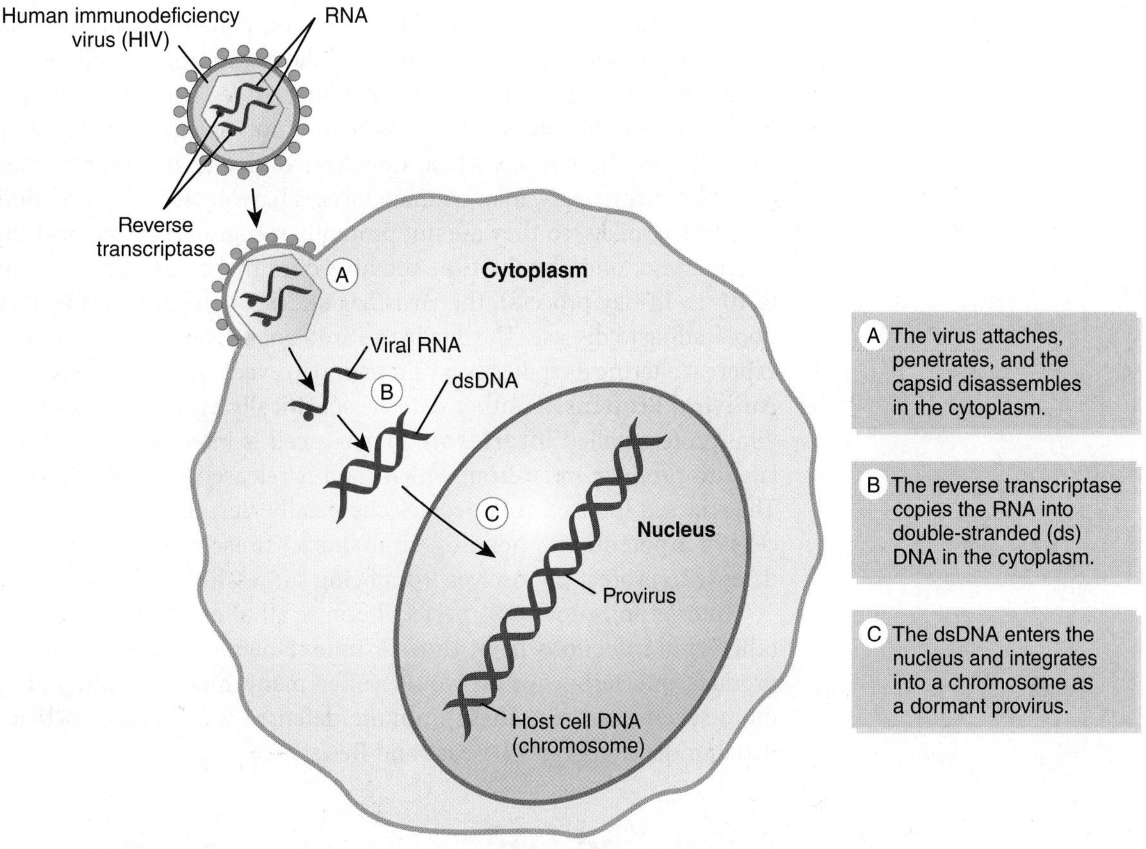
Human immunodeficiency virus (HIV)
RNA
Reverse transcriptase
A
Cytoplasm
Viral RNA
B
dsDNA
C
Nucleus
Provirus
Host cell DNA (chromosome)
A The virus attaches, penetrates, and the capsid disassembles in the cytoplasm.
B The reverse transcriptase copies the RNA into double-stranded (ds) DNA in the cytoplasm.
C The dsDNA enters the nucleus and integrates into a chromosome as a dormant provirus.

Fighting Back Against Viruses

If all viruses destroyed the cells they infected, all living cells would have been eliminated ages ago. So, infected host cells must have a way of fighting back. In fact, throughout the history of life on Earth, viruses and their host cells have tried to get one up on the other. How do host cells try to defend themselves against a viral attack?

Bacterial Defenses Against Phage

In response to phage infection, bacterial cells produce enzymes that specifically recognize viral genomes. These so-called **restriction enzymes** chop up the viral genome once it is in the cytoplasm. Some prokaryotes also contain in their chromosome short viral sequences called "clustered regularly interspaced short palindromic repeats" (**CRISPR**). Following phage infection, the CRISPR sequences are copied into RNA molecules, and if one of these molecules recognizes the viral DNA or RNA, the two join together. The virus genome-CRISPR complex then is recognized by another group of bacterial proteins called Cas. These proteins then chop up the complex. In both cases, without any viral instructions, virus replication is prevented, and no phages are formed. However, some phages have evolved modifications to their genome, making them resistant to the host's enzymes.

Human Host Defenses Against Viruses

Some viral diseases in humans are life threatening. AIDS and Ebola virus disease are two examples. Yet we do not usually die of colds, cold sores, and many other viral infections. Thus, the body must be able to defend itself from viral attack. The question is how.

- **Immune System Defenses.** One defense against viral attack (and all pathogen attacks) is our immune system. Molecular components of a virus, especially the capsid proteins and spikes, stimulate certain immune cells to produce huge numbers of highly specific **antibodies** that recognize the virus proteins. By attaching to the spikes or proteins on the viral capsid or envelope, antibodies neutralize the viruses, which now have difficulty attaching to their host cells. Not surprisingly, many human viruses have evolved ways of modifying the surface spikes, so they are not promptly recognized by the host immune system. Also, often by the time the host does make neutralizing antibodies (a 10- to 14-day process), the virus has undergone many rounds of replication, leading to disease. The flu viruses are a good example of a virus that is an expert at altering its spikes every flu season to evade quick antibody detection.
- **Antiviral Proteins.** Another defense specifically against viruses is an antiviral protein called **interferon**. When a cell is infected, the cell is stimulated to produce interferon, which then is released from the infected cell. The released interferon molecules chemically alert uninfected neighboring cells of a potential impending viral attack. Those cells then can put up a defense to protect themselves from being successfully infected.

  Interferon is not 100% perfect because all of us still get flus, colds, and other viral infections from time to time. However, without the ability to produce interferon, we all would suffer many more viral infections and more severe diseases. These immune defenses will be discussed in more detail in the chapter on Disease and Resistance.

6.4 New Viral Diseases: Where Are They Coming From?

Almost every year a newly emerging influenza virus descends upon the human population. Other viruses not even heard of a few decades ago, such as Ebola virus, Zika virus, and West Nile virus, are often in the news. Where are these viruses coming from?

Emerging Viruses

The United States and much of the world is at greater risk than ever from **zoonotic diseases**, which are diseases transmitted from other animals to humans (**TABLE 6.1**). Many of these are **emerging viral diseases**; that is, viruses appearing for the first time in a population or rapidly expanding their range with a corresponding increase in detectable disease. Many are transmitted by insects such as ticks, fleas, and mosquitoes. One example is West Nile virus disease, a mosquito-borne disease that marched across the United States between 1999 and 2009. It is now **endemic** across the continental United States.

Endemic: Referring to a constant presence of disease or persistence of an infectious agent at a low level in a population.

Mammals also are the source and transmitting agent of several viral diseases. Perhaps the most recent example was the Ebola epidemic in West Africa. The Ebola virus causing the epidemic hides out in bats. The bats then transmit the virus either directly to humans through a bat bite, or indirectly through bites to other animals (monkeys and apes), which if infected can transmit the virus to humans through some form of contact.

What about the near future? Scientists and microbiologists are wondering what diseases might emerge as climate change alters weather patterns. With

TABLE 6.1 Examples of Emerging Viruses

Virus	Human Contact
Influenza	Infected pig and avian populations
Dengue fever	Infected mosquitoes
Sin Nombre (hantavirus)	Infected deer mice or their dried urine/feces
Ebola/Marburg	Infected fruit bats
HIV	Infected chimpanzees and monkeys
West Nile	Infected mosquitoes
Nipah/Hendra	Infected flying foxes (bats)
Severe acute respiratory syndrome (SARS)-associated	Infected horseshoe bats
Chikungunya	Infected mosquitoes
Zika virus	Infected mosquitoes

possible warmer temperatures in the more temperate latitudes, insects like mosquitoes could expend their range. If infected with a viral pathogen, they might trigger new disease outbreaks.

But no matter how these viruses and zoonotic diseases are transmitted, what caused their emergence?

Jumping Viruses

One way "new" viruses arise is by a virus mixing genes from its genome with those from another virus to produce a new, unique combination of genes. Take for example influenza. Influenza viruses are notorious for swapping genes. The "swine flu" of 2009–2010 was the result of gene mixing between a strain of an avian flu virus, a human flu virus, and a swine flu virus. Viruses with the new combination jumped from pigs to humans, and a flu pandemic began.

Emerging viruses also arise from one of the driving forces of evolution—mutation. A **mutation** is a permanent change to the genetic information in a gene such that the mutation alters the genetic message. Although mutations often are lethal to the organism or virus, occasionally a mutation confers a benefit. In the case of flu viruses, beneficial mutations have resulted in new virus strains that, when they jump to humans, are resistant to the previous season's flu vaccine that the person might have received. With fast virus replication rates, it does not take long for the viruses with the beneficial mutation to spread, making treatment more challenging.

Even if a new virus has emerged, it must encounter an appropriate host to replicate and spread. Scientists believe the smallpox and measles viruses both evolved thousands of years ago in cattle and then jumped species to humans. Likewise, flu viruses today often originate in fowl and pigs and continue to jump to humans. HIV almost certainly evolved in apes or monkeys and then, as the result of a few mutations, the infecting virus jumped species to humans.

So, most emerging viruses are not "new" in the sense of appearing from nowhere. Rather they are the result of genome mixing and mutation between already present animal viruses that then jumped to the human species. What could facilitate such a jump to humans?

Today, population pressure is pushing humans into new uninhabited or less inhabited areas around the world. In these areas, potentially deadly animal viruses might be present. Increased agricultural expansion can bring wild rodents into contact with humans. If these animals are infected with a virus, it might be able to jump from rodent to human and lead to a disease outbreak.

An increase in the size of an animal population carrying a viral disease can trigger an outbreak. Prior to 1993, the deer mouse population in the Four Corners area of the United States (Arizona, Colorado, New Mexico, and Utah) had a low endemic infection with a virus called the hantavirus. Then, in the spring of 1993 the American Southwest experienced a wet season, providing ample food for deer mice. Their population quickly expanded, including those mice infected with the hantavirus. Rodent activities deposited mouse feces and dried urine in areas of human habitation. When aerosolized, the infected rodent excrement was inhaled by unsuspecting humans. In fact, 14 people died in the four Corners area from a respiratory illness caused by the hantavirus. This potentially fatal respiratory disease now is referred to as hantavirus pulmonary syndrome.

So, emerging viruses are not new. They simply are evolving from existing viruses and, through human contacts in the environment, are given the "opportunity" to spread (jump) to humans.

6.5 Tumors and Cancer: A Role for Some Viruses

Cancer is indiscriminate. It affects humans and animals, young and old, male and female, rich and poor. In the United States, the American Cancer Society projects that over 607,000 Americans will die of cancer in 2019. Cancer remains the second most common cause of death in the United States (after heart disease), accounting for nearly one of every four deaths. In addition, more than 1.7 million new cancer cases will be diagnosed in 2019. Worldwide, the World Cancer Research Fund International estimates more than 8 million people die of cancer each year and there will be more than 17 million new cancer cases diagnosed in 2018. Cancer is a very complex topic, so we will summarize the basics and then describe the association of some viruses with tumor and cancer development.

The Uncontrolled Growth and Spread of Abnormal Cells

Cancer starts with the uncontrolled reproduction of cells; that is, the frequency of cell division is greater for cancer cells than for normal cells. Cancer cells in some way escape controlling factors and, as they continue to multiply, they form an enlarging cluster of cells. Eventually, the cluster grows into an abnormal, large mass of cells called a **tumor**. Normally, the body will respond immunologically to a tumor or surround the tumor with layers of **connective tissue**. Such an "encapsulated" tumor is designated **benign** and is usually not life threatening. However, at times tumor cells evade immune attack, or they multiply too rapidly and break out of the capsule. They then can **metastasize** (spread) to other tissues and organs in the body. These tumors are called **malignant**, and the

Connective tissue: A group of cells that bind and support other tissues.

individual is said to have **cancer**. The term **oncology** (oncos = "a mass") is the study of tumors and cancer.

Cancer cells differ from normal cells in three major ways: they undergo cell division more frequently or for a longer period than do normal cells; they stick together less firmly than normal cells; and they often revert to an early stage in their development, often becoming formless cells dividing as rapidly as early embryonic cells. Moreover, cancer cells fail to exhibit contact inhibition; that is, they do not stop growing when they contact one another, as normal cells do. Rather, they overgrow one another to form a tumor. There appears to be no known boundary limiting the growth of a malignant tumor.

How can a mass of cells bring disease to the body? By their sheer numbers, cancer cells invade and erode local tissues, thereby interrupting normal functions and damaging organs. For example: a tumor in the kidney might block the tubules and prevent the flow of urine during excretory function, a brain tumor might cripple the organ by compressing the nerves and interfering with nerve impulse transmission, and a tumor in the bone marrow could disrupt blood cell production.

In addition, to satisfy their own metabolic needs, tumor cells rob the body's normal cells of vital nutrients. Others interfere with immune system functions so that microbial diseases might take hold. Ultimately, malignant tumors weaken the body until it fails.

The Involvement of Viruses

Many health experts believe that up to 65% of human cancers are the result of genetic abnormalities (mutations) in the body. Many of the remaining cancers result from chemical and physical agents called **carcinogens**. For example, chemical agents such as asbestos, nickel, certain pesticides and pollutants, and cigarette smoke can cause cellular changes leading to tumors and cancer (**FIGURE 6.9**). Among the known physical carcinogens are ultraviolet (UV) radiation and X rays. These physical agents damage DNA.

So, how are viruses involved? First, realize that most viruses and viral infections do not trigger tumor formation or cancer. Of all human cancers, oncologists believe up to 12% are a direct or indirect result of virus infections (**TABLE 6.2**). When these **oncogenic (tumor-causing) viruses** are transferred to test animals or cell cultures, an observable change occurs in the infected cells. Structural, biochemical, and/or growth patterns might be altered. However, the transformation of normal cells to abnormal tumor cells is a complex, multistep sequence of events, of which viruses play one part.

How Viruses Transform Cells

There are three known mechanisms by which oncogenic viruses transform normal cells into tumor cells.

- **Presence of a Viral Oncogene.** Some tumor viruses carry a so-called viral **oncogene** in their genome. On infection of a human cell, the protein produced by the oncogene converts a normal cell into a tumorous or cancerous cell (**FIGURE 6.10A**). These genes usually affect growth control or cell division.
- **Inhibition of a Tumor Suppressor Gene.** Cells normally contain several **tumor suppressor genes** (**TSGs**) whose function is to inhibit abnormal cell growth and tumor formation. Some oncogenic viruses will integrate their genome near a TSA gene, causing a loss of TSG function (**FIGURE 6.10B**). So, here the oncogenic virus influences tumor formation through the loss of host gene activity.

- **Activation of Host Genes.** Other oncogenic viruses physically interfere with host cell control. In this case, the virus inserts its genome near to or within a gene that normally controls cell division. Now under virus control, the human gene becomes active, and its protein products drive tumor formation (**FIGURE 6.10C**). However, these viruses insert their genome randomly within the DNA of the host genome. Consequently, the chance of insertion near a cell division–controlling gene is low.

FIGURE 6.9 The Onset of Cancer. Genetic factors, environmental agents, and some viruses are among the factors that can induce a normal cell to become abnormal. When the immune system is effective, it destroys abnormal cells and no cancer develops. However, when the abnormal cells evade the immune system, a tumor might develop and become malignant, spreading to other tissues in the body.

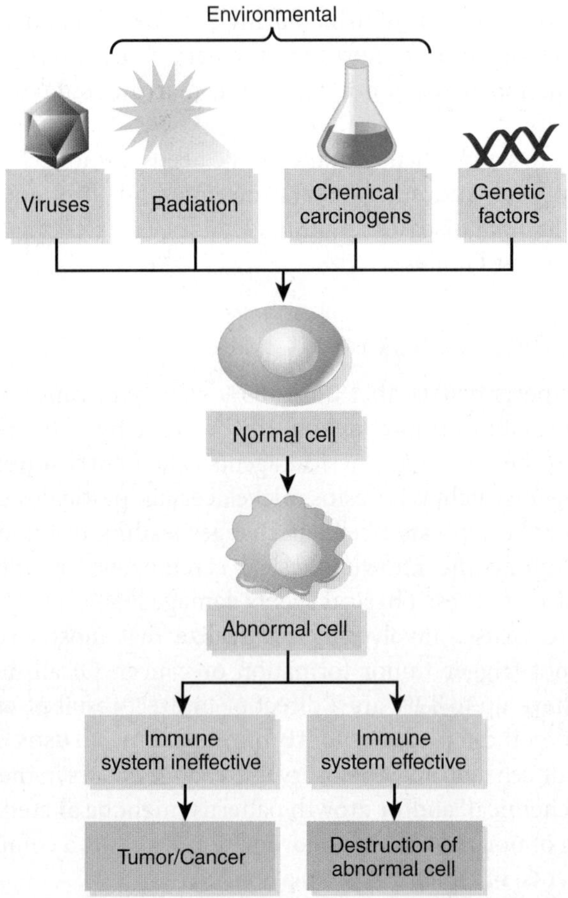

TABLE 6.2 Human Tumor Viruses and Their Effects on Cell Growth

Tumor Virus	Benign Disease	Effect	Associated Cancer
DNA Tumor Viruses			
Human papillomavirus (HPV)	Benign warts Genital warts	Encodes genes that inactivate cell growth regulatory proteins	Cervical cancer Penile cancer Oropharyngeal cancer
Merkel cell polyomavirus (MCPV)	Unknown	Being investigated	Merkel cell carcinoma

Epstein-Barr virus (EBV)	Infectious mononucleosis	Stimulates cell growth and activates a host cell gene that prevents cell death	Burkitt lymphoma Hodgkin lymphoma Nasopharyngeal carcinoma
Herpesvirus 8 (HV8)	Growths in lymph nodes	Forms lesions in connective tissue	Kaposi sarcoma
Cytomegalovirus (CMV)	Mononucleosis syndrome	Stimulates genes involved with tumor signaling	Salivary gland cancers
Hepatitis B virus (HBV)	Hepatitis B	Stimulates overproduction of a protein regulator	Liver cancer
RNA Tumor Viruses			
Hepatitis C virus (HCV)	Hepatitis C	Chromosomal aberrations, including enhanced chromosomal breaks and exchanges	Liver cancer
Human T-cell leukemia virus (HTLV-1)	Weakness of the legs	Encodes a protein that activates growth-stimulating gene expression	Adult T-cell leukemia/ lymphoma

FIGURE 6.10 Tumor Viruses and Tumor Initiation.

(A) Some oncogenic viruses carry an oncogene whose protein product directly causes uncontrolled growth.

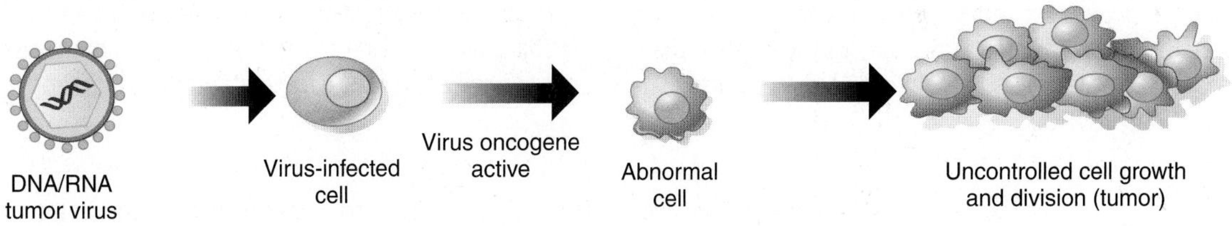

(B) Other viruses integrate their DNA next to a tumor suppressor gene (TSG), resulting in inactivation of the TSG and leading to uncontrolled cell growth.

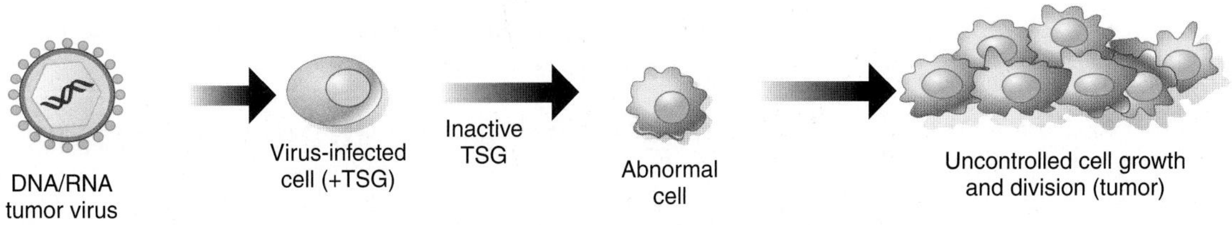

(C) Yet other viruses integrate their DNA next to a silent growth control gene, resulting in activation of the gene and uncontrolled growth.

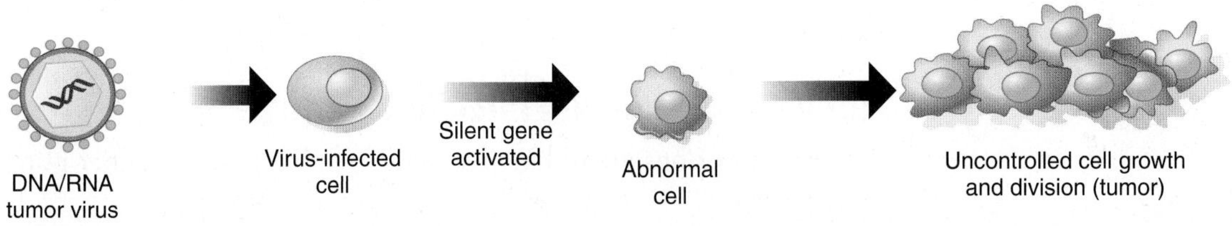

A CLOSER LOOK 6.3

The Power of the Virus

Ever since the discovery of viruses, scientists and physicians have wondered how they could use viruses to cure disease.

Almost 90 years ago, bacteriophages (phages) were experimented with as agents to cure human bacterial infections. Although the use of "phage therapy" for bacterial infections had some successes in the Soviet Union in the 1920s, the results were inconsistent. Then, with the advent of antibiotics in the 1940s as therapy for bacterial infections, much of the work on phage therapy declined.

At a time when we are seeing increased bacterial resistance to antibiotics, phage therapy using "phage cocktails" or phage enzymes is on the rebound. Although many of these phage therapy studies are in early clinical trials, the use of phage therapy holds great potential as a complementary approach for the treatment of acute and chronic bacterial infections and in support of human health.

There also are the viruses that infect humans. Can they be "tamed" in a form of "virotherapy" to fight human diseases like cancer (**FIGURE A**)? In 1997, a mutant herpesvirus was produced that was capable of only replicating in tumor cells. Because viruses make ideal cellular killers, perhaps these viruses would kill the infected cancer cells. Used on a terminal patient with a form of brain cancer, the virotherapy worked, as the patient has remained free of brain cancer. In 2018, scientists at the Duke Cancer Institute reported on their use of a genetically modified polio virus to attack a rare but aggressive and deadly brain cancer called glioblastoma. This is the disease that took the life of Senator Ted Kennedy and recently Senator John McCain. In patients taking part in the clinical trials, the modified polio virus was injected directly into the tumor. The glioblastoma patients showed a three-year survival rate of 21 percent compared with just 4 percent of patients given standard therapy. Additional "oncolytic" (cancer-killing) viruses are being developed. Today, phage therapy and virotherapy are in high gear with many early clinical trials in progress to examine the power of the virus to directly or indirectly attack bacterial infections or kill cancer cells. The future for viral therapy is looking bright.

FIGURE A False-color transmission electron microscope image of adenoviruses that have been used in some human virotherapy treatments. (Bar = 50 nm.)

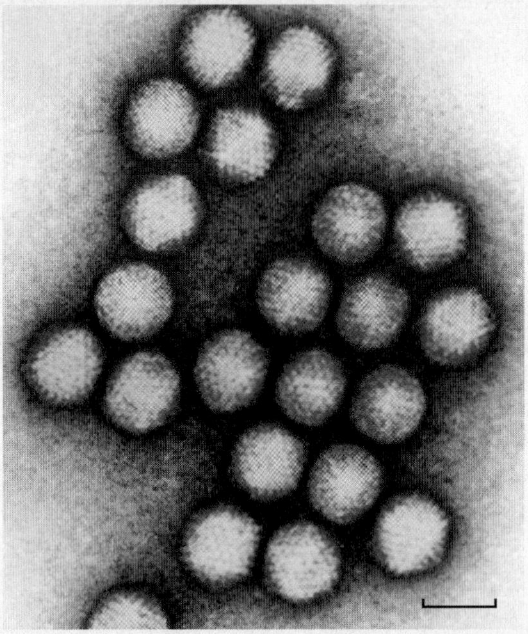

Courtesy of CDC.

In all three cases, virus infection leads to a disruption of normal growth control and a stimulation of cell division, leading to tumor formation. On the flip side, the natural ability of viruses to get inside cells and deliver their viral information has been manipulated by scientists to deliver genes to cure genetic illnesses. **A CLOSER LOOK 6.3** considers some uses of viruses for disease removal or cure.

6.6 Virus-Like Agents: Viroids and Prions

When viruses were discovered, scientists believed they were the ultimate microscopic infectious particles. It was difficult to conceive of anything smaller than viruses as agents of disease. However, this perception changed when scientists discovered even smaller disease agents of plants and animals. These "virus-like agents" are the viroids and prions.

Viroids and Prions Are Infectious Particles

Viroids and prions are unlike the viruses in that they are either pure RNA or pure protein.

Viroids

Viroids are infectious agents of some plants. The agents are composed of short pieces of RNA that lacks a capsid. In fact, the viroid RNAs are so small they are incapable of coding for any known proteins. When the particles infect plants, the RNA interferes with gene activity that controls plant development and results in stunted growth. More than two dozen crop diseases are associated with viroids.

Prions

Prions are a type of infectious particle that affects animals. These particles are composed only of protein; no nucleic acid is present. Prions cause several **neurodegenerative** diseases in mammals, including "mad cow disease" (MCD) in cattle and a very rare infection in humans called variant Creutzfeldt-Jakob disease (vCJD). So, without any genetic information, how do prions replicate and cause disease?

Prion: PRE-on

Neurodegenerative: Referring to the progressive loss of structure or function of neurons in the brain.

Basically, when the infectious proteins infect a neuron, they cause other proteins in the neuron to misfold into a nonfunctional shape. As they accumulate in the neuron, they eventually lead to the death of the cell. When enough nerve cells have been infected and die (this can take months or years), the animal suffers brain damage and eventually dies.

So, for vCJD, how does a person initially get infected with these prion proteins? A prion disease outbreak in the United Kingdom in the late 1990s to early 2000s illustrates the process. In 1996, a number of unusual cases of a neurodegenerative disease were reported in humans. These individuals exhibited symptoms such as personality changes, anxiety, depression, memory loss, blurred vision, and difficulty speaking. The human disease was named vCJD at that time.

Investigations soon discovered that the disease was originating from cattle that had mad cow disease. Much of the meat from the slaughtered animals had been made into meat products that then were introduced into the human food supply. The contaminated meat contained the prion proteins that, when consumed and absorbed, slowly made their way to the spinal cord and brain.

By the time the outbreak was over in 2008, 177 vCJD deaths in the United Kingdom were recorded. In the United States, there have been no reported cases of locally contracted vCJD and there have been only 19 infected animals identified with MCD in Canadian and American cattle. Importantly, protection measures are in place to quickly identify and remove any mad cow cattle and to ensure products from those animals are not used for human food or animal feed.

A Final Thought

When the author was a young assistant professor of biology, the question of whether viruses are "alive" came up for discussion at an informal social gathering. As several people had had some beer to drink, the conversation became quite animated supporting one or the other viewpoint on viruses being alive. Perhaps one of the best answers was, "Who cares! We treat viral diseases the same way regardless of whether they are alive or not."

But the question has continued to interest many microbiologists to this day. And now the question is put to you. Are viruses alive? Or are they inert chemical molecules that have life like abilities to replicate? Then again, are viruses neither totally inert nor totally alive, but somewhere on the threshold of life? From your "interactions" with viruses discussed in this chapter, and from your intimate contact with real viruses like those causing flus and colds, what is your thought?

Chapter Discussion Questions

What Was He Thinking?

From your reading of this chapter, identify and discuss five major points about the viruses that the author was trying to get across to you.

Questions to Consider

1. A textbook author referring to viruses once wrote: "Certain organisms seem to exist only to reproduce, and much of their activity and behavior is directed toward the goal of successful reproduction." Would you agree with this statement? Can you think of any creatures other than viruses that fit the description?
2. Researchers studying the bacteria that live in the oceans have long been troubled by the question of why bacteria have not saturated the oceanic environments. Based on the material in this chapter, how might the marine bacteria be kept in check?
3. An eminent immunologist once commented that viruses were "bad news wrapped in protein." What does this statement imply, what is the "bad news," and what is meant by "wrapped in protein?"
4. Bacteria can cause disease by using their toxins to interfere with important body processes, by overcoming body defenses, by using their enzymes to digest tissue cells, or by other similar mechanisms. Most viruses, by contrast, do not encode toxins and produce no digestive enzymes. How, then, do viruses cause disease?
5. When discussing the multiplication of viruses, virologists prefer to call the process replication, rather than reproduction. Why do you think this is so? Would you agree with virologists that replication is the better term?
6. How have revelations from studies on viruses, viroids, and prions complicated some of the traditional views about the principles of biology?

CHAPTER 7

Growth and Metabolism: Running the Microbial Machine

The Red Planet

Books have been written about it; movies have been made, even a radio play (the *War of the Worlds*) on Halloween night in 1938 about it allegedly frightened thousands of Americans. What is it? A Martian invasion of New Jersey!

In 1877 the Italian astronomer, Giovanni Schiaparelli, observed lines on Mars, which he and others assumed were canals built by Martians. It wasn't until well into the 20th century that this notion was disproved. Still, when we gaze at the Red Planet, we wonder: did life—microbial that is—ever exist there (see chapter opening image)?

Today, microbiologists have joined their other science colleagues in wondering if microbial life once existed on the Red Planet. Consider that on Mars temperatures can be far below 0°C (32°F), and the atmosphere contains little oxygen gas. In addition, the planet's surface is bombarded with ultraviolet (UV) radiation that can damage DNA.

To research the problem, investigators here on Earth built a device simulating the Martian environment. Microbes called **extremophiles** that are known to survive in extreme earthly environments (**TABLE 7.1**) were placed in this chamber. At the end of the experiment, the results suggested some members of the domain Archaea could grow in the Martian-like environment.

The National Aeronautics and Space Administration (NASA) has sent several spacecrafts to Mars to study the planet. Some findings suggest there are areas where salty seas once washed over the plains of Mars, possibly creating a life-friendly environment. The *Phoenix Mars Lander* uncovered water ice near

CHAPTER 7 OPENER This photo, taken by NASA's Mars rover *Curiosity,* shows Mount Sharp in the distance. In the center of the image is an area with clay-bearing rocks that scientists believe might help them determine if water was involved in forming Mount Sharp. All life needs water, because it is the medium in which growth and metabolism occur. If microbial life has existed on Mars, water would be necessary.

TABLE 7.1 Some Microbial Record Holders

Hottest environment (Juan de Fuca ridge)—121°C: Strain 121 (Archaea)
Coldest environment (Antarctica)—15°C: Cryptoendoliths (Bacteria and lichens)
Highest radiation survival—5 MRad, or 5000× what kills humans: *Deinococcus radiodurans* (Bacteria)
Deepest—3.2 km underground: Many bacterial and archaeal species
Most acid environment (Iron Mountain, CA)—pH 0.0 (most life is at least a factor of 100,000 less acidic): *Ferroplasma acidarmanus* (Archaea)
Most alkaline environment (Lake Calumet, IL)—pH 12.8 (most life is at least a factor of 1000 less basic): Proteobacteria (Bacteria)
Longest in space (NASA satellite)—6 years: *Bacillus subtilis* (Bacteria)
Saltiest environment (Eastern Mediterranean basin)—47% salt (15 times human blood saltiness): Several bacterial and archaeal species

the Martian soil surface. The Mars rover, *Curiosity*, detected simple organic molecules in sedimentary rock, suggesting more complex biological molecules might have existed on Mars in the distant past. Using a radar instrument on board the European Space Agency (ESA) *Mars Express* orbiter, researchers in 2018 reported evidence for a lake of liquid water about 20 km (12 miles) across under the planet's south polar ice cap (**FIGURE 7.1**).

Getting a conclusive answer for microbial life on Mars will require more studies and perhaps human travel to the red planet. Nevertheless, whether microorganisms are here on Earth in moderate or extreme environments, or on Mars, there are certain physical requirements they must possess to survive and grow. In this chapter, we explore the varied physical conditions for growth. We also will study some of the universal metabolic pathways microorganisms, and all life, possess to survive and grow. Much of the chapter emphasizes the role of carbohydrates in metabolism. Therefore, it might be worthwhile to review the material on the organic molecules (especially carbohydrates) presented in the chapter on Molecules of the Cell.

FIGURE 7.1 The South Pole of Mars. Scientists have proposed that a lake of liquid water exists beneath the ice near the south pole of Mars.

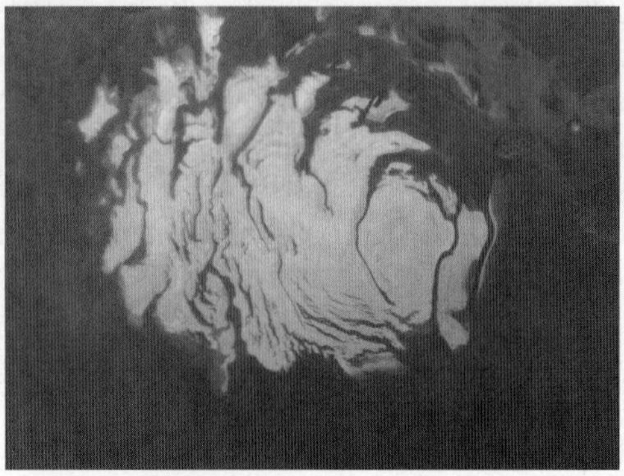

NASA/JPL/Malin Space Science Systems.

LOOKING AHEAD

After reading and completing this chapter, you will be able to:

7.1 Contrast microbial groups based on (a) temperature, (b) oxygen, (c) pH, and (d) salt requirements.
7.2 Identify the characteristics of enzymes and justify the need for adenosine triphosphate (ATP) energy by cells.
7.3 Distinguish between the stages of aerobic and anaerobic respiration.
7.4 Define and explain the importance of fermentation.
7.5 Summarize the two major reactions of photosynthesis.

7.1 Microbial Growth: Physical Factors

Microbes are metabolic machines with the potential for fast growth. Sometimes we suffer the consequences of microbial growth, such as when pathogens multiply in our bodies and cause disease. Conversely, industrial microbiologists can grow large quantities of microbes in large vats called **bioreactors** (**FIGURE 7.2**). In these controlled environments, some microbes can be coaxed into producing vitamins, amino acids, antibiotics, or other products valuable for human nutrition and medicine.

The growth of microbial populations is described in the chapter Exploring the Prokaryotic World. Here, the major physical factors that influence that growth are explained.

Water

Water is the medium for all chemical reactions in cells, so it should not be surprising that about 70% of a microbial cell is water (**FIGURE 7.3**). As mentioned in the chapter on Molecules of the Cell, water is the solvent in which solutes (e.g., salts, sugars) dissolve. Therefore, should a cell lose water to the environment, metabolism will slow down or stop, and the cell might die. Notable exceptions are bacterial and fungal spores. These dormant structures survive arid environments by temporarily shutting down metabolism.

Temperature

Another important factor for microbial growth is temperature. If the temperature is too low, the rate of the metabolism is lowered; if the temperature is too high, enzymes, which control metabolism, might be unfolded by the heat, and metabolism will cease.

Microbes have adapted to most environments on Earth; that means they have adapted to different temperatures. In Figure 7.4, notice that at the low and high end of the range, the microbes grow the slowest; that is, they survive but do not actively reproduce. Consequently, the optimal temperature will be somewhere in the middle of the range where they reproduce at their maximal rate. Based on these properties, microbes are assigned to one of three groups based on their temperature ranges (**FIGURE 7.4**).

Psychrophiles and Psychrotrophs

Those microbes growing at temperatures between −8°C (18°F) and 20°C (68°F) are said to be **psychrophiles** (*psychro* = "cold"; *phil* = "loving"); these "psychrophilic" microbes can be found in Arctic and Antarctic environments as

FIGURE 7.2 Growing Microbes. A pharmaceutical technician monitors a series of bioreactors to ensure a maximum yield of microbial product.

FIGURE 7.3 Water and Organic Compounds in Bacterial Cells. Bacterial cells are about 70% water. The other 30% are ions, small molecules, and organic compounds. Dry weight refers to the weight of materials after all the water has been removed.

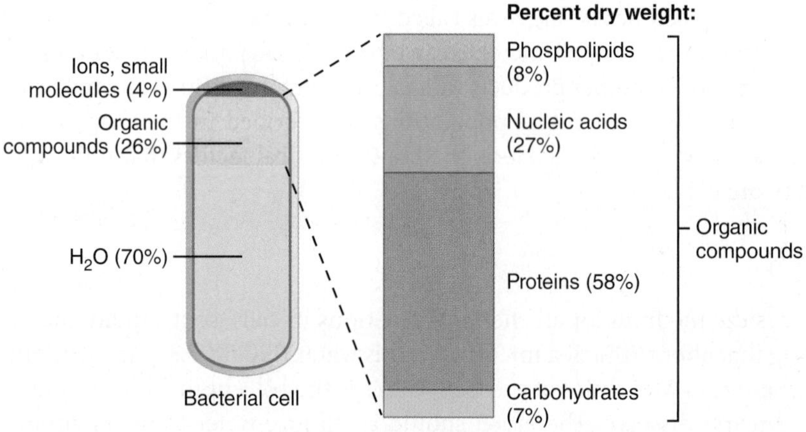

well as deep below the ocean surface. Considering 70% of Earth is covered by oceans having deep water temperatures below 5°C (41°F), there are lots of psychrophiles.

Another group of "cold-loving" microbes are the **psychrotrophs** (*troph* = "nourish"), which have a slightly higher temperature range than the psychrophiles. Psychrotrophs grow well at refrigerator temperatures (4°C–5°C/39°F–41°F) and can spoil foods. For example, streptococci grow in milk and deposit acid, causing the milk to become sour. Although these microbes do not represent a threat to health, excessive numbers in products like milk can make the product undesirable to the eye, nose, and taste buds. On the other hand, some dangerous "psychrotrophic" species can grow in refrigerated foods and deposit their diarrhea-inducing toxins in the contaminated food.

Mesophiles

Another group of microorganisms preferring warmer temperatures is the **mesophiles** (*meso* = "middle"). Their temperature range is between 10°C (50°F) and 45°C (113°F). Most human pathogens are mesophiles because human body temperature is about 37°C (98.6°F). Mesophiles also are found in aquatic and soil environments in temperate and tropical regions of the world. *Escherichia coli* is a typical mesophile.

Escherichia coli: esh-er-EE-key-ah KOH-lee

FIGURE 7.4 Growth Rates for Different Microorganisms in Response to Temperature. Temperature optima and ranges define the growth rates for all microbes. The growth rates decline quite rapidly to either side of the optimal growth temperature. Only some members of the Archaea can withstand extremely high temperatures.

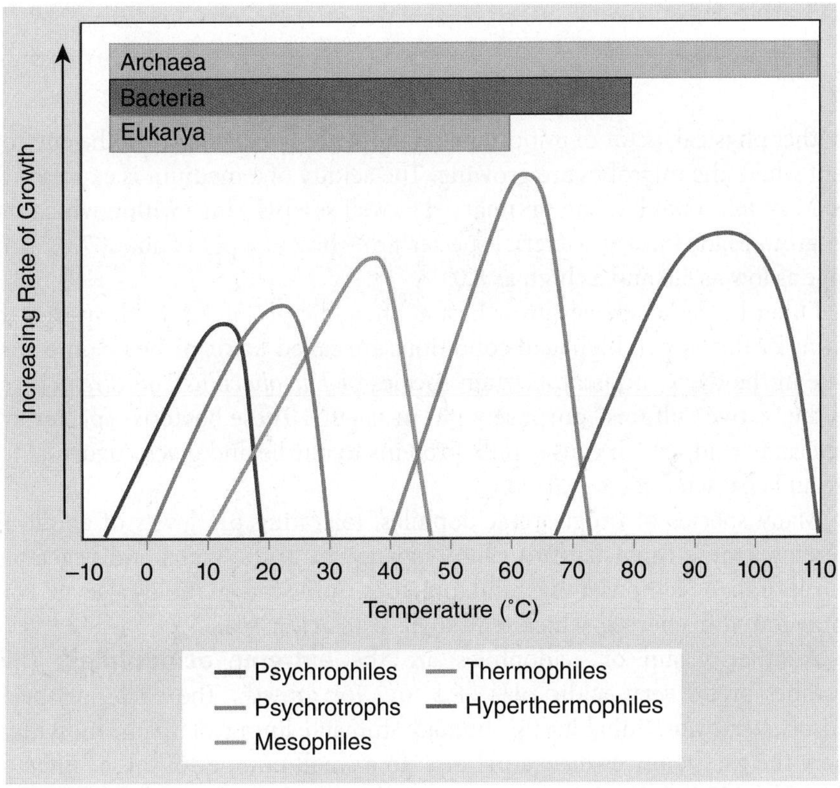

Thermophiles

Microorganisms tolerating and growing at high temperatures are called **thermophiles** (*thermo* = "heat"). These thermophiles grow at temperatures of 40°C (104°F) and higher, thriving in such varied environments as compost heaps, hot springs, and thermal vents on the ocean floor. Some species in the domains Bacteria and Archaea are thermophiles.

There also are archaeal species that can grow at temperatures at or above the boiling point of water (100°C/212°F), with some growing (perhaps just surviving) at an astounding 121°C (250°F). These **hyperthermophiles** are found in "hellish" environments, such as hot-water vents found along rifts on the floor of the Pacific Ocean (see Table 7.1). Here the water stays in a liquid form because of the extreme water pressure at these depths.

Oxygen

The growth of many microbes depends on a plentiful supply of oxygen gas (O_2; about 21% in air). Microbes requiring the gas for growth are said to be **aerobes**; humans also are "aerobic" organisms. In addition, there are species of microbes called **anaerobes** that live only in the absence of O_2; such organisms will be inhibited or die if the gas is present. Environments such as landfills (tightly packed with garbage), dense, muddy swamps, and the animal gut provide ideal environments for these anaerobic species.

Some pathogens also are anaerobes. For example, *Clostridium tetani*, the bacterial species causing tetanus, grows in the anaerobic, dead tissue of a wound.

Clostridium tetani: kla-STRIH-dee-um TEH-tahn-ee

In this environment, it produces powerful toxins that cause uncontrolled muscle spasms characteristic of the disease.

Several microbial species, including *E. coli*, can grow in the presence or the absence of oxygen gas. Microbes with this flexibility are referred to as **facultative** organisms.

Facultative: Referring to growth in the presence or absence of some substance, such as oxygen gas.

pH

Another physical factor of importance is the acidity or alkalinity of the environment where the microbes are growing. The acidity of a medium is expressed as **pH**. Most microbes have an optimal pH as well as a pH range within which they will grow. Many known bacterial species grow best at a pH of about 7.0, with a range as low as 5.0 and as high as 8.0.

pH: A measure of the hydrogen ion (H^+) concentration of an aqueous solution. Solutions with a pH less than 7 are said to be acidic and solutions with a pH greater than 7 are alkaline (basic). Pure water has a neutral pH of 7.

Other bacterial species grow best at an acidic pH of 3.0. Such species tolerating or thriving in high acid conditions are called **acidophiles**. Some are of value in the dairy industry. Certain species of *Lactobacillus* and *Streptococcus* are the "active cultures" purposely put in yogurt. These bacterial species produce lactic acid, which causes milk proteins to curdle and gives yogurt its texture and characteristic tart flavor.

Lactobacillus: lack-toe-bah-SIL-lus

Streptococcus: strep-toe-KOK-us

Many species of fungi are acidophiles, tolerating pH levels of about 5.0. Therefore, these fungi tend to cause spoilage in acidic fruits and vegetables, such as oranges, lemons, limes, and tomatoes. Some fungal molds also are commonly found in cheeses, which tend to have an acidic pH.

Another group of acidophiles are the **extreme acidophiles**. These microbes prefer very acidic pHs of 1 to 2 for growth. The bacterial species *Helicobacter pylori* inhabits the human stomach lining of many individuals where the gastric juices have a pH of 2. In a small number of these "infected" individuals, *H. pylori* might cause gastric ulcers and, among a small number of these individuals, might cause stomach cancer.

Helicobacter pylori: HE-lick-oh-bak-ter pie-LOW-ree

Another example of environmental extreme acidophiles are the bacterial and archaeal species living in the drainage water from some mines (**FIGURE 7.5**). Spain's Rio Tinto has an incredibly high iron-sulfur content due to waste from local mining activities. The river water has a pH of 1 to 2, which is the perfect environment for the growth of extreme acidophiles.

FIGURE 7.5 The Rio Tinto. The water in the Rio Tinto in southwestern Spain has a reddish-yellow-brown hue that comes from the reaction of its acidic waters with the iron in the soil. These unusual physical factors provide an environment for a diverse collection of extreme acidophiles.

FIGURE 7.6 Home to Halophilic Bacteria. The Great Salt Lake in Utah provides the high-salt environment favored by many halophilic bacteria. The pinkish color of the water is due to high numbers halophilic bacteria.

Other Physical Factors

In addition to temperature, oxygen gas, and pH, other common physical factors such as salt and radiation can influence the growth of a microbial population.

Salt

Microbes normally inhabiting salty environments are called **halophiles** (*halo* = "salt"). Examples of such "halophilic" species include the Halobacteria. Many of the species in this class are archaeal organisms that require at least 9% salt for growth (as compared to 1% for many nonhalophiles). Some halophiles can survive an incredibly high salt concentration of 47%. Although halophiles usually are thought of as inhabitants of the oceans, the salt concentration of the ocean is only about 3.5%. Halophiles are more commonly found in places like Utah's Great Salt Lake where the lake's salinity has ranged from a little less than 5% to nearly 27% (**FIGURE 7.6**).

Salt intolerance has been used in the food industry. By adding salt to cure meats and preserve other foods, the environment is so salty that most nonhalophiles dry out from the loss of water.

Radiation

One last physical factor is radiation, including ultraviolet (UV) light, X rays, and gamma rays. UV light is a component in sunlight and often is harmful to microbes inhabiting the surface of soil or water. However, bacterial species like *Deinococcus radiodurans* can survive extremely high levels of radiation (see Table 7.1). In fact, *D. radiodurans* has been referred to as "the world's toughest bacterium" by the Guinness World Records. Scientists hope to harness the metabolic powers of this microbe to help decontaminate radioactive waste sites, which are a continuing concern as a source of pollutants in the environment. If the scientists are successful, they will once again show how microbes can be a considerable benefit to society.

Deinococcus radiodurans: DIE-no-kok-us ray-dee-oh-DUR-anz

7.2 Microbial Metabolism: Enzymes and Energy

Metabolism refers to all the chemical reactions occurring in the cell during its growth. These chemical activities maintain the stability of the microbial cell and provide a dynamic set of chemical building blocks for the synthesis of new cellular materials. Indeed, much of the metabolism of human cells was first discovered and studied using microbial cells.

The Forms of Cellular Metabolism

There are thousands of different chemical reactions going on in cells at any moment. Many of these metabolic reactions fall into one of two general pathways: building (biosynthesis reactions) and disassembling (hydrolysis reactions) (**FIGURE 7.7**).

- **Biosynthesis Reactions.** "Biosynthesis" is a broad term that applies to any chemical process in which larger molecules are constructed from smaller building blocks. Metabolically, these are called **anabolic pathways** (**anabolism**). They require an input of energy because work must be done to construct large organic molecules, such as carbohydrates and proteins, from the building blocks.
- **Hydrolysis Reactions.** "Hydrolysis" is the term used to describe chemical reactions that break the large organic molecules into smaller pieces. These represent **catabolic pathways** (**catabolism**). They release energy, much of it in the form of unusable heat. The rest of the released energy will be available to produce **adenosine triphosphate** (**ATP**), the energy currency the cell uses to do work (see below).

The chemical reactions of metabolism occur in a highly organized and controlled manner. They do not occur spontaneously. Rather, each reaction is controlled by a specific enzyme.

Enzymes

Enzymes are large protein molecules that increase the chances that a chemical reaction will occur. Many enzymes can be identified by their names, which often end in "-ase." For example, "sucrase" is the enzyme that breaks down the sugar sucrose and "peptide synthase" refers to enzymes that, with the help of ribosomes, build amino acids into polypeptides. However, other enzymes do not have such descriptive names. Trypsin also is an enzyme that degrades protein.

FIGURE 7.7 Anabolic and Catabolic Pathways. Metabolism can be broken down into catabolic pathways that tear apart larger molecules (hydrolysis reactions) and release energy, and anabolic pathways that build larger molecules (biosynthesis reactions) from building blocks.

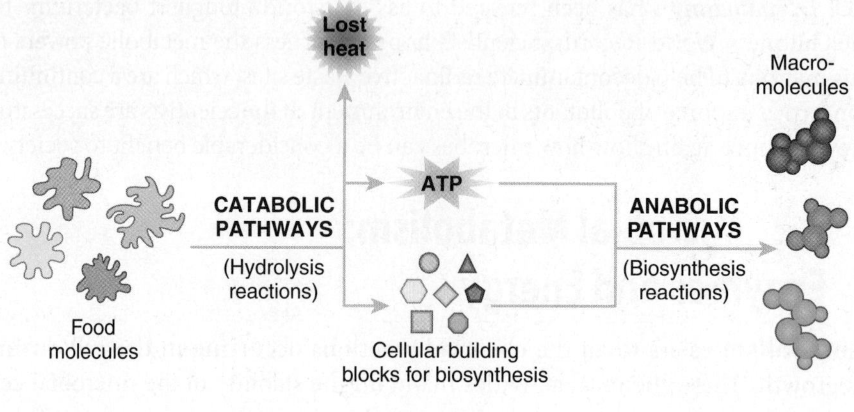

FIGURE 7.8 The Mechanism of Enzyme Action. Although this example shows an enzyme involved in an anabolic reaction, enzymes also control catabolic reactions.

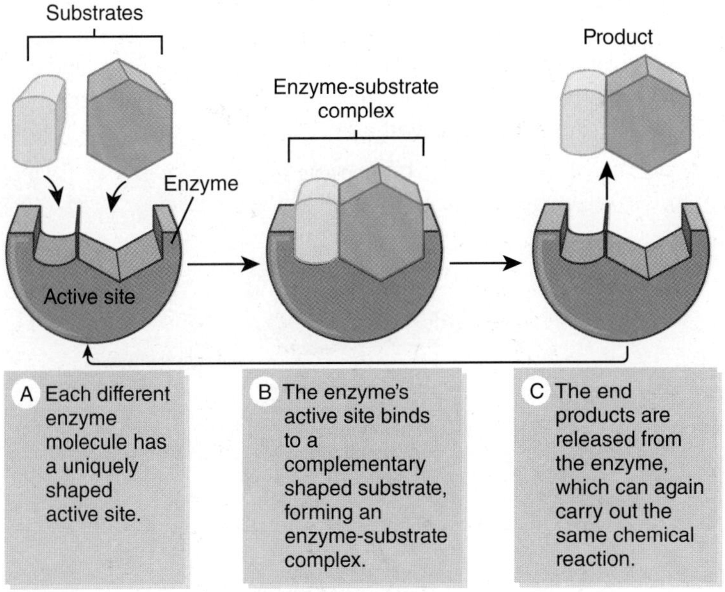

Substrates
Product
Enzyme-substrate complex
Enzyme
Active site
A Each different enzyme molecule has a uniquely shaped active site.
B The enzyme's active site binds to a complementary shaped substrate, forming an enzyme-substrate complex.
C The end products are released from the enzyme, which can again carry out the same chemical reaction.

Enzymes accomplish in fractions of a second what otherwise might take hours, days, or longer to happen spontaneously under normal environmental conditions. For example, without enzymes, it is highly unlikely that two reactants would randomly bump into one another in the precise way needed for a chemical reaction to link them together in the correct orientation.

Enzymes have several common characteristics that can be understood by referencing **FIGURE 7.8**.

Enzymes Are Highly Specific

An enzyme that functions in one chemical reaction will not be involved in another type of reaction. Consequently, each enzyme has a special pocket called an **active site** that has a unique three-dimensional shape. This shape is complementary to a reactant molecule (called a **substrate**). In the active site, the substrate is aligned in such a way that it is highly likely it will be affected in some way, resulting in the formation of one or more **products**.

Enzymes Are Reusable

In catalyzing a reaction, the enzyme itself is not changed. Once a chemical reaction has occurred, the enzyme releases the product or products and is ready to participate in another identical reaction. In fact, the same enzyme can catalyze the same type of reaction 1,000 to 1 million times each second if enough substrates are present.

Enzymes Are Required in Small Amounts

Because an enzyme can be used thousands of times to catalyze the same reaction, only small amounts of an enzyme are needed to ensure a fast and efficient metabolic reaction occurs.

Some Enzymes Are More Than Just Protein

Some enzymes in cells are not active unless they are associated with a small, nonprotein component called a **coenzyme**. Examples of two important coenzymes are **nicotinamide adenine dinucleotide** (NAD^+), which is derived

FIGURE 7.9 Cellular Respiration and Photosynthesis. Cell respiration and photosynthesis are complementary processes. Photosynthesis in the cell membrane of prokaryotes and chloroplasts of eukaryotes uses the sun's light energy to make chemical energy (glucose molecules; $C_6H_{12}O_6$). The energy in glucose molecules is extracted in cellular respiration at the cell membrane in prokaryotes and in the mitochondria of eukaryotes to make ATP. The glucose then is turned back into carbon dioxide (CO_2), which is used in photosynthesis.

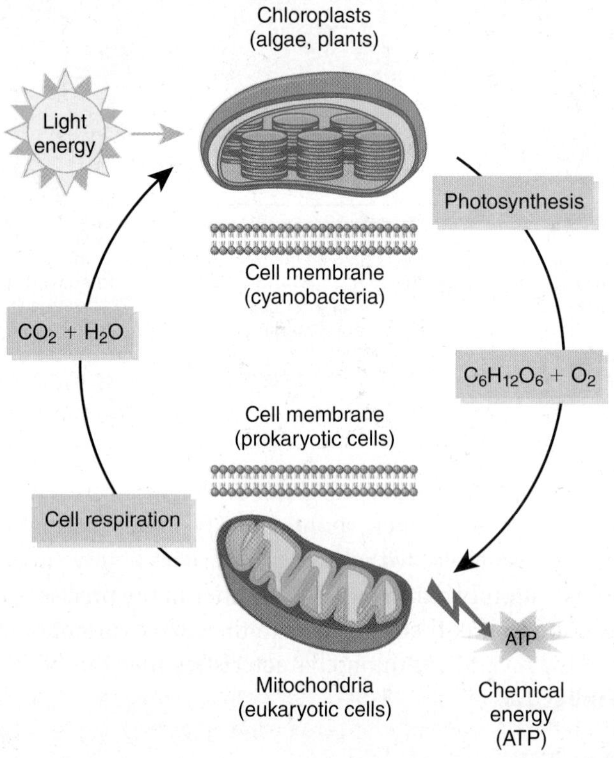

from vitamin B_3 (niacin), and **flavin adenine dinucleotide** (**FAD**), which is derived from vitamin B_2 (riboflavin). We will see the importance of these coenzymes later in this chapter.

Energy and ATP

Energy is the ability to do work or the capacity to cause change. Therefore, one of the major outcomes of catabolic pathways is the production of a steady supply of energy to perform the processes of life and to maintain order (see Figure 7.7). Physicists tell us the universe has a fixed amount of energy and this energy cannot be created or destroyed. However, energy can be converted from one form to another. An example is when logs are burned in the fireplace, the chemical energy in their molecules is released as heat energy and light energy. In the living world, organisms convert the chemical energy in carbohydrates and lipids into usable cellular energy; that is, into ATP (**FIGURE 7.9**).

ATP molecules are somewhat like portable batteries in that they can supply energy to any part of the cell where energy is needed. For example, ATP fuels the work needed to transport nutrients into the cell and to eliminate waste products from the cell. It is used in the synthesis of proteins and to provide movement of cells. ATP also energizes the reproduction process. However, unlike a portable battery, ATP energy needs to be used as it is generated. It cannot be stored.

ATP has been referred to as the "universal energy currency." As shown in **FIGURE 7.10**, the molecule contains three phosphate groups (phosphorus bonded to oxygen atoms). Much of ATP's energy is released when an enzyme breaks

FIGURE 7.10 The ATP Cycle. The energy released from the hydrolysis of ATP (1) is used to do cellular work (anabolic reactions). The energy released from cellular respiration (2) provides the energy to regenerate ATP.

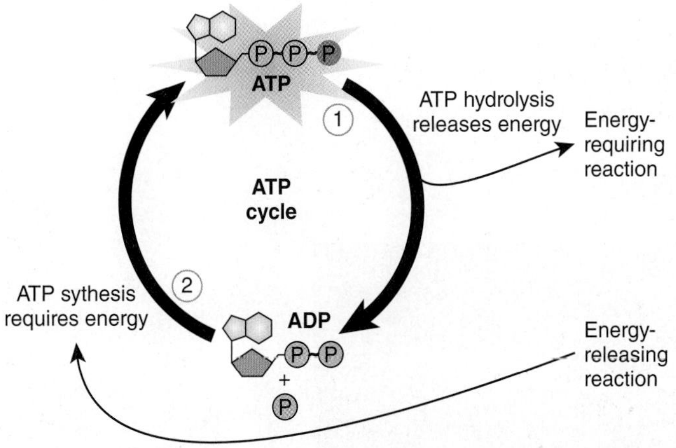

the high-energy bond connecting the terminal phosphate group to the rest of the molecule. That energy can then be used in an energy-requiring (anabolic) reaction. Most people are unaware of the importance of ATP in their lives—but without ATP, life as we know it could not continue.

7.3 Cellular Respiration: The Production of ATP

The step-by-step process by which organic compounds are broken down and their energy is released to generate ATP is called **cellular respiration**. This process is basic and essential to the metabolism taking place in all organisms. In prokaryotic cells, cell respiration occurs in the cytoplasm and the cell membrane, while in eukaryotic cells it occurs in the cytoplasm and mitochondria (see Figure 7.9).

Glucose is one of the most widely metabolized molecules for cellular respiration because the chemical bonds between carbon atoms are a major source of chemical energy. This energy can be used to generate ATP. As we will see, the cell respiration process can occur with or without oxygen gas.

Aerobic Respiration

The production of ATP using oxygen gas is called **aerobic respiration**. It occurs in three stages.

Stage 1: Glycolysis

The catabolism of glucose occurs through the process of **glycolysis** (*glycol* = "sweet," as in a sugar; *lysis* = break). **FIGURE 7.11** illustrates the pathway, so use the figure as a road map as we go through the key reactions. The process takes place in the cytoplasm where each of the 10 steps is catalyzed by a different enzyme.

During the "preparatory reactions" of glycolysis, each six-carbon (6C) glucose molecule is split into two 3C molecules. Then, in the "energy-harvesting reactions," the atoms in each of these 3C molecules are rearranged into a different 3C molecule called **pyruvate**. This is the end product of glycolysis.

At this point, the catabolism of glucose is complete. Notice in Figure 7.11 that for each glucose broken down, four ATP molecules are formed during the "energy-harvesting reactions." However, two ATP molecules were invested in the "preparatory reactions," so there is a net gain of two ATP molecules every time a glucose molecule is hydrolyzed into two pyruvates.

FIGURE 7.11 The Reactions of Glycolysis. Glycolysis is a metabolic process that converts glucose, a six-carbon (6C) sugar, into two 3C pyruvate products. In the process, two NADH coenzymes and a net gain of two ATP molecules are generated. Carbon atoms are represented by circles. Each enzyme-catalyzed reaction is represented by a numbered arrow.

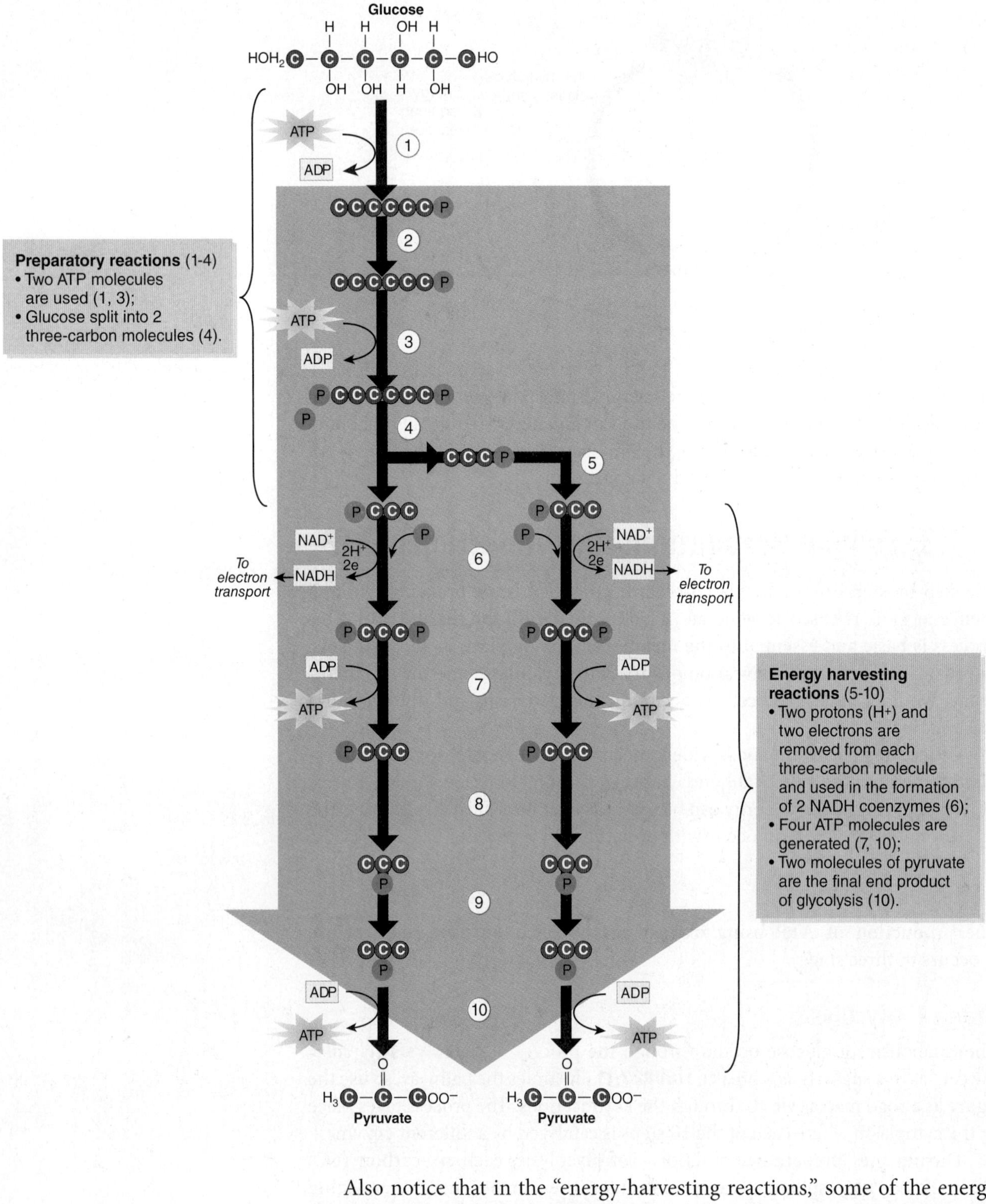

Also notice that in the "energy-harvesting reactions," some of the energy during the 3C rearrangements is transferred to the coenzyme NAD^+ forming NADH. Think of NAD^+ as an empty "dump truck" and NADH as a full dump truck carrying two electrons (e^-) and two hydrogen ions (H^+). We will see the importance of the coenzymes shortly.

The pyruvate molecules still have chemical energy locked up in the chemical bonds between carbon atoms, and this energy also can be harvested by further breaking down the pyruvates.

Stage 2: Citric Acid Cycle

Following glycolysis, the second stage of cellular respiration occurs. This stage is a cyclic pathway called the **citric acid cycle** (**FIGURE 7.12**). It is termed a "cycle" because the product formed at the end of the series of enzyme-controlled reactions serves as the starting point (the substrate) for another round of chemical reactions. In eukaryotic cells, the reactions of the citric acid cycle occur in the membranes of the mitochondria. In prokaryotic cells, which lack mitochondria, the reactions take place at the cell membrane.

The second stage begins with a "transition step" that connects glycolysis to the citric acid cycle. An enzyme removes a carbon atom from each pyruvate and releases that carbon as carbon dioxide (CO_2) gas (see box in Figure 7.12). The 2C remainder of each molecule bonds with a coenzyme called "coenzyme A." This change results in two 2C acetyl-coenzyme A (acetyl-CoA) molecules.

FIGURE 7.12 The Reactions of the Citric Acid Cycle. Each pyruvate first is broken down to a 2C molecule (transition step). This 2C molecule then enters the citric acid cycle and joins with a 4C molecule to form a 6C molecule. Each turn of the cycle releases CO_2, produces ATP, and forms NADH and $FADH_2$ coenzymes. Each enzyme-catalyzed reaction is represented by a lettered arrow.

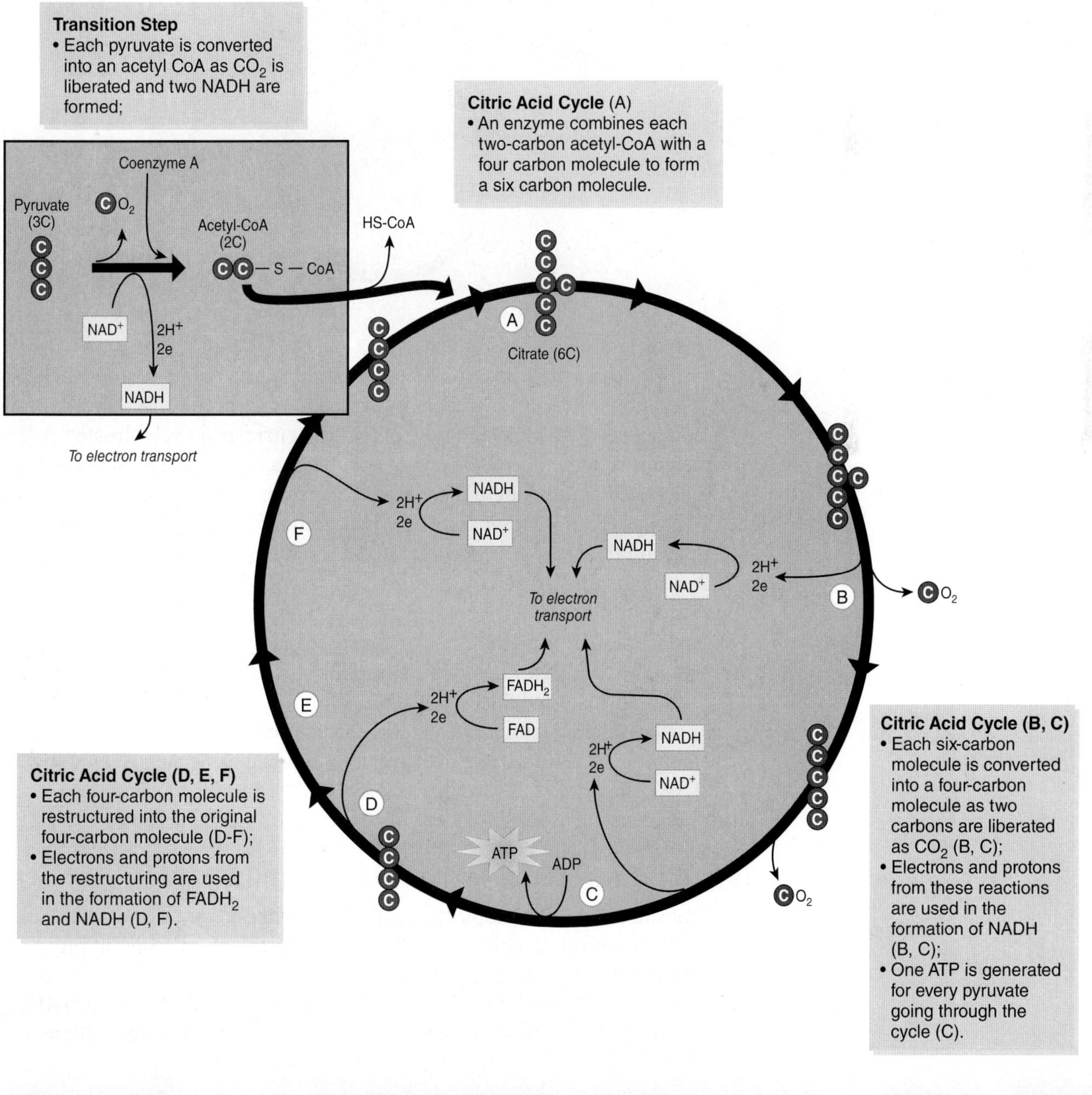

FIGURE 7.13 Summary of Glycolysis and the Citric Acid Cycle. Glycolysis and the citric acid cycle are the central metabolic pathways to extract chemical energy from glucose for generating a small amount of ATP and forming NADH and $FADH_2$ coenzymes.

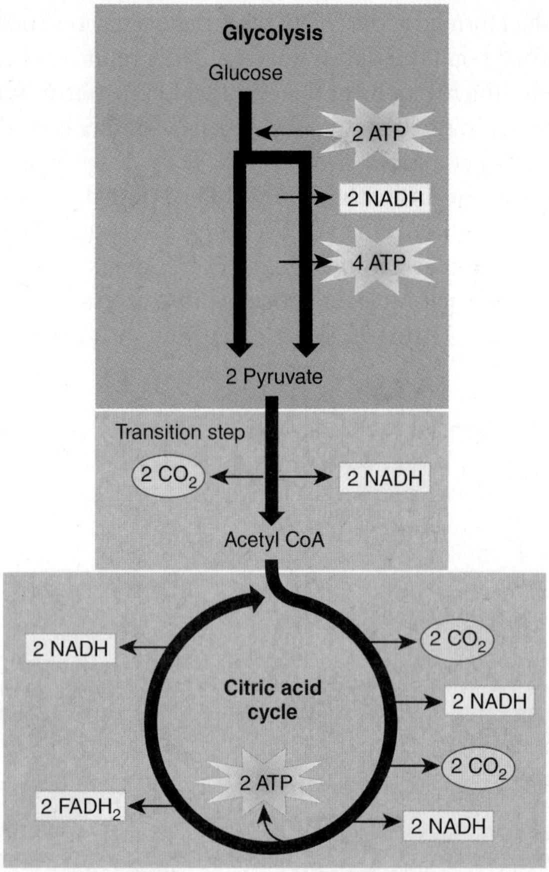

The importance of the transition step is that during the formation of each 2C molecule, another NADH molecule is formed.

Each acetyl-CoA molecule now enters the citric acid cycle. In step A, an enzyme connects the acetyl-CoA molecule with a 4C molecule to form a 6C molecule called citrate (or citric acid; thus, the naming of the cycle). In the next series of steps (B and C), separate enzymes convert the 6C molecule into a 5C molecule and then to a 4C molecule. Notice that the two carbon atoms are lost as CO_2 gas in the 6C to 4C conversion. Also, two ATP molecules are generated along with additional NADH and $FADH_2$ coenzymes.

As you follow around the cycle, you will see that the 4C molecule undergoes a series of chemical rearrangements (steps D and E) eventually reforming the 4C molecule that started the cycle (step F).

So, what is the significance of all these reactions of glycolysis and the citric acid cycle? Simply put, the breakdown of each glucose molecule generate 4 ATP molecules and 10 NADH and 2 $FADH_2$ coenzymes (**FIGURE 7.13**). The third and last stage of cellular respiration will use all the coenzymes to generate a substantial amount of ATP.

Stage 3: Electron Transport and ATP Synthesis

The final stage of aerobic respiration is called **electron transport**. This occurs in the mitochondria (in eukaryotes) and along the cell membrane (in prokaryotes; **FIGURE 7.14**). For aerobic respiration, this is the stage where O_2 is required.

In electron transport, the electron pairs carried by the NADH and $FADH_2$ are transferred to a series of compounds called **cytochromes**. These are iron-containing cell proteins that receive and pass along electron pairs in a biochemical "bucket brigade" fashion. The coenzymes and cytochromes are the vehicles through which the energy in electrons is released to produce ATP. Use Figure 7.14 as your road map (the steps are lettered A–G for easier tracking). Electron transport has two phases.

- **Electron Transfer.** First, the NADH and $FADH_2$ molecules pass their electron pairs to the first cytochrome (step A). In this way, NAD^+ and FAD molecules are renewed and returned to glycolysis and the citric acid cycle to be used again (step B). Then, the cytochrome "bucket brigade" involves the transfer of electron pairs from one cytochrome to the next. As transfer occurs, some energy is released (step C). The last cytochrome hands over the electron pairs to oxygen atoms, which now take on two H^+ from the surrounding environment to form a molecule of water (step D).

FIGURE 7.14 Electron Transport and ATP Synthesis in Bacteria. Originating in glycolysis and the citric acid cycle, coenzymes NADH and $FADH_2$ transport electron pairs to the electron transport chain in the cell membrane of microbes, which fuels the transport of H^+ across the cell membrane. The H^+ then reenter the cell through a protein channel in the ATP synthase enzyme. ADP molecules join with phosphates as protons move through the channel, producing ATP.

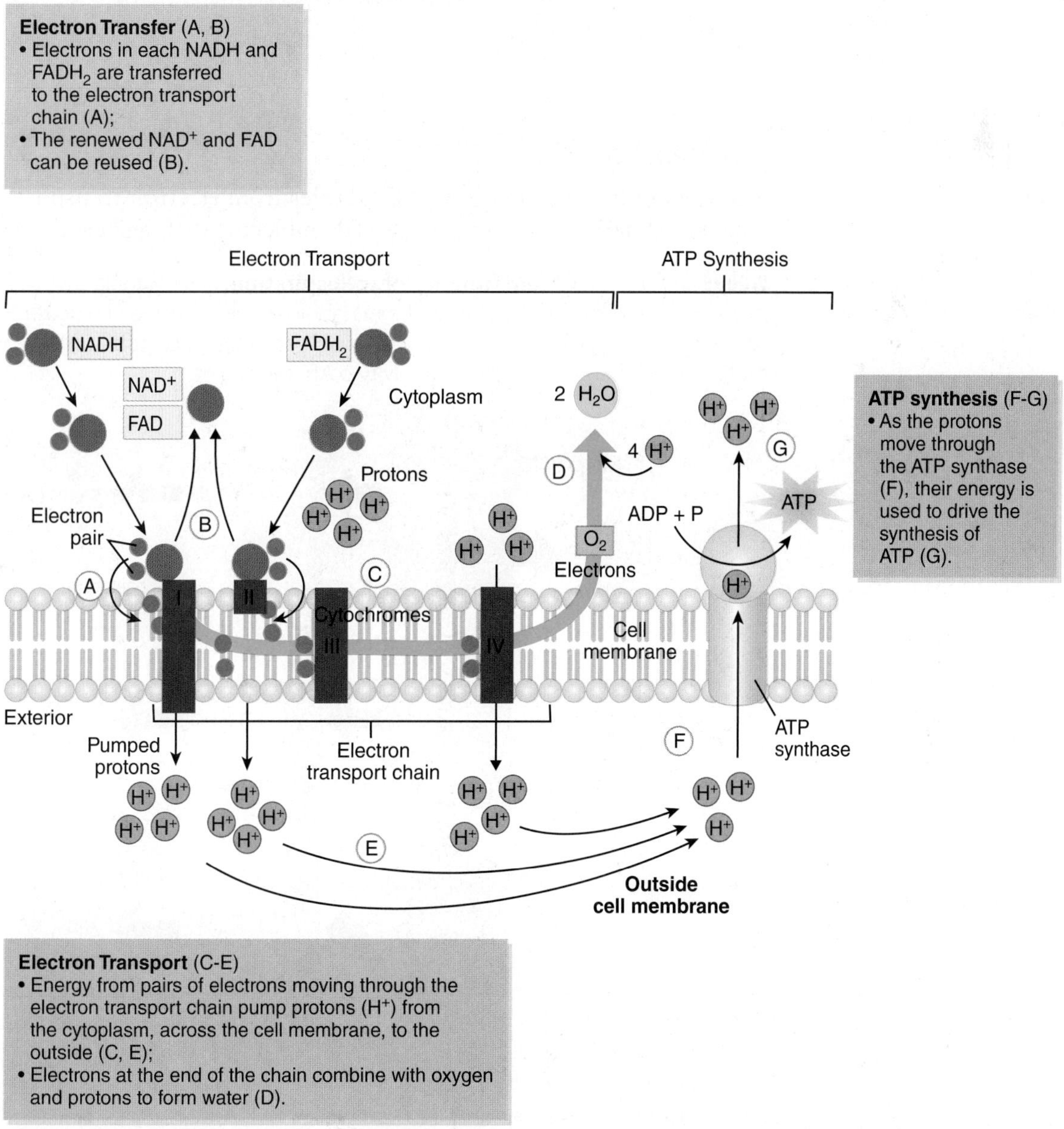

Let's pause for a moment and note the importance of oxygen in this process. In aerobic microbes, the oxygen atom is the only substance capable of accepting electrons at the end of the transport chain. If oxygen atoms are absent, the cytochromes could not pass on their electrons, and the series of electron transfers would back up and grind to a halt—the buckets would remain full of electrons, and the production of ATP would cease.

In the last phase of electron transport, we see the importance of moving all these electrons.

- **ATP Synthesis.** As the electrons pass among the various components of the electron transport chain, the energy released by the electrons is used to "pump" H^+ across the membrane (step E). This proton pumping results in a high concentration of H^+ on one side of the membrane and a low concentration on the other side. The unequal distribution results in H^+ flowing back through the membrane (step F). Importantly, this flow of H^+ occurs through special protein channels in the membrane containing the enzyme **ATP synthase**. During their flow, the H^+ provide enough energy to allow the ATP synthases to hook together ADP molecules and phosphate groups to form ATP molecules (step G).

  In summary, aerobic respiration uses glucose and oxygen gas to produce ATP.

$$C_6H_{12}O_6 + O_2 \rightarrow CO_2 + H_2O + ATP$$

The Final Score

The yield of life-sustaining ATP molecules from electron transport is considerable (**FIGURE 7.15**). For each NADH molecule that delivers its electrons

FIGURE 7.15 The ATP Yield from Aerobic Respiration. In a microbial cell, almost 38 molecules of ATP can result from the metabolism of a molecule of glucose. Each NADH molecule accounts for the formation of nearly three molecules of ATP; each molecule of $FADH_2$ accounts for roughly two ATP molecules.

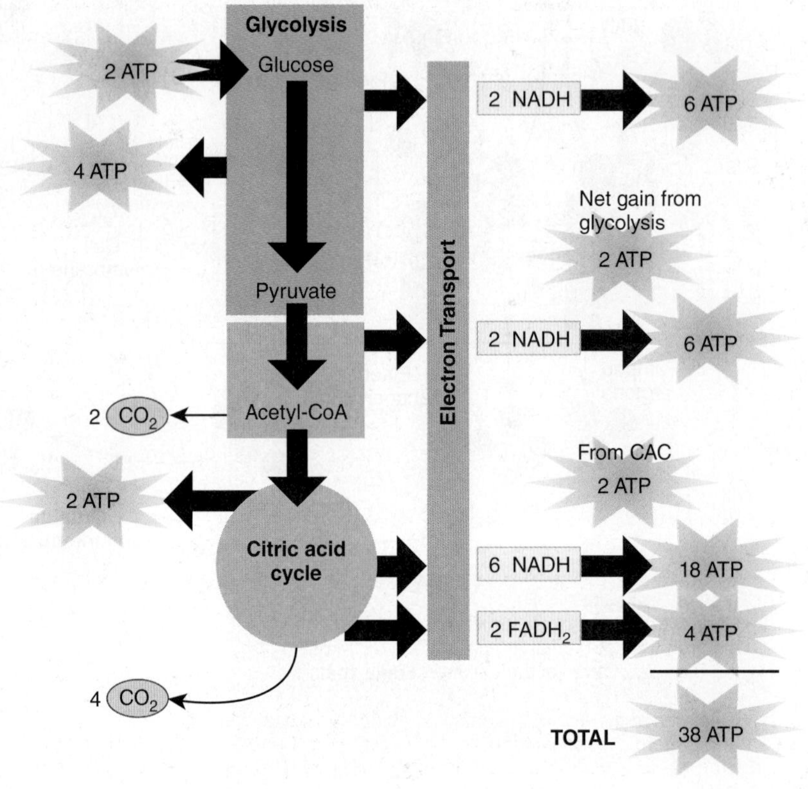

to the electron transport chain, enough energy is released to synthesize almost three ATP molecules. Each $FADH_2$ contributes enough energy to generate almost two ATP molecules in electron transport. So, from 10 NADH and 2 $FADH_2$ molecules about 34 ATP molecules are produced. Add to this the net gained two ATP molecules made directly in glycolysis and the two ATP molecules in the citric acid cycle, the final tally is some 38 molecules of ATP generated. Not bad for an initial investment of two ATP molecules!

We concentrated on glucose in this chapter because it is the richest source of chemical energy on Earth. However, other sugars, such as lactose and sucrose, and lipids also are used in microbial metabolism as sources of energy. These molecules use glycolysis and the citric acid cycle, although the starting materials might enter these pathways at different points.

It is interesting to note that oxygen was not always used for ATP synthesis on Earth, as **A CLOSER LOOK 7.1** explains.

A CLOSER LOOK 7.1

"It's Not Toxic to Us!"

It's hard to think of oxygen as a poisonous gas considering how many organisms, including humans, need it to survive today. Yet billions of years ago, oxygen was extremely toxic. One whiff by an organism and a cascade of highly destructive chemical reactions was set into motion. Death followed quickly.

Difficult to believe? Not if you realize that ancient ancestors of the members in the domains Bacteria and Archaea relied on anaerobic chemistry for their energy needs. The atmosphere was full of methane and other gases that they could use to generate energy. But no oxygen was present. And it was that way some 4 billion years ago.

Then, some 3 billion years ago, the cyanobacteria evolved. Floating on the surface of the oceans, the cyanobacteria trapped sunlight and converted it to chemical energy in the form of carbohydrates; the process was photosynthesis. However, there was a downside: oxygen was a waste product of the photosynthetic process—and it was deadly because the chemically active forms could disrupt cellular metabolism in other prokaryotes by "tearing away" electrons from essential cellular molecules. However, for the next few hundred million years the amount of oxygen gas in the atmosphere was minimal and was of no great consequence to other microorganisms.

About 2.4 billion years ago, there was a sudden and dramatic increase in the cyanobacteria population and in oxygen gas in the atmosphere. Unable to cope with the massive increase in the chemically active forms of oxygen being produced, enormous numbers of microbial species died and became extinct. Other species survive in oxygen-free environments, such as lake and deep-sea sediments where their present-day descendants still exist. The cyanobacteria survived in the open oceans because they evolved the enzymes to safely tuck away the chemically active forms of oxygen in a nontoxic form—that form was water.

FIGURE A **Stromatolites.** Shark Bay, Western Australia.

Among the survivors of these first communities were gigantic, shallow-water colonies called stromatolites. In fact, these structures—which look like rocks—still exist in a few places on Earth, such as Shark Bay off the western coast of Australia (see **FIGURE A**). These structures formed when ocean sediments and calcium carbonate became trapped in the microbial community. The top few inches in the crown of a stromatolite contain the photosynthetic cyanobacteria, whereas below them are other bacterial species that can also tolerate oxygen and sunlight. Buried beneath these organisms are other bacterial species that survive the anaerobic, dark niche of the stromatolite interior where neither oxygen nor sunlight can reach.

It still would be a couple of billion years before one particularly well-known species of oxygen-breathing creature evolved: *Homo sapiens*.

Anaerobic Respiration

As you discovered earlier in this chapter and in A Closer Look 7.1, many species of microbes live under anaerobic conditions; that is, in environments where there is little or no oxygen gas. These anaerobes still need to make ATP, so how do they do it without oxygen gas? It is through the process of anaerobic respiration.

Anaerobic respiration uses the glycolysis and the citric acid cycle pathways. However, anaerobic respiration employs something other than O_2 to accept electrons in the electron transport chain. For example, when it lives anaerobically, *E. coli* might use nitrate (NO^{3-}) as an electron acceptor in place of oxygen. This way, electrons in the electron transport chain keep moving and ATP production continues.

Other electron acceptors used by microbes include sulfate (SO^{4-}) and carbon dioxide (CO_2). When microbes use sulfate as electron acceptors, hydrogen sulfide (H_2S) gas is produced. Hydrogen sulfide has the odor of rotten eggs, and it gives a horrid smell to foods or soils where the microbes are living. When carbon dioxide is used as an electron acceptor, it is converted into methane gas (CH_4).

Anaerobic metabolism was a useful way of obtaining energy in the billions of years before oxygen filled Earth's atmosphere. Indeed, the anaerobes practicing this type of chemistry remind us that life today can exist in an oxygen-free environment. In fact, thousands of anaerobic species are found in the domains Bacteria and Archaea. They continue to exist deep in the soil, in marshes and swamps, in the ocean's depths, in the human gut, and in smelly landfills. Interestingly, sometimes the stink of gases like hydrogen sulfide can be an advantage for the microbe. **A CLOSER LOOK 7.2** explains.

7.4 Fermentation: A Metabolic "Safety Net"

Suppose you are a microbe and you find yourself in an environment without oxygen gas and without any of the other electron acceptors needed for anaerobic respiration. Without these, electron transport cannot function, NADH and $FADH_2$ coenzymes cannot be recycled, and ATP synthesis will quickly come to a halt. Death seems imminent—unless there is another way to generate ATP. In fact, some microbes have this ability, as they can carry out a process called fermentation.

Fermentation is an anaerobic process that some, but not all microbes can carry out. With fermentation, the pyruvate formed in glycolysis is not transformed into acetyl-CoA and sent through the citric acid cycle. Rather, pyruvate is converted by enzymes into other products, such as alcohols, acids, and CO_2. Fermentation thus provides a "safety net" by allowing some microbes to survive for a time in the absence of cellular respiration. Here is how fermentation works using yeast cells as an example.

Fermentation

Saccharomyces: sack-ah-roe-MY-seas

When yeast cells, such as *Saccharomyces*, metabolize glucose by glycolysis, they produce pyruvate (**FIGURE 7.16**). As the process proceeds, two ATP molecules is the net gain and two NAD^+ molecules are converted to NADH. To keep glycolysis going and producing two ATP molecules per glucose, the NADH coenzymes must be recycled to NAD^+. Normally, this would occur through the electron transport

A CLOSER LOOK 7.2

Microbes "Raise a Stink"

We are all familiar with body odor and bad breath. When one does not maintain a level of cleanliness or hygiene, bacterial species on the skin surface or in the mouth can grow out of proportion and, as they metabolize compounds like proteins, they produce noticeably unpleasant, smelly odors.

On a more environmental level, what about the smells often coming from rotting materials, such as in a landfill (see **FIGURE A**)? In fact, it is the microbes decomposing the garbage that produce the noxious odors. This got some marine biologists to thinking about such interactions in the oceans. When competing in marine environments with animal scavengers (such as crabs and fish) for food, do bacteria produce the odors to repel or deter animal species from consuming important food resources? Their hypothesis: decaying food resources become repugnant to scavenging crabs and fish.

To test their hypothesis, the research team baited crab traps near Savannah, Georgia, with menhaden, a typical bait fish for crabs. Some traps contained microbe-laden menhaden that had been allowed to rot for 1 or 2 days, while other traps contained freshly thawed samples having relatively few microbes. When the traps were inspected, those with fresh samples had more than twice the number of animals per trap than did the traps with microbe-laden menhaden. Lab studies with stone crabs showed they also avoided the microbe-loaded, rotting food, but readily consumed the freshly thawed, microbe-scarce menhaden.

This avoidance behavior by stone crabs was studied further. Some menhaden was allowed to rot in water without antibiotics. Other samples contained antibiotics in the water to prevent or inhibit microbial growth. Again, the crabs readily ate the antibiotic-incubated menhaden but avoided the menhaden without antibiotics. The presumption was that the presence of bacterial organisms in some way was responsible for the aversion.

Finally, the researchers took chemical extracts prepared from the microbe-loaded menhaden and mixed these with freshly thawed menhaden. Again, the crabs were repelled. Exactly what chemical compounds were responsible for the behavior were not evident from the study. However, it appears bacterial species not only act as decomposers in the environment, but they also can use chemical warfare—very successfully—to compete with relatively large animal consumers for mutually attractive food sources.

FIGURE A Smelly Landfill. A landfill can be a site where microbial decay is occurring, as the smell can confirm.

chain, but that process is not functional in this form of anaerobic metabolism. Therefore, pyruvate must be converted into some other product to reform NAD.

Fermentation End Products

For yeast cells, they have a gene that encodes the enzyme that converts pyruvate into ethyl alcohol (ethanol). In so doing, the chemical reaction regenerates NAD^+ for reuse in glycolysis, as shown in Figure 7.16A.

Fermentation by yeast cells has a special significance for humans because the end product is alcohol. If grape juice is the starting material, grape alcohol (wine) is the result and if barley is the starting material, barley alcohol (beer) results. In addition, CO_2 is liberated by *Saccharomyces* in the conversion of pyruvate to alcohol. This gas accounts for the bubbles appearing in champagne and beer and the ability of dough to rise when bread is made.

Numerous bacterial species also produce other fermentation end products (see Figure 7.16). For example, species of *Streptococcus* and *Lactobacillus* lack the enzymes to produce alcohol, but they do have the genes coding for enzymes to convert pyruvate into lactic acid. In the dairy industry, this acid converts condensed

FIGURE 7.16 Microbial Fermentation. Fermentation is an anaerobic process that converts NADH to NAD^+ by converting pyruvate into an end product. The presence of NAD^+ allows glycolysis to continue and generate a net gain of two ATPs for each glucose used.

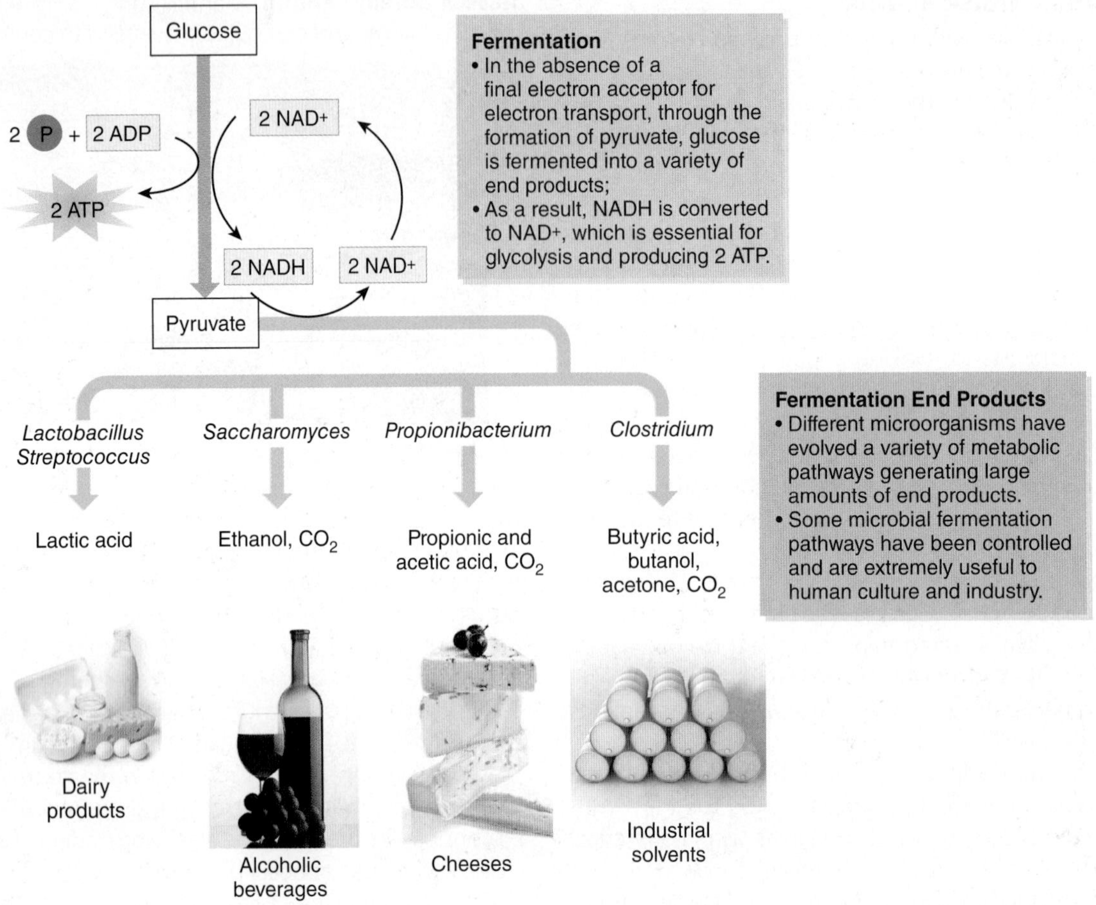

milk into yogurt. Many cheeses obtain their taste from the mixture of fermentation acids produced by other bacteria and molds during the ripening process. Industrially, such products as acetone, vitamins, many antibiotics, and numerous amino acids also are produced through microbial fermentation by other microbes. The chapter on Biotechnology and Industry details the processes behind these products.

In summary, microbes carrying out fermentation are just trying to stay alive until "better days" return and the electron acceptors for cell respiration again are present. Therefore, even though fermentation end products like alcohol and lactic acid might be important economically, they are of no significance to the microbes. Rather, by performing fermentation, the microbes have a steady supply of NAD^+ to run glycolysis and crank out a few life-saving ATP molecules.

7.5 Photosynthesis: An Anabolic Process

Cellular respiration and fermentation provide all living cells with the ATP energy they need for metabolism. However, much of the chemical energy to perform these processes comes from photosynthesis (see Figure 7.9). So, let's wrap up this chapter by briefly examining this process.

Photosynthesis is the anabolic activity that occurs on and near the cell membrane in the cyanobacteria and in the chloroplast of algae. These microbes have light-absorbing pigments to trap the energy in sunlight (light energy) and convert some of that energy into chemical energy (carbohydrates). Photosynthesis consists of the energy-trapping reactions and the carbon-trapping reactions **FIGURE 7.17**.

Energy-Trapping Reactions

In the **energy-trapping reactions** of photosynthesis, the energy in sunlight is harvested (trapped) by chlorophyll pigments that absorb light (step 1). The light "energizes" electron pairs, which are ejected from the chlorophyll molecules (step 2). The electrons move through an electron transport system, passing

FIGURE 7.17 Photosynthesis in Microbes. The energy-trapping reactions generate ATP and NADPH, while the carbon-trapping reactions use carbon dioxide gas (CO_2) along with the ATP and NADPH from the energy-trapping reactions produce glucose.

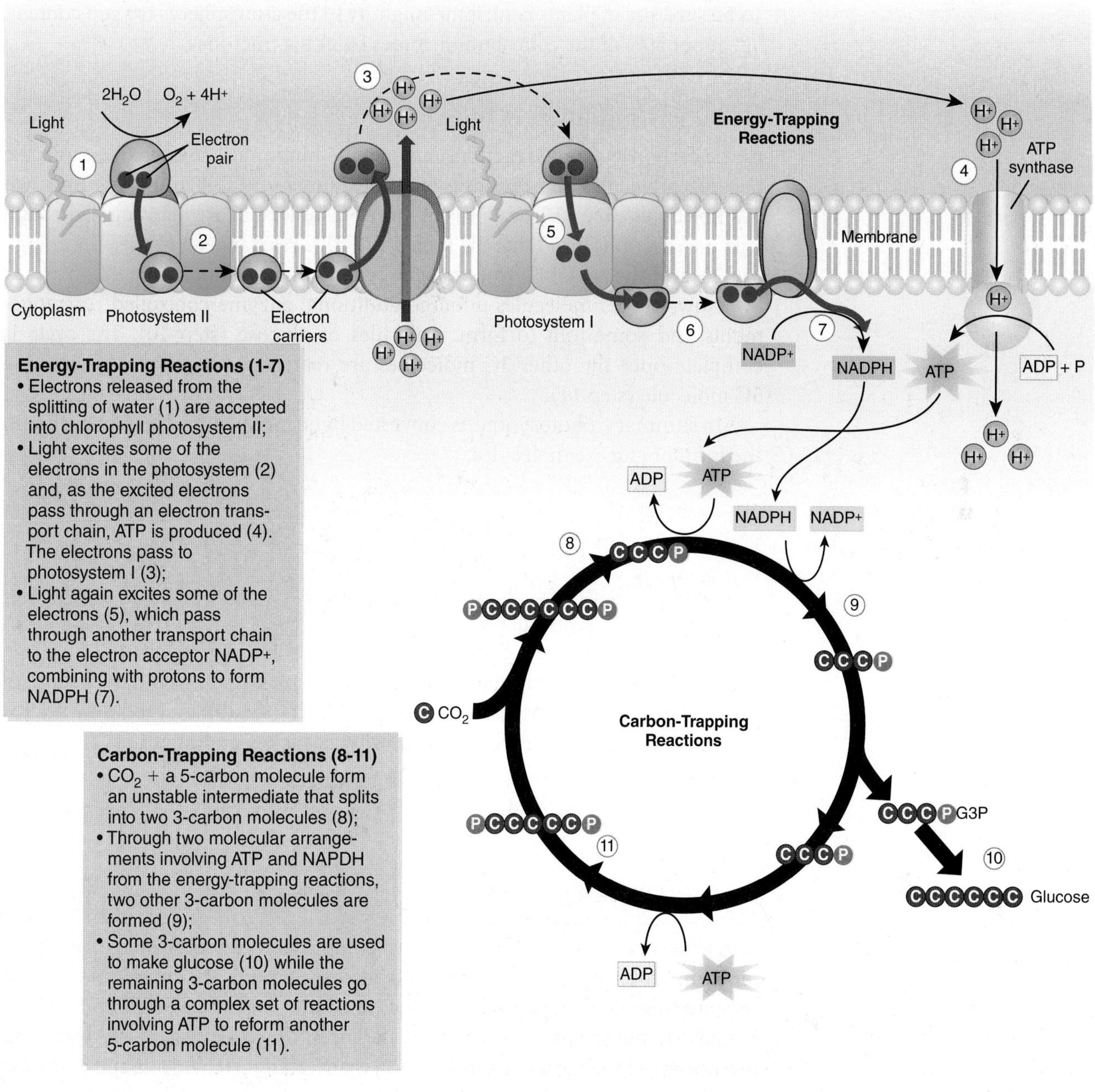

from one electron carrier to another, and ATP is formed in the process (steps 3 and 4). Once the electrons have passed through a series of cytochrome molecules, they are taken up by another coenzyme called **nicotinamide adenine dinucleotide phosphate** (**NADP**). The NADP molecule then takes on a H^+ and is converted to NADPH (steps 5-7). Thus, the two main products of the energy-trapping reactions are ATP and NADPH.

Before we leave the energy-trapping reactions, there is one extremely important reaction to mention. The energy-trapping reactions started with the loss of two electrons from a chlorophyll molecule (step 1). If the energy-trapping reactions are to continue, the chlorophyll must gain back the lost electrons. Notice in Figure 7.17A (step 1) that the cyanobacteria and algae obtain replacement electrons by splitting water molecules. Once the electrons have been incorporated, the remaining oxygen atoms from the water molecules are released to the atmosphere. Over billions of years of history, this oxygen has accumulated, and it now accounts for about 21% of the gaseous content of the atmosphere. It is the oxygen we breathe for cellular respiration that keeps our metabolism going. To be sure, green plants contribute mightily to the atmosphere's oxygen content, but about 50% of the contribution comes from the microbes.

Carbon-Trapping Reactions

In the carbon-trapping reactions, an enzyme binds a carbon dioxide molecule to a 5C molecule. The resulting 6C molecule is unstable and splits into two 3C molecules (step 8). The atoms in these 3C molecules are rearranged through enzyme reactions into other 3C molecules. This rearrangement uses the energy in ATP and the NADPH from the energy-trapping reactions (step 9).

Now, the 3C molecules undergo additional enzyme-controlled rearrangements and some join to form molecules of glucose (step 10). The cycle is complete once the other 3C molecules are rearranged back into the starting 5C molecule (step 11).

In summary, photosynthesis converted light energy into chemical energy in the form of glucose molecules.

$$6CO_2 + 6H_2O \xrightarrow{\text{Sunlight}} C_6H_{12}O_6 + 6O_2$$

These glucose molecules can be stored as starch and used later for glycolysis reactions and ATP production.

We have come to the end of our biochemical journey (whew!). It has been long and at times complicated, but if you have stuck it out, you have a sense of the chemical machinery of microbial life. It is not too difficult to know microbiology, but it is rather difficult to know what makes microbes tick. And that's what this chapter was all about—the chemical machinery hidden from the images we can see under a microscope or in a photograph.

A Final Thought

This chapter contains some of the most complicated concepts you will encounter in microbiology. Yet, the concepts also are among the most fundamental because this energy metabolism applies to all life—bacterial, plant, and human. Hopefully, you can step back and see the forest as well as the trees. Along with these observations comes the realization that certain mechanisms and pathways have been preserved in microbes, plants, and animals virtually intact through the billions of years of evolution. One of the great revelations of the 20th century was revealing the kinship of all living things. Indeed, one of the benefits of studying the metabolism of microbes is that one comes away with a better understanding of the biochemistry of all life, including humans.

Chapter Discussion Questions

What Was He Thinking?

From your reading of this chapter, identify and discuss five major points about microbial growth and metabolism that the author was trying to get across to you.

Questions to Consider

1. An organism is described as a facultative, halophilic mesophile. How might you translate this complex microbiological jargon into a description of the organism?
2. Like humans who expel carbon dioxide gas when exhaling, many microbial species also release carbon dioxide gas during metabolism. In both cases, where did the carbon dioxide gas originate in the human and microbial cells?
3. To prevent decay by bacterial species and to display the mummified remains of ancient Egyptian pharaohs, museum officials often place the mummies in sealed glass cases where oxygen has been replaced with nitrogen gas. Why do you think nitrogen is used?
4. One of the most important steps in the evolution of life on Earth was the appearance of cyanobacteria in which photosynthesis took place. Why was this critical?
5. A student goes on a college field trip and misses the microbiology exam covering microbial metabolism. Having made prior arrangements with the instructor for a make-up exam, she finds one question on the exam: "Discuss the interrelationships between anabolism and catabolism." How might you have answered this question?
6. ATP is an important energy source in all organisms, and yet it is never added to a microbial growth medium or consumed in vitamin pills or other growth supplements. Why do you think this is so?
7. From the material described in this chapter, why are vitamins like B_2 (riboflavin) and B_3 (niacin) essential to the human diet?
8. You have now studied many of the major physical factors that influence microbial growth and reproduction. Here on Earth, many microbes can survive these physical extremes (see Table 7.1). Could Martian microbes (assuming they are similar to those on Earth) survive the physical extremes on Mars? Here are some useful facts about Mars to help in your decision:
   - Surface temperature: Estimated to be from a warm 27°C (81°F) at the equator to −143°C (−225°F) at the winter polar caps.
   - Atmosphere: 95% carbon dioxide, but also present are nitrogen (2.7%), argon (1.6%), and oxygen (0.13%; 21% on Earth) gases. UV radiation is 3 times that on Earth.
   - Soil: Existence of water ice is confirmed; soil pH = about 7.7; several simple organic molecules identified, and chemicals have been found that could serve as nutrients for life forms, including magnesium, potassium, and chloride.
   - Salt: Dark, finger-like features seen on Mars could be the flow of salty water perhaps equivalent in salinity to Earth's oceans.

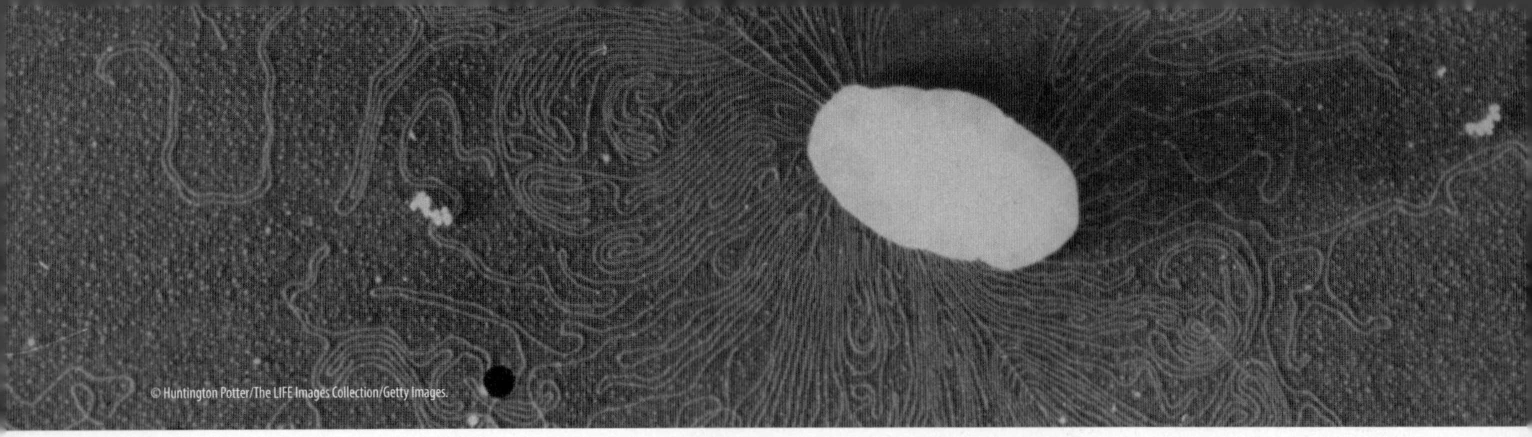

CHAPTER 8

The DNA Story: Chromosomes, Genes, and Genomics

Microbial Zombies?

Lobar pneumonia: A bacterial infection involving a large portion of one lobe or an entire lobe of the lungs.

Streptococcus pneumoniae: strep-toe-KOK-us new-MOH-nee-eye

In the early part of the 20th century, **lobar pneumonia** was deadly. Caused by *Streptococcus pneumoniae*, the disease was a common complication of the influenza pandemic of 1918–1919 (**FIGURE 8.1**). In fact, influenza and its complications took the lives of more than 50 million people worldwide. By 1928, lobar pneumonia still was a killer. In the United States alone, it killed more than 100,000 individuals that year.

With such a staggering mortality, Frederick Griffith, a British bacteriologist, was motivated to develop a vaccine to prevent pneumonia infections. His studies used two strains of *S. pneumoniae*. One strain produced large, smooth colonies (S strain) in agar culture, and the cells, when examined with the microscope, had a thick glycocalyx called a capsule. When Griffith injected S strain cells into mice, the animals soon died of pneumonia.

The second strain lacked a capsule and produced small, rough colonies (R strain) on agar. When injected into mice, the R strain cells were harmless, as no mice developed pneumonia. So, Griffith used heat to kill the pathogenic S strain cells, believing this might be a way to develop a vaccine. After these heat-killed cells were injected into mice, they all lived. None developed pneumonia, which made sense to Griffith.

Then, Griffith took the dead (heat-killed) S strain cells and mixed them with the live harmless R strain cells. The mixture was injected into mice. Griffith expected all the mice to live. Surprisingly, all the mice soon died from pneumonia. How could that be? Both the heat-killed S strain and the living R strain cells of *S. pneumoniae* were harmless as seen by his previous experiments.

CHAPTER 8 OPENER DNA is the molecule that codes the genetic information of all living organisms, including microbes, and some viruses. In this electron microscope image of an *Escherichia coli* cell, the cell has burst open, releasing its DNA. The DNA fiber seen in this image represents a single, circular molecule.

FIGURE 8.1 The Influenza Pandemic of 1918–1919. A digitally color-enhanced photo of an influenza ward at a U.S. Army Camp Hospital in France in 1918. Many influenza-infected patients ended up dying from pneumonia complications.

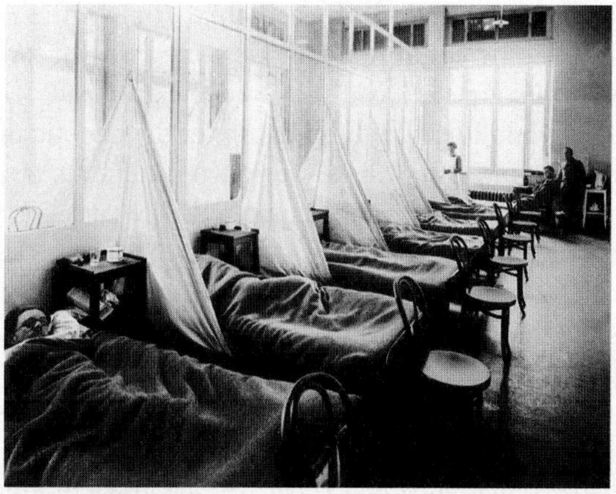

When Griffith cultured on agar a sample of the blood from the dead mice, he observed large numbers of living, S strain colonies. Moreover, the cells had a capsule! How could that be? All the S strain cells were dead when the experiment started. Did these bacteria return from the dead? Were they "zombie bacteria?" It all made no sense to Griffith.

In this chapter, we will uncover an explanation for Griffith's unresolved observations as we examine the role of **deoxyribonucleic acid** (**DNA**) and its functioning units, the **genes**. Then, we will explore how genes are decoded into the biochemical message in protein. Finally, we will see how scientists have used their understanding of DNA to develop the new discipline of genomics. It is a thrilling story of science and society, and one that is as current as today's headlines.

LOOKING AHEAD

After reading and completing this chapter, you will be able to:

8.1 Explain how the DNA molecule replicates itself.
8.2 Distinguish between transcription and translation.
8.3 Identify the different types of microbial genes that have been found in the human genome through comparative genomics.

8.1 DNA: The Hereditary Molecule in All Organisms

Until about 1952, scientists debated whether DNA or protein was the genetic material in living organisms. They knew **chromosomes** carried the genetic information and that it was composed of both DNA and protein. So, which component carried the genetic information? Many believed that DNA could not be the genetic material. How could the chemical complexity of any organism be determined by just the four nucleotides (letters) present in DNA. Proteins seemed a better possibility because it was composed of 20 amino acids (letters). By analogy, it would be impossible to build the entire English language from just four letters rather than the 26 in the English alphabet.

In 1928, Griffith published his work with the two strains of *S. pneumoniae*. He suggested that somehow R strain cells had transformed themselves into S strain cells. However, he could not provide a mechanism for the transformation. Then in 1952, two geneticists discovered the answer to Griffith's transformation experiment—and identified the nature of the genetic material.

Escherichia coli: esh-er-EE-key-ah KOH-lee

Alfred Hershey and Martha Chase were two geneticists trying to determine whether DNA or protein was the genetic material. They used *Escherichia coli* and **bacteriophages** (**phages**), viruses that infect bacterial cells. The phages use the bacterial cells as factories to make more viruses. Importantly, Hershey and Chase made use of the fact that the DNA in the core of the phage contains phosphorus atoms in its structure but no sulfur atoms. Also, the protein coat of the phage has sulfur atoms in some of its amino acids but no phosphorus atoms. They then produced phages that had radioactive phosphorus (called ^{32}P) in their DNA and radioactive sulfur (called ^{35}S) in their protein coat. These radioactive phages were mixed with *E. coli* cells and the investigators waited just long enough for new phages to be made in the infected cells (**FIGURE 8.2**).

When they examined the cells, only the ^{32}P radioactivity was detected inside the *E. coli* cells. The ^{35}S radioactivity remained outside the bacterial cells. Because the genetic information to make new phages would need to get into the

FIGURE 8.2 The Hershey–Chase Experiment. *Escherichia coli* cells and infecting phages were used to show that DNA is the genetic information in cells. ^{32}P is radioactive phosphorus and ^{35}S is radioactive sulfur.

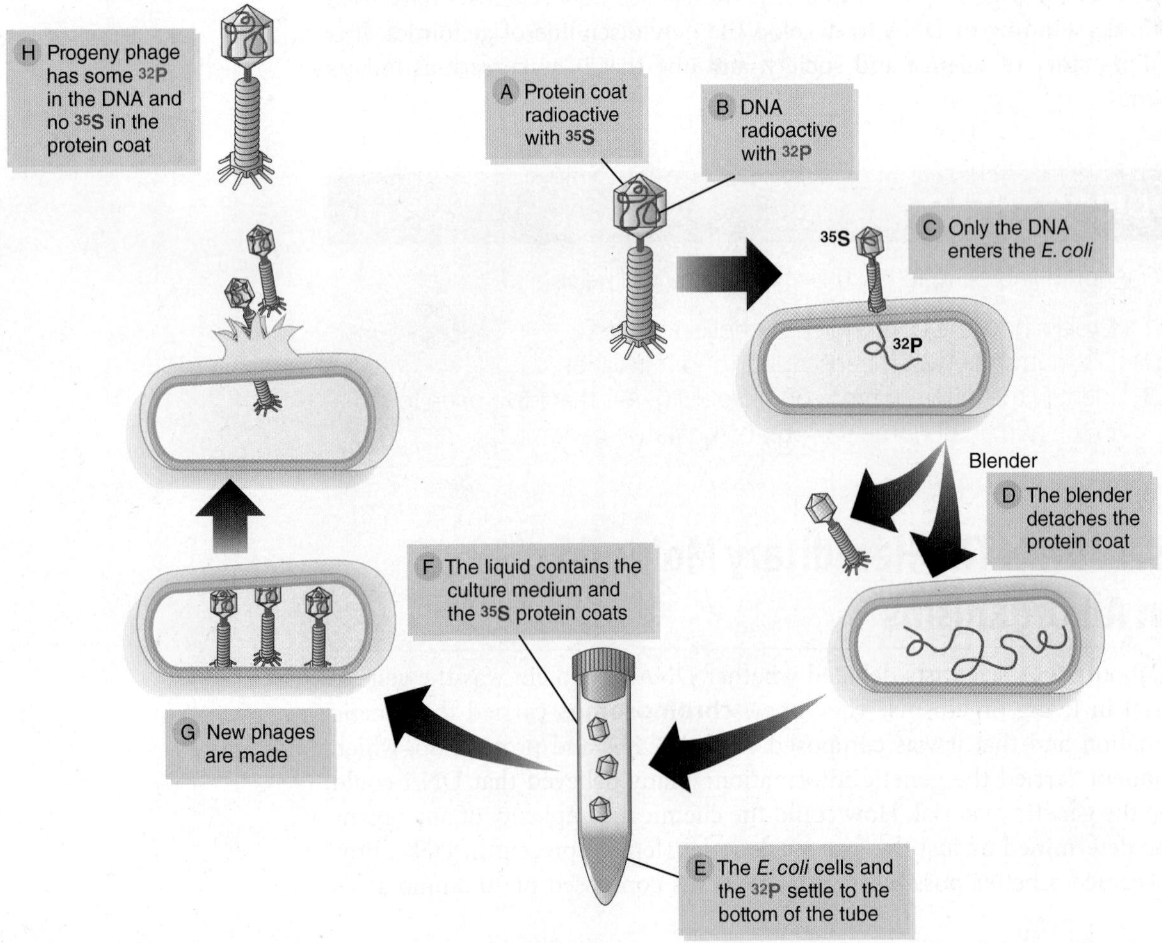

cytoplasm of the *E. coli* cells, Hershey and Chase concluded that virus DNA—and only DNA—is the substance responsible for producing more phages.

Certain experiments stand out as tipping points in scientific history. The Hershey–Chase experiments are an example. Their results had great influence on the biochemical thinking of the era because it solidified the understanding of DNA's role in heredity. Today, we know DNA is the hereditary information in all cells.

Their results also made it possible to interpret Griffith's transformation experiment. When heated, the S strain cells were destroyed and released their DNA as fragments into the liquid medium. Some fragments of that DNA were taken up by the R strain cells in the liquid, and the fragments were incorporated into the chromosome of R strain cells. That "new" DNA contained the genes to make a capsule. That was all that was needed to transform the harmless R strain cells into pathogenic S strain cells. No zombies needed! So, Griffith was ahead of his time in unknowingly showing that DNA carries the genetic information.

Once it was confirmed by other geneticists that DNA was the molecule of heredity, scientists needed to discover its chemical structure. By knowing the structure, that knowledge might explain how DNA directs and controls the metabolic and hereditary activities in cells.

The Double Helix

Since the 1920s, DNA was known to consist of three parts: a five-carbon sugar (deoxyribose), phosphate groups (PO_4), and a series of molecules called **nucleobases**, or just "bases," for short. The bases were **adenine (A)**, **thymine (T)**, **guanine (G)**, and **cytosine (C)** (**FIGURE 8.3A**).

DNA also was known to contain roughly equal proportions of phosphate groups, deoxyribose sugars, and bases. Therefore, scientists correctly concluded that DNA must be composed of these three components joined to one another as a unit. The unit was called a **nucleotide** with the base distinguishing one nucleotide from another (**FIGURE 8.3B**).

FIGURE 8.3 The Building Blocks of DNA. The structures of the four nucleobases found in DNA. The DNA nucleobases are bonded to a deoxyribose sugar, which in turn is bonded to a phosphate group. The complex is a nucleotide.

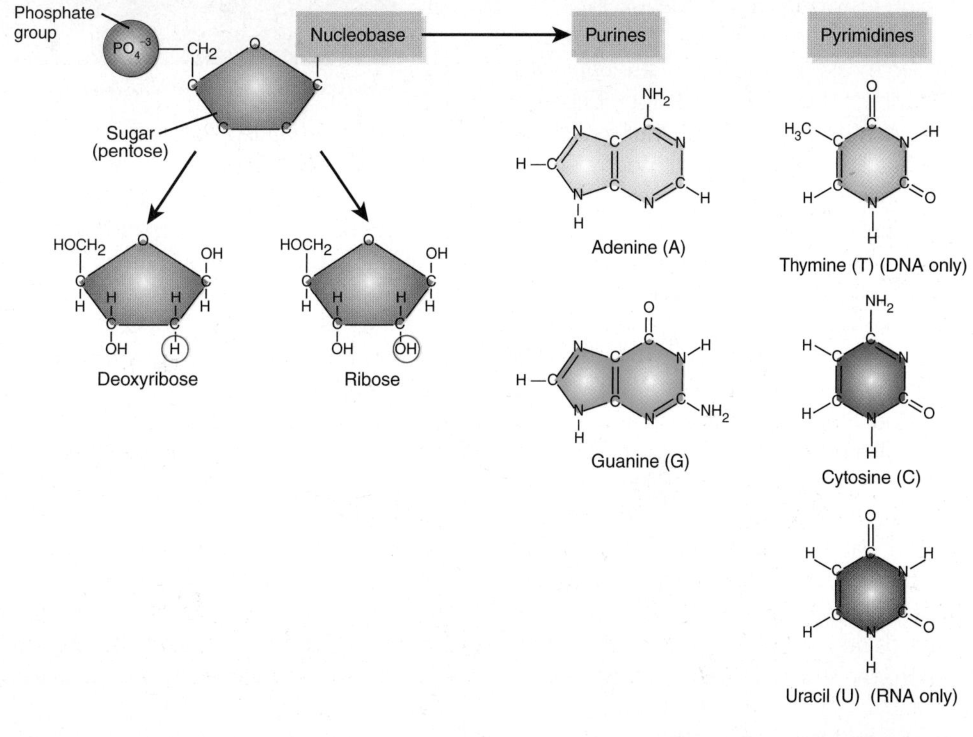

In the early 1950s, almost nothing was known about the spatial arrangement of the nucleotides in DNA. About this time, a new technique called X-ray diffraction became available. In X-ray diffraction, crystals of a chemical substance are showered with X rays. When the X rays are deflected (or diffracted) by electrons in the chemical substance, a characteristic pattern is recorded on a photographic plate. The pattern gives strong clues to the three-dimensional structure of the chemical substance.

Among the leading experts crystallizing DNA was the British biochemist Maurice Wilkins. Working with him was Rosalind Franklin, who used Wilkins' DNA preparation techniques to obtain clear X-ray diffraction patterns of DNA. **A CLOSER LOOK 8.1** presents the human-interest story concerning the race to discover the structure of DNA.

Watson and Crick concluded a base was joined to each deoxyribose sugar, hanging out as a side group from the chain, as **FIGURE 8.4A** illustrates. The whole

A CLOSER LOOK 8.1

The Tortoise and the Hare

We all remember the children's fable of the tortoise and the hare. The moral of the story was those who move along slowly and methodically (like the tortoise) will win the race over those who are speedy and impetuous (like the hare). The race to discover the structure of DNA is a story of collaboration and competition—a science "tortoise and the hare."

Rosalind Franklin (the tortoise) was 31 years old when she arrived at King's College in London in 1951. Having previously received a PhD in physical chemistry from Cambridge University, she moved to Paris where she learned the art of X-ray diffraction (see **FIGURE A**). At King's College, Franklin was part of Maurice Wilkins' group and she was assigned the task of using X-ray diffraction to work out the structure of DNA fibers. Her training and constant pursuit of excellence allowed her and a student, Raymond Gosling, to produce superb, high-resolution X-ray images of DNA.

Meanwhile, at Cambridge University, James Watson (the hare), an American postdoctoral student, was working with a British graduate student, Francis Crick, on the structure of DNA. Watson had a brash "bull in a china shop" attitude compared to Franklin's philosophy where you don't make conclusions until all the experimental facts have been analyzed.

Wilkins, falsely assuming that Franklin was his assistant, was more than willing to help Watson and Crick. Because Watson thought Franklin was "incompetent in interpreting X-ray photographs" and he was better able to use the data, Wilkins shared with Watson an X-ray image and report that Franklin had filed. From these materials, it was clear that DNA was a helical molecule. It is not known if Franklin also knew the structure at the time. Nevertheless, looking through the report that Wilkins shared, the proverbial "light bulb" went on when Watson and Crick saw what Franklin had missed: that the two DNA strands formed a double helix. This knowledge, together with Watson's ability to work out the base pairing, led Watson and Crick to their "leap of imagination." The structure of DNA was solved.

FIGURE A Rosalind Franklin inspecting the quality of a DNA sample.

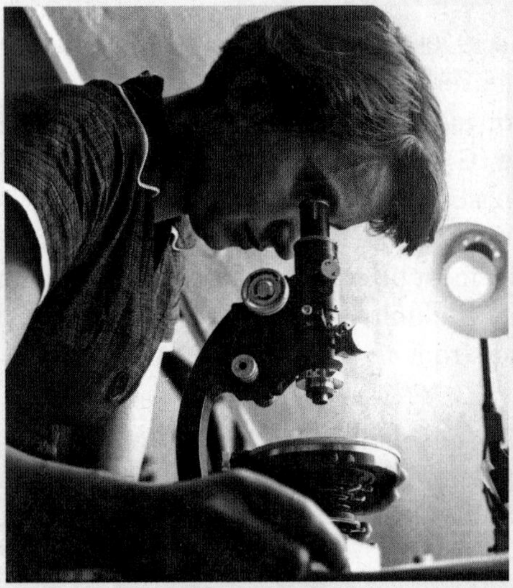

In her book entitled, *Rosalind Franklin: The Dark Lady of DNA* (HarperCollins, 2002), author Brenda Maddox suggests it is uncertain if Franklin knew DNA was a double helix. It was not in her character to jump beyond the data in hand. In this case, the leap of imagination won out over the methodical data collecting in research—the hare beat the tortoise this time. However, it cannot be denied that Franklin's data provided an important key from which Watson and Crick made the historical discovery.

In 1962, Watson, Crick, and Wilkins received the Nobel Prize in Physiology or Medicine for their work on working out the structure of DNA. Should Franklin have been included? The Nobel Prize committee does not make awards posthumously and Franklin had died 4 years earlier from ovarian cancer. So, if she had lived, did Rosalind Franklin deserve to be included in the award? But then who would have been left out? A Nobel Prize cannot have more than three recipients for any one award.

strand would represent a polynucleotide. Next, they knew from looking at Franklin's X-ray diffraction data that DNA was a **double helix**.

Through work done by others, Watson and Crick envisioned that for every A on one strand of DNA, there must be a T on the opposing strand (and vice versa). This would be the "complementary" base. Similarly, for every G on one strand, there must be a complementary C on the other strand (and vice versa). Weak chemical bonds formed between the complementary bases held the two strands together as the double helix (**FIGURE 8.4B**).

Watson and Crick expressed their hypothesis in a groundbreaking series of three papers published in 1953. The scientific community hailed their work as a great scientific leap forward, another tipping point in scientific breakthroughs. The published results made it possible for geneticists to postulate how DNA replicates and passes the genetic information on to the next generation.

FIGURE 8.4 The DNA Molecule. **(A)** The nucleotides in a segment of one strand **(polymucleotide)** of DNA. **(B)** A stylized model of DNA. The molecule consists of a double helix of two intertwined strands. Complementary bases in each strand extend out forming A-T and G-C base pairs. The structure to the right is what a DNA molecule might look like in three-dimensions.

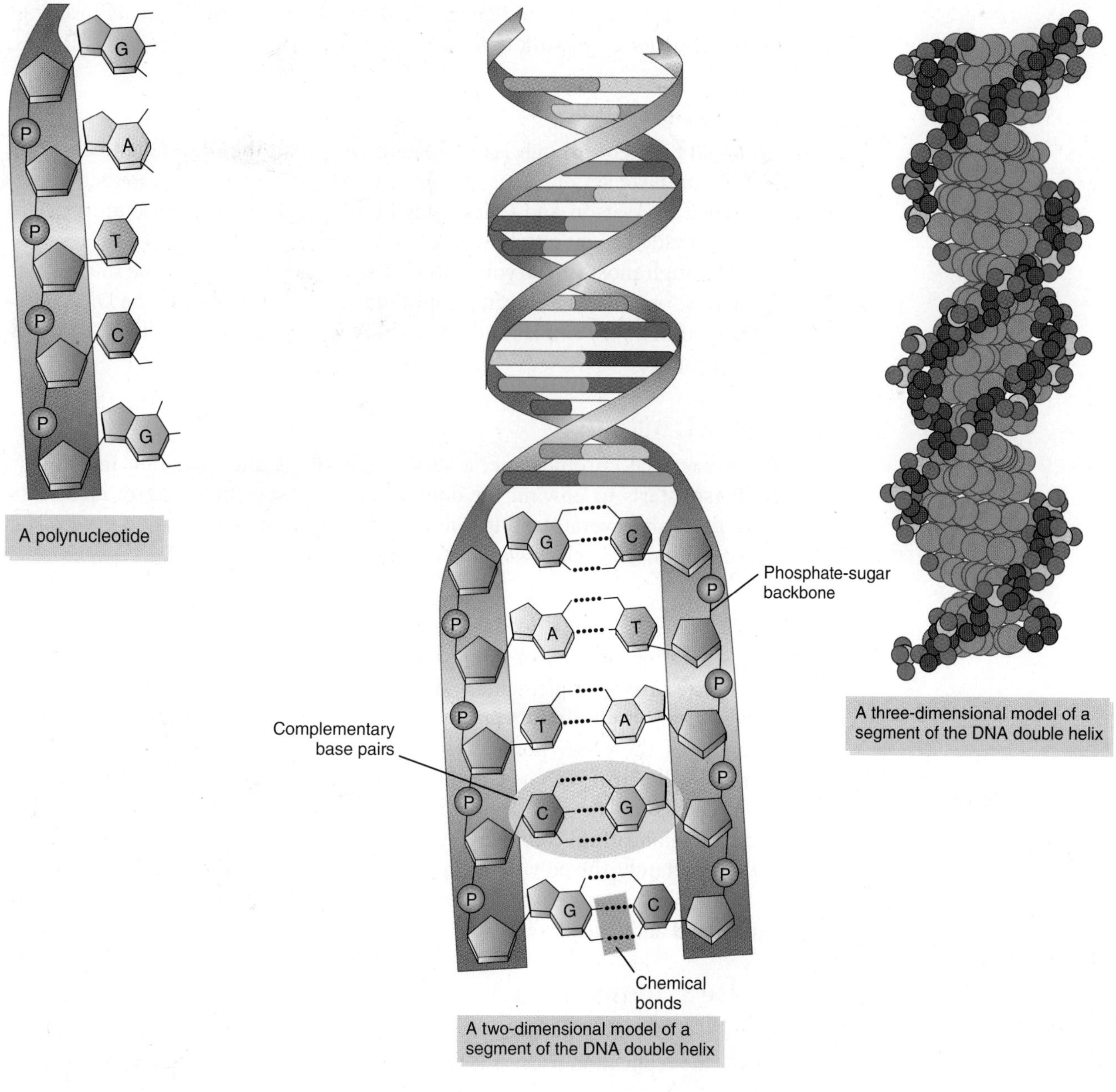

FIGURE 8.5 The Nobel Prize in Physiology or Medicine, 1962. Pictured in this 1962 photo are Nobel Prize awardees Professor Maurice Wilkins (far left), Professor Francis Crick (third from left), and Professor James Watson (second from right), each displaying their diplomas after formal ceremonies in Stockholm's Concert Hall. Note: the gentleman between Crick and Watson is author John Steinbeck, who was awarded the Nobel Prize in Literature.

In 1962,Watson, Crick, and Maurice Wilkins won the Nobel Prize in Physiology or Medicine for their work (**FIGURE 8.5**).

DNA Replication

When all cells go through asexual cell division, all the descendant cells must have a complete set of the genetic instructions. In one of the research papers published by Watson and Crick, they pointed out that the structure of DNA might provide insight into how DNA copies (replicates) itself.

Although most prokaryotic DNA is a circular molecule while the DNA in eukaryotic organisms is linear, prior to the cell division process, the DNA replicates itself in a very similar and precise way (**FIGURE 8.6**). Basically, it is a three-phase process.

Phase 1: Initiation

For a bacterial chromosome, a series of proteins and enzymes, including a **helicase**, starts to unwind the double helix at a specific point (in eukaryotic cells, there are several starting points). The initiation of replication proceeds in two directions, one clockwise and one counterclockwise.

Phase 2: Elongation

The separation of the two strands allows each strand to act as a template to synthesize a new complementary strand. An enzyme, called **DNA polymerase**, moves along each strand, matching and joining complementary bases. For example, if the DNA polymerase "sees" an A on one template strand, the enzyme will pair it with its complementary T. The DNA polymerases continue along each template strand and add more complementary bases. The result is a growing (elongating) complementary strand along each template.

Importantly, the polymerase also acts as a "proofreader"; that is, if an incorrect complementary base is added, usually the enzyme quickly removes the "typo" and adds the correct base.

Phase 3: Termination

Once the enzymes have reached the opposite side of the circular bacterial DNA, the replication process is terminated. The bacterial cell now contains two

FIGURE 8.6 DNA Replication. The complementary bases found in the two strands of DNA suggest each strand can serve as a template for a complementary strand. The three phases in the process are initiation, elongation, and termination.

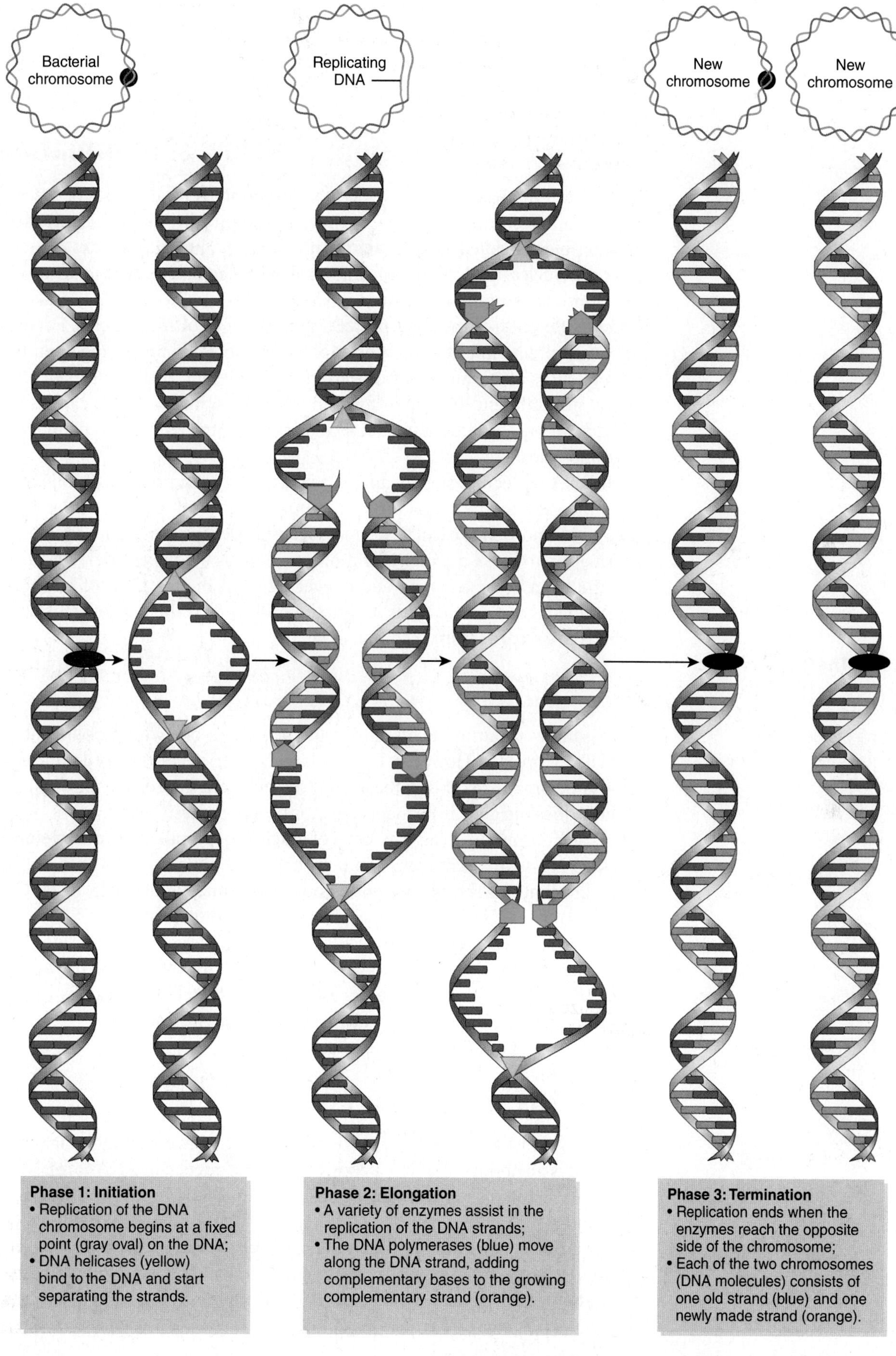
Bacterial chromosome
Replicating DNA
New chromosome
New chromosome
Phase 1: Initiation
• Replication of the DNA chromosome begins at a fixed point (gray oval) on the DNA;
• DNA helicases (yellow) bind to the DNA and start separating the strands.
Phase 2: Elongation
• A variety of enzymes assist in the replication of the DNA strands;
• The DNA polymerases (blue) move along the DNA strand, adding complementary bases to the growing complementary strand (orange).
Phase 3: Termination
• Replication ends when the enzymes reach the opposite side of the chromosome;
• Each of the two chromosomes (DNA molecules) consists of one old strand (blue) and one newly made strand (orange).

independent, circular chromosomes, which ensure that when the cell divides, each cell will have a complete set of genetic information.

This method of making new DNA molecules is called **semiconservative replication**, because one strand of each chromosome corresponds to the old, "conserved" (template) strand and represents half (*semi* = "half") of each DNA molecule.

8.2 Gene Expression: The Flow of Information

Proteins are the working and structural components of all cells. As described in the chapter on the Molecules of the Cell, chemically, they are composed of building blocks called **amino acids** (much as the building blocks of nucleic acids are nucleotides). Only 20 different amino acids are needed to produce the countless combinations found in the proteins of all cells. What makes one protein different from another protein is its sequence of amino acids. For example, two proteins might each consist of 33 amino acids linked together, but if the sequences of the amino acids (letters) are different, then the proteins (sentences) are different. Consider, for a moment, how many words can be composed from our 26-letter alphabet. For organisms, the alphabet is composed of 20 amino acids.

If DNA molecules specify amino acid sequences in proteins, it follows that some message in the DNA molecule must specify an amino acid sequence. Today, we know that each DNA message represents a gene and contains the information to make a protein (or more specifically a polypeptide). Let's look at how the DNA code gets read and converted from the language of nucleotides to the language of amino acids. In other words, how is the flow of information, called **gene expression**, accomplished?

FIGURE 8.7 The RNA Molecule. Ribonucleic acid (RNA) is a single-stranded molecule consisting of nucleotides containing a ribose sugar and the nucleobases adenine, guanine, cytosine, and uracil.

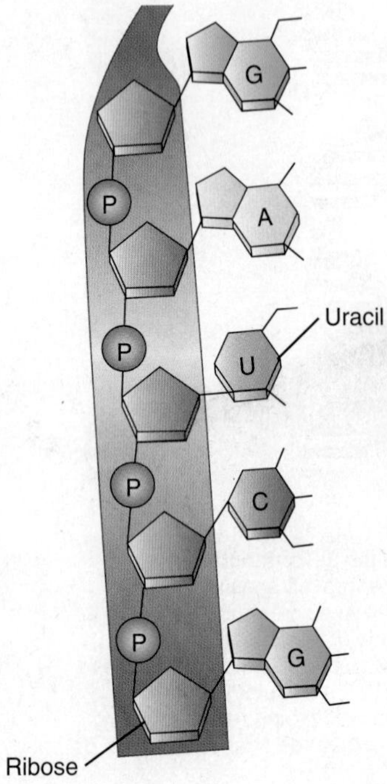

Ribonucleic Acid

The assembly of amino acids into proteins occurs on **ribosomes** located in the cell's cytoplasm. However, DNA is largely restricted to the cell's nucleus in eukaryotic organisms and to the area of the nucleoid in prokaryotic organisms. So, how does the gene information get to the ribosomes?

All cells contain a second type of nucleic acid called **ribonucleic acid (RNA)**. RNA is different from DNA in that RNA is (1) a single-stranded molecule; (2) contains the sugar ribose; and (3) has nucleotides with A, G, C and **uracil (U)** bases (**FIGURE 8.7**). Thymine bases are not present.

RNA is the "go-between" molecule that will get the gene information (DNA) to the ribosomes. The process of going from DNA to RNA is called **transcription** (**FIGURE 8.8**). The sequence of bases in each RNA then is read by the ribosomes to make a protein. This process is called **translation**.

Let's look at each of these processes.

Making RNA

If you transcribe a sentence, you make a copy of it. That essentially is what transcription is all about—making a copy of a gene sequence in RNA form.

The process of transcription begins with an uncoiling of the DNA double helix and a separation of the two DNA strands within a gene (**FIGURE 8.9**). The synthesis of RNA starts at one point, called the **promoter**, which is found on only one of the two separated strands of the gene. On that template strand, an enzyme called **RNA polymerase** will match and join complementary bases. For example, if the RNA polymerase "sees" an A on the template strand, the enzyme will pair it

FIGURE 8.8 Gene Expression. The flow of genetic information proceeds from DNA to RNA (transcription) and from RNA to protein (translation).

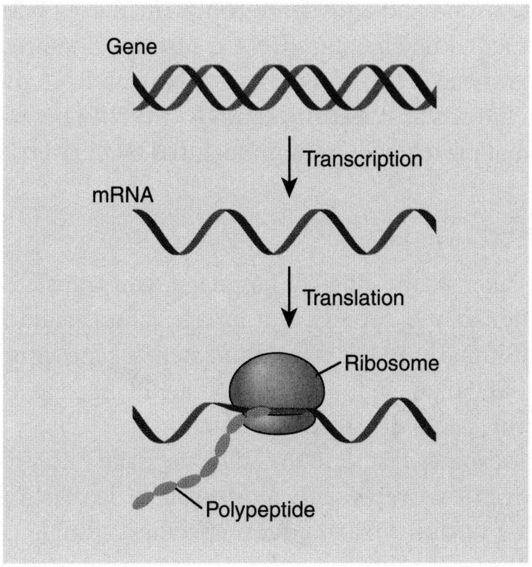

FIGURE 8.9 The Transcription Process. The enzyme RNA polymerase moves along the template strand of the DNA and synthesizes a complementary molecule of RNA using the base code of DNA as a template. The mRNA will carry the genetic message to the ribosomes, where translation occurs. Note that the nontemplate DNA strand of the gene is not transcribed.

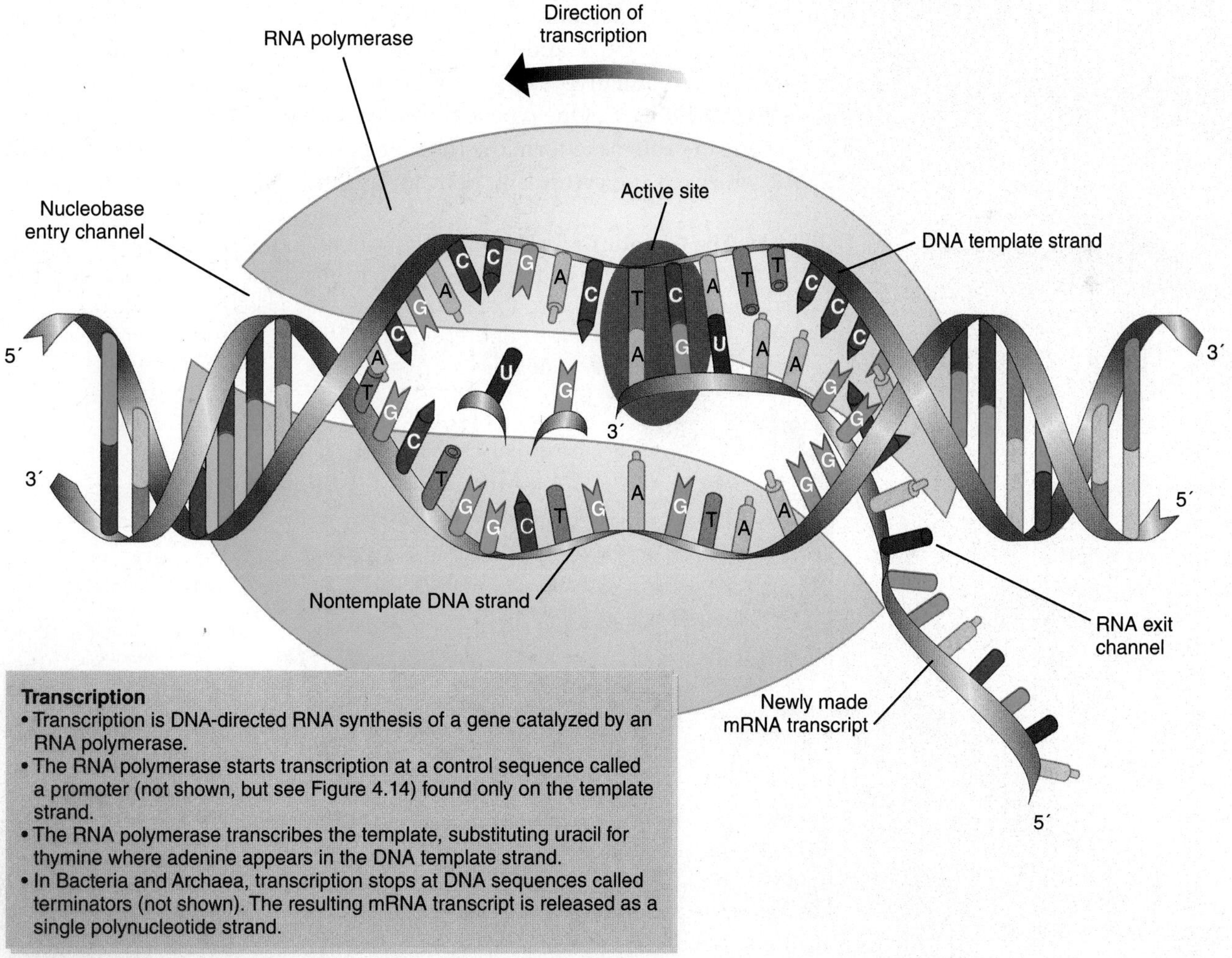

Transcription

- Transcription is DNA-directed RNA synthesis of a gene catalyzed by an RNA polymerase.
- The RNA polymerase starts transcription at a control sequence called a promoter (not shown, but see Figure 4.14) found only on the template strand.
- The RNA polymerase transcribes the template, substituting uracil for thymine where adenine appears in the DNA template strand.
- In Bacteria and Archaea, transcription stops at DNA sequences called terminators (not shown). The resulting mRNA transcript is released as a single polynucleotide strand.

with a complementary U on the new RNA (remember: RNA does not contain thymine nucleotides; G and C normally base pair). The RNA polymerase then moves along the template strand and adds more complementary nucleotides, forming a growing (elongating) complementary RNA transcript. So, through transcription, the base code in the gene is copied as a base code in an RNA molecule. The RNA so constructed is known as **messenger RNA** (**mRNA**) because it contains the "message" as to what protein will be manufactured by the ribosomes.

Making a Protein

If, as a native speaker of the English language, you translate a sentence from say Chinese to English, you convert that foreign language into one's own that is useful for others. Similarly, in the translation step of gene expression, ribosomes convert a nucleotide language into an amino acid language in the form of proteins that are useful (necessary!) for cell function.

So, how do ribosomes translate the RNA message into protein? The answer to this is that each mRNA represents a **genetic code** existing in sets of three-letter blocks called **codons**. Each codon specifies a single amino acid in the protein to be synthesized.

Using the set of four nucleotides in three-letter blocks (codon), 64 possible codons exist (for example, AUA, GAU, GCG, CAA, and so on), which is more than enough to specify the 20 amino acids (**FIGURE 8.10**). The genetic code is nearly universal because the same codon usually specifies the same amino acid regardless of whether the organism is a bacterium, a mushroom, or a human. Also note in Figure 8.10 that there is a redundancy in the genetic code. Often there is more than one codon specifying the same amino acid. For example, codons UCU, UCC, UCA, and UCG all code for the amino acid serine.

There are two other types of RNA that are transcribed from DNA but never translated (**FIGURE 8.11**). One type is **ribosomal RNA** (**rRNA**), which combines with a group of proteins to form the ribosomes. The other type is **transfer RNA** (**tRNA**), which, in the cytoplasm, delivers specific amino acids to a ribosome

FIGURE 8.10 The Genetic Code Decoder. The genetic code embedded in an mRNA is decoded by knowing which codon specifies which amino acid. Use the column to the left to find the first letter of the codon; then find the second letter from the top row; finally, read up or down from the right-most column to find the third letter. The three-letter abbreviations for the amino acids are given.

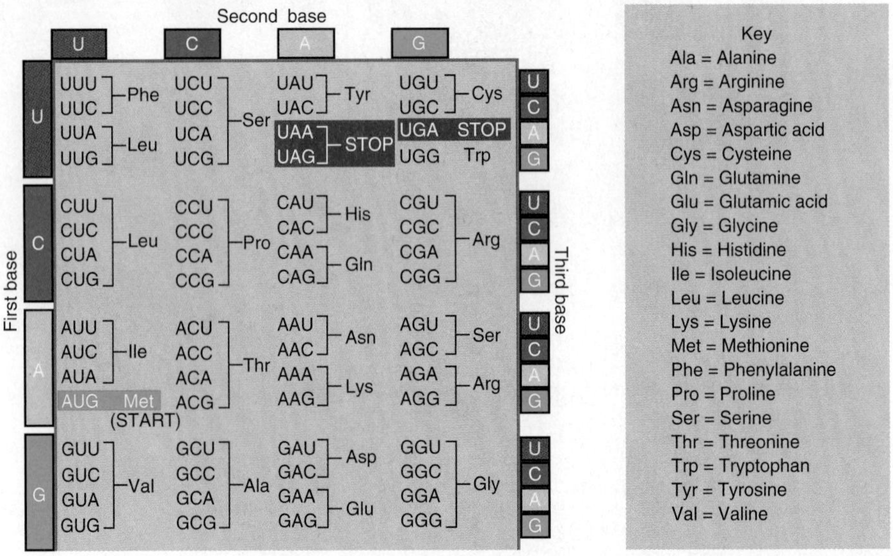

for use in translation. The amino acid binds at one end of the tRNA molecule, as shown in Figure 8.11. At another location on the tRNA, there is a sequence of three bases called the **anticodon**. The matching of complementary codon (on mRNA) and anticodon (on tRNA) is essential for the correct positioning of an amino acid during translation.

As translation occurs, only a portion of the mRNA molecule contacts the ribosome at a given time. **FIGURE 8.12** shows a moment in the translation process.

While the mRNA-ribosome complex is forming, amino acids are binding to their specific tRNA molecules in the cytoplasm. Then, the tRNA molecules transport the amino acids to the ribosome/mRNA complex. Next, as one

FIGURE 8.11 The Transcription of the Three Types of RNA. Different genetic sequences in the DNA contain the information to produce mRNAs, rRNAs that along with protein builds the ribosomes, and tRNAs that carry one of the 20 amino acids needed for translation.

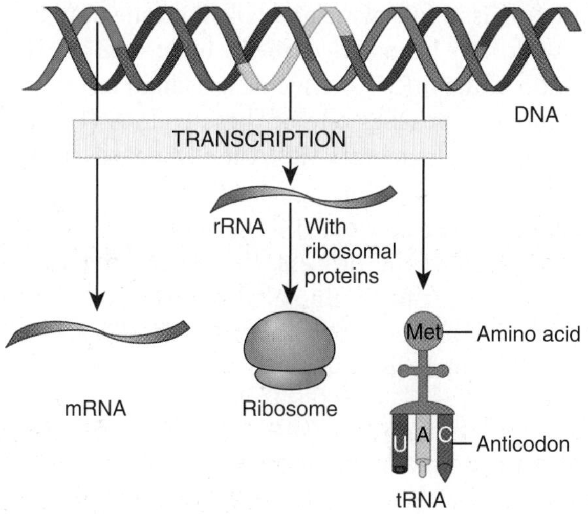

FIGURE 8.12 The Translation Process. The messenger RNA moves to the ribosome, where it is met by transfer RNA molecules bonded to different amino acids. The tRNA molecules align themselves opposite the mRNA molecule and bring the amino acids into position. A peptide bond forms between adjacent amino acids on the growing protein chain, after which the amino acid leaves the tRNA. The tRNA returns to the cytoplasm to bond with another amino acid molecule.

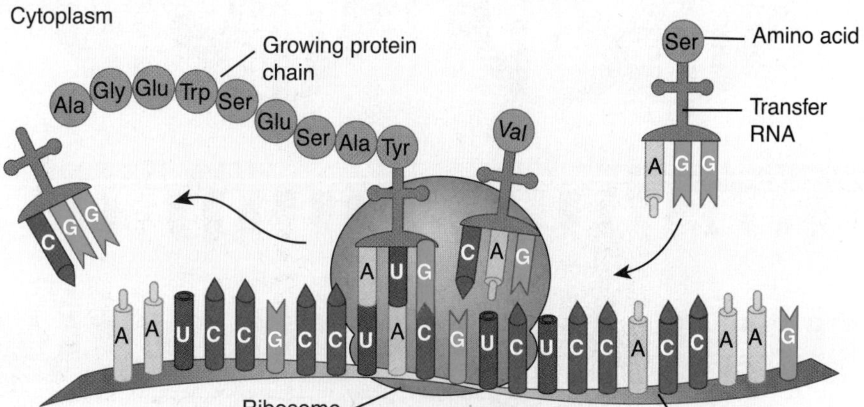

codon of the mRNA molecule is read by the ribosome, the tRNA with the complementary anticodon pairs up. The pairing brings that amino acid into position. Within the ribosome, a second tRNA molecule (holding a second amino acid) complementary pairs its anticodon with the second codon. Two tRNA molecules now have positioned their amino acids next to each other. Next, an enzyme joins the two amino acids to form a dipeptide (two amino acids joined together). The first tRNA molecule then is released from its amino acid and moves back into the cytoplasm to bind to another identical amino acid.

The ribosome now moves to the next codon of the mRNA molecule. The codon attaches its complementary anticodon on a tRNA molecule, which brings a third amino acid into position. As before, an enzyme chemically joins the first two amino acids to the newest amino acid to form a tripeptide (three amino acids in a chain). The second tRNA is then released back into the cytoplasm. And on it goes, codon after codon pairing with its anticodon, and amino acid after amino acid is positioned and joined to the growing peptide chain. The genetic message of DNA is being translated via mRNA into an amino acid sequence. Translation is in full swing and will continue until hundreds or thousands of amino acids have been added one by one to the growing chain.

The final codon of the mRNA molecule is a chain terminator, or "stop" signal. When the ribosome reaches one of these codons (UAA, UAG, or UGA; see Figure 8.10), no complementary tRNA molecule exists and thus no amino acid is added to the chain. The stop signal activates release factors that free the amino acid chain from the ribosome and translation concludes.

With the conclusion of the process, the genetic message encoded in the DNA has been expressed as the amino acid sequence in a polypeptide or protein. This extraordinary process, summarized in **FIGURE 8.13**, is one of the key underpinnings of microbial life—indeed, of all life.

FIGURE 8.13 Summary View of Protein Synthesis. The DNA molecule unwinds, and one strand is transcribed as a molecule of messenger RNA. The mRNA then operates as a series of codons, each codon having three bases. During translation, a codon specifies a specific amino acid for placement in the growing polypeptide. Note that codons 6 and 7 are identical and specify the same amino acid, alanine. Similarly, codons 2 and 5 are identical and encode glycine.

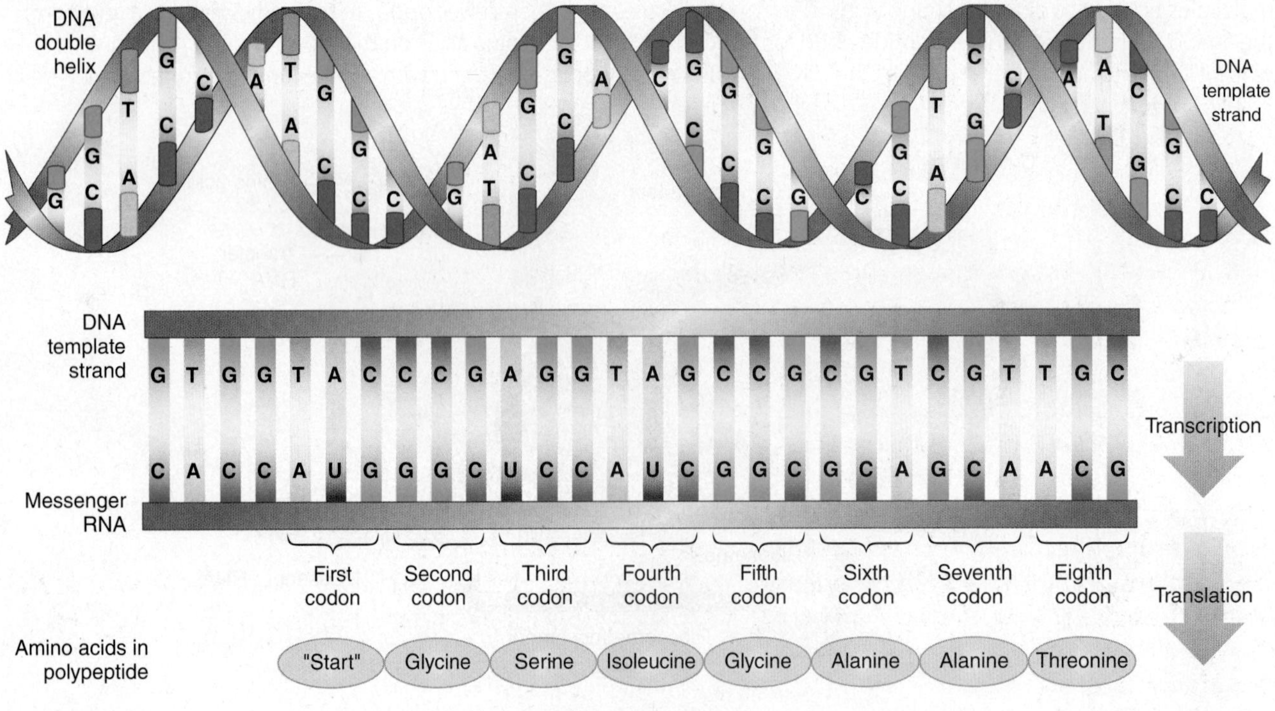

Gene Expression: Controlling the Flow of Information

The last topic of gene expression to address is how genes are regulated. It would be a waste of energy for cells or organisms to transcribe all their genes all the time whether the protein products were needed or not. So, there are many ways that the transcription of genes can be "turned on" and "turned off" (regulated) so their protein products are expressed only when needed. Here is one well-studied example of gene regulation in bacteria.

French biologist François Jacob and biochemist Jacques Monod investigated how genes in *E. coli* can be controlled through regulation of transcription. In 1965 they were awarded the Nobel Prize in Physiology or Medicine, along with microbiologist André Lwoff, for their groundbreaking work on the operon model for gene regulation. *E. coli* and other prokaryotic species contain sections of DNA called **operons**. Each operon is composed of three types of DNA sequences (**FIGURE 8.14A**).

- **Structural Genes.** One or more **structural genes** provide the information for proteins (through the gene expression process described earlier).
- **Operator.** A sequence of bases called the **operator** is next to the structural genes. The operator controls the expression of the structural genes and is not transcribed.
- **Promoter.** Mentioned earlier in the chapter, the **promoter** is located next to the operator. The promoter is where the RNA polymerase binds to start transcription of the structural genes. Again, the promotor is not transcribed.

Also, important, but not part of the operon, is a **regulatory gene** that controls the operator. It is always transcribed and translated.

The best way to understand the operon model is to observe how the *lac* (lactose) operon works. This operon controls the ability of the bacterial cell to produce the enzymes needed for the hydrolysis of lactose.

The *lac* Operon

In *E. coli*, the disaccharide sugar lactose can be used as an energy source if it is broken down into its monosaccharides glucose and galactose. However, when lactose is absent from the environment, there is no need for *E. coli* to produce lactose-digesting enzymes. Therefore, the microbe's regulatory gene encodes an mRNA to form a **repressor protein** that binds to the operator (**FIGURE 8.14B**). Although the RNA polymerase can still bind to the promoter, its movement down the DNA is blocked by the repressor protein attached to the operator. Therefore, the RNA polymerase cannot reach the structural genes (Z, Y, and A) and the cell does not produce the lactose-digesting enzymes.

Suppose lactose now enters the microbe's environment. How does the cell "turn on" the three structural genes needed for lactose digestion? In this context, lactose is known as an inducer because it will induce or "turn on" transcription of the structural genes. Because the repressor protein is always made, the inducer (lactose) binds to the repressor protein and changes the repressor protein's shape. With an incorrect shape, the repressor protein cannot bind to the operator (**FIGURE 8.14C**). With the operator sequence now unrestricted, the RNA polymerase can pass across the operator and transcribe the three structural genes. The mRNA is then translated, the enzymes soon appear, and digestion of lactose occurs.

FIGURE 8.14 The Bacterial Operon and Gene Regulation.

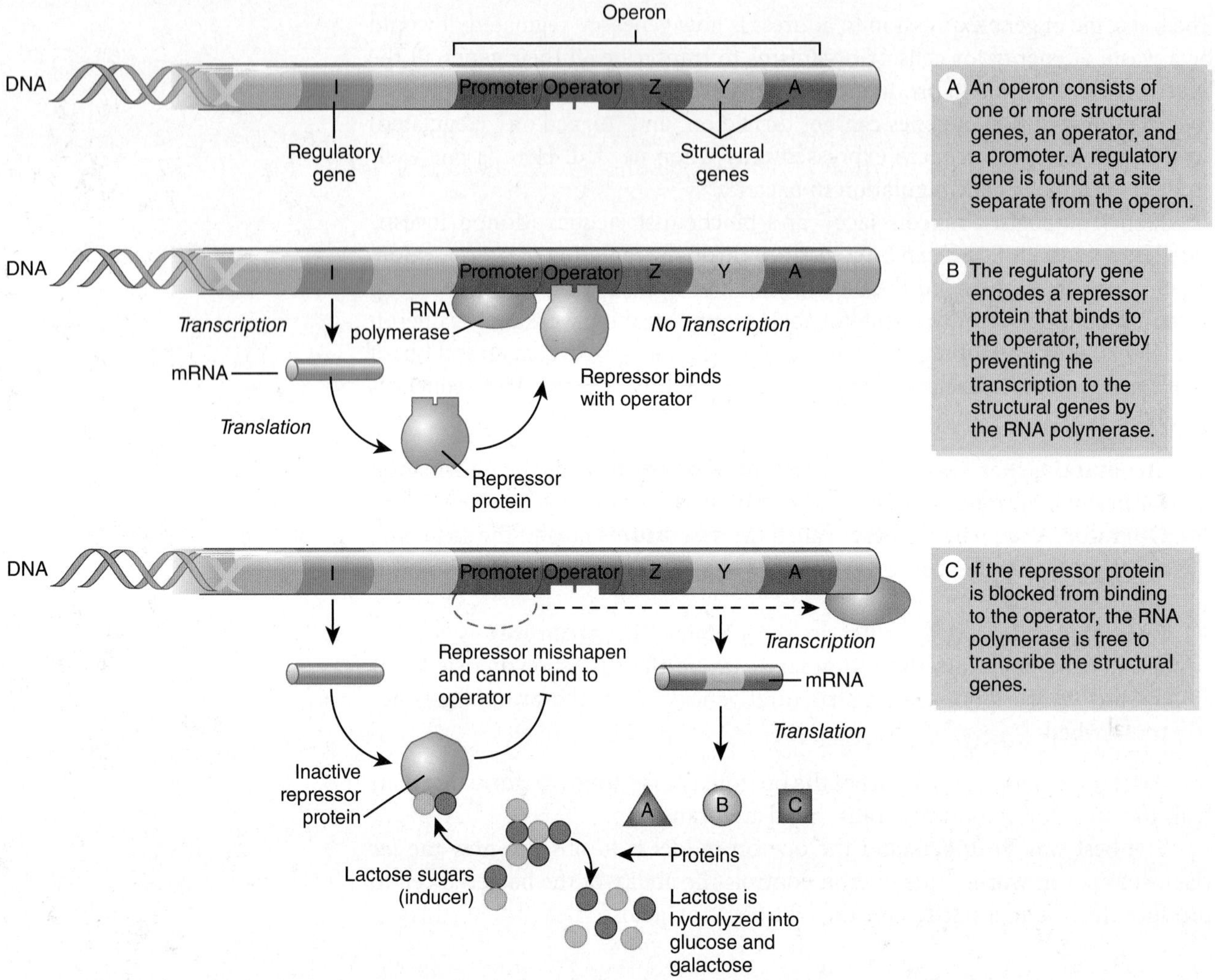

When the lactose has been totally digested, the lack of any more lactose allows repressor proteins to retain their original shape. Once again, they can bind to the operator, effectively shutting off gene expression of the *lac* operon.

8.3 Genes and Genomes: Human and Microbial

Genome: The complete set of genetic information in an organism or virus.

The **genome** making up any organism comprises the basic blueprint for that organism. Researchers now can compare large stretches of DNA within an organism and between different organisms through DNA sequencing. Such comparisons can yield an enormous amount of information about the biology of an organism, its responses to environmental influences, and its evolution.

DNA (gene) sequencing involves figuring out the order of nucleotides (A, G, C, and T) in a gene, a DNA segment, or an entire genome. The human genome consists of about 3 billion nucleobases (equivalent to 750 megabytes of computer data) and some 22,000 genes. By comparison, an *E. coli* cell has 4.7 million bases and 4,300 genes. Therefore, being substantially smaller, microbial genomes have been easier and much faster to sequence.

Many Microbial Genomes of Prokaryotes and Eukaryotes Have Been Sequenced

In 1976, sequencing technologies were used to sequence the genome of a bacteriophage. In 1995, the first complete genome of a free-living bacterium was sequenced, and in the following year the first complete genome sequence of a eukaryotic microbe, the common baker's (or brewer's) yeast, *Saccharomyces cerevisiae*, was reported.

Saccharomyces cerevisiae: sack-ah-roe-MY-seas seh-rih-VIS-ee-eye

With advances in sequencing technology, in April 2003, a publicly financed, $3 billion international consortium of biologists, industrial scientists, computer experts, engineers, and ethicists completed one of the most ambitious projects to date in the history of biology. The **Human Genome Project** (**HGP**) had succeeded in sequencing the **human genome**. The 3 billion nucleobases and some 22,000 genes in a human cell were identified and strung together (sequenced) in the correct order.

A logical extension of the HGP was to discover what microbes are resident in the human body. So, in 2007, the **Human Microbiome Project** (**HMP**) was started to identify and characterize the **human microbiome**. Five major body sites (nasal passages, oral cavity, skin, gastrointestinal tract, and urogenital tract) were selected for analysis. A second phase, known as the **Integrative Human Microbiome Project** (**iHMP**) was launched in 2014 and completed in 2016. Its aim was to get a better understanding of the roles these microbes play in human health and disease. Consequently, by the end of 2018, some 100,000 microbial sequencing projects had been completed or were being carried out (**FIGURE 8.15**).

Human microbiome: The communities of microorganisms (and their genes) living in association with the human body.

Having all these microbial sequences and the human genome sequence, it became of great interest to compare genes not only between different organisms but also between different domains of life.

Microbe and Human Genomes

One of the most active areas involving DNA sequencing is **genomics**, which is concerned with the structure, function, and evolution of genomes. The field of **comparative genomics** identifies DNA sequence similarities and differences between the genomes of different species or subspecies. Comparing sequences

FIGURE 8.15 Genomics Activities. The sequencing of bacterial genomes has far outpaced those of the archaeal and eukaryotic organisms.

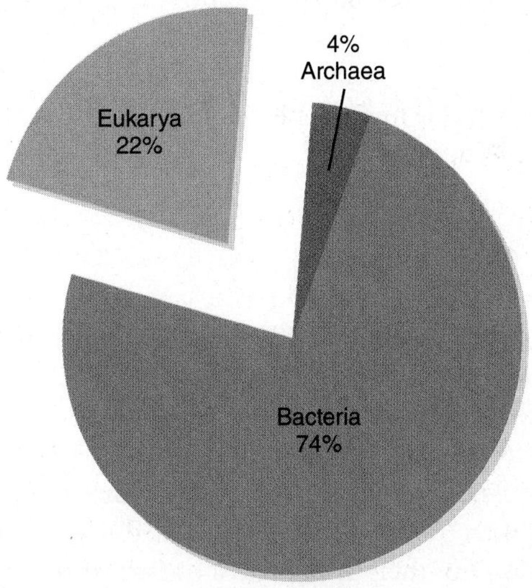

Data from Genomes OnLine Database. Retreived from http://www.genomesonline.org

FIGURE 8.16 Origins of Human Gene Sequences. Comparative genomics points to a long history of shared ancestry (percentages) with microbes.

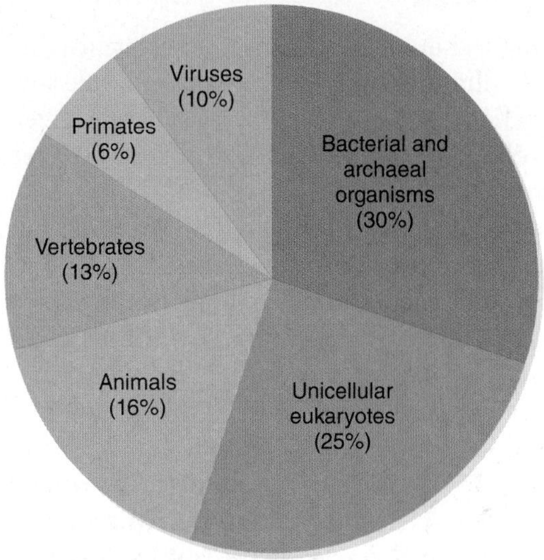

Data from Genomes OnLine Database. Retreived from http://www.genomesonline.org

of similar genes can provide evidence on how genomes have evolved over time. It also gives clues to the gene relationships between microbes and other organisms, including humans.

Humans don't look like microbes, but comparative genomics indicates that these organisms all share many common genes (**FIGURE 8.16**). In fact, genome comparisons indicate that as many as 6,600 of the 22,000 human genes (30% of the human genome) are essentially identical to those found in members of the domain Bacteria. About 6,000 human genes (25% of the human genome) are similar to genes found in fungal yeasts. However, most of these common genes also are found in other animal species as well. This suggests that these genes are so important, they have been preserved and passed along from organism to organism, and generation to generation, throughout evolution. They are life's oldest and most essential genes. As just one example, several researchers suggest some genes coding for brain signaling chemicals and for communication between cells did not evolve gradually in human ancestors. Rather, these genes became part of the animal lineage eons ago directly from microbial organisms. This claim remains highly controversial, though, and more work is needed to sort out the relationships of the microbial genes to animal and human evolution.

As an interesting side note, scientists have wondered if there is a minimum number of genes needed to maintain a cell's metabolism and growth. **A CLOSER LOOK 8.2** examines this question.

Virus and Human Genomes

Another discovery coming from comparative genomics concerns viral genomes. Several studies suggest that almost 10% of our genome (as well as parts of other animal genomes) is composed of fragments of viral DNA (see Figure 8.16). The infectious viruses that inserted this DNA into our ancestors' genome have been termed human **endogenous retroviruses** (**ERVs**). As described in the chapter on Viruses, **retroviruses** are RNA viruses that can reverse transcribe themselves; that is, after infection, their single-stranded RNA is copied into double-stranded DNA, which can then insert into a human chromosome.

The ERV "infections" presumably occurred in sex cells (egg and sperm) over the millions of years of primate evolution. During this period, the viral

A CLOSER LOOK 8.2

The Little Microbial Cell That Could

In 1930, a children's book entitled *The Little Engine That Could* was published in the United States. The premise of this story is that if you try really hard and work hard, you can accomplish your goal. This idea, skewed somewhat, can be applied to microbes as well.

Different organisms have different numbers of genes. In general, the more complex the organism, the more genes it has. However, there are exceptions. Humans have about 22,000 genes, but tomatoes have almost 32,000 genes. The tiny, almost microscopic water flea has 31,000 genes. In the microbial world, *E. coli* has about 4,300 genes, but many prokaryotic species have larger or smaller numbers. A species of *Streptomyces* has over 7,800 genes while *Pelagibacter ubique* has less than 1,400 genes. Especially interesting are the small genomes of **obligate symbionts** or parasites that have a reduced number of genes (see **Figure**).

Focusing on the free-living species, no matter how many genes the organism has, the genes can be subdivided into three sets.

- **Core Genes.** This set of genes is essential for cell function. It includes those genes involved with metabolism, growth, and reproduction.
- **Variable Genes.** Another set of genes represents those that are not essential for growth and reproduction, but, at times, might provide useful functions. In many cases, these variable genes contain information making a microbe a more dangerous pathogen.
- **Unique Genes.** The third set of genes codes for proteins that are unique to perhaps one strain of a species.

So, in a free-living microbe how small can a genome be? What is the minimal set of core genes?

One of the smallest genomes discovered is that of *Mycoplasma genitalium*. It has only 525 genes. Are all these genes necessary for its survival and reproduction? To find out, geneticists have taken *M. genitalium* cells and started removing genes one by one and then examining whether the organism can still grow and reproduce.

By removing the unnecessary genes, the scientists have now generated an artificial species (a little microbial species that could) having but 473 genes. With just 473 genes, this species can still grow and reproduce. Dissecting these genes: 48% are needed for gene expression and genome preservation, 18% for cell membrane structure and function, and 17% for cytoplasmic metabolism. The remaining genes (17%) have unknown functions. So, one would assume that this organism is approaching the minimal number of core genes—but it is still a little microbe that could.

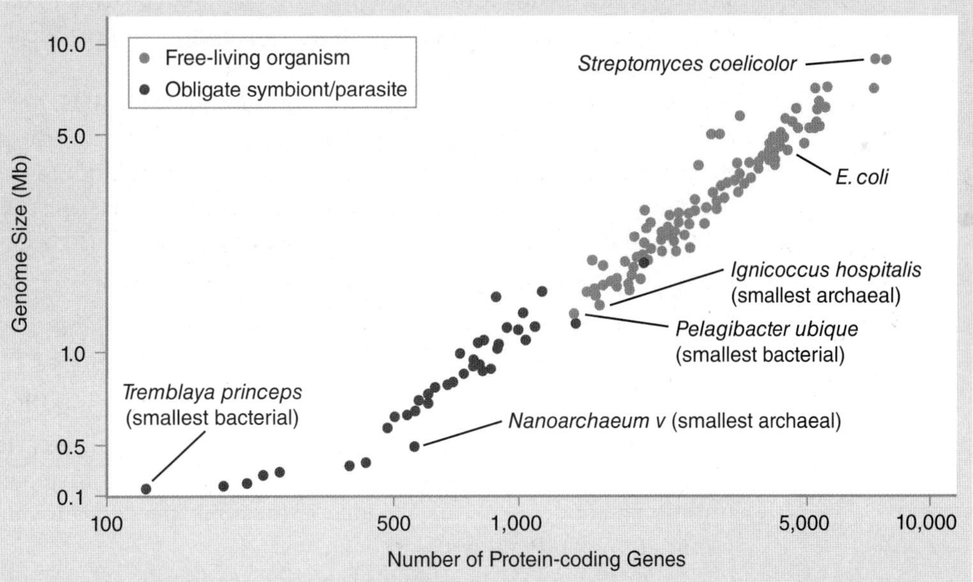

genome segments integrated into the primate lineage genome and have been transmitted to future generations ever since. These then are the 100,000 remnants discovered in human DNA today.

Many ERVs have accumulated harmful **mutations** during primate evolution and no longer cause disease in the human body. However, some ERVs can "reawaken" and act like genes or have evolved new functions. For example, at least six human ERV genes interact in the normal functioning of the human placenta. Another ERV gene codes for one protein among many needed for a fertilized egg to develop into a fetus. The ERV protein allows the cells in the outer

Obligate symbiont: A microbial species that must live in a host for its survival and reproduction.

layer of the placenta to fuse together. In fact, cell fusion in general might have evolved from ERV genes that slipped into the animal genome and now function as a necessary gene for cell fusions (e.g., muscle, skin, bone, and even sperm-egg fertilization) that are part of the development of complex multicellular life.

Mutation: A permanent change in an organism's DNA.

Amyotrophic lateral sclerosis (ALS): A progressive neurodegenerative disease that affects nerve cells in the brain and the spinal cord.

Other ERVs might play roles in cancer and neurodegenerative diseases of the spinal cord. One ERV is expressed at high levels in the brains of patients suffering from **amyotrophic lateral sclerosis** (**ALS**) and might play a role in this neurodegenerative disease.

Recently, even more virus "signatures" (nucleic acid sequences), very different from the ERVs, have been found in the human genome. In one scenario, these so-called **endogenous viral elements** (**EVEs**) have been proposed to have crossed from rats and infected evolving primates. In primates, the viral genes at some point managed to infect the genomes of egg and sperm cells, allowing the viral genes to be passed from generation to generation. Then, over time, the viral genes became disabled and could no longer produce new viruses.

All these studies strongly suggest that the human genome is a hybrid of eukaryotic, prokaryotic, and viral genetic information (see Figure 8.16). Undoubtedly, microbial and viral genes have affected gene expression and human evolution in ways that have yet to be completely understood.

Microbial Genomics Will Advance Our Understanding of the Microbial World

Microorganisms have existed on Earth for almost 4 billion years. Over this long period of evolution, they have become established in almost every environment on Earth. They make up a significant percentage of Earth's biomass. They are the smallest organisms on the planet, yet they influence—if not control—some of the largest events. Nonetheless, with few exceptions, a great deal remains to be understood about the microbes.

However, with the introduction of **microbial genomics**, microbiology is in another Golden Age. By analyzing and comparing the genome of a microorganism, remarkable scientific discoveries are being made toward understanding the workings and interactions of the microbial world. A few potential consequences for society from the understanding of microbial genomes are outlined below.

A Safer Food Supply

Microorganisms play important roles in our foods, both as contamination and spoilage agents. Therefore, understanding how they get into the food product and how they produce dangerous foodborne toxins will help the agriculture industry produce safer foods (**FIGURE 8.17**). However, a major limitation with traditional food safety surveillance is the time needed and number of tests required to identify a food pathogen.

In 2012, the U.S. Food and Drug Administration (FDA) initiated a project to identify the microbial genomes of foodborne pathogens. Called the GenomeTrakr network, the FDA, in collaboration with state, federal, and international food safety labs, now has sequence information on more than 317,000 strains of bacteria (**FIGURE 8.18**). Consequently, when a foodborne outbreak occurs, the sequence information in the network can speed up identification of the pathogen. It also can pinpoint which ingredient in a multi-ingredient food is contaminated. For example, in 2016, a multistate foodborne outbreak of *Listeria* occurred. Most infections by the bacterium occur by consuming improperly processed deli meats and unpasteurized milk products. So, what food product was to blame in this

Listeria: lis-TEH-ree-ah

FIGURE 8.17 Food Safety. Maintaining clean and safe foods to eat is an increasingly difficult problem. Microbial forensics can help to rapidly detect potential pathogen contamination.

Courtey of CDC.

FIGURE 8.18 GenomeTrakr. The GenomeTrakr network has sequenced more than 317,000 pathogen isolates, regularly sequencing over 5,000 isolates each month.

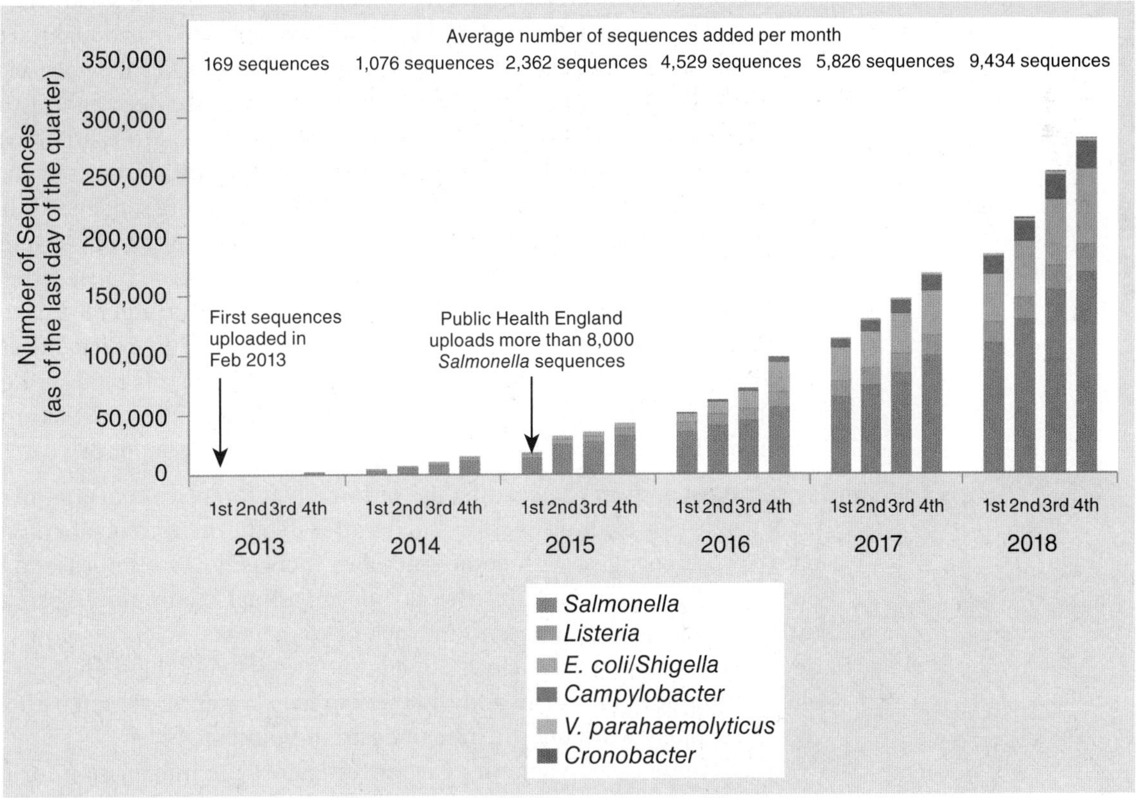

Courtesy of FDA.

outbreak? Using GenomeTrakr, the FDA identified a brand of frozen vegetables as the source of the *Listeria* contamination. Working with the Centers for Disease Control and Prevention (CDC), the FDA matched the *Listeria* genome sequence in the frozen vegetables with the *Listeria* sequence in people who had become ill.

Microbial Forensics

Bioterrorism: The use of microbes to cause fear in, or to inflict death or disease upon, a large population.

The history and threat of using microbes as agents of **bioterrorism** have led many researchers to look for rapid and efficient ways to detect the presence of such microbial bioweapons.

Many of the diseases caused by these potential biological agents cause no symptoms for at least several days after exposure. When symptoms do appear, initially they are not noticeable or are flu-like. Therefore, it can be difficult to distinguish between a natural disease outbreak and the intentional release of a potentially deadly pathogen. Such concerns have led to a relatively new and emerging area in microbiology called **microbial forensics**. This discipline attempts to quickly recognize patterns in a disease outbreak and identify the responsible pathogen. The discipline also looks for ways to control the pathogen's spread and to discover the source of the pathogenic agent.

Here are a few examples where microbial forensics has been critical to the identification and source of pathogens.

- **Swine Flu.** In 2009, some individuals suggested the swine flu pandemic that year was the result of an accidental release of the virus from a lab doing vaccine research. Investigators used microbial forensics to conclude that there was no accidental release of swine flu viruses. The genome sequences of the outbreak strain and research lab strain did not match.
- **Cholera Outbreak.** In 2010, a cholera outbreak occurred in Haiti. Because there had not been any cholera in Haiti for more than 100 years, where then did it come from? Some believed it was transmitted unintentionally by United Nations (UN) peacekeepers from Nepal. Many peacekeepers had arrived earlier that year as part of the UN relief effort after the 2010 earthquake. Genome sequencing of the cholera strains in Haiti and in the Nepalese peacekeepers confirmed their sequence similarity. In fact, Nepal had been fighting its own cholera outbreak at the time the peacekeepers were sent to Haiti.
- **Hepatitis C Outbreak.** In 2012, a cluster of 29 hepatitis C cases occurred in patients in a hospital in New Hampshire. Investigators sequenced the viral genome from the patients. They also took blood samples from hospital staff and looked for similar sequences in the samples. When completed, the forensic scientists traced the origin of the outbreak to a hospital medical technician who had hepatitis C. The report said that this virus-infected technician was injecting himself with the clinic's narcotics using hospital syringes. He then reused the now hepatitis C-contaminated needles for patient medications. The accused pleaded guilty and was sent to prison for 39 years.

One last and very interesting use of microbial forensics concerns an organism's death. For example, within 2 days after death, the microbial community on a corpse changes, with some microbes increasing in number, and others decreasing in number. After a week, the microbial community has changed further, partially in response to the lack of oxygen gas. Scientists now want to develop a "microbial death clock"; that is, they want to use the changes in microbial communities on a dead body as a way to help law enforcement pathologists more accurately work out the time of death in homicides.

In all these cases, genomics gave investigators the information for tracing an infecting microbe to an institution, place, or person(s) of origin. It is becoming more useful to medical diagnoses and to society as a whole.

Metagenomics Is Identifying a Previously Unseen Microbial World

Existing within, on, and around every living organism, and in most all environments on Earth, are microorganisms. Yet, some 98% of these species will not grow on any known growth medium. These organisms are referred to as "viable, but noncultured" (VBNC). This means that most microbes found in the soil, in oceans, and even in the human body have never been seen or grown—and certainly not studied. Yet the genes of these organisms represent a wealth of genetic diversity.

Advances in DNA sequencing technologies have created a new field of research for analyzing this uncultured majority. The field is called **metagenomics** (*meta* = "beyond"), and it refers to identifying the individual genome sequences within a mixture of genomes from multiple microorganisms in a community (the **metagenome**); that is, "beyond" what can be cultured.

Metagenomics today is stimulating the development of advances in fields as diverse as medicine, agriculture, and energy production. Here is how the metagenomics process is carried out.

Samples of the desired community of microorganisms are collected. These samples could be from soil, water, or even the human digestive tract. The DNA is extracted from the entire sample, which produces DNA fragments from all the microbes in the samples. The fragments can be amplified in amount so there is enough material to sequence. Using computer-based technologies, the sequenced segments are pieced back together like a jigsaw puzzle, each final sequence being unique to one of the microbial species in the original mixed sample. When completed, microbiologists have a broad spectrum of the microbial community. Researchers then can use the reassembled genomes to compare genetic relationships among organisms in the sample. Thus, the role of each microbe in that community can be studied without having to grow the organism in culture.

Today, metagenomics is one of the most intensively studied fields for understanding the relationships between and within microbial communities. The human microbiome projects mentioned above were driven by metagenomics because most of the microbes in humans cannot be grown in culture. On a more global scale, information from metagenomics studies is uncovering the diversity of microbes that might help solve some of the most complex medical, environmental, agricultural, and economic challenges in today's world (**TABLE 8.1**).

A Final Thought

For many decades, scientists pondered the nature of the gene. Try as they might, they could not imagine how DNA could be involved. They thought it too simple ("too stupid" some said) for the sophisticated function of controlling an organism's metabolism and growth.

But by the 1950s, compelling experiments pointing to DNA as anything but stupid. Then, when the structure of DNA was discovered, a flurry of brilliant research started snowballing and scientists cracked the code that makes DNA so smart. They were identifying genes and examining their structures, while researching the proteins those genes encoded. Then, they were ready to map the genome—human and microbe. What they have accomplished has made heads spin in the scientific community.

Deciphering the human genome has been compared to landing an astronaut on the moon. The identification of the human microbiome worked out through

TABLE 8.1 Some Uses for Metagenomics

Challenge	Example
Medicine	Understanding how the microbial communities that inhabit our bodies affect human health could provide genome information that can be used to develop new strategies for diagnosing, treating, and preventing infectious diseases.
Ecology and the Environment	Exploring how microbial communities in soil and the oceans affect the atmosphere and environmental conditions could help in understanding and predicting the effects of climate change.
Energy	Harnessing the power of microbial communities might result in sustainable and ecofriendly bioenergy sources.
Bioremediation	Adding to the arsenal of microorganism-based environmental tools can help in monitoring environmental damage and cleaning up oils, groundwater, sewage, nuclear waste, and other biohazards.
Agriculture	Understanding the roles of beneficial microorganisms living in, on, and around domesticated plants and animals could enable detection of diseases in crops and livestock and aid in the development of improved farming practices.
Biodefense	The addition of metagenomics tools to microbial forensics will help to monitor pathogens, create more effective vaccines and therapeutics against bioterror agents, and reconstruct attacks that involve microorganisms.

the HMP was no less awesome. The milestone of sequencing the genomes of humans and microbes is but the stepping-off point on an incredible journey that will take decades to complete. It is safe to assume scientists will not be turning off their sequencing machines any time soon. Today, society is crossing the threshold of knowledge so powerful that we will wonder how we ever got along without it.

Chapter Discussion Questions

What Was He Thinking?

From your reading of this chapter, identify and discuss five major points about DNA and genes that the author was trying to get across to you.

Questions to Consider

1. Try to put yourself in Griffith's position in 1928. Genetics is poorly understood, DNA is virtually unknown, and bacterial biochemistry has not been clearly defined. How would you explain the remarkable results of his experiments?
2. What is meant by saying that DNA replication is semiconservative?
3. In one sense, DNA is a relatively simple molecule, having only four different subunits (nucleotides). And yet, DNA can specify extraordinarily complex proteins having at least 20 different amino acids in chains of many thousands or more. How does DNA accomplish this seemingly impossible task?
4. It has been said that the names of Watson and Crick are permanently etched in the history of 20th-century biology as the "discoverers"

of DNA. Knowledgeable students of biology know, however, that Watson and Crick did not "discover" DNA. Which names should be remembered along with theirs, and why?

5. In this chapter, there has been a lot of information about microbial genes and genomes. Please explain the difference between the genes that a cell has and its genome.
6. You maintain that enzymes make proteins, but your classmate says that proteins make enzymes. Which of you is correct, and what is the basis for your conclusion?
7. Metagenomics is an important area in microbiology today. From your reading in this chapter, what is meant by the following statement: Metagenomics aims to access the genomic potential of an environmental sample?
8. You have two samples of DNA. Sample A contains the following percentages of nucleotides: 23.1% A, 26.9% C, 26.9% G, and 23.1% T. By contrast, sample B has 32.3% A, 32.3% C, 17.7% G, and 17.7% T. Which sample is double-stranded DNA, and which is single-stranded DNA. Why?

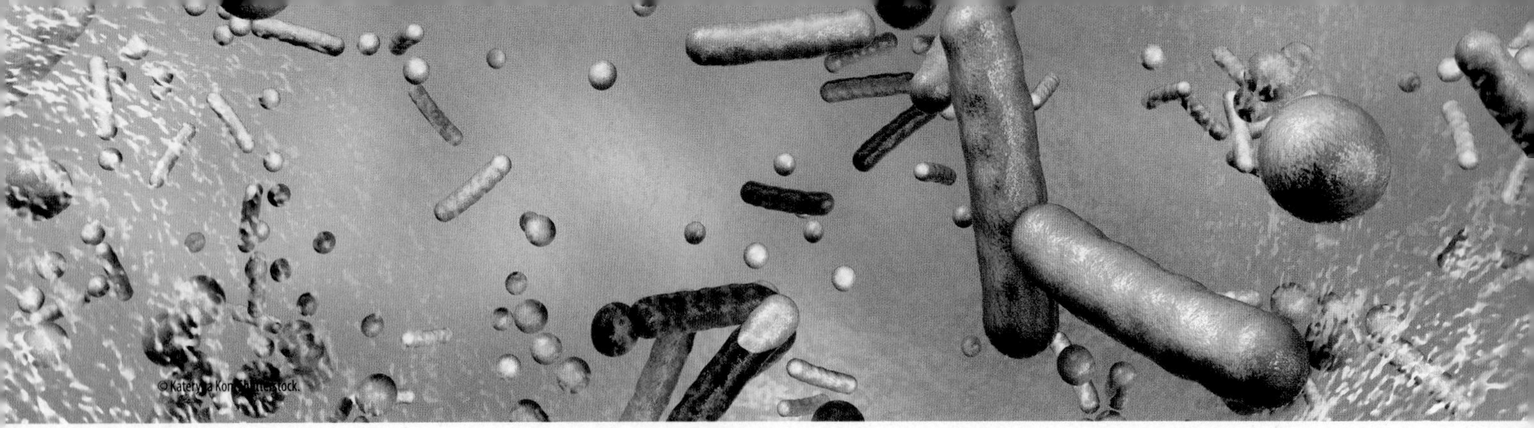

CHAPTER 9

Microbial Genetics: From Genes to Genetic Engineering

An Outbreak of Bacterial Dysentery

Shigella: shih-GEL-lah

In 1969, an extremely serious form of bacterial dysentery broke out in the Central American country of Guatemala. Bacterial dysentery can be a life-threatening and highly contagious infection caused by any of several species of *Shigella*. The infection by these gram-negative cells affects the human large intestines (colon), resulting in watery diarrhea containing blood or mucus (**FIGURE 9.1**). Waves of intense abdominal pain and stomach cramps also are common. The disease, called shigellosis, is spread by *Shigella*-contaminated food or water. In Guatemala, the outbreak spread quickly, with high infection rates in all age groups. However, the mortality rates were highest in young children due to dehydration.

Electrolyte: A mineral or salt in blood and other body fluids that regulates nerve and muscle function.

During the *Shigella* outbreak, patients were given salt tablets or oral salt solutions to help replace lost water and **electrolytes**. In addition, some patients also received one of four different antibiotics known to kill the pathogen. However, as the months passed, physicians became increasingly frustrated in their attempt to treat the disease. They discovered that none of the four antibiotics now was effective and were unable to eliminate the *Shigella* infection. Somehow, the pathogen had changed genetically to become antibiotic resistant. Without useful antibiotics, physicians lost a major weapon to stop shigellosis. By the third year of the outbreak, more than 112,000 people had been affected and at least 12,000 died.

The outbreak of shigellosis illustrates what can happen when antibiotic resistance (ABR) emerges in a population or community. At the time of the

CHAPTER 9 OPENER One of the biggest challenges facing microbiology, global health, and society today is antibiotic resistance. This occurs when bacteria that were susceptible to the drugs become resistant to these medicines. According to the Centers for Disease Control and Prevention, every year, more than two million people in the United States get infections that are resistant to antibiotics and at least 23,000 people die as a result.

FIGURE 9.1 Bacterial Dysentery. Bacterial dysentery is a painful intestinal infection that can be caused by a group of bacteria called *Shigella* (inset; bar = 10 μm.). Some people who are infected might have no symptoms but still pass the *Shigella* bacteria to others through sharing food or water contaminated with the pathogen.

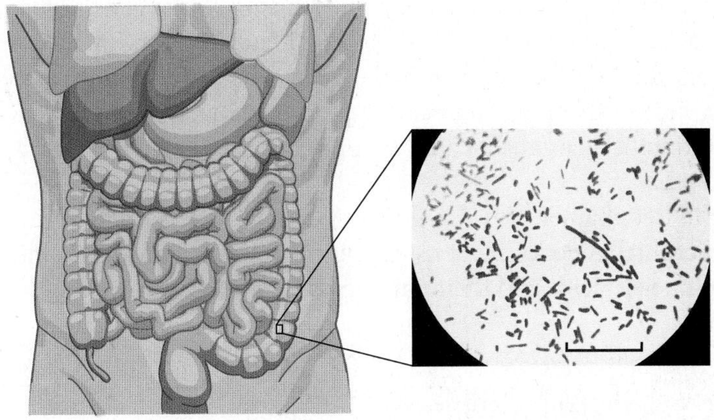

Guatemala outbreak, it was unusual to find infectious agents with resistance to one drug, much less four. Now, you might say, "Well that was 1969. Things must have changed in the last 50 years in terms of global ABR." In fact, things have changed—the problem of ABR is much worse today and, as part of **antimicrobial resistance** (**AMR**), represents one of the most pressing issues facing medicine and society on a global scale. The Centers for Disease Control and Prevention (CDC) says,

Antimicrobial resistance (AMR): The ability of a microbe to resist the effects of medication that once could successfully treat the microbe.

> *"People infected with antimicrobial-resistant organisms are more likely to have longer, more expensive hospital stays, and may be more likely to die as a result of the infection."*

In this chapter, we will learn how bacterial organisms like *Shigella* can acquire antibiotic resistance, as we explore the concept of microbial **genetics**. We will see how a DNA molecule naturally can be altered by mutations in its genes or through the acquisition of new genes. In addition, we will examine the topic of genetic engineering that purposely alters genes of microbes. The belief is that this technology can be used to improve the quality of people's lives and society.

Genetics: The branch of biology that involves the structure, function, and transmission of genes and the variation of inherited characteristics among similar or related organisms.

LOOKING AHEAD

After reading and completing this chapter, you will be able to:

9.1 Contrast chromosomal DNA and plasmid DNA found in bacterial cells.

9.2 Define mutations, and distinguish between base-substitution mutations and frameshift mutations.

9.3 Compare the three forms of gene recombination that can occur through horizontal gene transfer.

9.4 Explain the basics of genetic engineering, and discuss the implications for recombinant DNA molecules.

9.1 Bacterial DNA: Chromosomes and Plasmids

Modern-day prokaryotic species enjoy the fruits of all the genetic changes their ancestors have undergone over the nearly 4 billion years of their existence. In addition, a new generation of bacteria in some cases can be produced in just 20 minutes. Therefore, one can appreciate how a useful genetic change, such as antibiotic resistance, could be propagated quickly through a bacterial population.

Later in this chapter, we will study mutations. However, let's begin our briefing on microbial genetics by considering the bacterial chromosome. A quick review on nucleic acids and their parts in the chapter on The DNA Story might be helpful.

The Bacterial Chromosome

Most of the genetic information, collectively called the **genome**, in a bacterial cell is located within a single, circular chromosome. It exists in a region of the cytoplasm called the **nucleoid** that lacks a surrounding membrane. The bacterial chromosome occupies about half of the total volume of the cell and, extended its full length, measures about 1.5 mm long, which is about 1,000 times the length of the bacterial cell. This presents a problem. How can a 1.5-mm-long chromosome possibly fit into a 2.0-µm *Escherichia coli* cell? The answer is that the chromosome takes on a structure having numerous tightly packed loops. Each loop consists of about 10,000 nucleotides and is attached to other loops at anchorage points to form a "flower" structure (**FIGURE 9.2**). This very tight packing of the DNA accounts for its explosive release when the cell membrane and wall are broken.

Escherichia coli: esh-er-EE-key-ah KOH-lee

The chromosome of *E. coli* has been one of the most extensively studied. The chromosome contains some 4,300 genes (some viruses, by contrast, have as few as 7 genes and a human cell has some 22,000 genes). Other prokaryotic species might have more or fewer genes than *E. coli.*

FIGURE 9.2 Bacterial DNA.

(A) The loops in the structure of the bacterial chromosome, viewed head on. The loops in the DNA help account for the compacting of a large amount of DNA in a relatively small bacterial cell.

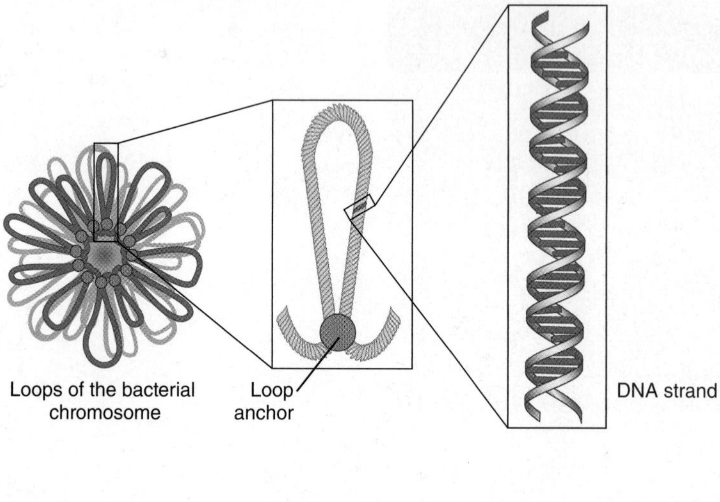

(B) A transmission electron microscope image of an *E. coli* cell immediately after disruption. The tangled mass is the organism's DNA. (Bar = 1 µm.)

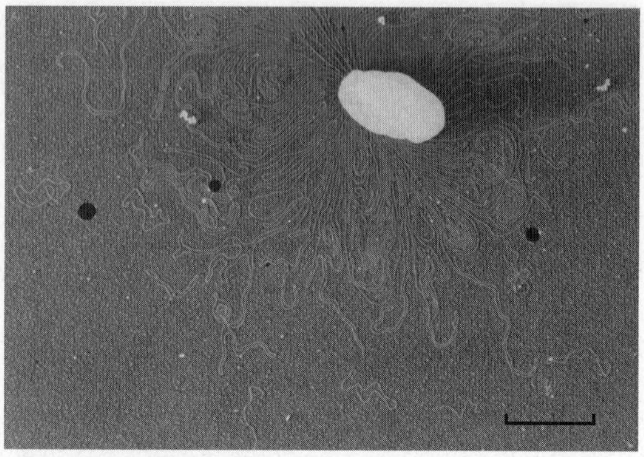

Plasmids

Besides the chromosome, many prokaryotic cells contain one or more small, circular DNA molecules called **plasmids**. These are found in the cytoplasm, distinct from the nucleoid (**FIGURE 9.3**). Plasmids contain about 2% of the total genetic information of the cell, and they multiply independently of the chromosome. Unlike the bacterial chromosome that contains all the essential genes for growth and reproduction, plasmids contain useful but not essential genes for the everyday existence of a bacterial cell. For example, many plasmids carry genes coding for ABR. These and other plasmids often can transfer their genetic material to another cell, as we shall see later in this chapter. By moving from cell to cell, these "traveling plasmids" might have allowed the *Shigella* species to pick up the ABR described in the chapter opener.

Plasmids also can contain genes encoding substances, called **virulence factors**. Many enzymes and toxins that are harmful to host cells are coded by genes on plasmids. These substances make the bacterial cell a more dangerous pathogen. In addition, some plasmids code for proteins that are toxic to other bacterial species. This gives the toxin producer an advantage in the soil where the competition for nutrients can be challenging. Later in this chapter, we will learn how microbiologists and biologists have used plasmids in the process of genetic engineering.

Virulence factor: A molecule on or produced by a pathogen that increases its ability to invade or cause disease in a host.

Genetic Diversity

With rare exceptions, humans go through their lives with virtually the same genes; what is inherited from one's parents is essentially what they have throughout life. This is not necessarily the case in the prokaryotic world because the genetic information coded by the bacterial plasmids and chromosomes is subject to change. Such changes are the basis for genetic diversity.

FIGURE 9.3 Bacterial Plasmids. A false-color transmission electron microscope image of bacterial plasmids. The plasmids are small, closed loops of double-stranded DNA. (Bar = 20 nm.)

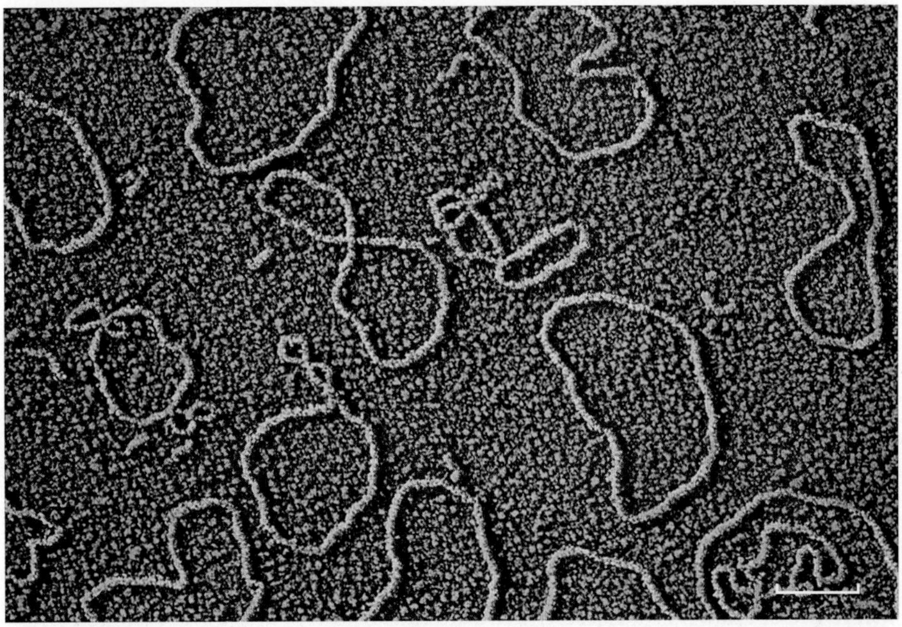

Genetic diversity: The total number of genetic characteristics in the genetic makeup of an organism or virus.

Natural selection: The process that results in the survival and reproductive success of individuals or populations best adjusted to their specific environment.

Genetic diversity among all organisms and viruses results from evolution, and the driving force is **natural selection**. For example, suppose a population of *Shigella* cells finds itself in an environment where the antibiotic ciprofloxacin (cipro) is present. Any cipro-resistant cells will survive and reproduce. The cells susceptible to the antibiotic will be killed or at least will be reduced in number. Consequently, natural selection has favored those cells resistant to cipro. Their growth in cell numbers will make them the dominant form in the population.

So, where did the antibiotic-resistant cells in the population come from in the first place? ABR can arise from changes to the cell's genome. Such changes are the result of mutation and/or genetic recombination. In both cases, the changes can have a substantial impact not only on the bacterial population but on human society as well. A review of replication and gene expression in the chapter on The DNA Story might be helpful to understand this material clearly.

9.2 Gene Mutations: Subject to "Change Without Notice"

A **mutation** is a permanent change to an organism's DNA. Such a change often involves disruption of the nucleotide sequence in a gene. This action might result in putting the wrong amino acids in a protein molecule being formed. Importantly, mutations are random events. Their occurrence cannot be predicted as to when and where they will occur. Such mutations are most easily identified when the mutation causes a change in an observable physical or biochemical characteristic called the **phenotype** of the organism. For example, this could be a change in the physical shape of a microbe or an alteration to its nutritional needs. It also might be detected by the sudden appearance of AMR in a population of once susceptible cells. A microbe with a normal, nonmutated characteristic is called the **wild type**, while an organism with the mutation is called the **mutant**.

Mutations Affect Genes

Nucleotide: In DNA, a molecule consisting of a five-carbon sugar bonded to a phosphate group and a nucleobase (A, T, G, or C).

Sometimes mutations are the result of a single **nucleotide** change in the sequence of bases in a gene. Mutations also are the result of losing or gaining a nucleotide in a gene. These changes are referred to as **point mutations** because the sequence of bases has changed at a single point. Such mutations fall into one of two groups.

Base-Pair Substitution

A mutation known as a **base-pair substitution** arises when a single nucleotide is incorrect (**FIGURE 9.4A**). Often these mutations are said to be "silent mutations" because the change in no way affects the structure or function of the protein made; there is no change in the amino acid sequence. Remember, some amino acids are coded by more than one codon. A "missense mutation" occurs if the substitution creates a faulty or less functional protein, or a protein with a new function. Some point mutations also represent a "nonsense mutation" where the change has stopped the production of the protein; the change generated a stop codon.

FIGURE 9.4 Categories and Results of Point Mutations. Mutations are permanent changes in DNA and their presence is identified by the mRNA and protein product. At the top is an illustrative normal sequence of bases in the template strand of a gene.

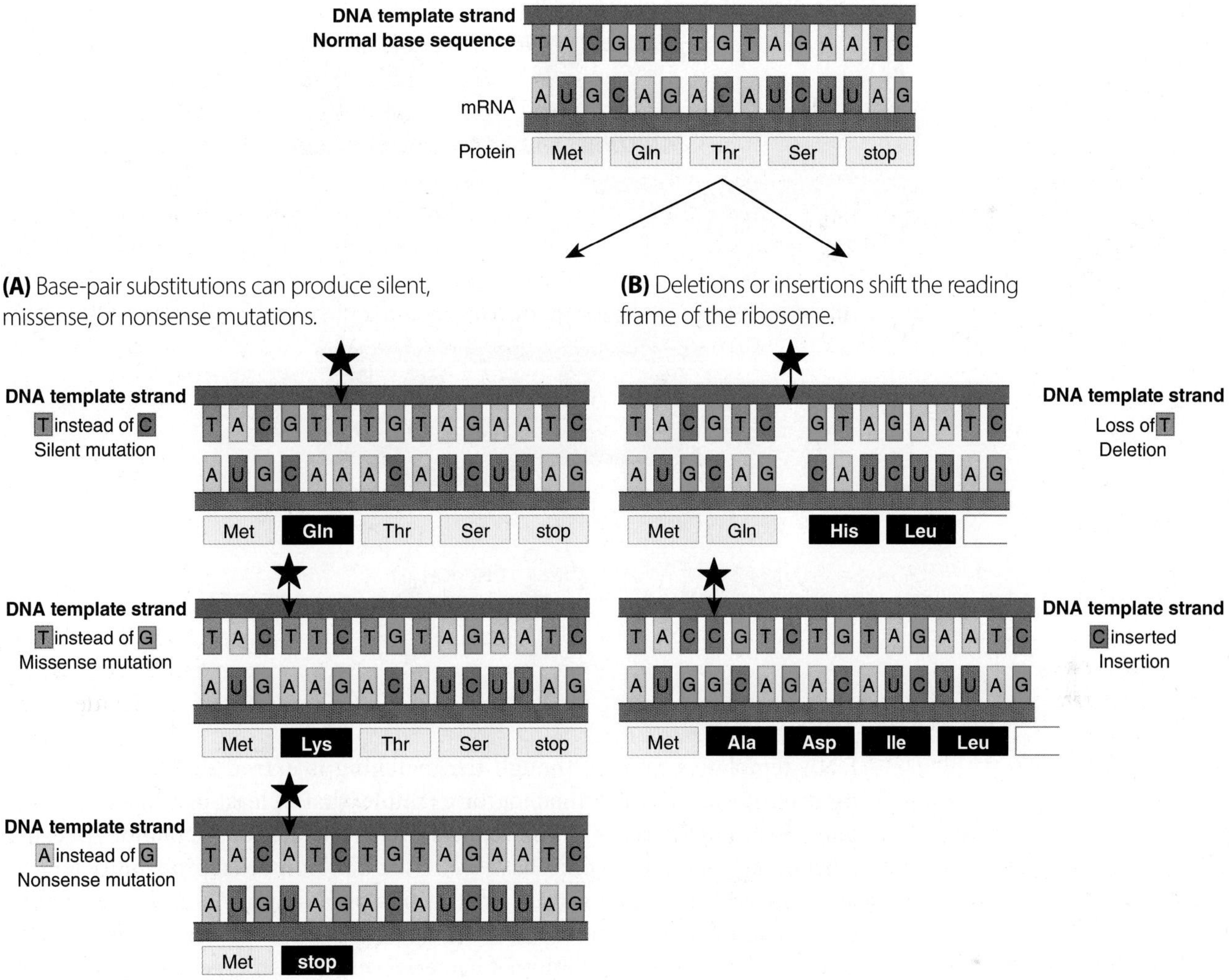

Frameshift Mutation

Mutations also might be the result of a loss or gain of a nucleotide (**FIGURE 9.4B**). A **frameshift mutation** changes the reading frame of the messenger RNA (mRNA). Remember, ribosomes always read the mRNA as "three-letter words" or codons. For example, if you had to read a sentence as three-letter words and a letter was missing, or an additional letter was added, a shift in the reading frame might produce a nonsense result.

Normal reading frame THE FAT CAT ATE THE RAT
Deletion (loss of the "F") THE ATC ATA TET HER AT
Insertion (addition of an "A" after "FAT") THE FAT ACA TAT ETH ERA T

Clearly, the shift in reading frame makes no sense. The same is true for a protein made; it probably will be useless, as many incorrect amino acids are present.

Causes of Mutations

Mutations can arise in several ways. In the environment, **spontaneous mutations** often are the result of DNA replication errors. During DNA replication, DNA polymerases add new complementary nucleotides to each

FIGURE 9.5 Ultraviolet Light and DNA.
(A) When cells are exposed to ultraviolet (UV) light either naturally or through experiment, the radiation can affect the cell's DNA.

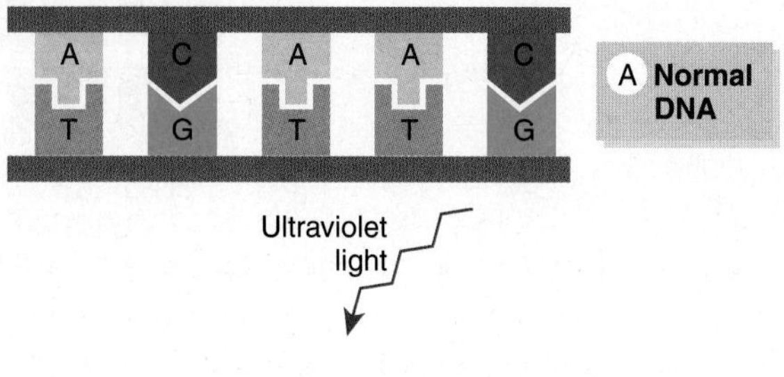

(B) UV light has caused adjacent thymine molecules to pair and distort the DNA strand.

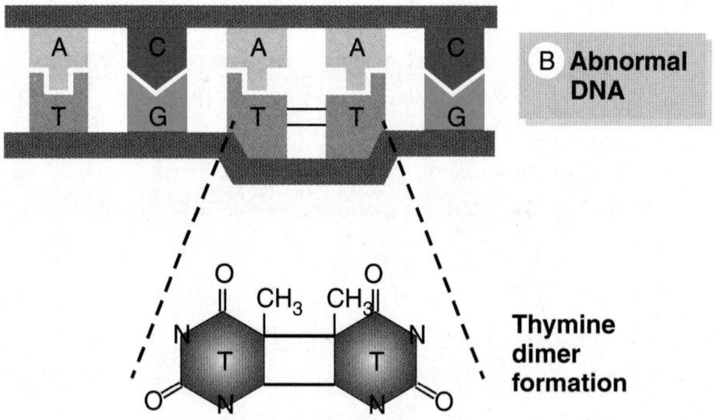

DNA template strand. Although the matching is extremely accurate, errors do occur. It has been estimated, for example, that at least one incorrect base pairing occurs for every billion replications in a bacterial population. Because a bacterial population could have well over a billion cells, probability tells us at least one mutant cell should exist in this population. As in the cipro antibiotic example above, natural selection might favor the mutant. That cell will multiply, and a new population of bacterial cells will emerge, all of which have the new characteristic.

Mutations also arise from DNA's exposure to **mutagens**, which are physical agents or chemical substances causing mutations in cells. Ultraviolet (UV) light, a component of sunlight, is a physical mutagen. When soil-dwelling bacterial cells absorb UV light, the mutagen might cause thymine or cytosine nucleobases next to one another in the same strand of DNA to bind together (**FIGURE 9.5**). In this abnormal arrangement, DNA replication and/or transcription is blocked because the polymerases cannot pass the distorted bases. Such a harmful change might be fatal to the cell.

Many chemicals in the environment also are mutagens. Aflatoxin, a chemical produced by a mold that infests peanuts and grain, and benzopyrene, a component of industrial smoke and cigarette smoke, are two such mutagens. They can cause the loss of nucleotides from, or addition of extra nucleotides into, the DNA during replication. Again, the change in nucleotide sequences might be deadly to the cell.

You might ask, "Could a bacterial cell, such as in the *Shigella* case in the chapter opener, become resistant to an antibiotic through a mutation?" Many antibiotics work by binding to an essential bacterial protein, disabling the protein's function. However, if the bacterial cell has a "beneficial" mutation, perhaps that mutation altered the protein just enough so that the antibiotic cannot bind. The mutant cell now survives, and its reproduction generates a whole population of resistant cells. This is one possible scenario for the *Shigella* example.

9.3 Genetic Recombination: Sharing Genes

Normally, cases of bacterial dysentery can be treated effectively with antibiotics. So, *Shigella*'s ability to become resistant to all the usable antibiotics was an alarming aspect of the Guatemala outbreak. Oral rehydration solutions helped with treatment for dehydrated patients, but antibiotics were needed for many cases of severe, life–threatening shigellosis. The question then is how could a bacterial species like *Shigella* acquire AMR to four different antibiotics? How had the organism developed **multidrug resistance** (**MDR**)? As we will see below, MDR might have involved sharing genes between bacterial species.

Years before the Guatemala outbreak began, species of *E. coli* had been identified with the same AMR as would be found in the Guatemala shigellosis cases. Could this be coincidence? Probably not because mutations are random events. The chances of two separate species developing identical mutations is about as likely as winning the lottery twice in the same week. Rather, the medical experts believed resistance to the four drugs must have been transferred directly from *E. coli* to *Shigella*.

To see if this could happen, investigators mixed together liquid suspensions of MDR *E. coli* and drug-susceptible *Shigella*. Within a few days, they found that the *Shigella* cells now were resistant to the same four drugs as the *E. coli* cells. Presumably, at some point patients with MDR *E. coli* infections also became infected with *Shigella*. Once together, transfer of MDR from *E. coli* to *Shigella* had taken place. How might MDR resistance have been acquired?

Besides mutations causing genome changes in a cell, bacterial species can undergo **genetic recombination** to modify their genome. This is a type of genetic transfer in bacteria where DNA from a donor cell is transmitted to (recombined with) the DNA from a recipient cell. In fact, the process is widespread in the bacterial world. For example, biochemists have calculated that *E. coli* has acquired nearly 20% of its genome directly from other bacterial species. A nonpathogenic species, *Thermotoga maritima*, has acquired about 25% of its genome from other prokaryotes in both the Bacteria and Archaea domains.

Thermotoga maritima: ther-moh-TOE-gah mar-eh-TEA-mah

In normal cell reproduction, the hereditary information coded in DNA is passed down from one generation to the next by **vertical gene transfer** (**FIGURE 9.6A**). If a nonlethal mutation occurs in a cell's DNA, the mutation will be passed to the next generation during the binary fission process. This type of gene transfer is common in all organisms. In humans, for example, it is thought that the mutation leading to breast cancer can be passed from mother to daughter during conception. Similarly, genes can be passed from one bacterial cell generation to the next through vertical gene transfer.

However, the genetic recombination process in bacteria involves the physical transfer of DNA directly from a donor cell to a recipient cell. This mechanism, called **horizontal gene transfer**, can happen in one of three ways: conjugation, transduction, and transformation (**FIGURE 9.6B**). Let's look at each.

Bacterial Conjugation

In the recombination process called **conjugation**, two live bacterial cells come together and the donor cell transfers some of its genetic material to the recipient cell (**FIGURE 9.7**). Specifically, the donor is known as an F^+ cell because it has an **F factor**, which is a plasmid containing about 20 genes mostly associated with conjugation. The recipient is known as an F^- cell because it lacks an F factor.

In gram-negative species, like *E. coli* and *Shigella,* to the process of conjugation begins with the F^+ cell forming a **conjugation pilus** that contacts the recipient cell. The pilus shortens, bringing the two cells close together.

FIGURE 9.6 Genetic Recombination.
(A) Vertical gene transfer involves genes (such as gene *A*) being passed down to succeeding generations of bacterial cells.

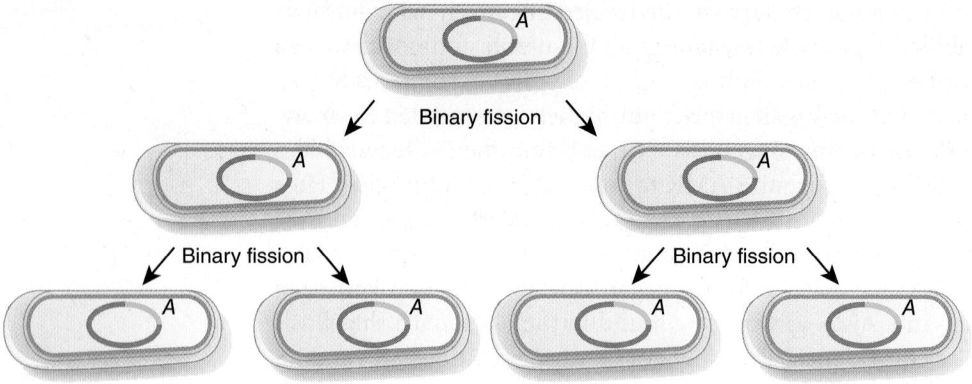

(B) Horizontal gene transfer involves genes (such as gene *A*) being passed directly to another cell through conjugation, transduction, or transformation.

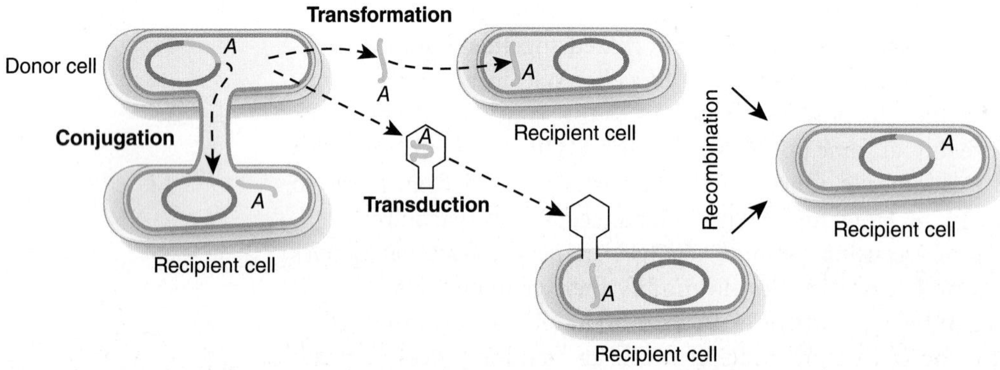

FIGURE 9.7 Bacterial Conjugation. When the F factor (plasmid) is transferred from a donor (F^+) cell to a recipient (F^-) cell, the F^- cell becomes an F^+ cell by acquiring a copy of the F factor.

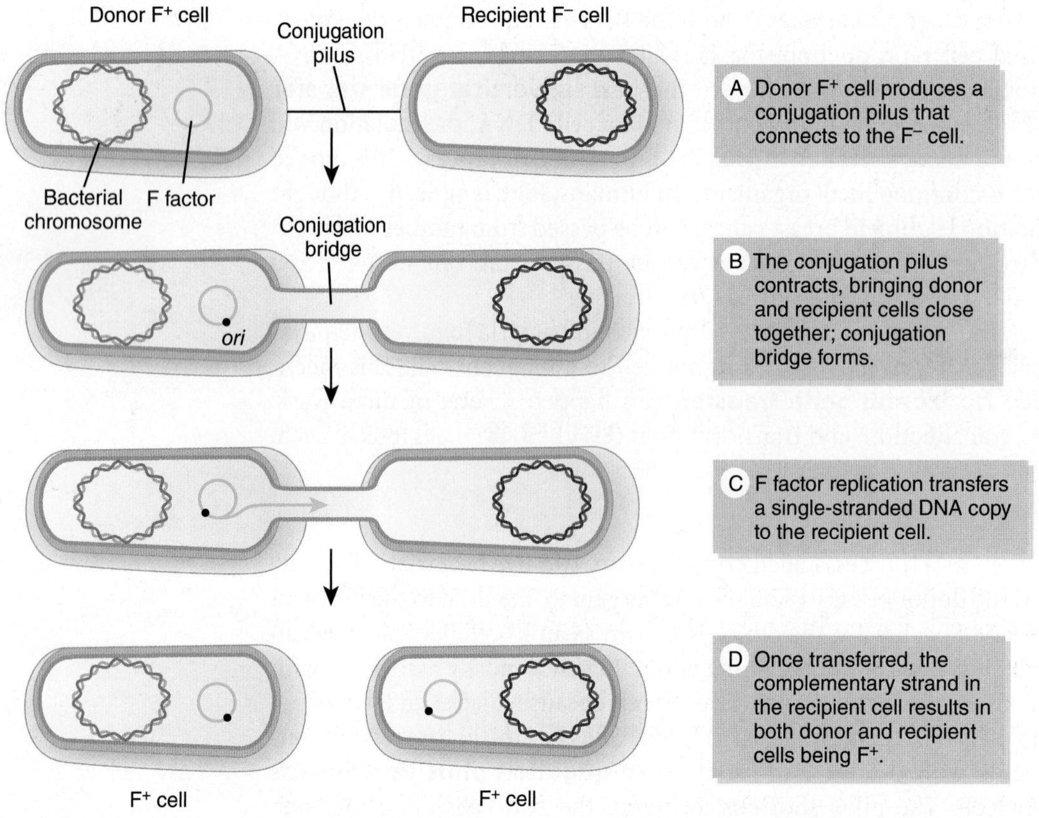

A conjugation bridge then connects the two cytoplasms. Once connected, an enzyme starts copying one strand of the plasmid DNA. That single strand of DNA then passes through the bridge and into the recipient cell. When the DNA arrives in the recipient cell, enzymes synthesize a complementary strand, and a double helix is formed.

By acquiring a copy of the F factor, the recipient cell now is an F^+ cell. Meanwhile, back in the donor cell, a complementary new strand of DNA forms and unites with the leftover strand of the original F factor. So, the donor cell has not changed in any way.

Let's think about this with the shigellosis outbreak. This form of conjugation could have transferred the MDR genes from *E. coli* to *Shigella*. If an F^+ plasmid in *E. coli* contained the four genes for AMR, contact with *Shigella* cells could result in transfer of the genes. In fact, resistance to multiple antibiotics can be acquired through this form of genetic recombination. Add to this that other types of plasmids might contain genes for disease factors like bacterial toxins or enzymes, conjugation can increase the disease potential of a bacterial population.

Transduction

A second recombination process involving horizontal gene transfer is **transduction**. This process involves the help of a bacterial virus **bacteriophage** (**phage**) to transfer the bacterial DNA to a recipient cell. As described in the chapter on Viruses, during phage replication, the virus DNA penetrates the bacterial host cell and immediately starts the replication of new phages in the bacterial cytoplasm. When this happens, the bacterial cell serves as a biochemical factory to produce hundreds or thousands of new phages.

During phage replication, the bacterial chromosome is cut into many small fragments. Sometimes during the virus replication process, some newly forming phages mistakenly contain a bacterial gene or a small DNA fragment along with an incomplete set of viral genes (**FIGURE 9.8**). In either case, these are called "defective particles" because they do not carry a complete set of viral genes. The defective particles can infect another bacterial cell but they cannot replicate in the cell. On infection, the bacterial DNA (and any viral DNA) of the defective particle might recombine with the recipient cell's chromosome.

Look again at Figure 9.8. Suppose the defective particle contained a gene for resistance to cipro. The defective particle will infect and recombine its DNA into the recipient's chromosome. Then, cipro resistance will be a permanent part of the recipient's genome. Transduction is a rare event because phages seldom make the "mistake" of picking up bacterial DNA when they replicate. Therefore, it is unlikely that transduction events passed MDR to *Shigella* in the Guatemala shigellosis cases.

However, in the oceans where there are unimaginable numbers of bacterial cells and even greater numbers of phages, transduction can be a significant form of genetic recombination, as **A CLOSER LOOK 9.1** points out.

Transformation

The third form of genetic recombination is **transformation**. During this form of horizontal gene transfer, a fragment of DNA released from a dead bacterial cell is acquired by a living recipient cell (**FIGURE 9.9**).

Let's consider an example of how transformation can occur. Suppose a patient has been treated for an infection of the colon caused by a pathogenic strain of *E. coli*, such as *E. coli* O157:H7. The physician used an antibiotic called gentamicin, and the susceptible bacterial cells were killed by the drug.

FIGURE 9.8 Transduction. Defective phages (particles) can transfer a few bacterial genes to a recipient (transduced) cell.

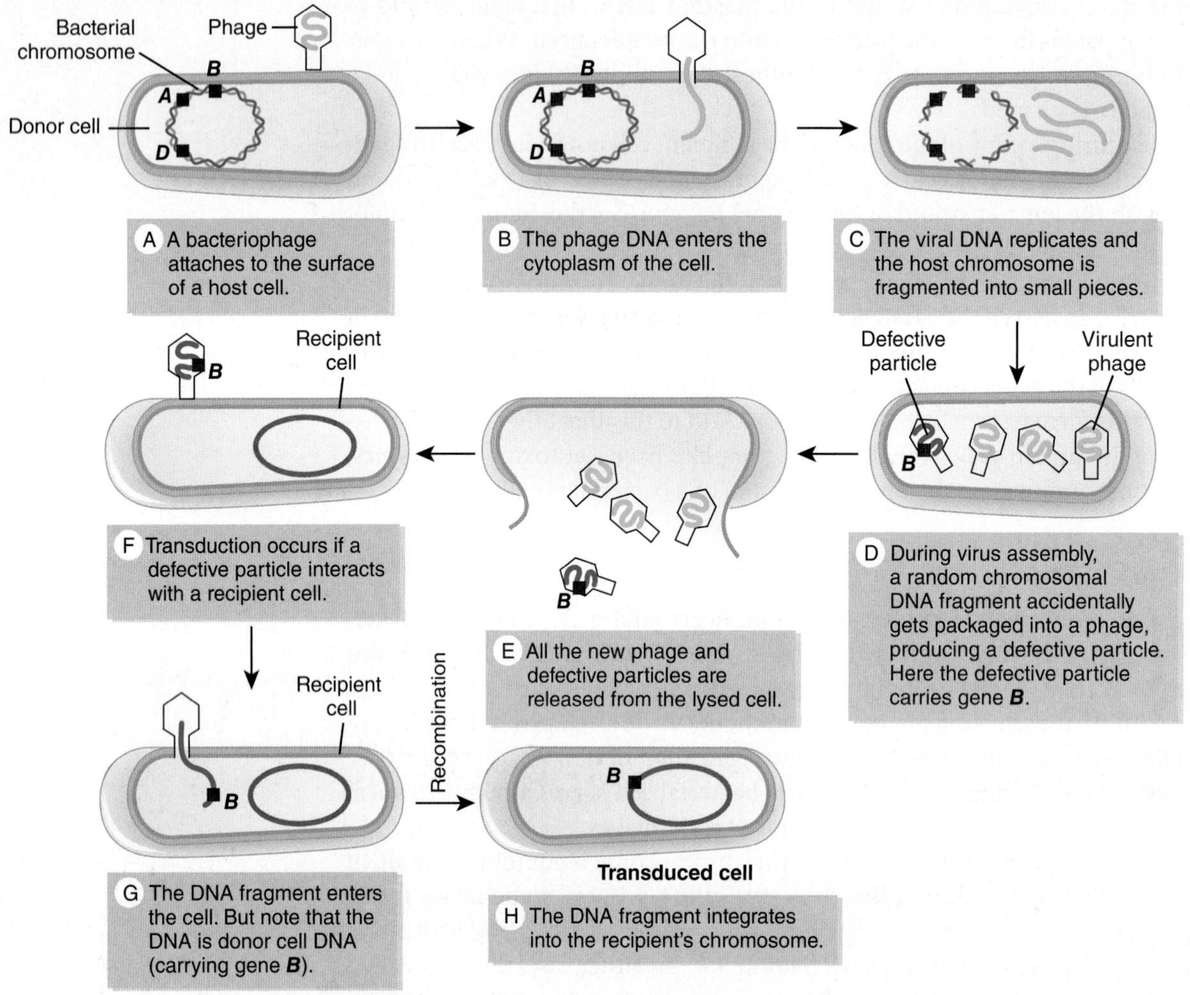

Now, suppose this strain of *E. coli* had a gene enabling the cells to resist the antibiotic cipro; that is, the gene would have protected the microbes against cipro if the physician had prescribed that antibiotic.

However, the physician chose gentamicin, and the bacterial cells were killed. Cell killing though resulted in fragments of the *E. coli* DNA being strewn about in the intestinal tract of the patient.

Now, through drinking *E. coli*-contaminated water, a few *Shigella* cells enter the patient's intestine. The *Shigella* cells can acquire fragments of the *E. coli* DNA present in the intestine. So, the *Shigella* cells pick up the gene for resistance to cipro. They then recombine the new DNA into their own DNA and become resistant to cipro. Next, they pass out of the intestine in the feces and accumulate in soil or water, where they reproduce to a sizable bacterial population, all of which have resistance to cipro.

Over the course of years, cells of this *Shigella* species conceivably could acquire resistance to numerous antibiotics through transformation. Perhaps, during that fateful period of 1969, such an MDR *Shigella* species found its way into a food or water supply. Consumption of the contaminated food or water by the unsuspecting people of Guatemala then lead to shigellosis and the misery and death described in the chapter introduction.

This is just a hypothetical example, but it shows a way that *Shigella* could acquire MDR through transformation. Health experts will never know exactly

A CLOSER LOOK 9.1

Gene Swapping in the World's Oceans

We are aware of the immense numbers of microorganisms in and around us. However, we are less familiar with the massive numbers of microbes in the world's oceans. Microbial ecologists estimate there are an estimated 10^{29} prokaryotes in these marine environments. More impressively, there are some 10^{30} bacteriophages in the oceans that can infect these oceanic microbes.

During the transduction process, bacteriophages sometimes by mistake carry pieces of the bacterial chromosome (rather than viral DNA) from the infected cell. These defective phages then transfer the DNA to another recipient cell. However, these transduction events are rare, occurring only once in every 100 million (10^8) virus infections. That might seem insignificant until the number of phages and susceptible bacteria in the oceans are factored in. Virologists calculate that if a transduction event happens only 1 in every 10^8 infections, there still are about 10 million billion (10^{16}) such gene transfers per second in the world's oceans. That is about 10^{21} transduction events per day!

Microbiologists do not understand what all this genetic recombination means on the global scale. What we can conclude is there's an awful lot of gene swapping going on!

Courtesy of Dr. Jeffrey Pommerville.

FIGURE 9.9 Bacterial Transformation. Transformation is the process in which a DNA fragment from the environment binds to a live recipient cell, passes into the recipient, and incorporates into the recipient's chromosome.

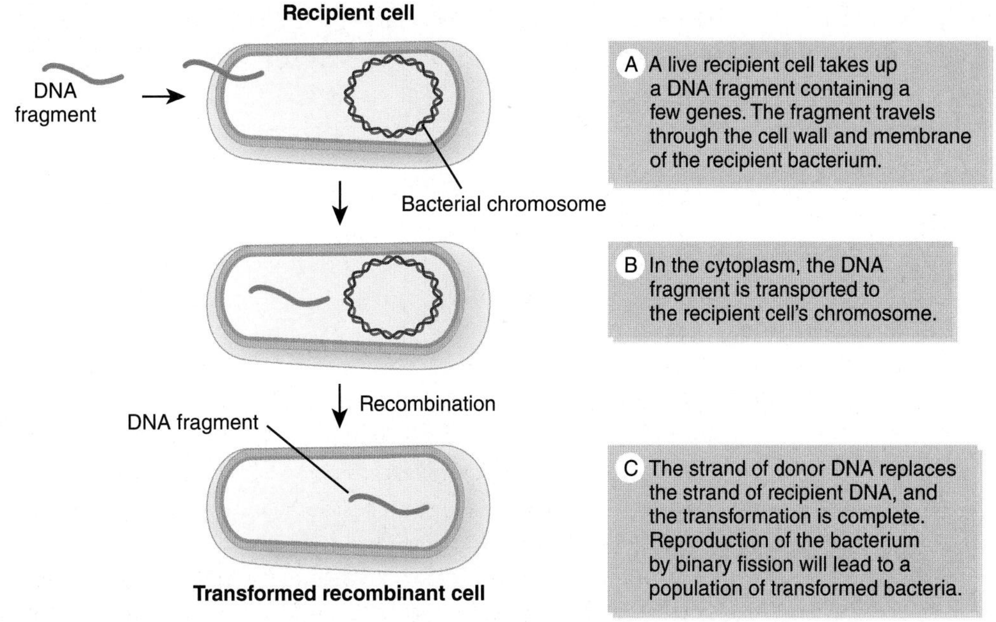

how MDR was acquired by *Shigella*. Most likely, it would have been conjugation or transformation.

Other Ways of Transferring Genes Horizontally

Another possible form of genetic transfer involves movable genetic elements called **transposons**. Transposons are small segments of DNA that can move from one position to another in the bacterial chromosome. Transposons have been called "jumping genes" because the segments simply jump from one chromosomal site to another, from a plasmid to a chromosome, or from a chromosome to a plasmid.

Because transposons can be found in plasmids, the transposons also can be transferred to a recipient cell through a horizontal gene transfer process like conjugation (**FIGURE 9.10**). Therefore, if a transposon in a plasmid carries information for MDR, plasmid transfer will carry the transposon to the recipient and confer to that cell the antibiotic resistance. There, the transposon might then jump to the recipient's chromosome and become a permanent part of its genome.

Antibiotic Resistance in Today's World

An alarming number of bacterial species have evolved antibiotic resistance, and many of these exhibit MDR. This latter group, the so-called **superbugs**, have become one of the most serious problems facing the medical community and

FIGURE 9.10 Transposon Transfer.

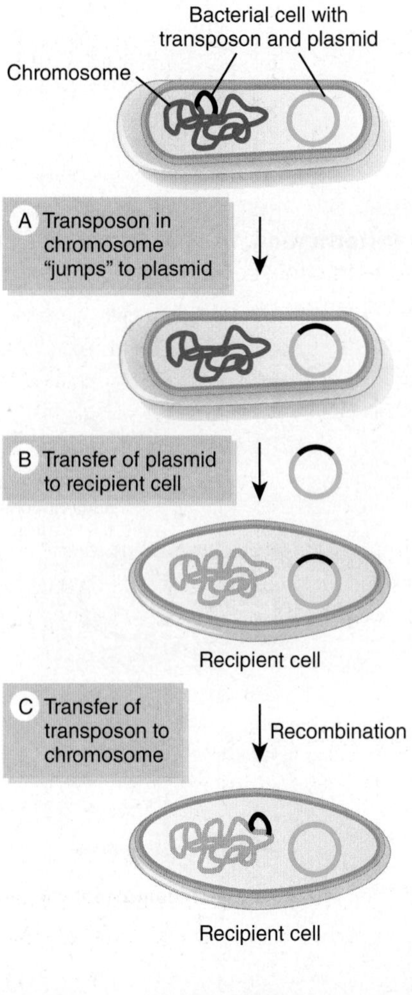

society today. Although the topic of antibiotic resistance will be presented in the chapter on Controlling Microbes, here it is worth mentioning tuberculosis. This disease is high on the list of concerns regarding antibiotic resistance.

Tuberculosis (TB) has been around for thousands of years, continuing to evolve and resist medicine's best drugs. The disease is caused by *Mycobacterium tuberculosis,* a small, aerobic, nonmotile rod that spreads from person to person through airborne respiratory droplets. Over the centuries, TB has continued to be a killer in the human population. Just in the last two centuries it killed an estimated 1 billion people. In 2017, the CDC reported more than 9,000 new cases in the United States, while the World Health Organization (WHO) reported that globally 10.4 million people fell ill with TB and 1.4 million died. TB remains the top infectious killer and is the main cause of deaths related to AMR.

Mycobacterium tuberculosis: my-koh-back-TIER-ee-um too-ber-cue-LOH-sis

For decades, TB has been treated with so-called first-line drugs that can cure the disease. Now, many *M. tuberculosis* strains are resistant to these drugs as well as other drugs used to replace the original ones. Such patients are said to have multidrug-resistant tuberculosis (MDR-TB). Almost half of these cases are in India, China, and the Russian Federation. Alarmingly, an increasing number of MDR-TB cases now have evolved into extensively drug-resistant tuberculosis (XDR-TB), meaning almost all antibiotics used to treat TB are useless. Few treatment options remain for these individuals, and a successful outcome is uncertain.

Physicians still have a few useful TB treatment alternatives by using drug combinations. Still, they are grappling with the possibility that one day soon nothing will be left in the antibiotic arsenal to treat TB patients. Importantly, the crisis of AMR is all due to evolution by mutations and genetic recombination, as we will consider in the chapter on Controlling Microbes.

9.4 Genetic Engineering: Intentional Gene Recombination

During the 1970s, researchers discovered they could mimic the processes of mutation and genetic recombination in the laboratory. They could alter the gene content of DNA. Soon, the scientists were cutting and joining DNA from different organisms, and they were removing and inserting genes into cells. They were breaking new ground in pure and applied science.

The technology is called **genetic engineering**, and it aims to change the genetic makeup of a cell or an organism. Genetic engineers do this by removing genes from a cell or by introducing new DNA prepared in the laboratory. Today, the technology provides possibilities of genome alteration that can benefit the environment and society. Few scientists could have imagined any of this a generation ago.

In this concluding section, we explore how genetic engineering emerged from studies in microbial genetics. Another chapter on Biotechnology and Industry describes the fabulous fruits of this technology called **biotechnology**.

Restriction Enzymes and Recombinant DNA

To carry out genetic engineering, several molecular tools and special techniques are required. Here are the basics of the process.

Restriction Enzymes

Many bacteria produce enzymes called **restriction endonucleases**. Genetic engineers use these enzymes as "biochemical scissors" to cut a DNA molecule at

FIGURE 9.11 Restriction Endonucleases.
(A) A restriction endonuclease is an enzyme that cuts through two strands of a DNA molecule at a specific site.

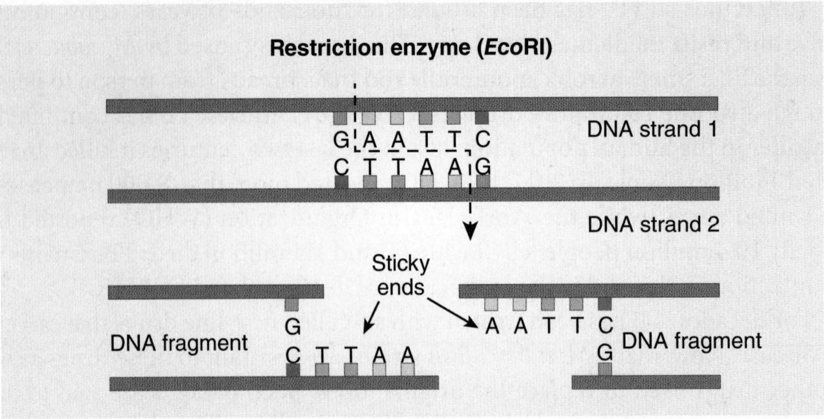

(B) The recognition sites recognized by three endonucleases. Note that *Eco*RI [in (A)] and *Hind*III [in (B)] leave "sticky ends" that can attach to other fragments with complementary "sticky ends."

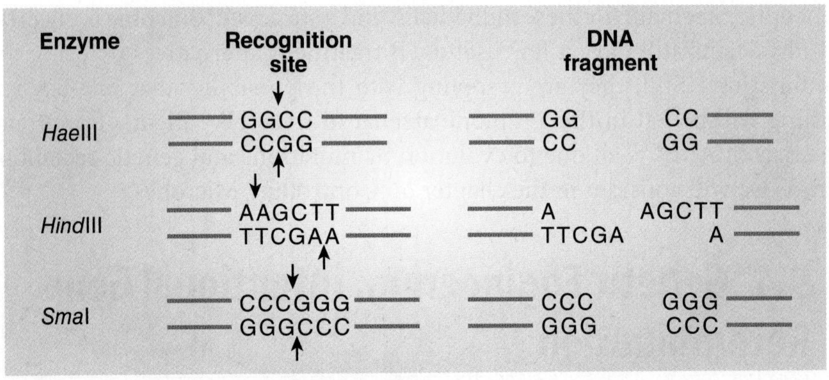

specific points into small fragments (**FIGURE 9.11A**). The cutting often produces "**sticky ends**," which represent unpaired DNA bases with the potential to complementary base pair. For example, in **FIGURE 9.11B**, enzyme *Hind*III would cut any DNA molecule at sequence AAGCCT, regardless of whether the DNA was from a plant, animal, bacterium, or virus.

Recombinant DNA

Recombinant DNA refers to two or more pieces of DNA from different organisms and/or viruses that have been joined together into one DNA molecule. As a simple example, let's use the endonuclease *Eco*RI to cut open the DNA molecule from a virus and then insert that viral DNA into an *E. coli* plasmid that also had been cut open with the same endonuclease (**FIGURE 9.12**). Notice that sticky ends are produced when the two DNA molecules are cut. When the two DNA molecules are mixed, the sticky ends complementary base pair and an enzyme called **DNA ligase** is added seal the molecule. In doing so, a recombinant DNA molecule has been produced containing DNA from two different sources.

Using Recombinant DNA Molecules

Having formed this recombinant DNA molecule, genetic engineers can transfer the molecule mechanically into bacterial cells, mimicking the natural process of transformation. In doing so, they can move foreign genes into new

FIGURE 9.12 Construction of a Recombinant DNA Molecule. In this construction, two unrelated DNA molecules (bacterial and virus) are joined to form a single recombinant DNA molecule.

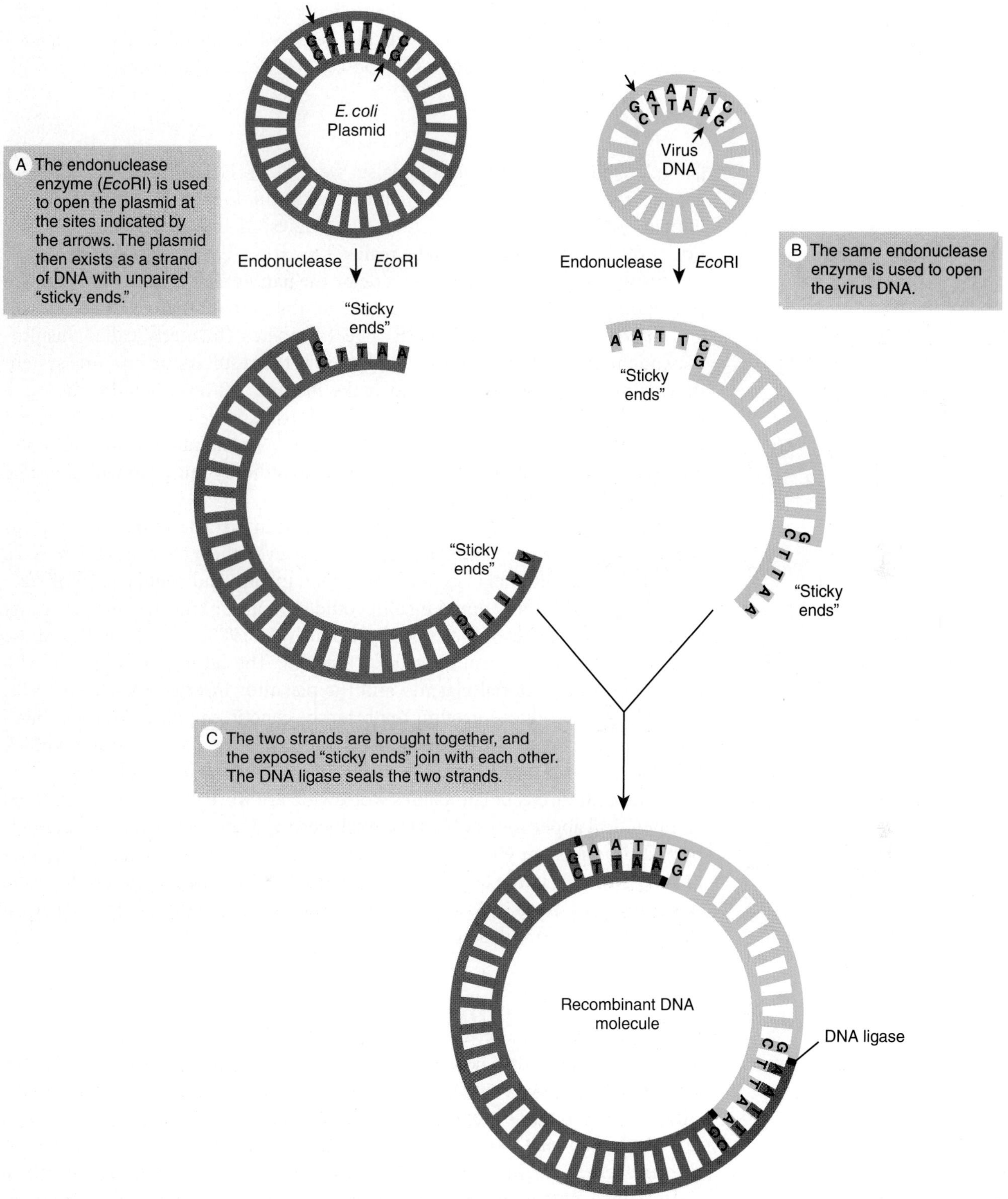

organisms. In this case, viral genes have been introduced into the bacterial cells using the plasmids as the gene carriers. In the cells, the genes on the plasmids will start transcription. The bacterial cells therefore act as tiny factories, producing the protein coded by the genes on the plasmids.

This process of moving genes from one source to another to produce a recombinant DNA molecule is called **recombinant DNA technology**. Importantly, the technology has great value to many aspects of society, including science, agriculture, industry, and medicine. Let's look at one medical consequence of this technology that illustrates the power of genetic engineering.

Diabetes

Genetic engineers and other biologists were quick to see the implications of genetic engineering and recombinant DNA technology. Could they correct a human genetic disorder using the technologies? If they could, the potential power of the technology would be realized.

Insulin is a protein hormone made by the pancreas. It is needed to regulate the blood glucose level and ensure cells get the glucose they need for growth and survival. In individuals with type I diabetes (formerly called insulin-dependent or juvenile diabetes), the disease is the result of the immune system destroying the pancreas' ability to make insulin. Without insulin, there are increased levels of glucose in the blood and urine, which can cause increased fatigue and long-term damage to body organs. To maintain a normal blood glucose level, these diabetics need to get the insulin by injection with a needle or with an insulin pump.

Before 1982, diabetics received purified insulin extracted from the pancreas of cattle and pigs, or even human cadavers. However, these sources often posed a problem because animal insulin is not human insulin and could trigger allergic reactions. In addition, animal insulin could contain unknown disease-causing viruses that had infected the animal previously. In 1982, Eli Lilly marketed the first synthetic human insulin, called Humulin. The human insulin gene was transferred to a bacterial plasmid and the plasmids inserted into *E. coli* cells. The cells then produced insulin. Importantly, genetic engineered human insulin is identical to the insulin naturally made inside the human body. **A CLOSER LOOK 9.2** describes how the engineering achievement was accomplished.

Today, hundreds of companies worldwide are working on the commercial and practical applications of genetic engineering. Many of the genetically engineered products are either proteins expressed by the recombinant DNA in a host organism or in the target organism (**FIGURE 9.13**). The protein products are numerous and diverse. Again, many of these are discussed in the chapter on Biotechnology and Industry.

A Final Thought

You decide you'd rather be 6 foot 2 than 5 foot 1. So, you go out and get some "tall genes"—maybe you pick them up at a used-gene shop or order them online. "That's crazy," you say. "People can't get new genes or change their genetic makeup."

Maybe you can't shop for such genes (yet), but with the advances being made through genetic engineering and recombinant DNA technology, it is possible to change one's phenotype, such as one's height. Medically, one can receive human growth hormone (HGH) that, like insulin, has been produced in bacterial cells. Today, with medical supervision, genetically engineered HGH therapy is available for children with **idiopathic** short stature; in fact, such therapy seems to be effective in reducing the deficit in height as they grow to adults. But what constitutes short stature, and who should be treated medically? Should growth hormone be made available to parents who want their "normal-height" child to grow very tall so he or she can be a basketball or volleyball player? These are

Idiopathic: Referring to a disease or disorder that has no known cause.

A CLOSER LOOK 9.2

Bacteria to the Rescue

Today, people with type I diabetes receive regular injections of insulin to control their blood glucose level. The basic steps in producing genetically engineered insulin are outlined in **FIGURE A**. Note: The segment of DNA containing the human insulin gene (B) can be obtained in several ways (not described here). Also, there are several methods to identify the bacterial cells (F) producing insulin (again, not described here).

At the industrial scale, the bacterial cells can be grown in large, sterile stainless-steel vessels containing all the nutrients and optimal conditions for bacterial growth; the cells divide and produce a large population of genetically identical cells all containing the plasmid (F). When the insulin has been produced, the protein is purified and packaged for distribution (G).

FIGURE A Genetic Engineering. An outline of the steps to engineer the insulin gene into *Escherichia coli* cells.

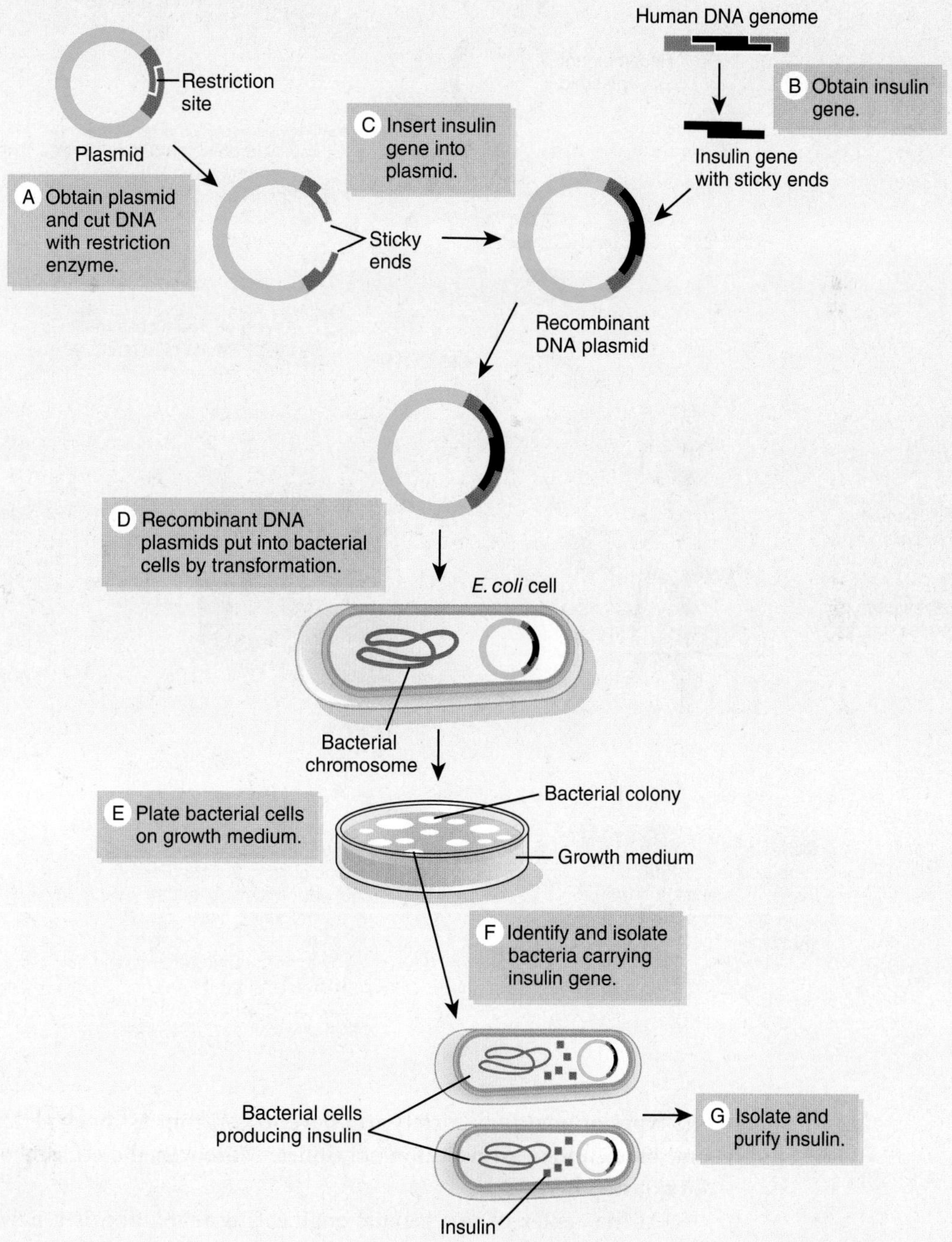

FIGURE 9.13 Developing New Products Using Genetic Engineering. Genetic engineering is a method for inserting foreign genes into bacterial cells and obtaining chemically useful products.

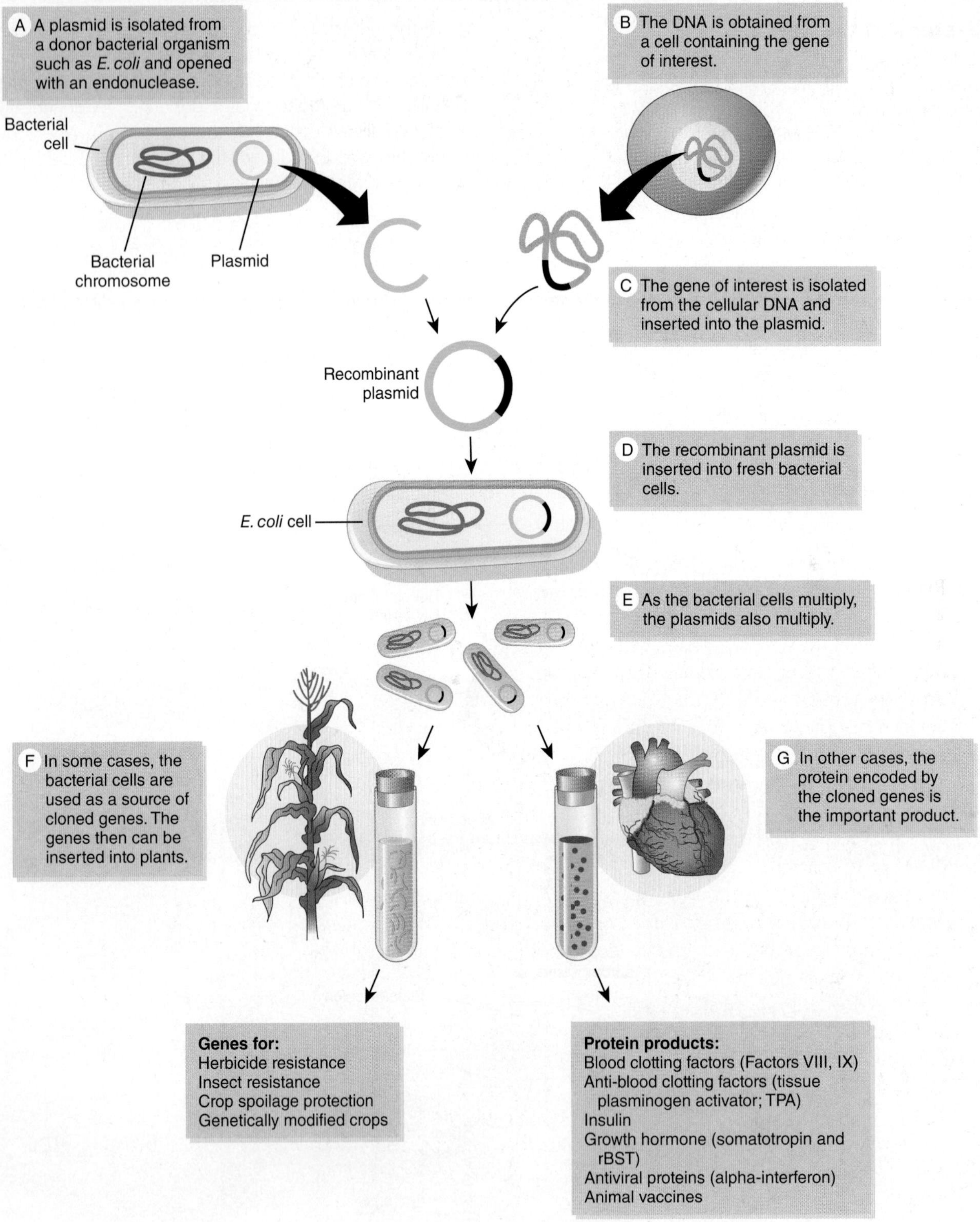

the types of questions society will have to confront as the genetic engineering and biotechnology revolution continues. Moreover, the ethical considerations do not stop here.

At the center of this genetic engineering revolution is a new tool to edit genes and genomes in any organism. This gene editor, called CRISPR, was discussed briefly in the chapter on Viruses. In simple terms, CRISPR is a natural process used by many bacterial species to add "immunity genes" to their genome to protect themselves from phage attack. Genetic engineers and other biologists now have taken this process and modified it, so they can add or change genes in other organisms. CRISPR has been used to reverse antibiotic resistance, making such bacterial cells again susceptible to antibiotics. It has been used to alter mosquito genomes, so the insects cannot transmit diseases like malaria. For humans, and other animals and plants, such "genome engineering" is envisioned as a tool to correct genetic defects or to prevent organisms from evolving undesirable traits. In short, it could correct mutations, which means it could alter the course of evolution by limiting natural selection. Such power has its risks as it could be used for malevolent purposes as well as beneficial ones. In early 2019, Chinese scientists announced the birth of twin girls whose genomes were edited by CRISPR. Consequently, in the coming years, such technology needs to be discussed and carefully evaluated.

Chapter Discussion Questions

What Was He Thinking?

From your reading of this chapter, identify and discuss five major points about microbial genetics that the author was trying to get across to you.

Questions to Consider

1. Explain how a mutation would most easily recognized.
2. The author of an introductory biology textbook writes about the development of antibiotic resistance. She says: "The speed at which bacteria reproduce ensures that sooner or later a mutant bacterium will appear capable of resisting the antibiotic." How might this mutant bacterial cell have come into existence? Do you agree with the statement? Does this bode ill for the future use of antibiotics?
3. Which of the recombination processes (transformation, conjugation, or transduction) would be most likely to occur in the natural environment? What factors would encourage or discourage your choice from taking place?
4. In 1976, an outbreak of pulmonary infections among participants at an American Legion convention in Philadelphia led to the identification of a new disease, Legionnaires' disease. The bacterium responsible for the disease had never before been known to be pathogenic. From your knowledge of bacterial genetics, can you postulate how it might have acquired the ability to cause disease?
5. Some scientists believe mutations are the single most important event in evolution. Do you agree? Why or why not?
6. In 2011, researchers discovered bacterial cells possessing human genes. They found that more than 10% of a *Neisseria gonorrhoeae* population, the species responsible for gonorrhea, contained a human DNA fragment. This fragment was not found in other species of *Neisseria*. How might a bacterial cell have acquired human DNA?
7. It is not uncommon for students of microbiology to confuse the terms reproduction and recombination. How do the terms differ?
8. A classmate asks you to explain the difference between antibacterial resistance (ABR), antimicrobial resistance (AMR), and multidrug resistance (MDR). How would you answer his question?
9. In your opinion, are the prospects of gene technologies (genetic engineering and genome engineering) exciting and promising or are they alarming and worrisome? Give examples to support your answer.

Neisseria gonorrhoeae: nye-SEER-ee-ah gah-nor-REE-eye

CHAPTER 10

Controlling Microbes: In Our Surroundings and Close to Home

Sanitation, Hygiene, and Antibiotics

Until the early 20th century, disease outbreaks and epidemics were common. Many believed these diseases were caused by uncontrollable vapors called miasmas in the air and garbage in the streets. Such perceptions in the early 1800s led an English lawyer and social organizer named Edwin Chadwick to pioneer efforts to safeguard public health. What developed was the Great Sanitary Movement. This program attempted to control filth, odor, and contagion in cities by building sewers, removing garbage, and providing clean water. Although it was more a social crusade than medical program, the movement represents the start of modern public health and sanitation. In fact, the *British Medical Journal* in 2007 polled its readers asking: "What has been the most important medical milestone since 1840?" Sanitation was the top vote getter.

Consequently, today, in the community, we expect sanitary public bathrooms, clean streets, and safe drinking water. Yet, in many parts of the world, unsanitary conditions still exist (**FIGURE 10.1**). Closer to home, we try to keep microbes in check by cooking and refrigerating foods. We clean our kitchen counters and bathrooms with disinfectant chemicals (see chapter opening photo). For personal hygiene, we wash our hands, take regular showers or baths, brush our teeth with fluoride toothpaste, and use an underarm deodorant. In most cases, this daily routine is performed in the hopes of preventing infection and disease.

CHAPTER 10 OPENER Household and industrial cleaning products are diverse and formulated for every cleaning need. Many are especially formulated to disinfect surroundings, such as the kitchen counter or the bathroom toilet bowl in our homes, the food-processing area in a meat factory, or the floors in a hospital ward. The aim with all these products is to maintain sanitary conditions.

FIGURE 10.1 Lack of Sanitation. A young boy in Sierra Leone is washing clothes with water that is trickling out of a wastewater pipe.

Courtesy of CDC.

Louis Pasteur put forward and Robert Koch validated the germ theory of disease in the late 1800s. Their work and that of other microbe hunters added considerably to the understanding of disease and disease transmission. However, this awareness did little to help the infected patient. Then, in the 1940s, thanks to earlier efforts by Paul Ehrlich and Alexander Fleming, antibiotics like penicillin burst on the scene, and a revolution in medicine began. Doctors now could cure patients with bacterial infections. As a matter of fact, discovery of antibiotics was a close second to sanitation in the British journal poll mentioned above. Accordingly, even closer to home, if we still get sick and develop an infectious disease within the body, we often depend on antibiotics or other antimicrobial drugs to pull us through.

In this chapter, we will examine various physical and chemical methods used to control microbes in the environment. Then, we will turn our attention to antimicrobial drugs and especially antibiotics. These medications are at the core of our healthcare delivery system for treating infectious diseases. Applied on a broad scale, sanitation and antimicrobial therapy together make up a major deterrent to infection and disease.

LOOKING AHEAD

After reading and completing this chapter, you will be able to:

10.1 Compare five ways physical methods are used to control microbial growth in the environment and at home.

10.2 Describe several chemical agents used to control microbial growth.

10.3 Explain the mode of action of penicillin and identify the four other bacterial targets for antibiotics.

10.4 Assess the challenge of antimicrobial resistance to society today, and identify the abuses and misuses of these medications.

10.1 Physical Methods of Control: From Hot to Cold

The Citadel, a novel by A. J. Cronin, follows the life of a young British physician named Andrew Manson. In the 1920s, Manson arrives at a small coal-mining town in Wales and soon encounters an epidemic of typhoid fever. When his first patient dies of the disease, Manson becomes troubled. At first, he believes there is nothing he can do to alter the course of the epidemic. Then, in a moment of insight, he realizes microbes succumb to the effects of intense heat. So, he builds a huge bonfire into which go all his patients' bed sheets, clothing, and personal effects. To his delight, the epidemic soon disappears.

Heat Disrupts Cellular Protein Structure

Most microbes populate environments within a specific range of temperatures. Although they survive at the limits of these ranges, they cannot live beyond those points. As described in the chapter on Molecules of the Cell, at high temperatures cellular proteins unravel; they lose their three-dimensional shape. Enzymes become inoperative, and the microbes die as their chemical reactions grind to a halt.

Heat has long been known as a fast, inexpensive, and reliable way of controlling microbes. Often the heating process is designed to achieve **sterilization**. This term implies that all forms of life and viruses have been destroyed or removed. In microbiology, sterilization is an absolute term because an object cannot be "partially sterilized." It is either devoid of life and viruses, or it remains "contaminated" with them. However, in commercial food preparation, **commercial sterilization** is used for canned foods. This form of sterilization applies just enough heat to eliminate all pathogens. However, the food product could still contain some nonpathogenic microbes. It is assumed that cooking (heating) will kill any remaining microbes.

Moist Heat Methods

Among the most widely used methods for controlling microbes is moist heat. The intense heat of the steam generated by boiling water at 100°C (212°F) kills most microbes in 10 minutes. A notable exception is bacterial endospores that require 2 hours or more of heating (**FIGURE 10.2**). Often pressure is used to raise the temperature of steam above 100°C (212°F). As the pressure increases, the temperature of the steam rises, and the destruction of microbes increases proportionally. An apparatus called the **autoclave** (**FIGURE 10.3**) increases steam pressure above atmospheric pressure, allowing the temperature of the steam to increase from 100°C to 121°C (250°F). At this temperature, metal instruments, glassware, microbial media, hospital and laboratory equipment, and virtually anything else that can withstand the high temperature and pressure, will be sterilized. Important exceptions are plastics that would melt and certain chemical substances (e.g., vaccines, antibiotics) that would be rendered inactive.

As consumers, probably the most well-known commercial sterilization process familiar to us is **pasteurization**. The process uses moist heat to destroy any pathogens that might have been introduced into products such as milk, fruit juices, and beer. Pasteurization only reduces the number of spoilage microorganisms in these liquids. Most products today are subjected to the **flash pasteurization method**, because it is quick and fast (**TABLE 10.1**). As an alternative, some products today are being subjected to **ultra-high temperature** (**UHT**)

FIGURE 10.2 Temperature and the Physical Control of Microorganisms.
Notice that materials containing bacterial endospores require longer exposure times and higher temperatures for killing.

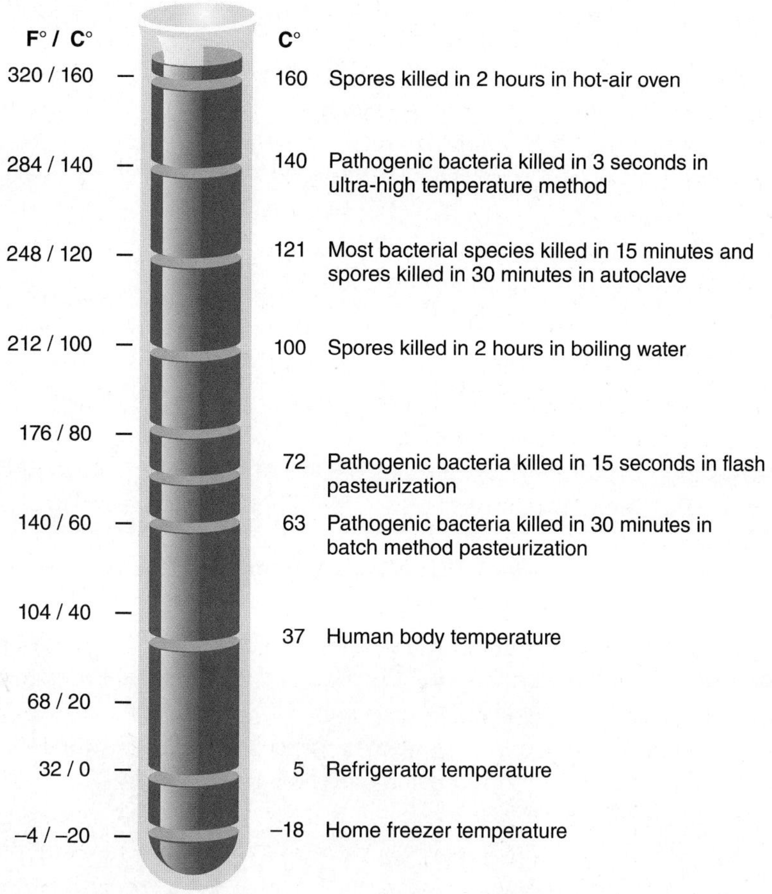

FIGURE 10.3 Operation of an Autoclave.
(A) An autoclave can be used to sterilize liquid materials.

(B) Steam enters through the port (A) and passes into the jacket (B). After the air has been exhausted through the vent, a valve (C) opens to admit pressurized steam (D) that circulates among and through the materials, thereby sterilizing them. At the end of the cycle, steam is exhausted through the steam exhaust valve (E).

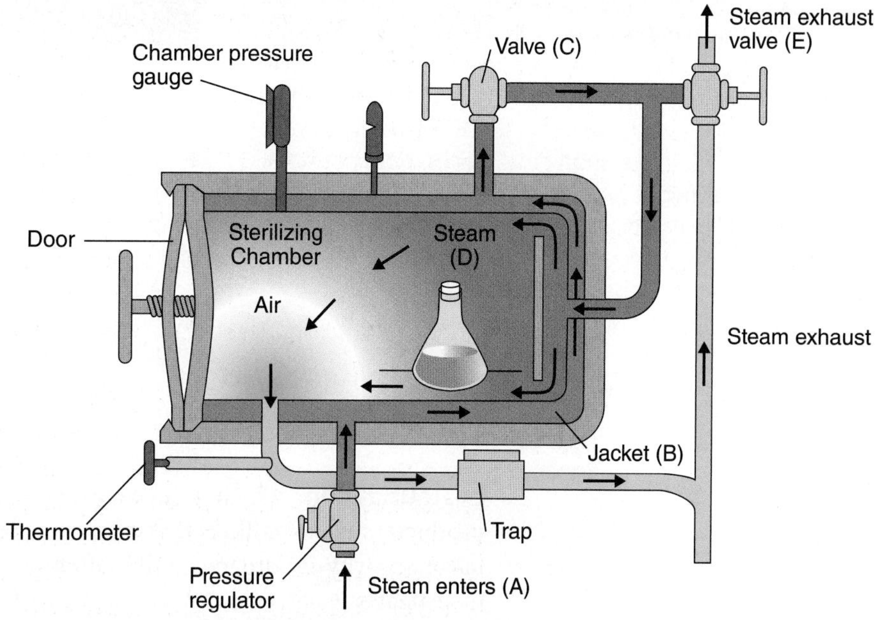

TABLE 10.1 Pasteurization Methods

Method	Temperature	Processing Time	Shelf Life
Batch (vat) pasteurization (low temperature long time; LTLT)	63°C (145°F)	30 minutes	5–7 days (refrigerated)
Flash pasteurization (high temperature short time; HTST)	72°C (162°F)	15 seconds	14–16 days (refrigerated)
Ultra-pasteurization (ultra-high temperature; UHT)	140°C (280°F)	1–3 seconds	2–4 weeks (refrigerated)* * Up to 9 months unrefrigerated if using previously sterilized containers under sterile conditions.

A CLOSER LOOK 10.1

Does Milk Stay Fresher Longer If It's Organic?

If you ever go shopping for milk at the local market, especially one specializing in "natural and organic foods," you might notice that the "Best by" date expires much sooner on a carton of regular milk than on a carton of organic milk. In fact, the date on the organic milk might be some 3 weeks longer than on the regular milk, which typically is 5 to 16 days from the store delivery date. So, being organic, does that ensure a longer shelf life?

The fact that the milk is organic has nothing to do with its longer shelf life. Labeling the milk as "organic" only means the cows on the dairy farm were not given antibiotics or hormones like bovine growth hormone (BGH), which stimulates a cow's milk production. The reason it has a longer shelf life is due to the pasteurization process. Organic milk is subjected to the ultra-high pasteurization (UHT) process, where the milk is heated to 140°C (280°F) for 3 seconds (see **FIGURE A**). This kills all the microorganisms that might be in the liquid—it is sterile. Most regular milk today is subjected to the flash pasteurization method where the milk is held at about 72°C (162°F) for 15 seconds. This "high temperature, short time" process does not kill all microbes that might be in the milk; only the pathogens have been eliminated. Because there are bacterial species that can survive refrigerator temperatures, the milk can spoil if left on the refrigerated shelf too long.

Regular milk also could be subjected to UHT. However, it usually is not because it must travel only a short distance to market; organic products, on the other hand, are not often produced throughout the country, so they have further to travel to reach the consumer. Therefore, UHT preserves the product longer. Although appearing more frequently in the United States, room-temperature Parmalat milk is a product of UHT and can be found commonly in Europe, Mexico, and other parts of the world. So, if shelf life is important to you, simply look for products treated by UHT.

FIGURE A UHT Means Longer Shelf Life. A carton of organic milk.

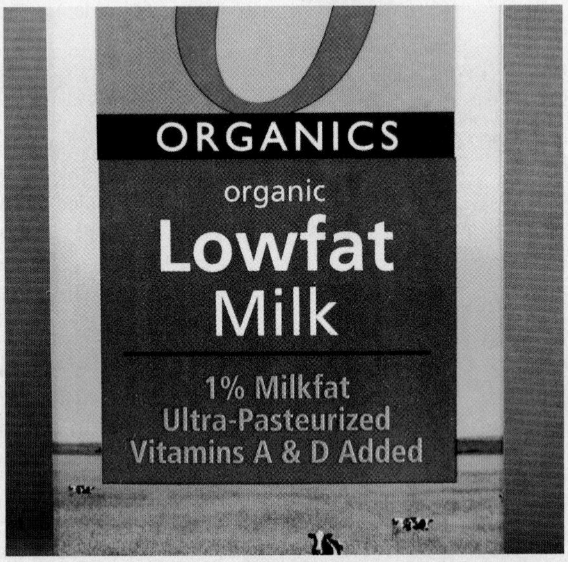

Courtesy of Dr. Jeffrey Pommerville.

pasteurization, which is even a faster process. The drawback with these products, such as milk, is that some people say the product can have a burned taste. So, why is "organic milk" often subjected to UHT? **A CLOSER LOOK 10.1** investigates.

FIGURE 10.4 Incineration Is an Extreme Form of Dry Heat. Healthcare workers burn mattresses used by patients infected with the Ebola virus during the 2014–2016 epidemic in West Africa.

Dry Heat Methods

The other method of heat uses dry heat. Dry heat penetrates less rapidly than moist heat, so the sterilizing temperatures tend to be higher and the exposure times longer. To achieve sterilization, a temperature of 160°C to 170°C (320°F to 338°F) must be applied to a population of microbes for a period of 2 or more hours.

A direct flame will incinerate microbes very rapidly. This form of dry heat can generate temperatures greater than 1,500°C (2,700°F), which burns material to ashes. In the 14th century, the bodies of disease victims were burned to prevent spread of the Black Death (the Plague). Like Cronin's *The Citadel*, today disposable hospital gowns and other burnable materials might be incinerated, as was done during the 2014–2016 Ebola epidemic in West Africa (**FIGURE 10.4**).

Although heat is a valuable physical method for controlling or eliminating microorganisms, sometimes it is impractical to use. No one would suggest removing the microbial population from a tabletop by using a blowtorch! In addition, heat-sensitive solutions cannot be sterilized in an autoclave. In instances such as these, and numerous others, a heat-free method must be used to eliminate or reduce microbial populations.

Radiation Damages Genetic Material

Various types of radiation have destructive effects on microbes (**FIGURE 10.5**). **Ultraviolet radiation** (**UV light**), for example, interacts with the DNA, causing adjacent thymine or cytosine bases to bond together. This alteration disrupts DNA replication and transcription such that the damaged organisms can no longer produce critical proteins or reproduce, and eventually die. UV light is a useful sterilizing agent for dry surfaces, such as a tabletop or flat instruments. It can be used in a closed environment, such as an operating room, to lower the microbial population of the air. UV light does not penetrate liquids or solids effectively.

Other kinds of radiation, including X rays and gamma rays, also are useful for sterilization. These forms of high energy radiation are about 10,000 times more active than UV light. They represent **ionizing radiation**, which emits ions that break each DNA strand into smaller fragments. If the affected cell cannot repair the damage, cell death will occur.

Ionizing radiations also are used for preserving foods. Today, **irradiation** of foods is used in more than 40 countries for over 100 food items. These include spices, potatoes, onions, cereals, flour, and fresh fruit. Irradiation reduces potential foodborne illness by killing any pathogens in poultry, red meats (beef,

Irradiation: The process by which an object is exposed to radiation.

FIGURE 10.5 The Ionizing and Electromagnetic Spectrum of Energies. The complete spectrum is presented at the bottom of the chart, and the ultraviolet and visible sections are expanded at the top. Bactericidal refers to the killing of bacteria.

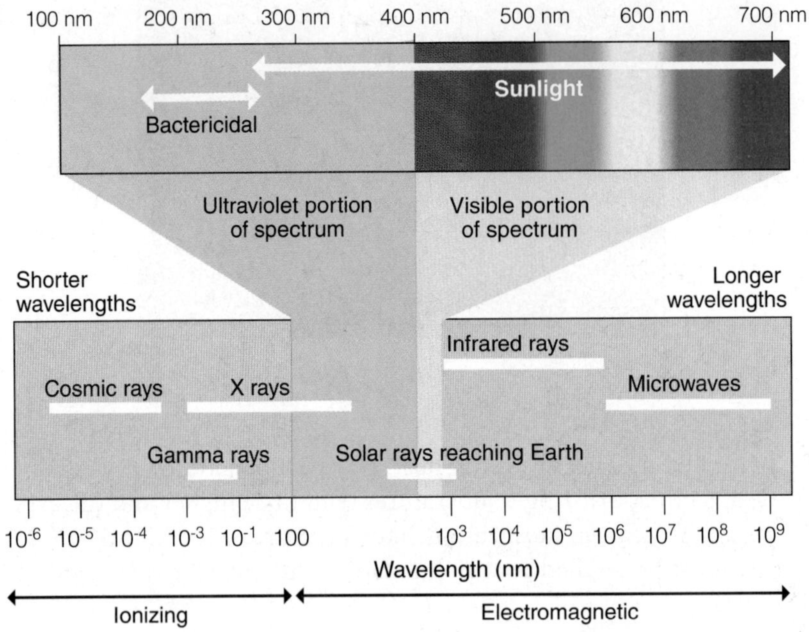

FIGURE 10.6 Salted Fish. These fish in a Malaysian fishing village are being preserved by salting, which removes the water.

lamb, and pork), fresh and bagged spinach, and iceberg lettuce. The irradiation level used on American products is called a **pasteurizing dose**. This treatment is not intended to eliminate all microbes in the food. Rather, like pasteurization, the dose will eliminate the pathogens. Importantly, the irradiated food is not radioactive. The radiation passes completely through the object and therefore is of no danger to the consumer.

Drying Removes Water

Another non-heat method for controlling microbial populations is by drying. Humans have used drying to preserve foods, such as meat and fish, for centuries by drawing out the water with salt (**FIGURE 10.6**). Likewise, dried fruit can be preserved by using sugar, rather than salt. Evaporative drying is used today to prepare foods such as cereals, grains, and numerous other products for storage in the home pantry. Whether the water is lost by evaporation, or by using salt or sugar, the lack of water means metabolism in microbes cannot occur. It then follows that microbes cannot grow or reproduce.

Hikers, campers, and backpackers often carry foods prepared by **freeze-drying**, or more technically, **lyophilization**. In this process, the food is frozen and then the water is **vaporized** by vacuum pressure. The result is an extremely light and dry product that can be rehydrated when needed.

Vaporized: Referring to a solid being converted directly into vapor (gas).

Filtration Traps Microbes

Microbes can be removed physically from a liquid solution by the process of **filtration**. During filtration, liquid passes through a porous filter while trapping microbes in its submicroscopic pores (**FIGURE 10.7**). For heat-sensitive solutions, such as vaccines and antibiotics, filtration is a useful method to obtain a microbe-free solution. Many beverages, such as "cold-filtered beer," are filtered rather than pasteurized because the heating process might affect the quality or taste of the product.

Air also can be purified to remove most microorganisms. Most of us have some type of air filter in our home air conditioning/heating system. These filters trap dust and microscopic particles, like smoke, mold spores, and pollen. Hospitals and facilities that need to maintain a high standard of air quality use **high-efficiency particulate air (HEPA) filters**. These filters consist of a mat of randomly arranged fibers that trap particles, microorganisms, and spores.

Low Temperatures Slow Microbial Growth

Low temperatures are a physical method to control, but not eliminate, microbial populations. At low temperatures, such as in a refrigerator (5°C/41°F) or freezer (−18°C/0°F), enzyme activity diminishes. This means microbial reproduction

FIGURE 10.7 The Principle of Filtration. A liquid containing bacterial cells is poured into a filter, and a vacuum pump helps pull the liquid through and into the flask below. The bacterial cells, being larger than the pores of the filter, become trapped on the filter surface. Inset: A scanning electron microscope image of bacterial cells trapped on the surface of a filter. (Bar - 5 µm).

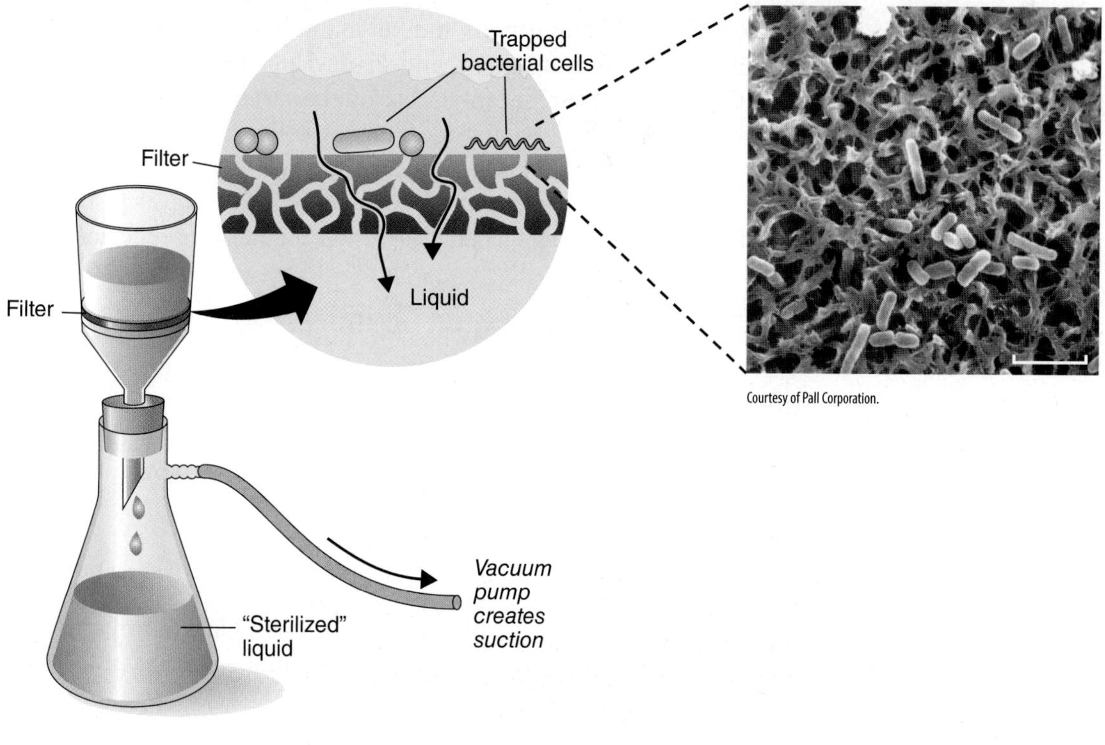

slows considerably (see Figure 10.2). Consequently, refrigerating or freezing foods preserves the foods but does not sterilize the foods. Frozen foods should not be thawed and refrozen because during the thawing process, the bacteria have time to start multiplying.

10.2 Chemical Methods of Control: Antiseptics and Disinfectants

The practices of sanitation and chemical control of microbes are not new. The Bible often refers to cleanliness and prescribes certain dietary laws to prevent consumption of contaminated foods. The Egyptians tried to prevent decay of mummies by using resins and aromatic chemicals for embalming. Other ancient peoples burned sulfur as a way to deodorize and cleanse materials.

Medicinal chemicals came into widespread use in the 1800s. During the American Civil War, solutions of iodine were valuable in preventing infections in soldiers. The chemical control of microbes received a considerable boost in the 1860s with the work of Joseph Lister. As a British surgeon, he was puzzled as to why more than half his amputation patients died—not from the surgery—but rather from postoperative infections. Hearing of Pasteur's germ theory, Lister hypothesized that the surgical infections his patients developed resulted from germs in the air. So, in 1865 Lister started using a carbolic acid spray in surgery and on surgical wounds. The results were spectacular. Most of the surgical amputations healed without infection. His technique would revolutionize medicine and the practice of surgery. Importantly, it ushered in the practice of **antisepsis**, the use of chemical methods to prevent infections of external living surfaces, such as the skin.

General Principles of Chemical Control

Chemical controls are designed to interrupt the spread of microbes. The chemicals are applied in such diverse locations as the hospital environment, the food-processing plant, restaurants, and the typical household. Most chemicals do not sterilize. Rather, they accomplish **disinfection**, which represents a method to lower microbial populations and kill many, but not necessarily all, pathogens. Specifically, **disinfectants** are chemicals developed for use on inanimate (lifeless) objects. **Antiseptics** are meant for use on the surface of the body (**FIGURE 10.8**).

No single chemical agent is ideal for controlling all microbes under all conditions. However, if an ideal chemical agent were to exist, it would possess an elaborate array of characteristics as summarized in **TABLE 10.2**. With such stringent requirements, it is not surprising that an ideal disinfectant or antiseptic does not exist.

A Survey of Some Common Chemical Agents

The chemical agents currently in use for controlling microorganisms range from very simple liquid substances to very dangerous gases. We will survey several common groups of chemical agents and indicate how they are applied to control microorganisms. **A CLOSER LOOK 10.2** looks in your pantry to identify some common but surprising antiseptics.

Alcohols and Peroxides

Among the widespread chemical agents for microbial control are the **alcohols**. The most widely used is ethyl alcohol (ethanol), usually in a 70% solution.

FIGURE 10.8 Some Uses for Antiseptics and Disinfectants.
(A) Antiseptics are used on body tissues, such as on a wound or before piercing the skin to take blood.

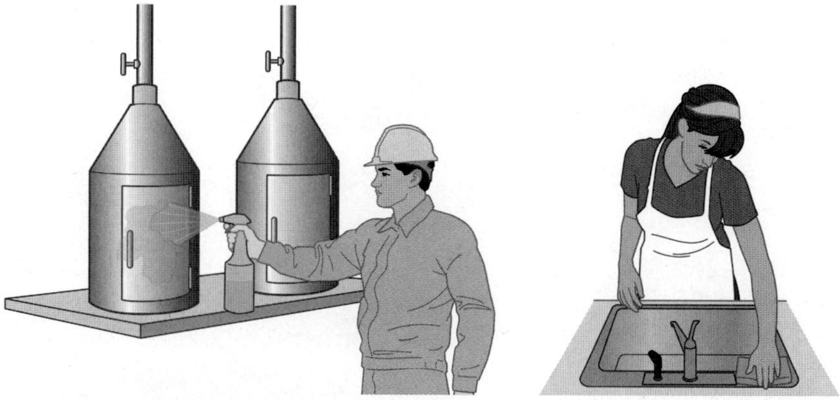

(B) Antiseptics are used on inanimate objects, such as equipment used in an industrial process or tabletops and sinks.

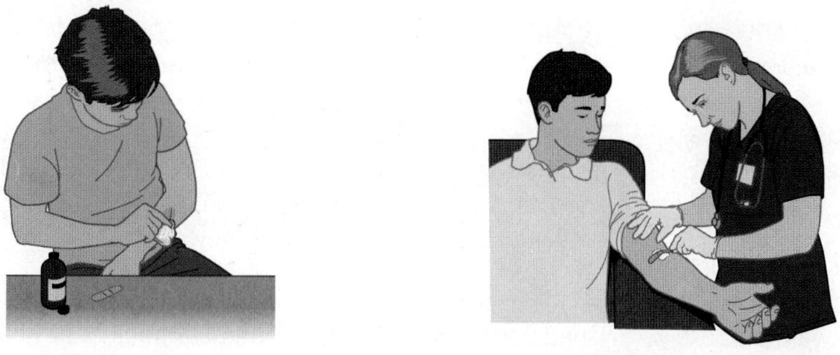

TABLE 10.2 Characteristics of an Ideal Chemical Agent

Characteristic
Kills all microbes.
Dissolves in water.
Remains stable on standing for extended period.
Is nontoxic to humans and animals.
Does not combine with organic matter other than in microbes.
Has strongest action on microbes at room or body temperature.
Penetrates surfaces efficiently.
Does not corrode or rust metals or damage or stain fabrics.
Is available in useful quantities and at reasonable prices.

A CLOSER LOOK 10.2

Antiseptics in Your Pantry?

Today, we live in an age when alternative and herbal medicine claims are always in the news. Often, many of these claims seem unbelievable. Regarding "natural products," are there some that have genuine medicinal and antiseptic properties?

Garlic

In 1858, Louis Pasteur examined the properties of garlic as an antiseptic. During World War II, when penicillin and sulfa drugs were in short supply, garlic was used as an antiseptic to disinfect open wounds and prevent infection. Since then, numerous scientific studies have tried to discover garlic's antiseptic powers.

Many research studies have identified a sulfur compound called allicin as one key to garlic's antiseptic properties. When a raw garlic clove is crushed or chewed, allicin gives garlic its characteristic smell and taste. Laboratory studies suggest that this compound can interfere with the microbes causing the common cold, flu, sore throat, and sinus and respiratory infections. The findings indicate that the compound blocks key proteins that bacteria and viruses need to invade and damage host cells.

Cinnamon

Cinnamon might be an antiseptic that could control pathogens in fruit beverages. In several studies, commercially pasteurized apple juice was purposely contaminated with typical foodborne pathogens and viruses. Cinnamon then was added to some of the contaminated juices. The investigators discovered that more pathogens were killed in the cinnamon blend than in the cinnamon-free juice.

In another study carried out by a different research group, 10% Saigon and Ceylon cinnamons deactivated 99.9% of viruses, suggesting that the spice might help eliminate or prevent virus infections.

Honey

For more than three decades, Professor Peter Molan, has been studying the medicinal properties of and uses for honey. Its acidity, between 3.2 and 4.5, is low enough to inhibit many pathogens. Its low water content (15% to 21% by weight) means that it "drains water" from wounds, depriving pathogens of an ideal environment. In addition, two proteins in honey interfere with the integrity of the bacterial cell wall and cell membrane.

PHOTO A Natural Antiseptics. Honey and cinnamon are two substances with potential antimicrobial properties.

Courtesy of Dr. Jeffrey Pommerville.

Apparently, the antibacterial properties of honey depend on the kind of nectar (the plant pollen) that bees use to make honey. Manuka honey from New Zealand and honeydew from central Europe are thought to have useful levels of antiseptic potency. In 2011, the United States Food and Drug Administration (FDA) approved wound dressings containing manuka honey.

Licorice Root

Dried licorice root has been used in traditional Chinese medicine for centuries. Scientists have identified two substances in licorice root capable of killing the two most common bacterial species causing tooth decay and one species responsible for gum disease. But don't run out and start eating lots of licorice candy. The licorice root extract originally in candies has been replaced by anise oil. It has a similar flavor but no antiseptic properties.

Wasabi

The green, pungent, Japanese horseradish called wasabi is a spicy condiment for sushi. However, there are some natural chemicals in wasabi called isothiocyanates. In lab experiments, these chemicals inhibit the growth of the bacterial species involved with tooth decay. Interesting, but any antiseptic properties of wasabi will need to be proven in human clinical trials.

So, there appear to be products that might have antimicrobial properties—and there are many more than can be described here.

Venipuncture: The piercing of a vein to take blood, to feed somebody intravenously, or to administer a drug.

Topical: Refers to a surface area, especially the skin.

The alcohol disrupts protein structure and dissolves lipids in cell membranes. It is the active ingredient in many popular hand sanitizers, and it is used as an antiseptic to treat the skin before a **venipuncture**. In these cases, the alcohol mechanically removes bacterial cells from the skin surface. Isopropyl (rubbing) alcohol also is effective as a **topical** antiseptic.

Hydrogen peroxide is a mild household antiseptic. The liquid commonly is used on the skin to prevent infection of minor scrapes and abrasions. When applied to the affected area, the solution foams and bubbles. This is because the enzyme catalase in the damaged tissue breaks down hydrogen peroxide to oxygen gas (the bubbles) and water. The furious bubbling loosens dirt, debris, and dead tissue, and the oxygen gas is effective against anaerobic bacterial species. Hydrogen peroxide breakdown also produces a reactive form of oxygen that is highly toxic to microorganisms and viruses. However, the chemical is not recommended as an antiseptic for open wounds because its action can harm the damaged tissue and delay healing.

Halogens and Heavy Metals

Halogens are extremely reactive elements. Two common halogens are iodine and chlorine. Iodine can be used as a **tincture** (2% iodine in ethyl alcohol) or as an iodine-detergent compound. The detergent loosens microbes from the skin surface and the iodine kills them.

Tincture: A low concentration of a chemical dissolved in alcohol.

Chlorine is used as a gas, liquid, or solid to reduce the microbial content in water, such as in swimming pools. Like iodine, chlorine reacts with proteins, is effective against all types of microbes, and might be used in antiseptics or disinfectants depending on the preparation and concentration. Chlorine also is used in the form of sodium hypochlorite in a 5% concentration in household bleach (Clorox). Other uses for chlorine are shown in **FIGURE 10.9**.

Three **heavy metals** have been used as the active ingredient in chemical antiseptics and disinfectants. Silver is employed as a general antiseptic in wound dressings and creams. In these cases, the heavy metal damages enzymes

FIGURE 10.9 Some Practical Applications of Disinfection with Chlorine Compounds. Different chlorine compounds have been used for water disinfection and for countertop and floor disinfection. As antiseptic solutions, chlorine is effective for topical applications.

in membranes. Mercury is another heavy metal that disrupts bacterial metabolism. Although mercury can be toxic, it has been used in some antiseptics, such as Mercurochrome and Merthiolate, for treating minor cuts and scrapes on the skin. As a topical antiseptic, the mercury compounds have been replaced by other safer chemical agents.

Copper, in the form of copper sulfate, is used to control cyanobacteria in swimming pools and to restore the clarity of the water. Copper is very toxic to bacterial cells. In recent studies, hospital fixtures containing copper alloys were found to contain 70% to 90% fewer microbes than did non-copper fixtures. Using such fixtures on high touch surfaces reduced the risk of acquiring an infection by 40%.

Soaps and Detergents

Soaps and detergents are strong wetting agents and surface tension reducers. They work their way between microbes and a surface and "loosen" the microbes, so they can be removed with the wash water. They also dissolve microbial cell membranes by reacting with its lipids. The membranes then become leaky and cell death results.

The most useful detergents to control microorganisms and many viruses are derivatives of ammonium chloride. Called **quaternary ammonium compounds** (**quats**), they are used as disinfectant agents for industrial equipment and food utensils. They also are effective as disinfectants on hospital walls and floors. Quats are found in some contact lens cleaners and mouthwashes. If you use a mouthwash, look on label and if you cannot pronounce the chemical names of the active ingredients, they are probably quats. So, just shake the bottle—if it foams, it contains quats.

Phenols

Phenol (**carbolic acid**) was among the first chemical agents used for microbial control. It was employed by Lister in his antisepsis surgical treatment described earlier in this chapter. Phenol derivatives are the active ingredients in Lysol and some handwashes.

A very widespread phenol derivative is **triclosan**. This chemical destroys bacterial cell membranes by blocking lipid synthesis. The chemical is the active ingredient in numerous products, such as antibacterial soaps, lotions, mouthwashes, toothpastes, toys, food trays, underwear, kitchen sponges, utensils, and cutting boards. Whether triclosan is better than plain soap and water at preventing the spread of pathogens has been questioned. In late 2016, the United States Food and Drug Administration (FDA) banned triclosan and 18 other ingredients in antibacterial soaps and other home antibacterial products because of the lack of evidence supporting the effectiveness of the chemicals.

Ethylene Oxide

The chemical agents described above usually are used in liquid form. However, a few agents can sometimes be used in a gaseous form for sterilization.

Ethylene oxide is a gas that is used to sterilize paper, wood, metals, rubber, and plastic products. For hospitals, catheters, artificial heart valves, heart-lung machine components, and optical equipment can be sterilized with the gas. It is one of the few chemical compounds that is effective at killing bacterial endospores. The National Aeronautics and Space Administration (NASA) even employs the gas to sterilize interplanetary spacecraft.

10.3 Antimicrobial Drugs: Antibiotics and Other Agents

For many centuries, physicians believed heroic measures were necessary to save patients from the ravages of infectious disease. They prescribed frightening courses of purges and bloodlettings. At other times, they were given enormous doses of strange chemical concoctions, immersed in ice-water baths, or simply starved. These treatments probably complicated an already bad situation by reducing the natural body defenses to the point of exhaustion. In fact, the death of George Washington in 1799 is believed to have been due to a streptococcal infection of the throat. His condition probably was made worse by the bloodletting treatment used to remove almost 2 liters of his blood within a 24-hour period.

Thankfully, a revolution in medicine took place during the 1940s, when antibiotics burst on the scene. Doctors were astonished to discover they could kill bacteria in the body without doing substantial harm to the body. Physicians found they now could successfully alter the course of disease using "magic bullets" called antibiotics.

Magic Bullets

When the germ theory of disease emerged in the late 1800s, physicians gained a better understanding of the causes of infectious disease. However, the identification of the infectious agents did not change the treatment for infected patients. Tuberculosis continued to kill one out of every seven people. Streptococcal pneumonia remained a fatal experience. Malaria still took a heavy toll in human life.

Against this backdrop, scientists and physicians dreamed of chemical substances capable of killing microbes in the body without damaging the body. In 1910, the German chemist Paul Ehrlich was searching for such a magic bullet. In his hunt, he and his Japanese assistant Sahachiro Hata identified an arsenic-based compound that showed promise against sexually transmitted infections like syphilis. In 1932, another German investigator, Gerhard Domagk, discovered another magic bullet that was effective against several common bacterial infections in the human body. To bring all this work together, in 1941 Selman Waksman, a Jewish-American microbiologist, suggested the term "antibiotic" be used for this new group of chemicals. Today, an **antibiotic** refers to those antimicrobial substances naturally produced by a few mold and bacterial species, or **synthetically** produced, that inhibit the growth or kill other bacterial species.

Synthetically: Referring to a chemical that is made in a research lab, rather than produced naturally.

Penicillin Is a Game Changer

In 1928, Alexander Fleming, a Scottish physician and microbiologist, was performing research on staphylococci that can cause boils, sore throats, and abscesses (**FIGURE 10.10**). Before going on vacation, he spread staphylococci on plates of nutrient agar and set them aside to incubate. On his return, he found one plate was contaminated by a green mold, which was identified as ***Penicillium***. Fleming's attention was drawn to the clear area around the mold, an area where the staphylococci were unable to grow. Unsure what was happening, he cultivated the *Penicillium* in broth and added a small drop of the broth to a culture of staphylococci. The staphylococci soon died. Other gram-positive bacteria also were susceptible to the *Penicillium* mold, so he called this magic bullet **penicillin**.

Penicillium: pen-ih-SIL-lee-um

By 1940, a group led by Howard Florey and Ernst Chain at the University of Oxford had isolated the penicillin molecule and started to produce larger quantities of the drug. Human clinical trials with the drug began in 1941 and

FIGURE 10.10 Fleming and Penicillin. Alexander Fleming noticed on some culture plates that bacterial growth was inhibited by the fungus *Penicillium*.

© Peter Purdy/Hulton Archive/Getty Images.

A CLOSER LOOK 10.3

Hiding a Treasure

Their timing could not have been worse. Howard Florey, Ernst Chain, Norman Heatley, and others of the team had rediscovered penicillin, purified it, and proved it useful in treating patients with bacterial infections. But it was 1939, and German bombs were falling on London. This was a dangerous time to be doing research into new drugs and medicines. What would they do if there was a German invasion of England? If the enemy were to learn the secret of penicillin, the team would have to destroy all their work. So, how could they preserve the vital fungus yet keep it from falling into enemy hands?

Heatley suggested each team member would rub the mold on the inside lining of his coat. The *Penicillium* mold spores would cling to the rough coat surface where the spores could survive for years (if necessary) in a dormant form. If an invasion did occur, hopefully at least one team member would make it to safety along with his "moldy coat." Then, in a safe country, the spores would be used to start new cultures, and the research could continue.

FIGURE A Two *Penicillium* colonies growing in agar culture.

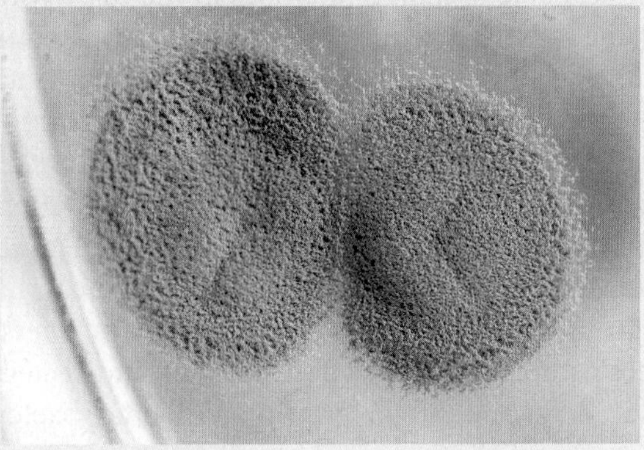

© Rattiya Thongdumhyu/Shutterstock.

Of course, a German invasion of England did not occur, but the plan was an ingenious way to hide the treasured organism.

The whole penicillin story is well told in *The Mold in Dr. Flory's Coat* by Eric Lax (Henry Holt Publishers, 2004).

it was soon apparent that people's lives could be saved with penicillin therapy. However, England was at war; German bombs were falling on London, and the researchers feared for their lives. **A CLOSER LOOK 10.3** describes how the research group concealed the *Penicillium* mold from a possible German invasion and moved their work to the United States.

Soon, American companies were producing large amounts of penicillin, and the Age of Antibiotics was under way. Penicillin was such a game changer to the field of infectious disease that Fleming, Florey, and Chain were awarded the 1945 Nobel Prize in Physiology or Medicine.

Antibiotics Work in Different Ways

Dozens of new antibiotics were developed in the late 1940s to early 1960s. The excitement was so intense that by 1969 the perception among many in the medical community was that antibiotics soon would defeat all infectious diseases. Unfortunately, that would not be the case, as we will see later in this chapter.

One of the unique characteristics of antibiotics was that they target different structures of bacterial cells. Let's look at a few of the antibiotics and identify the structures they target. Use **FIGURE 10.11** as a road map for this section.

Inhibition of Cell Wall Formation

Of all the antibiotics that have been discovered or synthesized in the lab, penicillin has remained the most widely used. Since its introduction in the 1940s, numerous derivatives of the natural penicillins have been developed (**FIGURE 10.12**). These "**semisynthetic**" penicillins are more stable, better tolerated, and less likely to cause drug resistance than the natural ones. All the penicillins, though, share the same fundamental structure. They have a chemical complex called a beta-lactam ring at the core of the molecule. For this reason, these antibiotics often are called the **beta-lactams**.

Semisynthetic: Referring to a chemical that has been chemically modified from its natural form.

As described in the chapter on Exploring the Prokaryotic World, penicillin and its derivatives block the formation of the protein cross-links within the cell wall of gram-positive bacteria. Without these links, the wall is weak and gives way to internal pressures, causing the microbe to swell and burst. Such antibiotics that only work against a select group of bacteria are referred to as **narrow-spectrum antibiotics**.

FIGURE 10.11 The Targets for Antibiotics. There are several cellular targets that different antibiotics affect.

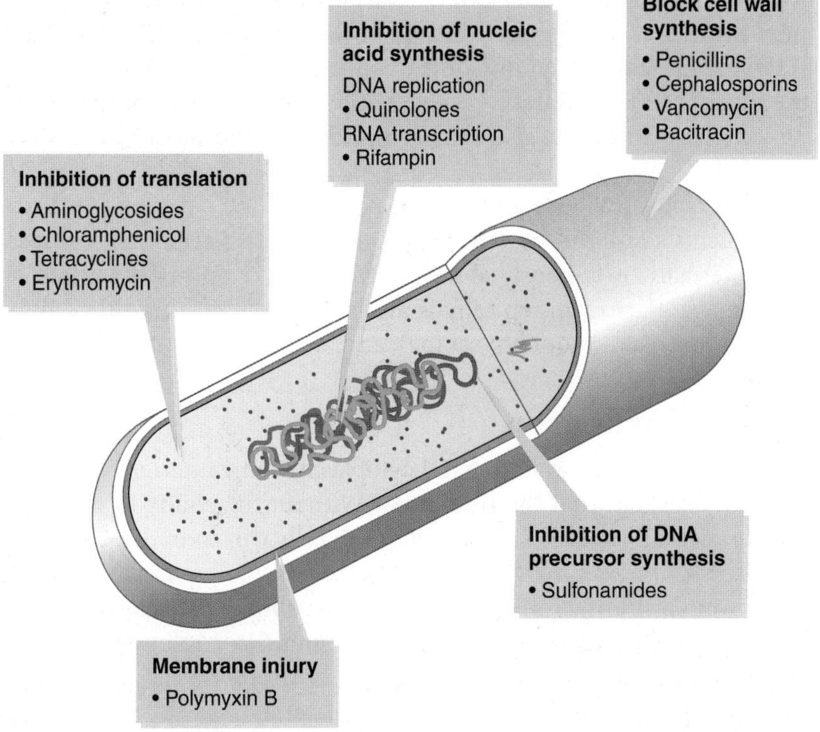

FIGURE 10.12 Some Members of the Penicillin Group of Antibiotics. The beta-lactam ring is common to all the penicillins. Semisynthetic penicillins are formed by changing the side group (highlighted in blue) on the molecule.

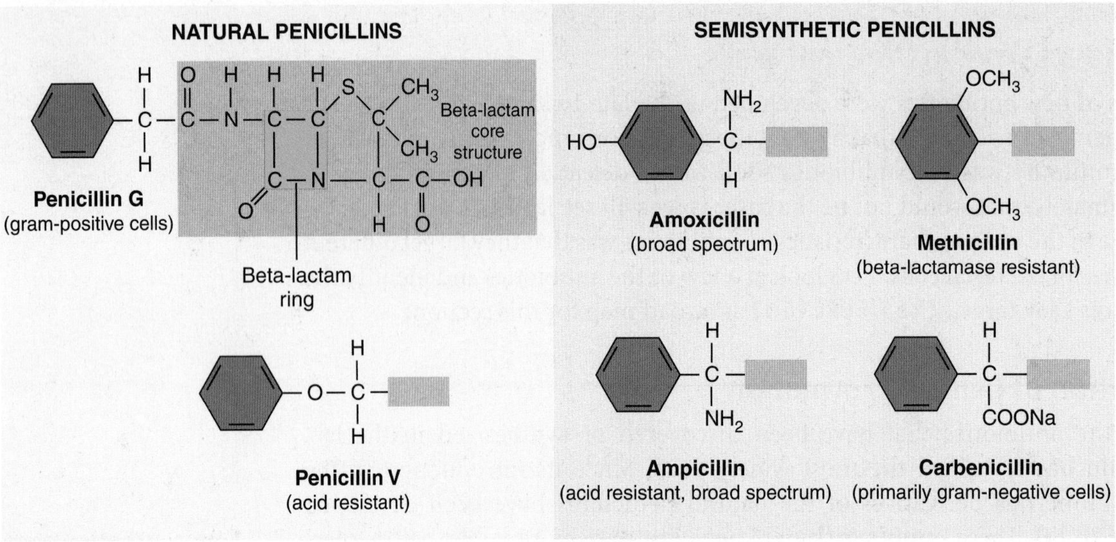

Cephalosporium: sef-ah-low-SPOH-ree-um

Bacillus: bah-SIL-lus

Streptomyces: strep-toe-MY-seas

Over the decades of use, many bacterial species have developed resistance to penicillin. These organisms have acquired the gene that codes for an enzyme called **penicillinase**. Bacteria producing this enzyme can break the beta-lactam ring and disable the antibiotic.

There are other antibiotics that also interfere with the synthesis of the gram-positive cell wall. One group is the **cephalosporins** that are derived from a mold called *Cephalosporium*. These antibiotics contain a slight modification to the beta-lactam ring that makes them resistant to the penicillinases. Among the more prescribed cephalosporins are cephalexin (Keflex) and ceftriaxone (Rocephin).

Vancomycin (Vancocin) is an antibiotic used to treat serious infections caused by antibiotic-resistant gram-positive bacteria. It is considered an antibiotic of last resort because it might be the only effective drug left for treatment. Vancomycin is not routinely used for trivial situations because of its damaging side effects in the ears and kidneys. Rather, it is reserved for the most serious antibiotic resistant infections. Unfortunately, bacterial resistance to this antibiotic also has been observed, and substitute drugs have been sought.

Another wall-inhibiting antibiotic is bacitracin. This drug is a mixture of several proteins produced by *Bacillus* species. It is restricted for use because, if taken internally, it is poorly absorbed from the intestine and might cause kidney damage. Bacitracin typically is available in therapeutic skin ointments (Polysporin) to treat minor cuts, scrapes, and burns.

Inhibition of Translation

Many antibiotics work in the bacterial cytoplasm. Some interfere with the protein translation abilities of the ribosomes. Without the ability to manufacture proteins to maintain metabolism, the affected cells will soon die. Such antimicrobials are considered **microbiostatic** because they kill microbes.

One group of translation inhibitors is the **aminoglycosides** that are produced by numerous species of the soil bacterium ***Streptomyces***. These antibiotics, including gentamicin (Garamycin) and streptomycin, are effective in treating infections caused by gram-negative bacteria.

Many antibiotics inhibiting translation are referred to as **broad-spectrum antibiotics** because they are effective against infections caused by both gram-positive and gram-negative bacteria. Examples include chloramphenicol (Amphenicol) and tetracycline derivatives, such as doxycycline (Oracea). Because doxycycline has few side effects, many physicians prescribe this valuable antibiotic in trivial situations (such as for acne). However, its over prescription has stimulated antibiotic-resistant bacteria to emerge.

Other Antibiotics Have Different Modes of Action

There are many other natural and synthetic antibiotics that target other critical sites in the bacterial cell.

Inhibition of Nucleic Acid Synthesis

Besides those antibiotics that act on the translation stage of gene expression, there are other drugs that act on transcription and DNA replication. Without RNA and DNA, bacterial cell metabolism will cease, and cell death will result.

Rifampin (Rifadin) is a synthetic antibiotic prescribed for tuberculosis patients. The antibiotic acts by interfering with the RNA polymerase enzyme during the transcription process. Taking rifampin can give an orange-red color to urine, tears, and other body secretions. It also can cause liver damage.

A group of antibiotics called the quinolones are synthetic drugs. They are designed to block DNA replication. Among the most prescribed antibiotics in the United States are the fluoroquinolones, such as ciprofloxacin (Cipro). These drugs are effective for urinary and intestinal tract infections and to treat gonorrhea.

The **sulfonamides** (**sulfa drugs**) are a group of synthetic antibiotics that slow down growth by interfering with the materials needed for DNA replication and RNA transcription. Such chemicals that slow the growth of microbes are said to be **microbiostatic**. Newer formulations of sulfonamides often are a combination of two drugs. For example, co-trimoxazole (Bactrim) contains two sulfa drugs and is used to treat infections of the urinary tract, lungs, and ears. This combination of drugs is an example of **synergism**, meaning the two drugs together are more effective than either drug alone. Consequently, the development of antibiotic resistance to both drugs is less likely.

Membrane Injury

Polymyxin B is a protein produced by some *Bacillus* species. The drug pokes holes in cell membranes that makes the cells leaky and cell death results. Like bacitracin, it generally is prescribed for skin infections involving cuts, abrasions, and burns produced by gram-negative bacilli.

Antiviral, Antifungal, and Antiprotistan Drugs

Antibiotics are effective only on bacterial infections. That is why individuals should not take, and physicians should not prescribe, antibiotics for viral infections. The drugs will not work. However, over the decades, scientists have developed other types of **antimicrobial drugs** that are effective against viral, fungal, and protistan infections.

Antiviral Drugs

A variety of antiviral drugs have been developed that target a specific stage in viral replication. As described in the chapter on Viruses, these infectious agents go through a replication process involving five stages. These are attachment, penetration, biosynthesis, assembly, and release. By blocking one of these stages, virus replication can be disrupted. Most of the antiviral drugs have been designed and developed against acquired immunodeficiency syndrome (AIDS). Fewer drugs have been created for herpes virus infections, such as fever blisters and shingles, and for influenza.

Importantly, unlike antibiotics that can cure a patient of a bacterial infection, there is only one antiviral drug intended to cure a viral disease. This is for hepatitis C infections. "Cure" means the hepatitis C virus cannot be detected in the blood 3 months after treatment is completed. As of today, there are no other antiviral drugs capable of curing a patient of a viral infection. Rather, the drugs only lessen the symptoms or make the disease of shorter duration.

Antifungal Drugs

Candida albicans: KAN-did-ah AL-bih-kanz

The physician has relatively few antimicrobial drugs for treating fungal infections. Most are for topical use only. One antifungal drug is nystatin (Mycostatin), which is effective against yeast infections caused by *Candida albicans*. Other useful antifungal drugs include griseofulvin (Grisovin) for ringworm and athlete's foot. Another class of antifungal drugs is the imidazoles. Many of these compounds, such as clotrimazole (Lotrimin) and miconazole (Monistat), also are effective against infections of the skin surface. For invasive and often life-threatening fungal infections, a few imidazoles and amphotericin B (Amphocin) are used. The latter drug is quite toxic in the body, so it is a last resort antifungal medication only.

Antiprotistan Drugs

Plasmodium: plaz-MOH_dee-um

Protistan-caused parasitic diseases afflict many individuals living in the developing nations of the world. One of the biggest disease threats is malaria, which is caused by several species of *Plasmodium*. The 2018 World Malaria Report published by the World Health Organization (WHO) estimates there were more than 219 million new malaria cases worldwide and over 435,000 deaths, mostly (90%) in Africa. The current therapeutic drug, artemisinin, can swiftly reduce the number of *Plasmodium* parasites in the blood of malaria patients. However, the parasite has become resistant to several other antimalarial drugs, and resistance to artemisinin has been detected in Southeast Asia.

10.4 Antimicrobial Resistance: A Growing Challenge

Antimicrobial resistance (AMR) refers to the ability of bacteria, protists, fungi, and viruses to become unaffected by antimicrobial drugs (such as antibiotic, antiviral, antifungal, and antiparasite medications). As a result, infections might persist in the body, increasing the risk of spread to others. AMR can affect anyone of any age around the world.

AMR is a phenomenon that develops naturally, often as a result of mutation. For example, some bacterial and fungal species that live in the soil produce antibiotics as chemical warfare agents. By producing these compounds, the producers limit the growth and spread of their bacterial competitors. Consequently, other bacterial species have evolved ways to defend themselves and survive by becoming resistant to the chemical agents. So, it is important to realize that AMR is part of everyday life in the microbial world. It is when these resistance mechanisms get into human pathogens, often through the misuse of antimicrobial drugs in human and animal populations, that the problems of resistance arise.

Since the late 1960s, an alarming number of microbes and viruses of human concern have acquired resistance to one or more antimicrobial drugs. As a result, hundreds of thousands of people around the globe, including in the United States, die every year from infections that have become untreatable. These drug-resistant pathogens are especially dangerous to those individuals whose body's natural defenses are very weak. This includes patients in intensive care units and burn wards, young children, older people, and individuals with weakened immune systems. Professor Dame Sally Davies, Chief Medical Officer for England, in her first annual report to the Department of Health and Social Care, has said

> *the development and spread of AMR "...is arguably as important as climate change for the world."*

Antibiotic resistance (**ABR**) is a subset of AMR and applies to bacteria that have become resistant to antibiotics. The remainder of the chapter will focus on this critical societal issue.

Antibiotic Resistance and Superbugs

ABR is of most concern when it develops in a human pathogen. To make matters worse, many of these organisms are resistant to many antibiotics. They are referred to as being **multidrug resistant** (**MDR**) and are called **superbugs**. This means standard medical treatments become ineffective and infections persist. This predicament is not only dangerous for the affected patient but also for society, as these superbugs can spread to others in the community, nation, or even globally (**FIGURE 10.13**). In the United States, the Centers for Disease Control and Prevention (CDC) estimates that 23,000 Americans die every year from MDR infections. A British report, The Review on Antimicrobial Resistance, estimates that worldwide 700,000 people die each year from such superbug infections.

The development of ABR in human societies is a result of microorganisms being exposed to antimicrobial drugs for a prolonged time. Like microbes in the soil, human pathogens can evolve beneficial mutations by chance, and some of these mutations might provide a resistance mechanism (**FIGURE 10.14**). For example, suppose a gene in a bacterial cell undergoes a mutation that results in a change to the structure of the ribosomes. If that mutant cell now is exposed to an antibiotic that targets the ribosomes, the organism will be resistant to that antibiotic and often several similarly acting antibiotics.

What is more prevalent in today's world is the ability of many bacterial species to undergo genetic recombination resulting in ABR. As described in the chapter on Microbial Genetics, genes randomly are swapped between species through the process of **horizontal gene transfer**. In the process, a bacterial cell with an antibiotic resistance gene could transfer a copy of the gene to another bacterial cell previously susceptible to the antibiotic. That recipient cell now has resistance to the antibiotic.

For example, recently a new and alarming form of antibiotic resistance has arisen through **conjugation**. This is one of the three horizontal gene transfer processes. Some bacteria have been identified that carry a gene called *NDM-1* on a plasmid. This gene codes for an enzyme that disables essentially all beta-lactam antibiotics. In fact, bacterial pathogens carrying the gene have been referred to as "nightmare superbugs," because they usually have resistance to many other types of antibiotics as well. As a result, an individual who is infected with one of these MDR pathogens has few treatment options. In 2017, a woman in Nevada died of

FIGURE 10.13 The Global Threat from Antibiotic Resistance. Many bacterial species around the world are becoming resistant to antibiotic medicines.

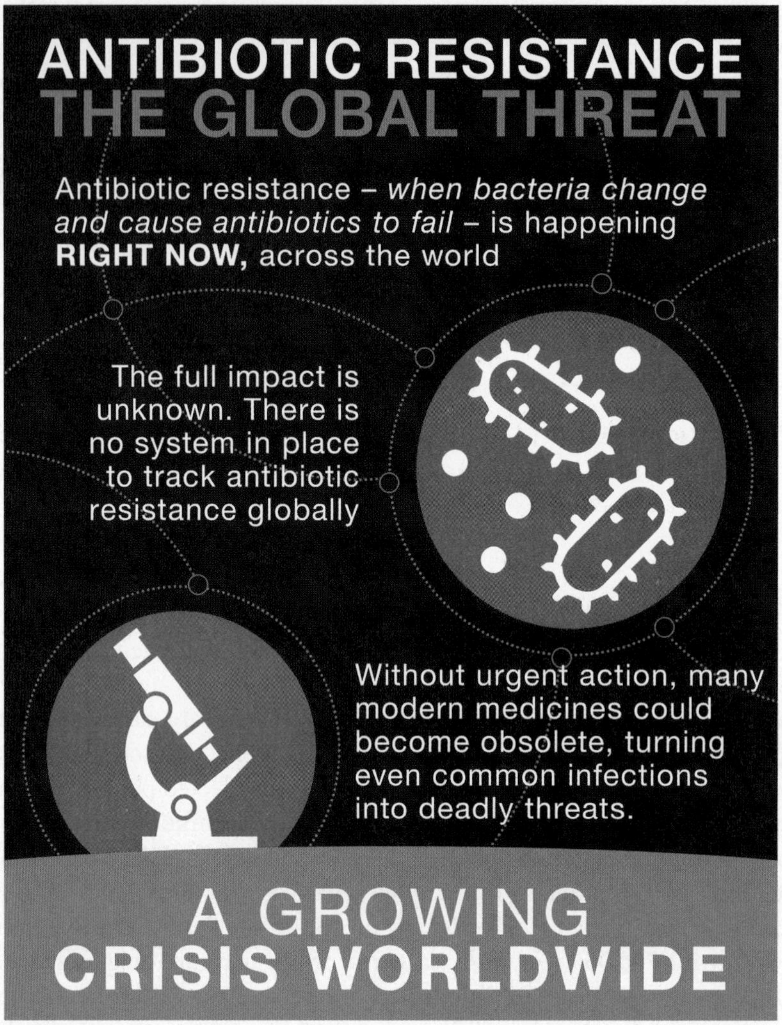

Courtesy of CDC.

an incurable infection due to such a nightmare superbug she picked up in India. That superbug was resistant to all 26 antibiotics available in the United States.

Today, one of the biggest causes for ABR (and AMR as well) is the excessive and often unnecessary prescribing by medical professionals of, and demanding by the public for, antimicrobials for human illnesses. As the WHO has stated, ABR is a global concern—and here are some reasons why.

1. *ABR Kills.* Infections caused by resistant microorganisms fail to respond to the standard treatment, resulting in prolonged illness in the patient and greater risk of death. In fact, the death rate for patients with ABR infections is about twice the rate of that in patients with nonABR infections. In addition, ABR reduces the effectiveness of treatment, meaning patients remain infectious for a longer time, which increases the risk of spreading resistant microorganisms to others.
2. *ABR Threatens a Future That Could Be a Post-Antibiotic Era.* If AMR continues to spread, it is very possible and probable that the infectious diseases once controlled and cured with antibiotics will become untreatable and uncontrollable. If we run out of effective drugs, we are looking toward a post-antimicrobial era. Speaking at a conference

FIGURE 10.14 The Possible Outcomes of Antibiotic Treatment.

(A) Ideally, with a complete course of antibiotics, all pathogens will be destroyed.

(B) If there are some resistant cells in the infecting population, they will survive and grow without any competition.

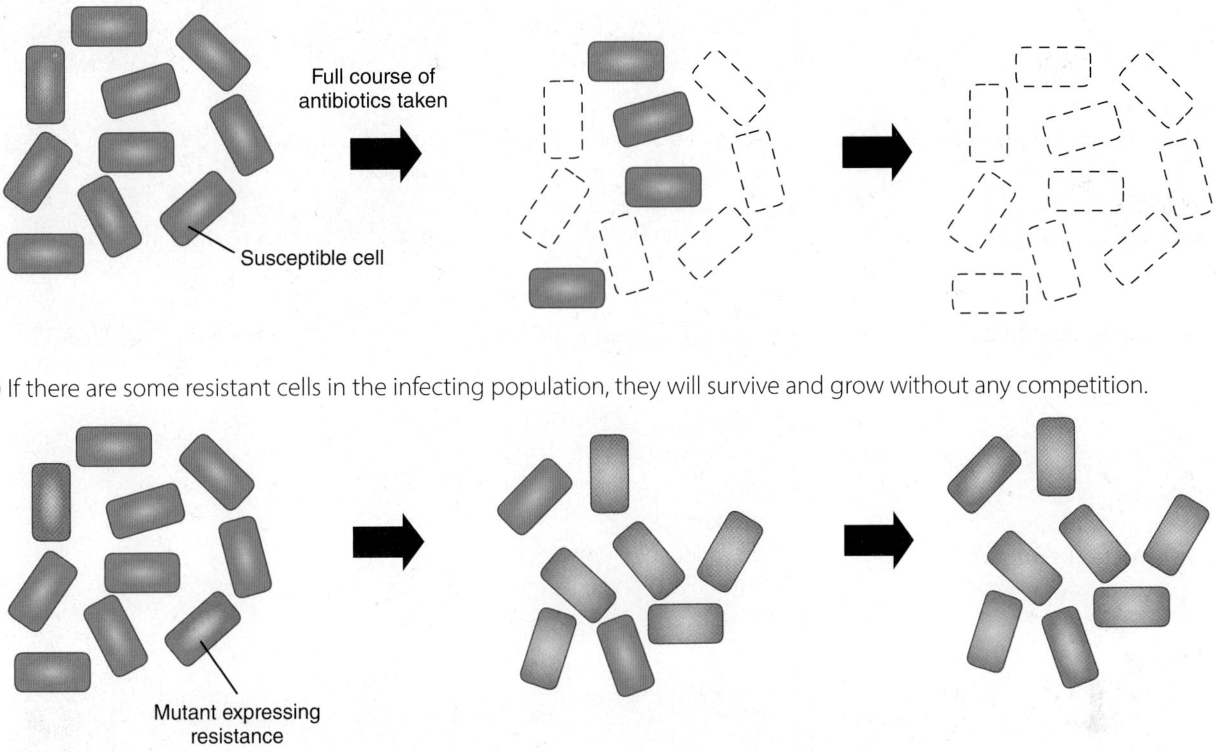

called "Combating antimicrobial resistance: Time for action," Margaret Chan, Director-General of the WHO, said of antibiotics

> "...the pipeline is virtually dry...the cupboard is nearly bare."

Recently, there have been a few glimmers of hope with the development of a few new antibiotics. Let's hope it is not too little too late.

Very simply, ABR to a large part is society driven. It is the result of abuse and misuse of antibiotics by the medical community and by the public. In fact, certain human actions have greatly accelerated the emergence and spread of ABR.

Antibiotic Abuse and Misuse in Medicine

The rise in antibiotic resistance is partly the result of the abuse of antibiotics. For example, drug companies promote antibiotics heavily and patients pressure doctors for quick cures. Physicians sometimes misdiagnose infections, or they might write prescriptions to avoid ordering costly tests to pinpoint a patient's illness.

Hospitals are another source for the emergence of ABR. Sometimes physicians use unnecessarily large doses of antibiotics to prevent infection during and following surgery. This runs the risk of destroying much of the intestinal human microbiome. Now, the natural microbial defense is destroyed, and an antibiotic-resistant pathogen can thrive (see Figure 10.14). The result is a **superinfection**, which can be extremely hard to control and cure. **TABLE 10.3** lists most of the microbes known to exhibit ABR.

Antibiotics also are abused in developing countries where they often are available without prescription. Some countries permit the over-the-counter sale of potent antibiotics, and large doses promote resistance to develop.

TABLE 10.3 Pathogens and Diseases Associated with AMR (See Appendix A for organism pronunciations.)[1]

Pathogen	Diseases
Bacteria	
Acinetobacter baumannii[2]	Pneumonia, skin and wound infections, and meningitis
Bacillus anthracis	Skin, intestinal, and respiratory diseases
Clostridium difficile	Diarrhea
Enterococcus faecium	Urinary tract infections, blood and heart infections, intestinal infections, and meningitis
Group B streptococci	Blood infections of newborns, elderly
Klebsiella pneumoniae	Pneumonia, bloodstream infections, wound or surgical site infections, and meningitis
Mycobacterium tuberculosis	Tuberculosis
Neisseria gonorrhoeae	Gonorrhea
Neisseria meningitidis	Childhood meningitis
Salmonella enterica (Typhi)	Typhoid fever
Shigella dysenteriae	Diarrhea
Staphylococcus aureus	Wide variety of diseases
Streptococcus pneumoniae	Pneumonia, blood infections, and sinus and middle ear infections
Viruses	
Influenza viruses	Influenza
Human immunodeficiency virus (HIV)	AIDS
Fungi	
Candida albicans	Oral thrush, vaginitis, and body infections
Protists	
Plasmodium species	Malaria

[1] Data are from the CDC.
[2] **Bold** = Currently cause most U.S. hospital infections.

Antibiotic Abuse in Livestock

A major cause for the emerging public health crisis of antibiotic resistance is the abuse and misuse of antibiotics in animal feeds. Globally, more than 50 million kilograms (55,000 tons) of antibiotics are produced each year. Some estimates suggest that up to 80% of these antibiotics are not for clinical purposes. Rather, they are dispensed as low, nontherapeutic amounts on large commercial feedlots to promote the faster growth of healthy cattle, swine, and poultry (**FIGURE 10.15**).

By using these antibiotics unselectively, many feedlots are becoming breeding grounds for the evolution of antibiotic-resistant bacterial species in these animals. Some of these antibiotic-resistant strains have found their way into people through the consumption of animal food products. If MDR pathogens are present, the pathogen can directly cause illness in the consumer or they might horizontally transfer the resistance genes of other bacteria in the human intestine. In 2015, bacteria carrying resistance to the antibiotic polymyxin E (Colistin) were discovered in livestock, meat, and humans in China, where the antibiotic is used in livestock feed. Soon, polymyxin resistance mysteriously made its way to Denmark, where it was discovered in several samples from animals and food. In the United States, the FDA has issued new voluntary rules (FDA Guidance #213) to curtail through 2020 the proliferation

FIGURE 10.15 Antibiotic Use, United States. This chart shows the estimated annual use of antibiotics.

Other = pets, aquaculture, and agricultural crops.

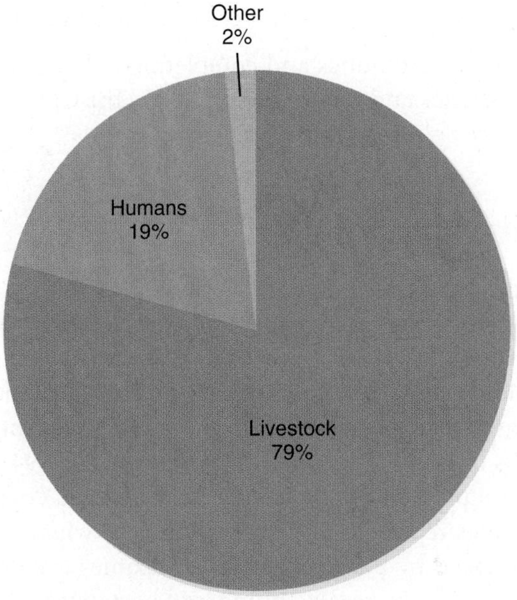

Data from U.S. Food and Drug Administration.

of antibiotics in animals for production purposes. During this period, the FDA will work with the industry to help phase out the use in food animals of those antibiotics used as human medicines. The FDA also wants to bring the therapeutic uses of animal-feed antibiotics under the control of licensed veterinarians.

Controlling the Antibiotic Resistance Problem

It will never be possible to eliminate ABR. However, it can be controlled through a two-pronged approach.

First, as already described, governments and industry need to preserve the antibiotics still available by using them prudently. Preventing abuse and misuse of antimicrobial drugs is something society can do to stem the spread of resistance. In the case of antibiotic resistance, this means taking and sustaining the following four steps, as suggested by the CDC.

1. *Do Not Ask for Antibiotics.* For illnesses not treatable with antibiotics (e.g., colds, flu, most fevers, any viral infection), do not ask a physician for an antibiotic. In fact, the physician should not prescribe antibiotics for such illnesses.
2. *Do Not Stop Taking a Prescribed Antibiotic.* If an antibiotic is prescribed correctly for a bacterial illness, take the full course of therapy. Do not stop taking the antibiotic as soon as you start "feeling better." There might still be some pathogens "hanging around" in the body, so you need to make sure they all are eliminated.
3. *Do Not Save Antibiotics for a Future Illness.* If for some reason you have "extra" antibiotic pills, do not save them for a future illness. If at some time in the future you have similar symptoms of an illness, do not assume it is the same illness you had previously and take the leftover antibiotics. Also, if you have a very different illness, do not take the leftover antibiotics. They probably will not work.

4. *Do Not Share Your Antibiotics with Another Person.* If you know someone who appears to have similar symptoms of an illness you had, don't give that person your antibiotics. You don't know if it is the same infection, and you don't know if the person might be allergic to the antibiotic.

Appropriate use of antibiotics and completion of antibiotic treatment will limit antibiotic resistance and reduce its spread. As the CDC has stated in their Antibiotic Resistance Solutions Initiative:

> *"Get smart. Know when antibiotics work."*

The second part of the approach to controlling drug resistance is more challenging. You might ask that if there are fewer effective antibiotics, scientists and microbiologists in pharmaceutical companies need to discover and develop new drugs to which bacterial are susceptible. Excellent idea but it hasn't really happened as of late. For example, between the late 1940s and the 1960s, nearly all major groups of antibiotics were discovered and developed. However, the following decades represent an "innovation gap" where no new antibiotic classes were introduced by pharmaceutical companies (**FIGURE 10.16**). Importantly, during this gap, an increasing number of bacterial species became resistant to many of these antibiotics.

There are at least a couple of reasons for the lack of development of new antibiotics and other antimicrobial drugs. First, due to the perception in the late 1960s that antibiotics would cure all infectious diseases, pharmaceutical companies and research organizations stopped or severely slowed the development of new antibiotics, even though it was obvious many bacteria and bacterial pathogens were becoming resistant to them.

Second, the cost to develop and bring an antibiotic to market is staggering. For a new antibiotic, the costs can be close to $1 billion over 10 years. Pharmaceutical companies say that the relatively large research cost and time for development is lost if the medication does not make it to market. Therefore, pharmaceutical companies find there is a much higher financial reward in developing and marketing medications for the treatment of long-lasting illnesses. These include medications for conditions like depression, hypertension, diabetes, cancer, high cholesterol, and arthritis. Also, unlike antibiotics, which are usually given for a short 5- to 14-day period and then discontinued, medications for long-lasting illnesses might be taken for a lifetime. These medications therefore are more profitable. So, the development of new antibiotics and antimicrobial agents has declined at a time when there is a pressing need for them. In Margaret Chan's conference speech mentioned above, she also said:

> *"A post-antibiotic era means, in effect, an end to modern medicine as we know it. Things as common as strep throat or a child's scratched knee could once again kill. Some sophisticated interventions, like hip replacements, organ transplants, cancer chemotherapy, and care of preterm infants, would become far more difficult or even too dangerous to undertake. At a time of multiple calamities in the world, we cannot allow the loss of essential antimicrobials, essential cures for many millions of people, to become the next global crisis."*

Perhaps the antimicrobial development pendulum is starting to swing back. In the last few years, a few new antibiotics have entered clinical trials. Still, it

FIGURE 10.16 The Innovation Gap. This time line shows when major antibiotics were introduced and the gap that existed in the development of new antibiotics.

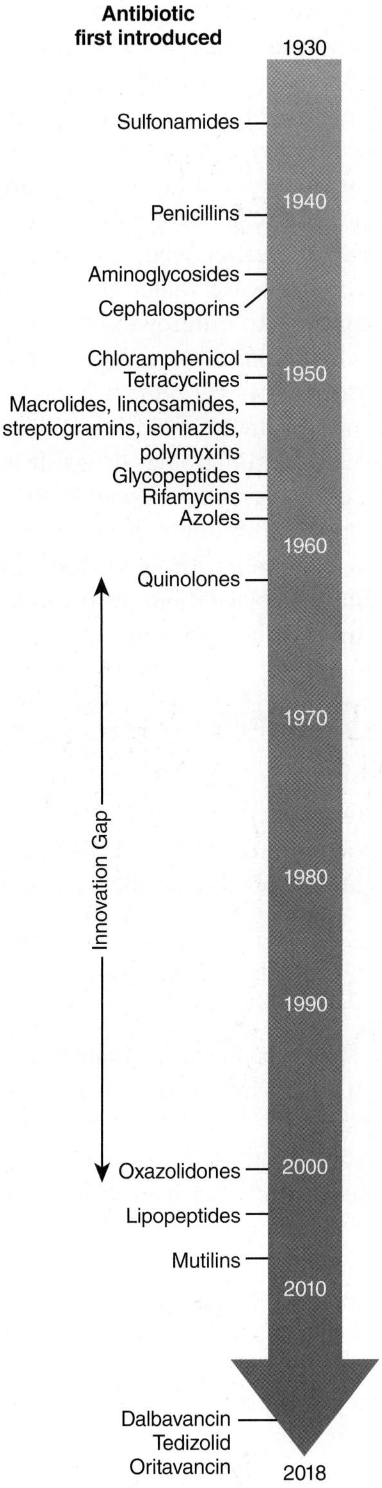

is time to accelerate the development and supply of new antimicrobial drugs. Although these medications are not the perfect magic bullets once perceived by Ehrlich, new drug discovery might provide many new antimicrobial agents to fight pathogens and infectious disease.

A Final Thought

In the past 100 years, there were two periods in which the incidence of disease declined sharply. The first was in the early 1900s. The new understanding of the disease process led to numerous social measures, such as water purification, care in food production, control of insects, milk pasteurization, and isolation of infectious patients. These sanitary practices reduced the possibility that pathogens could reach their human targets and be spread through a population.

The second period began in the 1940s with the development of antibiotics and blossomed in the years thereafter, when physicians found they could treat established cases of infectious disease. Major health gains were made as serious infections came under control. An outgrowth of these successes was the belief that science can cure any infectious disease. A shot of this, a tablet for that, and then perfect health, Unfortunately, that often is not the case.

Scientists might show us how to avoid infectious microbes. Doctors can control certain diseases with antimicrobial drugs. It is society's role to make sure antimicrobials are used correctly. However, the ultimate body defense against infection relies on the immune system to provide the strong cellular and chemical defenses to overcome pathogens. The antimicrobials supplement natural defenses; they do not replace them. In the end, it is well to remember that good health comes from within, not from without.

Chapter Discussion Questions

What Was He Thinking?

From your reading of this chapter, identify and discuss (1) three major points about controlling microbes with physical and chemical methods and (2) three essential points about antimicrobials that the author was trying to get across to you.

Questions to Consider

1. Instead of saying "food has been irradiated," processors indicate it has been "cold pasteurized." Why do you believe they use this term? Do you think itis appropriate?
2. Of all the sterilization methods described in this chapter, why do you think none of these methods has been widely adapted to the sterilization of milk? Which, in your opinion, holds the most promise?
3. While on a camping trip, you make camp near the stream where, as a young child, you once swam and from which you drank freely. Fearing contamination of the stream, you decide it would be wise to use some form of disinfection before drinking the water. The nearest town has only a grocery store, pharmacy, and post office. What might you purchase? Why?
4. In the United States, almost 5 million children under the age of five attend day care centers, nurseries, or preschools. With so many in these organized activities, the possibilities for infectious disease transmission among children is considerable. Under what circumstances might antiseptics and disinfectants be used to preclude the spread of potentially dangerous microorganisms?
5. Look at the label on a bottle of disinfectant or antiseptic you have at home. Does it provide any of the information listed in Table 10.2?

6. One of the novel approaches to treating gum disease is to impregnate tiny vinyl bands with antibiotic, stretch them across the teeth, and push them beneath the gumline. Presumably, the antibiotic would kill bacteria that form pockets of infection in the gums. What might be the advantages and disadvantages of this therapeutic device?
7. The antibiotic issue can be argued from two perspectives. Some people contend that because of side effects and ABR, medical use of antibiotics will eventually be abandoned. Others see hope for a recently reported "superantibiotic," a type of "magic bullet" that would attack bacteria in such a way that the microbe should not be able to generate resistance. What arguments can you offer for either view? Which do you see as more likely to be true?
8. Why do physicians often prescribe a broad-spectrum antibiotic when the cause of the disease is unknown? What are some drawbacks of this practice?
9. While at a party, several of your friends are talking and one of them says she has been feeling ill lately, having headaches, a low-grade fever, and fatigue. Another person in the group tells her he has some antibiotics left over from an illness he had several months ago. He offers to give her the antibiotics. What should be your response in this circumstance, and what should you say to both individuals?
10. You purchase a household cleaning product that states, "Kills 99% of household bacteria." Being somewhat skeptical, you decide to test the product. Based on what you have learned about microbes and how they can be grown in culture, design a simple experiment that would determine if the product's statement is true. Assume you have access to all needed microbiological materials.
11. Paul Ehrlich coined the term "magic bullet" to refer to drugs that could kill bacteria in the body without doing substantial harm to the body. From your reading about antimicrobials and antibiotics in this chapter, do these drugs fit that definition? Explain your reasoning.

© Vrx/Shutterstock

CHAPTER 11

Microbial Crosstalk: Uncovering the Social Life of Microbes

The Case of the Disappearing Squid

In the shallow coastal waters off the Hawaiian Islands in the Pacific Ocean dwells the very small (3 cm in length), pear-shaped Hawaiian bobtail squid (**FIGURE 11.1**). Being so small, it can be easy prey for predators like seals. Therefore, during the day, the squid burrows into the sand on the sea floor to rest and avoid capture. However, when it comes out at night to feed on crabs and shrimp, the moonlight shining down will illuminate the squid. This would produce a silhouette, making it easily seen from below by predators. How can the squid hunt at night and evade capture? It needs to disappear!

To prevent any silhouette being cast by the moonlight, the squid employs the help of bacteria. Not just any marine bacteria, but rather one species called *Aliivibrio fischeri*. This species has taken up residence in a special light organ on the underside of the squid. At night, this community of bacterial cells glows continuously; that is, it exhibits **bioluminescence**, like the glow naturally produced by fireflies. In the case of the squid, the light mimics the intensity of the moonlight, which means there is no silhouette. Depending on the phase of the moon, the squid can control how much light is emitted to match the moonlight's intensity. The squid can swim through the shallow water camouflaged and undetected by predators.

Aliivibrio fischeri: alee-ee-VIB-ree-oh FISH-er-ee

Come the next morning, the squid expels about 90% of the bacteria in the light organ. The remaining bacterial cells will reproduce and increase in number during the day as the squid hides out in the sand, ready for another night's light show.

CHAPTER 11 OPENER The human microbiome consists of trillions of microbial cells (and their genomes) that exist permanently on and in the human body. Different parts of the body have diverse but distinct communities of microbes that communicate with one another and with both nearby and distant human cells.

FIGURE 11.1 The Hawaiian Bobtail Squid. The bobtail squid is native to the central Pacific Ocean, where it occurs in shallow coastal waters.

By this point, you have learned quite a bit about the types of organisms and viruses that inhabit the microbial world and the daily activities they exhibit as they grow and reproduce. In this last chapter to Part I, we will explore one more very surprising characteristic of most microbes—their social lives. Social behaviors are not characteristics that were thought to be associated with microbes. However, microbiologists have discovered that microbes communicate with one another within microbial communities, between microbial communities, between microbes and animals (such as the bobtail squid), and between microbes and humans (see chapter opening image). Most surprising, microbiologists now realize that the composition and activities of the microbes in the human body carry out key biological processes for the normal functioning of our digestive system and immune system. These microbes even play a role in our cognitive capabilities and emotional states.

LOOKING AHEAD

After reading and completing this chapter, you will be able to:

11.1 Explain what is meant by saying microbes are social creatures.
11.2 Describe the development and composition of a biofilm and illustrate three examples of bacterial quorum sensing.
11.3 Discuss the reason why humans and microbes form a symbiotic relationship and identify the human body systems where a microbiome (bacterial biofilm) is found.
11.4 Provide four examples for the role of the gut microbiome in health and disease through "talking" back and forth with the host.
11.5 Explain what it means to say that a human is "more than a human."

11.1 Microorganisms: Their Rich Social Life

Microbes live remarkably social lives. These minuscule life forms can avoid, attack, support , or ignore their neighbors, depending on the situation. In fact, these social interactions within tightly bound communities are the norm in the microbial world. Like any society, be they ant colonies, beehives, or human communities, microbes live in a cliquish world where they hang out with their friends, cooperate with one another, but at the same time, keep an "eye" on everyone else in the community.

The understanding of the social lives of microbes has expanded in the past few decades. The traditional view of microbes—and perhaps your impression from previous chapters—was that they are relatively independent, single-celled organisms. They lack the type of cooperative social behavior characteristic of ant and bee colonies and mammals. However, a rapidly expanding body of research has completely overturned this idea. Today, it is startling to realize that just like social animals, bacteria display a variety of social behaviors involving cooperation and communication. Some even display individual self-sacrifice for the good of the group.

11.2 Bacteria Lead Social Lives: Biofilms and Cell Communication

What do pond scum, a slimy sink drain, a middle-ear infection, dental plaque, and the human gut have in common? They all are composed of active microbial metropolises that exhibit a rich social life. In these communities, living and "working" together, and "listening" and "talking" through chemical communication, are critical for their survival.

Biofilms Represent Multicellular, Social Communities

In their natural environments, microbes grow and multiply. Interactions with other microbes once were thought to be rare and brief. Occasionally, they might aggregate together and cause an infectious disease, or they might produce colonies if grown on an agar plate (**FIGURE 11.2**). This view of microbial life is incorrect.

Up to 95% of bacterial organisms live in highly organized communities called **biofilms**. These groups of cells represent multicellular associations of one or more types of microorganisms. A mature biofilm, which might contain millions or billions of cells, can form on an environmental surface like a rock or the inside surface of a drainpipe. In the human body, biofilms form on the surface of a contaminated indwelling or implanted medical device like a knee **prosthesis** or **catheter** (**FIGURE 11.3**). Biofilms also form the various communities that are part of the human microbiome.

Prosthesis: An artificial body part, such as a leg, hip, or knee replacement.

Catheter: A thin, flexible tube that carries fluids into or out of the body.

The best studied biofilms are those composed of bacterial cells. Therefore, most of the following pages will focus on their community organization and social behavior.

A biofilm forms when free-living bacterial cells attach to a surface. They then start to excrete a sticky film consisting of proteins, sugars, and other substances (**FIGURE 11.4**). As the biofilm matrix grows, the cells become embedded in this self-generated coating. When the biofilm is mature, it behaves like living

FIGURE 11.2 Bacterial Colonies. In a sense, each bacterial colony growing on an agar surface can be viewed as a biofilm.

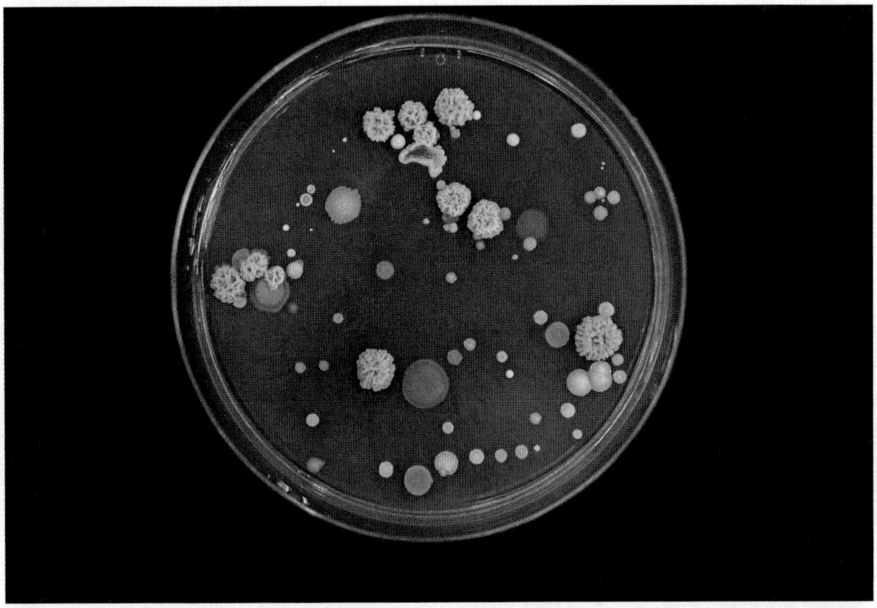

Courtesy of Dr. Jeffrey Pommerville.

FIGURE 11.3 Bacterial Biofilm. This scanning electron microscopic image shows bacteria within a biofilm on the surface of a catheter. The sticky biofilm matrix holds the cells together.

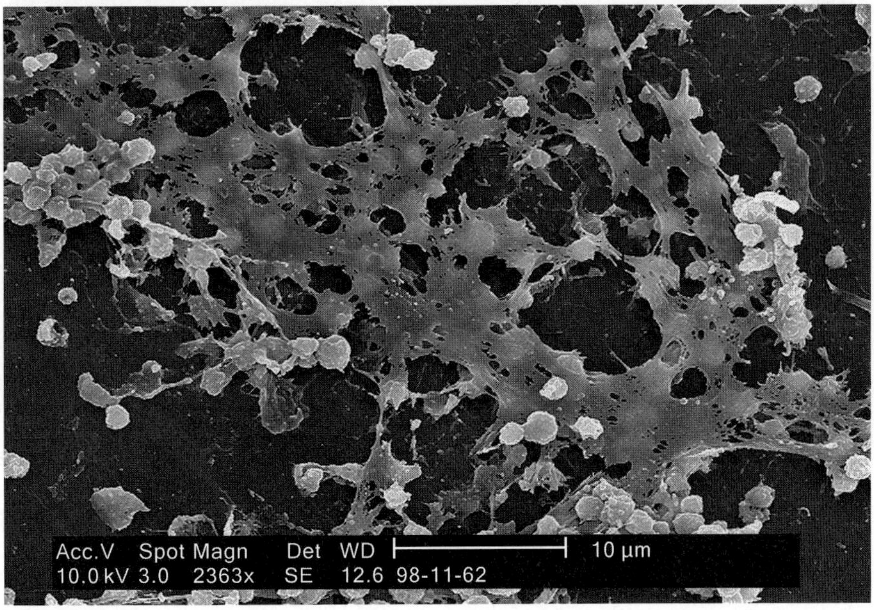

Courtesy of CDC.

human tissue. The community of cells contains a primitive circulatory system made of water channels to bring in nutrients and eliminate wastes.

Consequently, a biofilm serves to:

1. **Protect the Community**. The cells within a biofilm are shielded from attack by predators (other bacteria and protists) and from the effects of environmental chemical agents. In the case of a human infection, the biofilm protects the cells from the immune system. For example, patients suffering from **cystic fibrosis** (**CF**) are susceptible to a

Cystic fibrosis: A genetic disorder that primarily affects the lungs and causes the buildup of thick mucus, which makes breathing difficult.

FIGURE 11.4 Biofilm Development. The biofilm life cycle is an example of cell cooperation in the development of a multicellular structure.

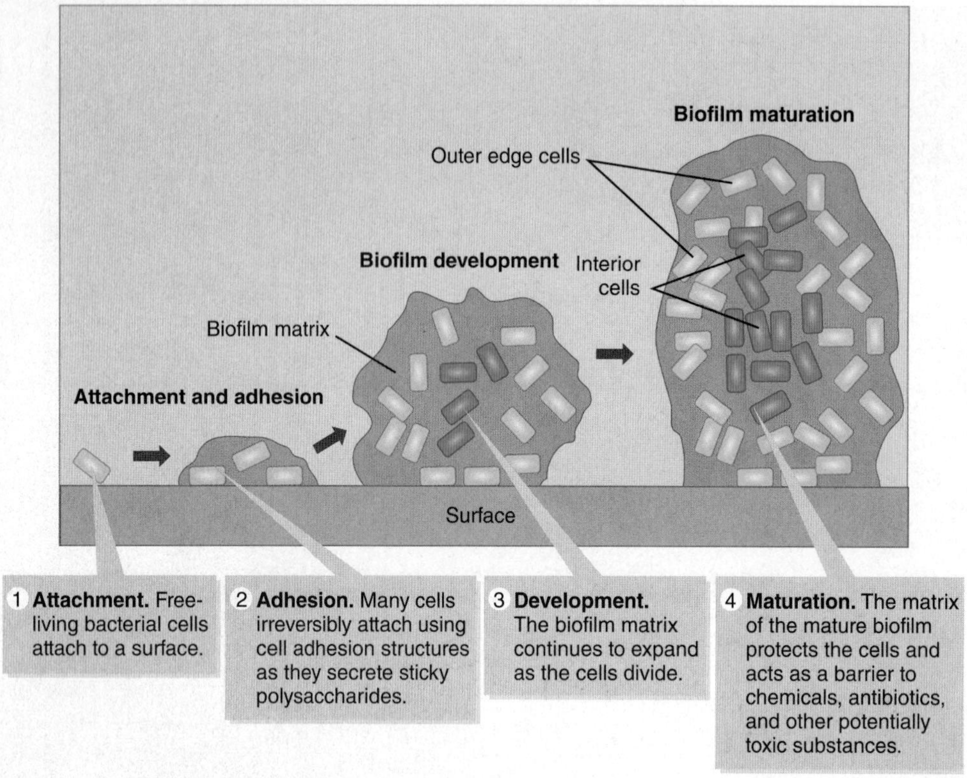
Biofilm maturation
Outer edge cells
Biofilm development
Interior cells
Biofilm matrix
Attachment and adhesion
Surface
1 Attachment. Free-living bacterial cells attach to a surface.
2 Adhesion. Many cells irreversibly attach using cell adhesion structures as they secrete sticky polysaccharides.
3 Development. The biofilm matrix continues to expand as the cells divide.
4 Maturation. The matrix of the mature biofilm protects the cells and acts as a barrier to chemicals, antibiotics, and other potentially toxic substances.

Pseudomonas aeruginosa: sue-doh-MOH-nahs ah-rue-gih-NO-sah

potentially fatal lung disease called cystic fibrosis pneumonia. Illness occurs if the bacterial pathogen *Pseudomonas aeruginosa* infects the lungs and forms a biofilm. The biofilm protects the cells from immune attack, and being in an impermeable film, the cells are protected from antibiotics and other antimicrobial agents.

2. **Trap and Process Nutrients**. The sticky biofilm matrix concentrates nutrients more easily than if the nutrients were dispersed in more dilute surroundings. This ability to trap chemicals has been manipulated in a positive way in sewage treatment plants. In these facilities, naturally occurring bacterial biofilms are used to break down solids in wastewater into cleaner, more environmentally friendly products. Likewise, a process called **bioremediation** uses naturally occurring biofilms, or artificially (lab) created ones, to break down chemical contaminants, such as those in toxic waste sites or oil spills (**FIGURE 11.5**).
3. **Provide Effective Chemical Communication.** The cells within a biofilm are closely packed together. This makes chemical communication among the cells in the biofilm much easier than if they were dispersed as single cells over a wider area. In biofilms, social networks can develop.

Cells in Biofilms Communicate by Social Networks

Millions of people today are use social networking to stay in touch with friends, family, or classmates or for business purposes to connect with clients or customers. What is astonishing is that microbes, such as bacteria, have been using a form of social networking for millennia. They too use it to stay in contact with their neighbors. However, their social communication is not electronic but rather chemical.

FIGURE 11.5 Bioremediation. EPA scientists testing the effectiveness of bioremediation on an Alaskan beach following the *North Cape* oil spill of 1996.

FIGURE 11.6 Cell Communication and Quorum Sensing.
(A) At low cell densities, few signaling molecules accumulate. **(B)** As the cell density increases (quorum sensing), these molecules reach a threshold concentration that, when detected by the cells, leads to a change in cell behavior through new gene activity.

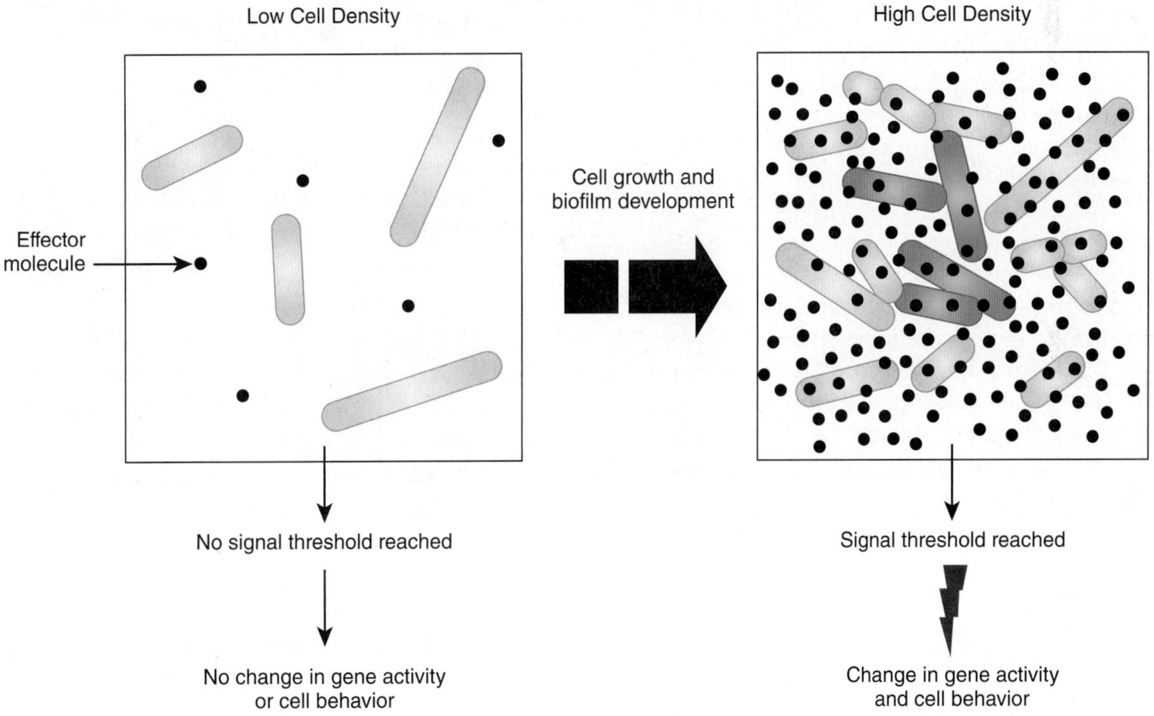

The bacterial cells within a biofilm chemically communicate with one another through a process called **quorum sensing** (**QS**). When a biofilm starts to develop, the cells secrete signaling molecules throughout the biofilm. The cells use these chemicals to monitor their population size within the community. When the signaling molecules reach a high concentration, or a "tipping point," the cells sense that there now are enough cells (a **quorum**) in the biofilm to execute some type of group behavior (**FIGURE 11.6**). This behavior might trigger infectious disease onset (e.g., cystic fibrosis pneumonia) or produce enzymes capable of degrading wastewater contamination in a sewage treatment

Quorum: The minimum number of "members" of a deliberative assembly necessary to conduct the business of that group.

plant. Whatever the behavior, often it is the result of new gene expression. Thus, by monitoring their numbers through QS, the behavior of cells in the biofilm is controlled. Genes are not active unless the cells "realize" there are enough individuals that the products produced will accomplish a task or display some behavior.

Not all the "bacterial chatter" is limited to signals between cells of the same species. Bacterial cells also can intercept communication signals from other microbial species, or, in the case of a human infection, even from cells of the immune system, as we will see shortly. **A CLOSER LOOK 11.1** surveys some microbial interactions.

Consequently, it is not uncommon for bacterial cells, like their counterparts in the plant and animal world, to have sophisticated and complex social communication networks to coordinate their behavior. Here are a few very diverse examples.

A CLOSER LOOK 11.1

Life in a Biofilm

The interactions within a biofilm depend on chemical interactions (signals) among the cells in the group. These signals can have immediate consequences. Some of the behaviors could be beneficial while others might be detrimental to the everyday life of that community. Here are some examples.

- *Microbial Cheating*. When the cells in a biofilm reach a quorum, the resulting response is not always "all for one and one for all." As in human and other animal societies, some members of a bacterial biofilm might cheat.

  Suppose the cells in a biofilm need to obtain iron from the environment. Iron is needed for energy (ATP) generation. Therefore, cells in a biofilm go through the process of secreting molecules capable of accumulating iron. This is an energy costly process. However, there are cells that seem to be "freeloaders" in the biofilm; that is, they do not participate in producing the iron scavenging molecules. In the end though, they still profit (get iron for nothing) from the work done by those "exhausted" cells that do produce these molecules.

  But is this "cheating," or is it an example of a more complex behavior? Perhaps this is a case of division of labor, which is typical in multicellular plants and animals. In these organisms, different cells have different functions for the good of the whole. So, the cheaters might have a different role to the benefit of the biofilm.

  Here are a few other examples of what microbiologists are discovering about chemical signaling through signaling molecules (see **FIGURE A**).

- *Signaling Molecules as Cues*. If exposed to low concentrations of a chemical, the chemical might influence the control of hundreds of genes involved in metabolism and organism behavior. This might cue the sender (species 1) to isolate itself in a biofilm, now separated from species 2.
- *Signaling Molecules as Synergistic Signals*. Some molecules act as a synergistic signal, triggering cooperation between different species. The result often is biofilm formation. Such signaling then benefits both the sender (species 1) and recipient (species 2).
- *Signaling Molecules as Competitive Manipulators*. A signal produced by one species can manipulate chemically another species to benefit the chemical producer. For example, species 1 produces a chemical that results in species 2 fleeing the environment. The manipulation of species 2 by species 1 lowers the competition without killing the competition. Some antibiotics produced by soil bacteria might work this way.

FIGURE A Possible roles of chemical signals.

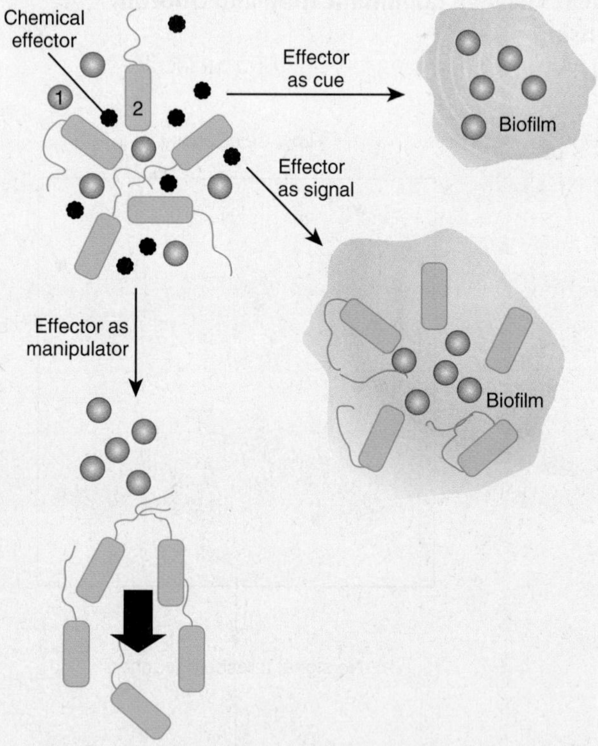

Data from Ratcliff, W. C., and Denison, R. F. 2011. *Science* 332(6029): 547–548.

Bacteria Behaviors Within a Biofilm

In a biofilm, bacterial cells undergo chemical "cross-talk" through QS, leading to some form of community behavior. Let's look at four distinct behaviors of bacterial social interactions, starting with the bobtail squid story described in the chapter opener.

Bobtail Squid Bioluminescence

Microbial bioluminescence is used by several marine animals as a defense weapon or as a lure for prey. In the chapter opener, the Hawaiian bobtail squid used bioluminescence to evade predators.

The *A. fischeri* cells do not produce light until the cell population reaches a quorum. Chemically sensing that point, the cells turn on new genes that trigger light production (**FIGURE 11.7**). Because bioluminescence is costly in terms of energy (ATP) production, the genes for bioluminescence only become active at high cell densities. Then, all the cells equally take part in the process.

Dental Plaque

After the common cold, dental caries (*cario* = "rottenness"), or tooth decay, is the most prevalent human infectious disease. On a tooth surface, cells of *Streptococcus mutans* can attach, multiply, and begin to form a biofilm (**FIGURE 11.8**). In ways that are not completely understood, through QS these cells prepare the area for a succession of up to 500 bacterial species. Importantly, these species communicate and interact in such a way that a dynamic biofilm commonly called **dental plaque** can form. Although most of the species in the biofilm do not cause dental caries, changes in the local environment can have a destructive effect. A diet high in sugary foods coupled with poor oral hygiene can shift the balance of microbes to one favoring the growth and metabolism of acid-producing *S. mutans*. These microbes then produce acids that can erode the protective tooth enamel over an extended period. This can increase the risk of dental caries. For that reason, dental plaque needs to be removed through regular flossing, brushing, and periodic dental cleaning.

Streptococcus mutans: strep-toe-KOK-us MEW-tanz

FIGURE 11.7 Bioluminescence. Cells of *Aliivibrio fischeri* were spread on an agar plate and allowed to grow (pale creamy yellow streaks on left plate). When the cells have reached a quorum, bioluminescence occurs (right plate). Note the areas (ovals) where there are insufficient cells to trigger bioluminescence.

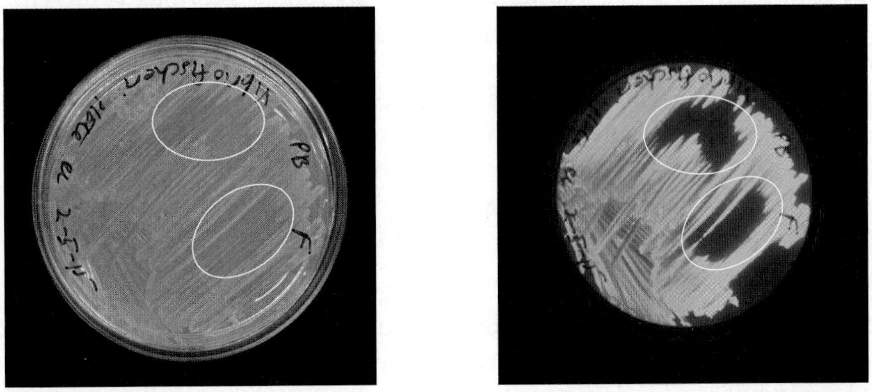

FIGURE 11.8 Dental Plaque Represents a Biofilm.

(A) This false-color scanning electron micrograph of a tooth surface shows dental plaque (pink) attached to the enamel of the tooth.

(B) In the drawing, a set of early colonizing species establish a biofilm at the tooth surface. Without removal, over time the bacteria set the stage for the development of a mature biofilm (dental plaque) with the addition of late colonizers.

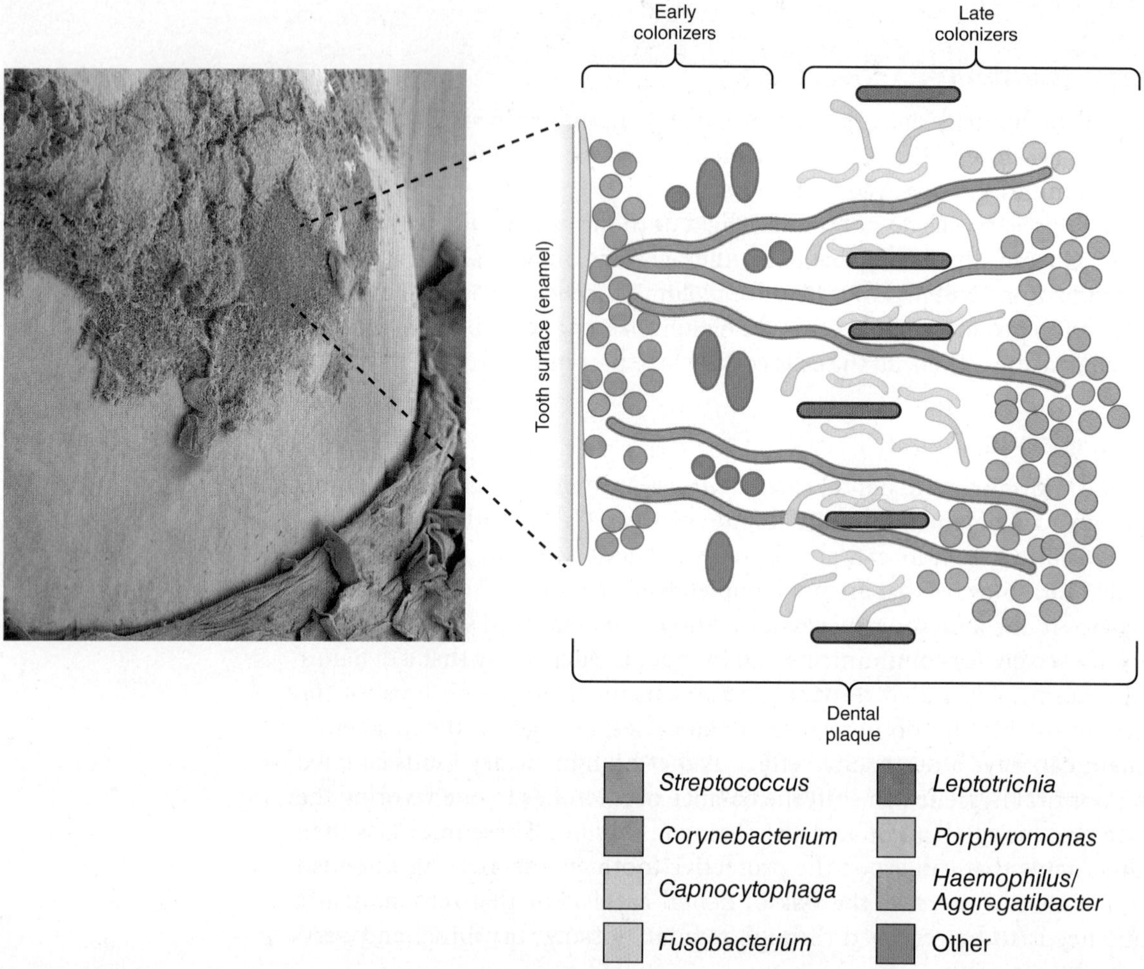

(B) Data from Welch, J. L. M., et al. 2016. PNAS doi:10.1073/pnas.1522149113.

Bacteria Behaviors Associated with Learning

When we think about learning, we tend to associate it with mammals and perhaps with some forms of "machine learning" by robots. Learning is the process of acquiring new, or modifying existing, knowledge, behaviors, skills, or values. It is not a behavior that has been associated with prokaryotic organisms. However, the assumption would be incorrect. Surprisingly, bacterial cells can "learn," but not in the cognitive way that humans learn. Rather, bacteria can anticipate and prepare for future conditions by modifying their existing "knowledge." Here are two examples.

Anticipating Change

Associative learning: A process in which a new response becomes associated with a stimulus.

Remember Pavlov's dogs? Ivan Pavlov was a Russian physiologist who is famous for his work in **associative learning**. In experiments carried out in the 1890s, Pavlov conditioned his dogs to associate a meal with a ringing bell. Over time, the dogs, hearing the bell (new stimulus) would anticipate the coming meal by salivating (new response), even before the meal arrived. Microbiologists have shown that bacterial organisms demonstrate a similar type of associative or conditioned learning. Here is one recent experiment.

Lab cultures of *Escherichia coli* cells were repeatedly shifted from 25°C (77°F) and 20% oxygen gas (O_2) to 37°C (98.6°F) and an eventual drop to 0% O_2. The bacterial cells over the weeks of the experiment "learned" to anticipate the coming O_2 drop by associating it with the temperature increase. When the cells sensed the temperature increase (new stimulus), they altered their metabolism (new response) in anticipation for the coming O_2 drop. In other words, the cells had associated the change in temperature (25°C to 37°C) with a coming drop in O_2 (20% to 0%).

Escherichia coli: esh-er-EE-key-ah KOH-lee

When bacterial cells experience repeated cycles of an environmental stimulus in Nature, similar associative responses probably occur. The cells associate the stimulus with past events and "plan ahead" for change. This is just another example of conditioned behavior that shows the enormous flexibility and adaptability present in the microbial world.

Dealing with Social Conflicts

Another fascinating example of bacterial cell behavior within a biofilm concerns the community's ability to resolve internal "social conflicts." Take, for example, the locations of cells within a biofilm (see Figure 11.4). Because of their proximity, the cells near the outer edge of the biofilm protect and shelter the cells in the interior. The outer-edge cells also have immediate access to nutrients needed for growth. Therefore, if these cells grow rapidly, few nutrients are left to "feed" the sheltered interior cells. Death would be forthcoming. Consequently, there is a social conflict here between outer and interior cells for nutrients and protection. In this case, it is resolved peaceably by QS.

The interior cells normally produce ammonium ions (NH_4^+). This is a small metabolite that the outer-edge cells need for growth and reproduction. Therefore, through QS, if the interior cells chemically sense that the cells at the outer edge are increasing too rapidly in number, the interior cells change metabolic behavior and reduce the secretion of NH_4^+. As a result, the outer-edge cells slow down their growth, allowing the inner cells to receive enough nutrients to stay alive.

Although this is described as a social conflict, it also is an example of a division of labor. In multicellular plants and animals, cells are specialized to carry out their own set of functions. However, they all combine to give an integrated set of functions that define the organism. This is true in a prokaryotic biofilm as well, where the biofilm consists of a cohesive set of cells that define a multicellular system. The example just described is not a rare behavior. Here is one more example involving altruistic behavior.

Microbes Exhibit Altruistic Behaviors

Altruism refers to a type of charitable behavior where the well-being of others is put before the well-being or survival of one's own. With altruistic behavior, the costs and benefits often are measured in terms of **reproductive fitness**. This refers to an organism's ability to transmit genes to the next generation in a way that ensures that generation can pass them on to their next generation. As with learning, this is a behavior that is associated with apes, humans, and other organized animal societies, such as bees and ants.

As with the other behaviors described above, microbiologists have discovered selfless acts (in human terms) or "charitable" behavior in bacterial populations. Clearly, there is no cognitive origin for this behavior in bacteria. However, like a robot that can be programmed to "learn" how to respond to a stimulus, so can bacterial cells. They can be genetically programmed to generate a specific response that outwardly looks like a learned behavior.

The Martyr Effect

The chapter on Controlling Microbes described how many bacterial species develop (or evolve) resistance to antibiotics. The perception could be reached that only the resistant bacterial strains survive because they in some way can overcome the antibiotic and multiply (**FIGURE 11.9A**). The susceptible cells lack resistance mechanisms and are killed.

Within populations of *E. coli* cells, there often are some antibiotic-resistant individuals. Research studies have shown that a few of these cells with the highest antibiotic resistance can act as "martyrs." If the population comes under antibiotic attack, some of the antibiotic-resistant cells will sacrifice their own well-being to improve the community's overall chance of survival.

This "bacterial altruism" is a result of chemical communication among community members. If the group is exposed to the antibiotic, some of the most resistant individuals produce a signaling molecule called indole (**FIGURE 11.9B**). Indole triggers the antibiotic-susceptible cells to activate membrane pumps to expel the antibiotic before it can kill them. The downside is that while indole might save most of the community, its production takes a toll on the fitness level of those individuals producing indole. Indole production requires a large expenditure of ATP, so the "martyrs," the most antibiotic-resistant cells, grow slower—and some might die.

FIGURE 11.9 Altruistic Behavior in a Bacterial Colony. (A) The traditional view of antibiotic resistance suggests that in the presence of an antibiotic, only the cells resistant to the antibiotic will survive and replicate. All the susceptible cells die.

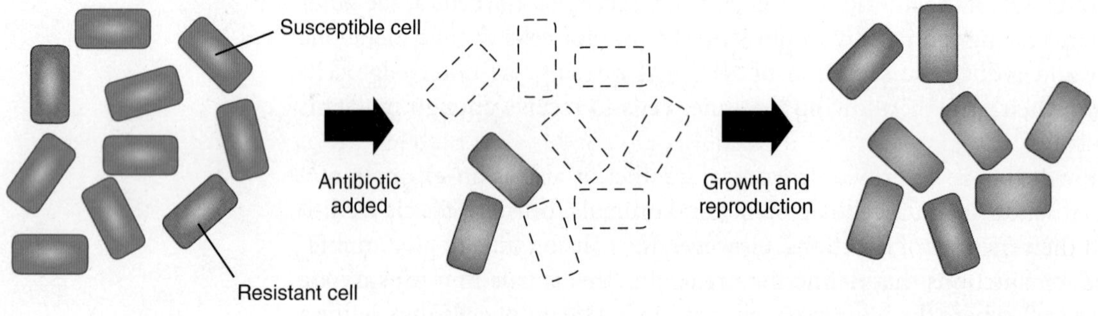

(B) Research now supports an alternative idea. In the presence of an antibiotic, some of the resistant cells produce a chemical mediator called indole that protects the antibiotic-susceptible cells from being killed by the antibiotic. As a result, the community survives.

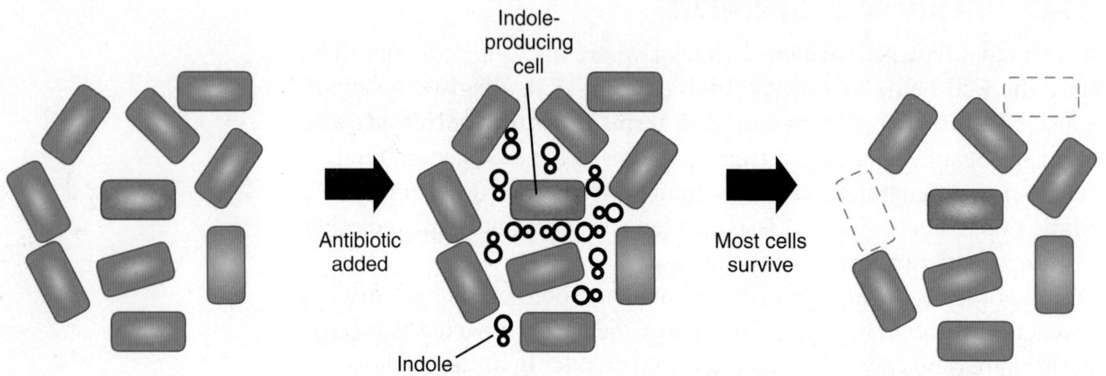

11.3 Biofilms Within the Human Body: The Human Microbiome

The Human Microbiome Project was completed in 2013. It provided the first in-depth look at the diversity of microbial communities that live in and on the human body. Since then, hundreds of additional research papers and reports have been published on the **human microbiome**. These studies are beginning to provide a glimpse into the varied roles these microbes play in human health and disease. Therefore, this section of the chapter will focus on these human biofilms, especially the gut microbiome.

There are about 10 trillion human cells that build the human body. There are another 30 trillion microbes that are distributed as biofilms among several different organs and tissues within the body. The majority of these microbes are bacterial. They have been detected by DNA sequencing methods because most species will not grow in culture. Behaving as **symbionts**, they have nutritional needs that have yet to be identified. Nonetheless, like the other biofilms discussed in this chapter, these human biofilms chemically "talk" with one another as well as with cells comprising the human body.

Symbiont: An organism that is very closely associated (living) with another, usually larger, organism.

Most of the microbes present in an adult are acquired at the time of birth. In a vaginal birth, delivery occurs through the mother's birth canal; that is, a fetus moves from the uterus (womb) out through the vagina. During the newborn's passage, it is exposed to the microbes normally found in the mother's birth canal. It then contacts microbes in the external environment. The body sites that will be colonized by microbes are those that have contact with the environment—namely, the skin, mouth, nose, ears, eyes, respiratory tract, gastrointestinal tract, and urogenital tract (**FIGURE 11.10**). Other internal organs, such as the heart, liver, kidneys, and brain, as well as the blood, remain sterile. These sites normally are free of any microbes and viruses unless an infection occurs.

Why a Symbiosis?

A **symbiosis** refers to two or more different organisms living in close association with one another. In this association, the **host** is the partner that is occupied by, or is infected with, the other partner, the symbiont. Why do many organisms, including humans, support symbiotic microbial communities, and why do these symbionts want to live in a biofilm with the host?

The symbiotic association with human and microbes has been coevolving for millennia. The human microbiome contains 100 times more genes than the human genome (22,000 genes). Consequently, both the host and symbiont have come to depend on the physical and biochemical/genetic properties of one another. Here are some examples.

Host Services

The human host provides an ideal environment for the symbionts to form a biofilm.

1. *Location*. The human body has numerous surfaces (e.g., skin, respiratory, digestive) on which microbes can establish a biofilm.
2. *Nutrients*. The environment surrounding the microbiome usually contains the preferred types of nutrients the organisms need for growth and survival.
3. *Protection*. The microbes are protected from potentially adverse conditions, like fluctuating temperatures, pH, and oxygen levels.

FIGURE 11.10 A Sampling of Microbes Composing the Human Microbiome. There are thousands of microbial species (primarily bacteria) on and within different systems of the human body. A few of the more prominent genera are listed. See **Appendix A** for organism pronunciations.

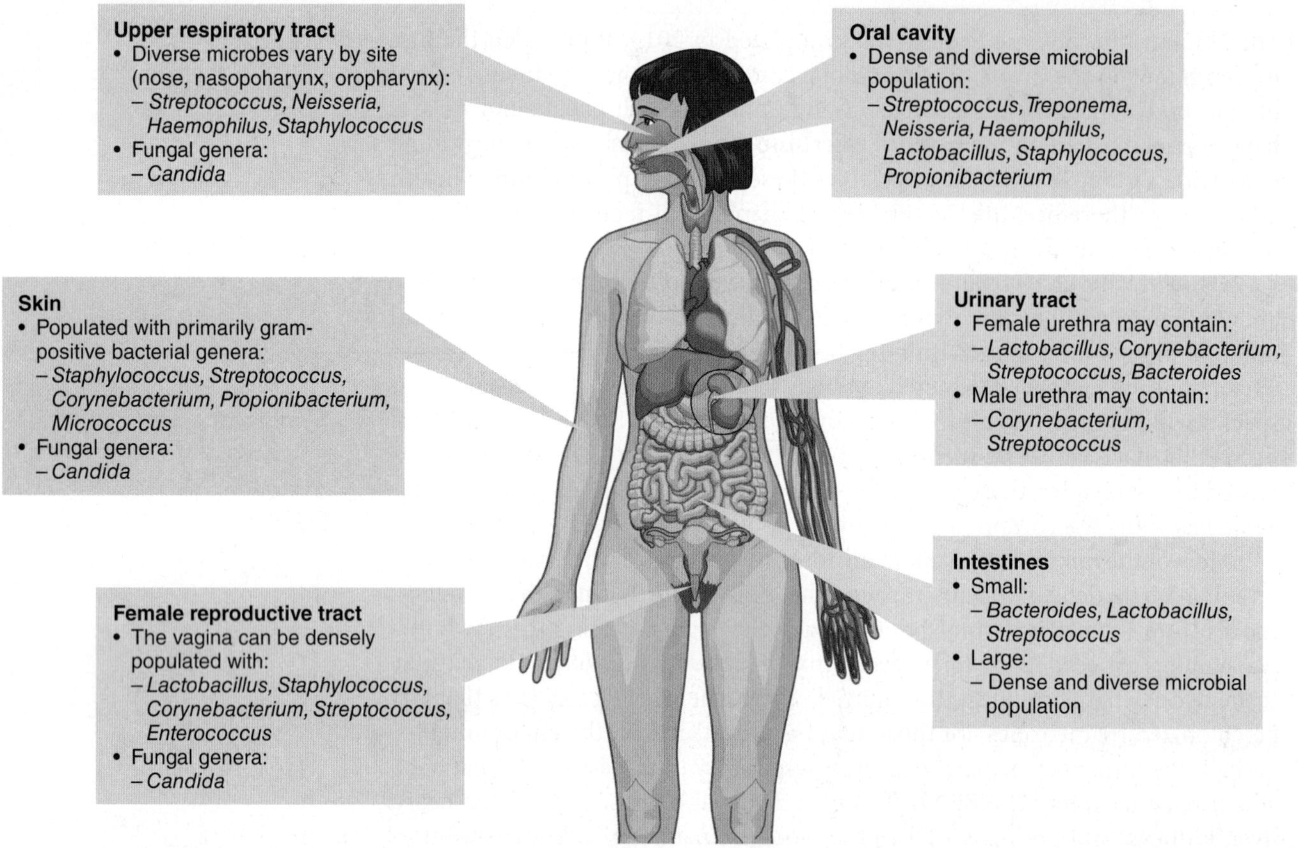

Symbiont Services

The symbionts provide metabolic capabilities that are lacking in, or adding benefit for, the host.

1. *Digestive Enzymes.* Microbes produce numerous enzymes that degrade the complex, indigestible carbohydrates (fiber) that humans often consume.
2. *Essential Nutrients.* Microbes supply essential vitamins and amino acids.
3. *Toxin Degradation.* Microbes produce enzymes that degrade toxins that are ingested in some foods.
4. *Defense Against Pathogens.* The resident species in a microbiome form a barrier to pathogen invasion. However, many also perform **microbial antagonism**. This behavior allows the symbionts to outcompete invading pathogens for space and nutrients. The microbes also produce antimicrobial substances that inhibit the growth of those pathogenic species. For these reasons, the human microbiome sometimes is referred to as a "second immune system."

This dependence between host and microbiome suggests that an imbalance in, or disturbance to, the microbial community might have harmful outcomes for the host. This so-called **dysbiosis** might result in illness or disease. The working premise, called the **disappearing microbe hypothesis**, proposes that over the past century there have been changes in diet, modern birth practices, pollution, and antibiotic use that have contributed to an imbalance among and loss of some bacterial species in the gut. This shift has altered the symbiotic relationship within the gut microbiome and the services that the microbes have provided. Therefore, numerous research investigations have examined how the symbionts promote good health in some cases and, in a dysbiosis state, to illness or disease.

Let's briefly examine the three body systems (skin, respiratory, and digestive) most intensively studied today.

The Skin Harbors a Resident Microbiome

The skin is the largest organ in the human body, covering about 2 square meters (21 square feet) on an adult. It is home to a complex skin microbiome comprised of trillions of bacteria. This biofilm remains relatively stable in healthy individuals despite constant exposure to environmental fluctuations, such as temperature, humidity, and body hygiene.

The skin microbiome is dominated by four bacterial phyla. These are the gram-positive Actinobacteria and Firmicutes and the gram-negative Proteobacteria and Bacteroidetes (**FIGURE 11.11A**). The phyla are described in more detail in the chapter on Exploring the Prokaryotic World. However, different areas of the skin create distinct physical environments (habitats) and, therefore, are home to separate and unique communities of microbes (**FIGURE 11.11B**).

Sebaceous Skin Sites

Bacterial diversity appears lowest at sebaceous (oily) sites, such as the forehead and back. *Propionibacterium acnes* inhabits hair follicles at these sites and can metabolize the **sebum** into products that are toxic to many other bacterial species including pathogens. In higher numbers, some *P. acnes* strains are associated with skin conditions like acne.

Propionibacterium acnes: pro-pea-OHN-ee-bak-tier-ee-um AK-nees

Sebum: An oily secretion of the sebaceous glands that keeps the skin and hair soft and moist.

Moist Skin Sites

Moist sites on the skin, such as the navel, groin, sole of the foot, back of the knee, and the inner elbow, are dominated by *Staphylococcus* and *Corynebacterium* species. Processing of sweat by the staphylococci and corynebacteria produces the characteristic body odor associated with sweat.

Staphylococcus: staff-ih-loh-KOK-us

Corynebacterium: KOH-ree-nee-back-tier-ee-um

Dry Skin Sites

The skin areas with the highest microbial diversity are the dry sites, such as the forearm, palms, and buttocks. These sites are dominated by *Staphylococcus epidermidis* and several gram-negative species in the Proteobacteria and Bacteroidetes phyla.

Staphylococcus epidermidis: staff-ih-loh-KOK-us eh-peh-DER-mih-dis

No matter what skin site, many of the resident species display microbial antagonism. A more recent and surprising discovery is that some members of the skin microbiome, including *S. epidermidis*, can produce anti-cancer substances. These substances, which have no effect on normal skin cells, can inhibit DNA replication in cancer cells. It is thought that such chemicals, if they can be identified and pharmaceutically produced, might be useful in helping prevent skin tumors.

The Healthy Respiratory Tract Harbors a Diverse Resident Microbiome

Hundreds of different bacterial species representing the same four phyla form the microbiome of the respiratory system (**FIGURE 11.12**). These microbes in healthy individuals cause no known illness or disease.

FIGURE 11.11 The Skin Microbiome.
(A) Members of four phyla compose most of the skin microbiome.

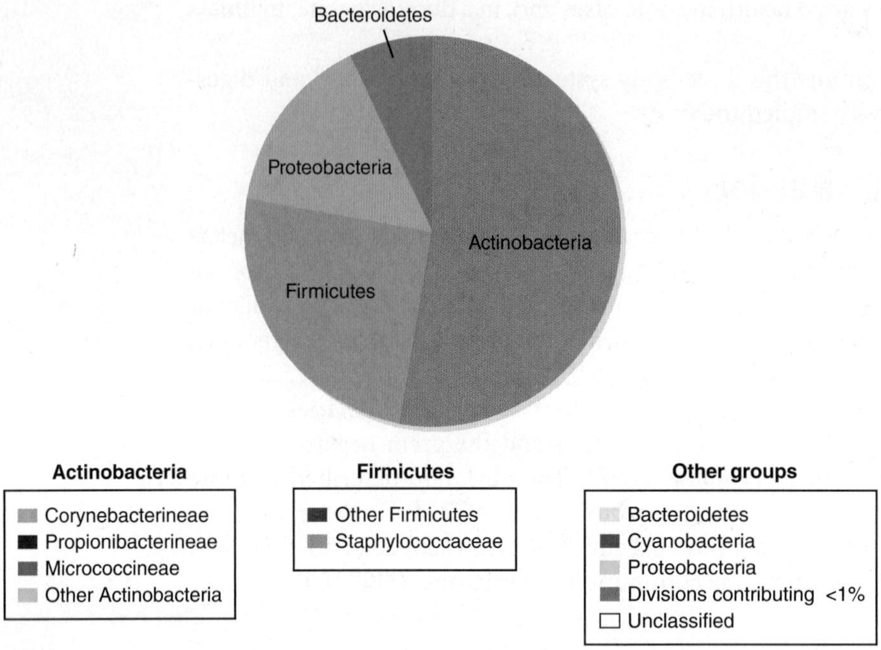

Data from Cho, I., and Blaser, M. J. 2012. *Nat Rev Gen* 13(4): 260–270.

(B) A topographical distribution of the skin microbiome. The presence of specific bacteria is governed by the microenvironment of each of the three skin sites. Lettered labels: blue = sebaceous sites; green = moist sites; red = dry sites.

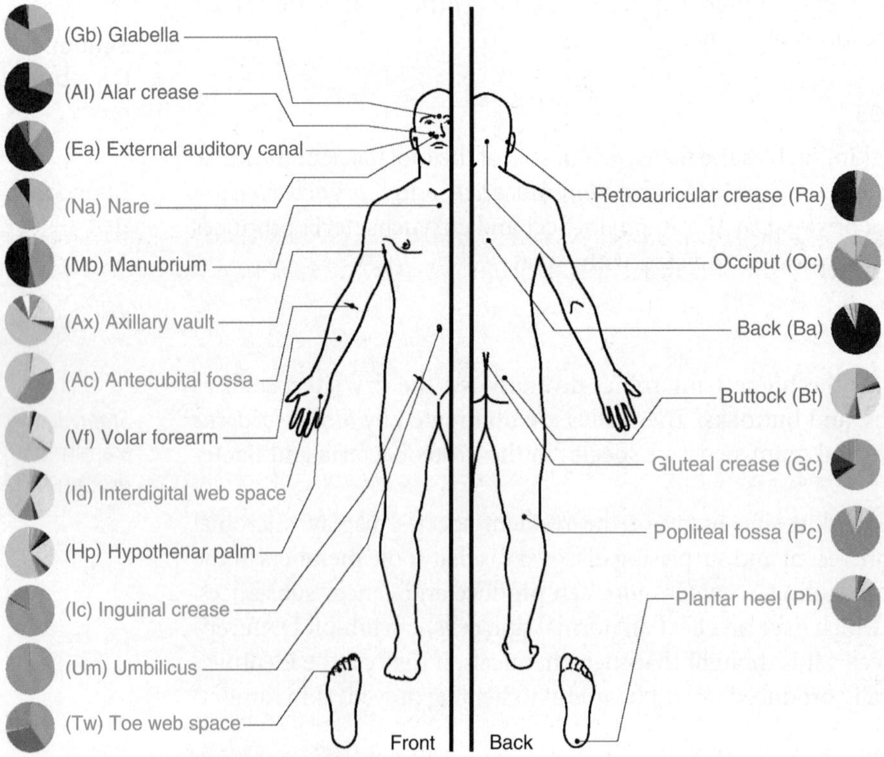

Courtesy of Darryl Leja, National Human Genome Research Institute (NHGRI).

Unique biofilms are present in different regions within the respiratory system. The surface of the nasal cavity, for example, mainly is colonized by members of the Actinobacteria (e.g., *Corynebacterium* and *Propionibacterium*) and Firmicutes (e.g., *Staphylococcus, Streptococcus*, and *Neisseria*). In the oropharynx,

Neisseria: nye-SEER-ee-ah

FIGURE 11.12 The Human Respiratory Microbiome. The overall compositional diversity of the respiratory tract is compared to that for the nasal cavity and the oropharynx.

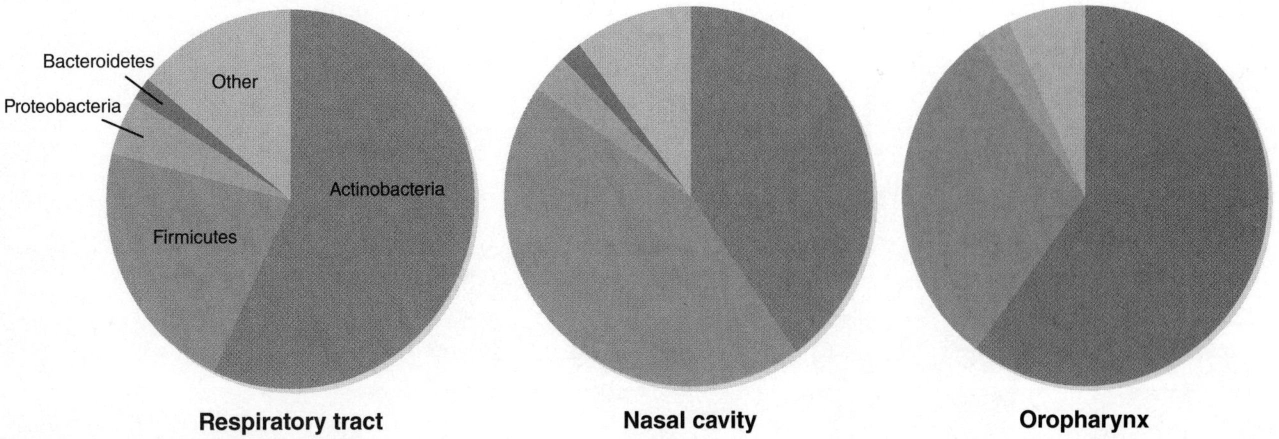

Data from Cho, I., and Blaser, M. J. 2012. *Nature Reviews Genetics* 13: 260–270.

which is the middle portion of the **pharynx**, a richer and less varied group of genera are present.

Pharynx: The part of the throat behind the mouth and nasal cavity.

The passage of aerosolized fluids, secretions, and microbes from the oropharynx appears to be the source for the microbes found in the microbiome of the lower respiratory tract (windpipe and lungs). However, the resident microbial community here is 1,000 times less dense than the microbiome in the nasal cavity and pharynx. Without the presence of strong microbial antagonism, infections of the lungs (e.g., pneumonia and tuberculosis) often are more serious than nasal or throat infections.

The Gastrointestinal Tract Harbors the Largest and Most Diverse Microbiome

About 98% of all bacteria of the human microbiome are found in the **gastrointestinal (GI) tract** (gut). Like the skin, differences in anatomy, physiology, and organization in each part of the GI tract generate unique habitats that are home to unique microbial biofilms (**FIGURE 11.13**). Species number and diversity are very high in the mouth (oral cavity), which is home to some 20 billion (2×10^9) bacterial cells comprising up to 700 different species. Species numbers and types drop off dramatically in the stomach, only to increase again in the small intestines. Numbers and diversity climb sharply in the **colon**, which contains the highest density of bacterial species in the body. It is home to almost 4 trillion cells distributed among 1,000 different bacterial species.

Gastrointestinal (GI) tract: A group of organs that includes the mouth, esophagus, stomach, pancreas, liver, gallbladder, small intestine, colon, and rectum.

Colon: Also called the large intestine, this region of the digestive system is where food that has not been digested in the stomach and small intestine is processed prior to defecation.

Clearly, from the above description, most studies of the gut microbiome until recently involved the cataloging of species. Although it is important to know what species are present, what is needed is an understanding of what these microbes do beyond the services mentioned above. Part of the hurdle is sorting through the hundreds of species that have been discovered in the gut. To make matters more complex, there appears to be no common set of microbes shared by all humans. In fact, only 10% to 20% of the gut bacteria in one person are found in another person. Nevertheless, all healthy humans carry out the same basic digestive reactions in the gut. Therefore, different gut microbiomes must still carry out a common metabolism. Simplistically, one individual might have bacterial species A that can digest plant fiber A. Another individual lacks species A, but has species B that also can digest plant fiber A.

FIGURE 11.13 The Digestive System Microbiome. There can be substantial differences in the microbiome composition within an individual at different anatomical sites. Inset: The number and diversity of bacterial species increases from the stomach to the colon. The small intestine is composed of the duodenum, jejunum, and ileum.

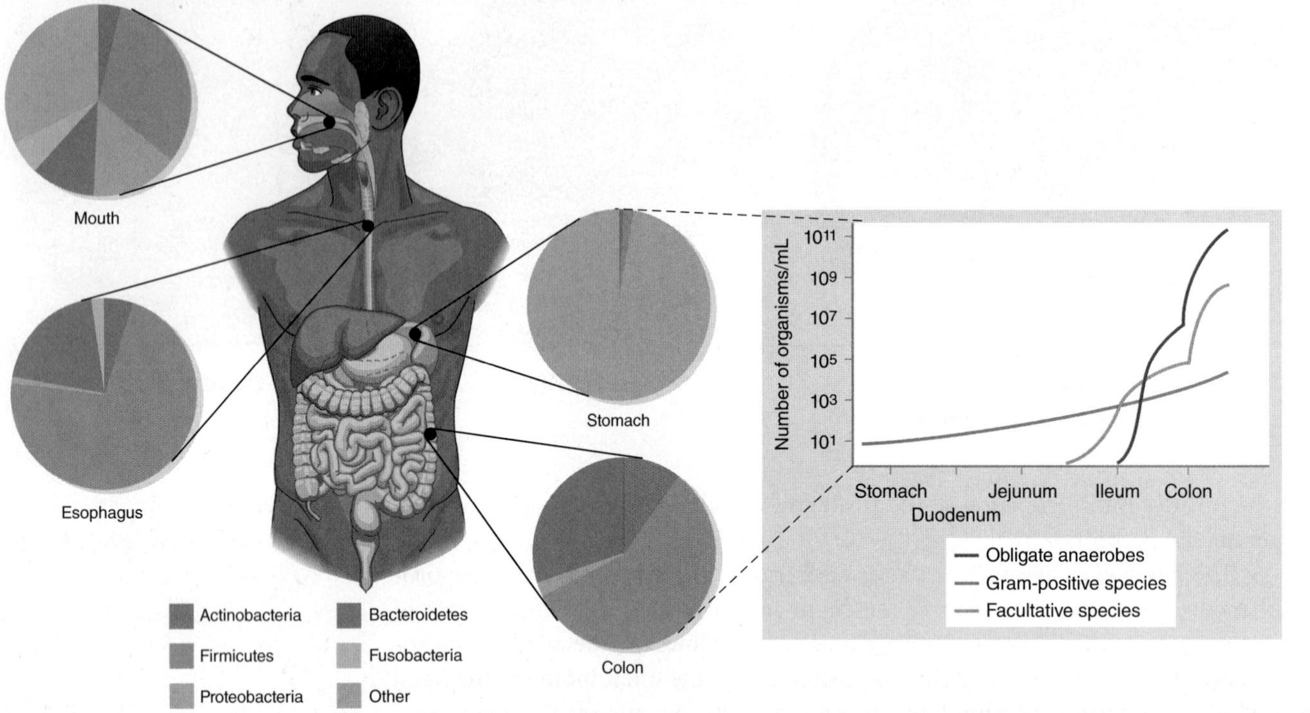

Data from Cho, I., and Blaser, M. J. 2012. *Nature Reviews Genetics* 13: 260–270.

Without a core gut microbiome, it is challenging to determine exactly what is a "healthy microbiome".

Despite these hurdles, microbiologists are beginning to understand some roles the gut microbiome plays in health and disease. They also are studying the gut microbiome's fascinating role in shaping human behavior. In these studies, it is not possible to invasively sample gut material from the intestine of living humans. Therefore, most human studies involve analysis of feces composing the **fecal gut microbiome**.

Let's look at a few of the better-studied functions. Realize that much of this work is preliminary and sketchy at best, and often the headlines move faster than the science. In the coming paragraphs, words like "might" and "could" and "if" are often used because the whole picture is not clear and in some cases still quite foggy.

Obesity and the Gut Microbiome

Obese: Someone who is 30 pounds or more overweight (according to the National Institutes of Health).

According to the most recent National Health and Nutrition Examination Survey, almost 20% of American children and nearly 40% of adults are **obese**. For a long time, obesity was associated with an unhealthy diet, a sedentary lifestyle, and perhaps some family genes. Although certainly important, microbiologists are discovering that some gut bacteria can alter how fat is stored. These bacteria also might control how the body balances glucose levels in the blood and how hormones can make us feel hungry or full.

Research studies reveal that the gut microbiome of lean individuals is much more diverse in types of bacteria than that of obese people. The microbes from lean individuals are very efficient at breaking down bulky plant starches and fibers into shorter molecules. The body can use these as a source of energy and as signals regulating multiple functions in the gut and beyond. A dysbiosis, and the missing microbes, might set the stage for obesity.

FIGURE 11.14 The Effect of Gut Microbes on Obesity.
(A) The fecal transfer of gut microbes from lean mice keeps the recipient mice lean, while the fecal transfer of gut microbes from obese mice causes the recipient mice to remain obese.

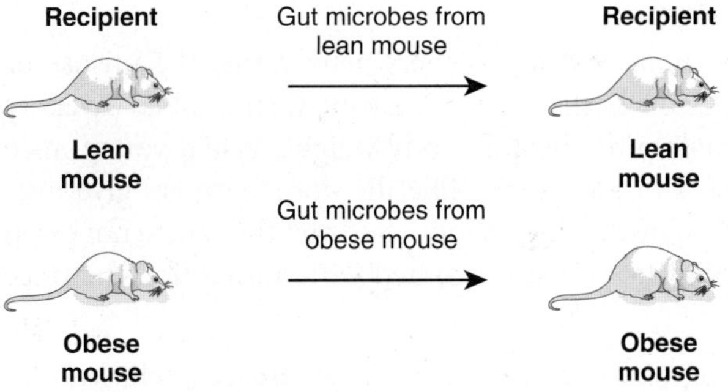

(B) If obese mice are given a fecal transfer of microbes from lean mice, the obese mice lose weight and become lean. Similarly, if lean mice are given a fecal transfer of microbes from obese mice, they gain weight and become obese.

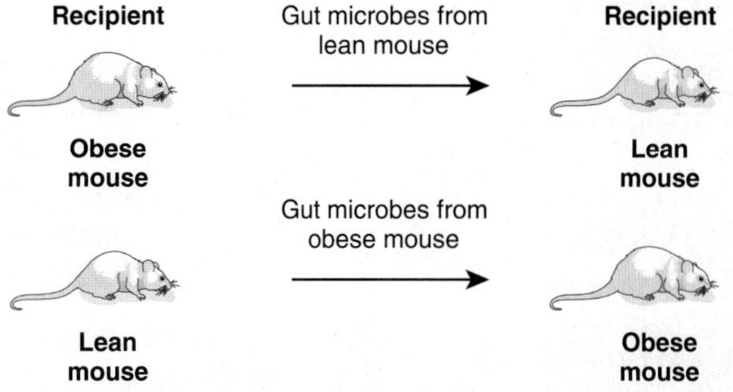

Gnotobiotic: Referring to an animal that is microbe free or in which only certain known species are present.

Many studies of the gut microbiome use **gnotobiotic** animals, especially mice, as models to substitute for humans. By reducing the number of bacterial species present in mice, it is easier for investigators to follow the behavior of just a few species. Also, mice are easier to study and control in experiments than are humans. In addition, some experiments using humans might not be possible or ethical to conduct.

To demonstrate that gut microbes can influence obesity, genetically identical gnotobiotic mice living in a microbe-free environment were used. The mice were given either intestinal microbes collected from lean mice or intestinal microbes from obese mice (**FIGURE 11.14A**). The mice in both groups ate the same diet in equal amounts. At the conclusion of the experiments, if lean mice received gut microbes from lean mice, the recipients remained lean. Similarly, if obese mice received gut microbes from obese mice, the recipients remained obese. When the mouse fecal microbiome was analyzed, the obese mice had a less diverse community of microbes in the gut than did the lean mice. Some microbe species had disappeared.

To further study this cause-and-effect relationship, gut bacteria from the lean mice were transferred to the guts of obese mice. The results showed that the mice given the "lean community" of bacteria lost weight. They were no longer obese (**FIGURE 11.14B**). Likewise, if lean mice received gut microbes from obese mice, the recipients gained weight. Again, the plausible explanation for these findings is that microbes from obese mice have experienced a dysbiosis. Bacterial species needed to perform key roles in maintaining a healthy body weight and normal metabolism are missing.

So, why would these microbes be absent? Here is what the experts believe.

1. *Change in Diet.* A typical "Western diet" is composed of highly processed foods that are high in fat and low in fiber. If mice are fed a diet high in fat and low in fiber (e.g., red meats, sugar-sweetened foods), the mice gain weight. The unhealthy diet lacked the nutrients needed to support a diverse gut microbiome. The bacteria needed for a diverse microbiome do not survive and disappear.

   The microbiome of mice fed a low-fat, high-fiber diet (e.g., fruits and vegetables) contained the needed diverse bacteria. The diet of these mice provided for the growth of the diverse gut microbiome.

2. *Food Intake*. The gut microbiome can influence the release of several key chemical transmitters that operate on the gut–brain axis (see below). Some of these transmitters might modulate food intake. Without a diverse gut microbiome, the control signals are disrupted or missing. The result might be higher food intake that would further intensify weight gain.
3. *Misuse/Abuse of Antibiotics*. Researchers believe that the misuse or abuse of antibiotics in children might destroy the microbial diversity needed to help maintain a healthy body weight. When young mice are given low doses of antibiotics, like the doses farmers give livestock, the mice develop about 15% more body fat than mice not given antibiotics. If a high-fat diet is combined with antibiotics, the mice became obese.

Lactobacillus bulgaricus:
lack-toe-bah-SIL-lus
bull-GAIR-ee-kus

Streptococcus thermophilus:
strep-toe-KOK-us
ther-MOH-fill-us

Bifidobacterium lactis:
bi-fih-doe-back-TIER-ee-um
LACK-tiss

Researchers are quick to point out that this is a new research field with far more questions than answers. What is needed is to identify the strains of bacteria associated with leanness, determine their roles, and develop treatments accordingly. **A CLOSER LOOK 11.2** discusses one possible treatment: probiotics and/or prebiotics.

A CLOSER LOOK 11.2

Probiotics and Prebiotics to the Rescue?

Could probiotics or prebiotics help maintain, or restore, a normal, healthy microbiome? Maybe.

One potential treatment to reverse dysbiosis is to "pump up" the good bacteria in the gut by using probiotics and/or prebiotics. **Probiotics** are foods or pill supplements that contain live microbes that might help maintain or reestablish the normal gut microbiome (see figure). One of the most familiar probiotic foods is yogurt. Yogurts containing *Lactobacillus bulgaricus, Streptococcus thermophilus*, and *Bifidobacterium lactis* are said to contain "live, active cultures." These bacterial species are believed to help maintain good gut health through the fermentation metabolism they carry out. Physicians often suggest patients eat yogurt after taking oral antibiotics, as these antimicrobial drugs can destroy much of the gut microbiome.

FIGURE A Probiotics in food and pill forms.

Other probiotic products vary widely among manufacturers, but no one can say with certainty which bacterial species are beneficial or how many microbes one must ingest. Some probiotic pill supplements, for example, contain only a handful of bacterial species, while others contain as many as 20 different species and strains. Many of these products simply are throwing bacteria at the "problem."

Prebiotics are high-fiber foods or pill supplements that are thought to influence the growth of the good gut microbes. In a sense, prebiotics are thought to act like "fertilizer" to stimulate the growth of a diverse gut microbiome. Again, which fibers and in what amounts they need to be consumed is an open question.

In addition, the United States Food and Drug Administration does not regulate the diet supplement industry. That means the makers of these "health" products don't have to show their products are beneficial or contain exactly what the label says. So, it is questionable whether they will increase the diversity of microbes in the gut.

So, what is the take-home lesson? Adding probiotics and prebiotics to the diet is questionable and probably will contribute little to balancing or maintaining good gut health. For most healthy people, probiotics and prebiotics will not do any harm. For individuals with gut problems, more research into the roles of specific bacterial species is needed before the true value of these supposedly health-promoting therapies can be evaluated. No one in the field believes that probiotics and prebiotics alone will keep a person's gut healthy. What can be said is that along with exercising and good eating habits, the gut microbiome certainly is a key partner to a healthy gut.

11.4 Microbiome and Host: "Talking" Back and Forth

There are two other fascinating fields being studied concerning the gut microbiome's role in human health and behavior. One area is the crosstalk between the gut microbiome and the immune system. The other area is the crosstalk between gut bacteria and the nervous system, specifically the brain activities that underpin human personality and cognition.

Immune System Function and the Gut Microbiome

The immune system is a complex network of cells and signaling molecules that, among other roles, protect the body from infectious disease. It usually works with almost flawless molecular precision to recognize and defend against pathogens. This control depends on a variety of **antibodies** and antimicrobial substances that can limit the spread of pathogens. The inner workings of the immune system are described in the chapter on Disease and Resistance.

Antibody: A protein produced in response to foreign substances, such as bacteria, that enter the body.

However, immune system function is not solely a human-controlled trait. Rather, the gut microbiome might be part of a complex network of identification and response to an invader. If so, a vigorous immune system is a consequence of complex interactions between human cells and microbial cells. What might the gut microbiome be doing?

The gut microbiome releases many metabolic products that promote proper development of the immune system. Research suggests the gut microbes assist in the secretion of antibodies against infecting pathogens (**FIGURE 11.15**). There also is experimental evidence that the gut microbiome helps control some allergic conditions (e.g., asthma, eczema) by controlling immune system responses to foreign substances. In all these interactions, if there is a dysbiosis, the result might be ill health.

One example is the development of the immune system in infants. Studies have reported that children delivered by Caesarean section are not exposed to the same microbes as infants born by vaginal delivery. Apparently, C-section-born

FIGURE 11.15 Effects of Dysbiosis on Immune System Function.
With a healthy immune system and balanced gut microbiome, normal immune behaviors occur. If the immune system is dysfunctional and/or the gut microbiome is unbalanced, various disease outcomes might occur.

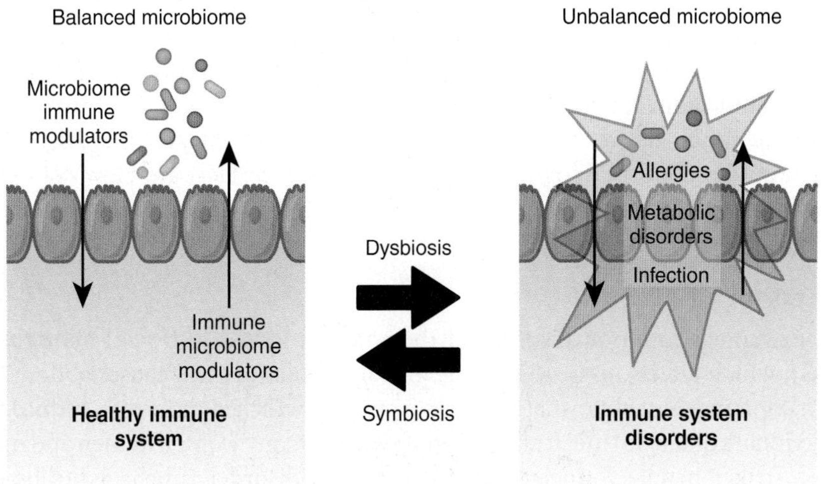

infants are missing essential microbes that normally would be picked up from the mother during vaginal delivery. These disturbances to the gut microbiomes could affect immune development. Some studies suggest that a dysbiosis during early life might make the child more predisposed to developing allergies and less fit to fight off infections.

The misuse of antibiotics in infants and young children is another example where the drugs can dramatically alter the gut microbiome. Studies examining the influence of antibiotics on infant immune system development also find there is a loss of essential bacterial species with excessive antibiotic use. It is proposed that with the disappearance of important bacterial species, immune system modulation might be disturbed.

In both examples concerning C-section delivery and antibiotic misuse, the cross talk between gut microbes and immune system does not occur as it should. Unfortunately, what these chemical conversations are communicating remains a mystery.

Nervous System Function and the Gut Microbiome

The human endocrine system is a collection of glands that produce hormones. These hormones help regulate metabolism, growth and development, tissue function, sexual function, reproduction, sleep, and mood. To carry out these roles, the system continuously communicates with different parts of the body to help maintain an internal steady state.

It should not come as a surprise, especially after reading this chapter, that over the thousands of years of coevolution between human and microbiome partners, there are cross-species chemical "conversations" going on, influencing, and sometimes affecting, the behavior of the endocrine system. For example, human blood is packed with metabolic products coming from gut microbes that cross the gut lining and enter the bloodstream. Do these microbial products influence gene expression and behavior in cells of the endocrine system? Human stress releases hormones, such as **epinephrine**, into the blood. Do such molecules move in the opposite direction and influence the gut microbiome?

Many of the chemical conversations involve the so-called **gut–brain axis** (**GBA**). This communication pathway connects the body's **central nervous system** with the **enteric nervous system** in the GI tract. The chemical communication involving the back-and-forth neural, hormonal, and immunological communication between the gut and brain has been known for a long time (**FIGURE 11.16**). Because research investigations suggest the gut microbiome is important in influencing this crosstalk, perhaps a better name for the communication link should be the "**gut–brain–microbiome axis**." As in the other examples involving gut health, when chemical communication is normal and proper, the gut and brain function optimally. Conversely, dysbiosis might disrupt the communication network. Some research experts even suggest the disconnect might initiate abnormal cognitive and social behaviors. Here are a few examples. Again, gnotobiotic mice often are the experimental animals used in the studies.

Epinephrine: A neurotransmitter (also known as adrenalin) that under stress is released by some neurons to stimulate the heart muscle, accelerate the heart rate, and increase cardiac output.

Central nervous system: That part of the nervous system that includes the nerves in the brain and spinal cord.

Enteric nervous system: The arrangement of nerve cells and supporting cells that govern the function of the gastrointestinal tract.

Irritable Bowel Syndrome

An extremely common disorder of the colon is **irritable bowel syndrome** (**IBS**), which affects up to 15% of the global population. IBS causes abdominal pain, bloating, and abnormal bowel movements, which significantly reduce the individual's quality of life. It affects about twice as many women as men and most often strikes people younger than 45 years. The disorder is believed to be the result of abnormal GI tract movements and a disruption in the communication

FIGURE 11.16 The Gut-Brain-Microbiome Axis. The **vagus nerve** represents the major neuroanatomical link, and the blood represents the major circulatory link between the gastrointestinal tract and the brain.

Vagus nerve: One of the cranial nerves that connects the brain with the abdomen.

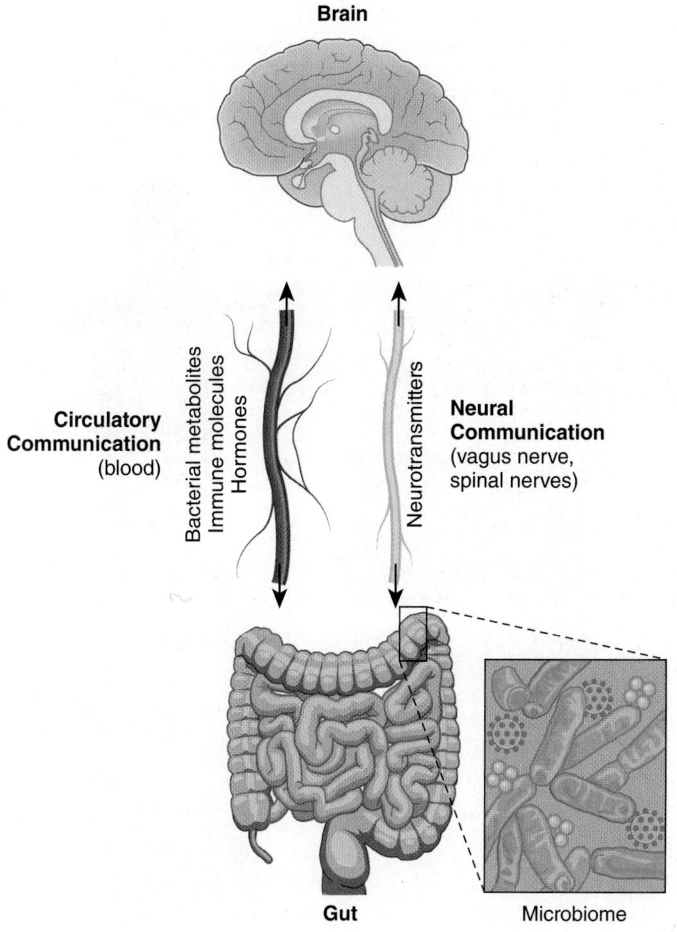

between the brain and the gut. Treatment options have long been limited to managing the symptoms.

Over the past several years, numerous studies have documented a disturbance in the diversity of bacteria in the gut microbiome of people with IBS. This has led researchers to postulate that the microbiome and dysbiosis in the gut–brain–microbiome axis might play a key role in the development of IBS and progression of the disorder.

There are several factors that might lead to microbiome disruption. These include antibiotic use, infection, diet, and stress. Any and all of these can affect the diversity of gut bacteria. Therefore, modulation of the axis with dietary changes, prebiotics, probiotics, select antibiotics, and stress-reduction strategies have been suggested as ways to try to alleviate the disorder.

There is a growing interest in an investigational treatment called **fecal microbiome transplantation** (**FMT**). For FMT therapy, a fresh or frozen stool (microbiome) sample from a healthy donor is introduced into the patient's colon with a **colonoscope**, **enema**, or feeding tube. The thought is that the new microbiome introduced will restore the balance of intestinal bacteria and return the patient's gut microbiome to a normal state.

Treatment with FMT has been a spectacular success when treating recurrent infections of the gut caused by antibiotic-resistant *Clostridioides difficile*. Using FMT, cure rates are up to 94% effective (**FIGURE 11.17**). These results have

Colonoscope: A long, flexible tube inserted into the rectum and colon that, for FMT, is used to deliver a fecal sample.

Enema: An injection of fluid into the lower bowel by way of the rectum.

Clostridioides difficile: kla-strih-dee-OY-deez DIF-fih-sil-ee

FIGURE 11.17 Fecal Microbiome Transplantation Therapy. Fecal microbiome transplantation therapy can be used to cure *Clostridioides difficile* infections or possibly reduce the effects of a gut microbiome imbalance, such as irritable bowel syndrome.

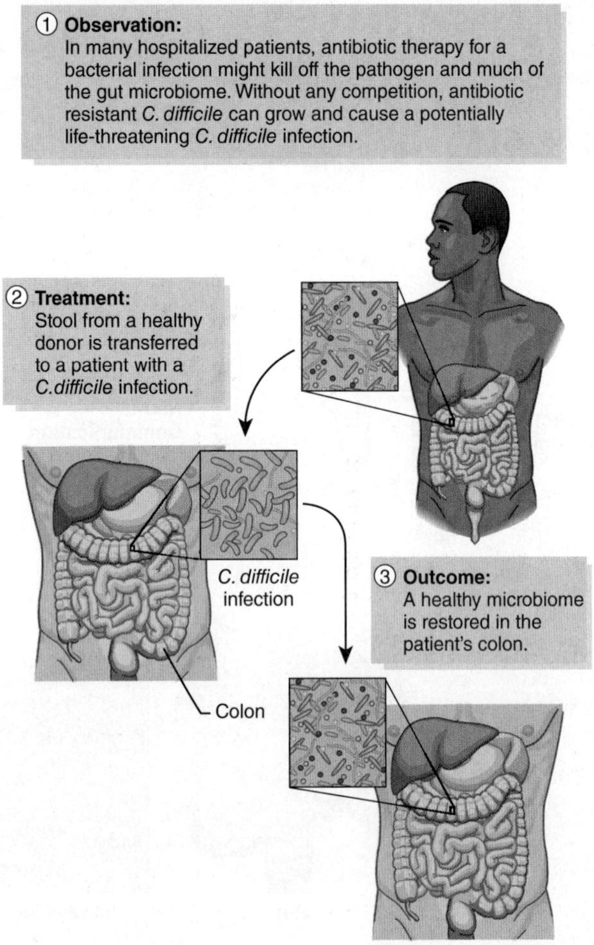

https://jamanetwork.com/journals/jama/fullarticle/2635633.

stimulated investigators to consider FMT for IBS patients. The assumption is that the introduced microbes would reestablish the normal communication links between gut and brain. With a return to normalcy, immune system control would also be modulated correctly. Among the small number of IBS patients receiving FMT so far, there has been about a 60% improvement in symptom relief. Although encouraging, more clinical trials involving IBS patients need to be carried out to evaluate the effectiveness of FMT.

Stress and the Gut Microbiome

Studies on the gut–brain–microbiome axis are shedding new light on how self-awareness, personality traits, and emotional state relate to the higher functions of the human brain. For example, behavioral studies on mice have found that changes in the gut microbiome affect cognitive function, social behavior, and stress-related responses. These would be akin to anxiety and depression in humans.

Recently, scientists have shown that injecting beneficial bacteria into gnotobiotic mice makes the rodents more resilient to stress. In the study, healthy mice were injected with a heat-killed soil bacterium called *Mycobacterium vaccae*. A control group was not injected with the bacterium. During a 19-day period, the mice were placed in a cage with a larger, more aggressive male mouse.

Mycobacterium vaccae: my-koh-back-TIER-ee-um VAK-keye

Both groups of mice then were watched for stress-related behavioral changes. The investigators reported that the bacteria-treated mice exhibited less anxiety or fear of the aggressive mouse as compared to the control group. So, what could *M. vaccae* be doing?

Further physiological studies found that injection of *M. vaccae* caused a group of nerve cells in the brains of the injected mice to secrete serotonin. This is a **neurotransmitter** known to modulate anxiety. The assumption then is that the production of serotonin by the injected mice was channeled through the gut–brain–microbiome axis. As a result, the neurotransmitter changed its behavior such that it was less stressed than the control mice.

Neurotransmitter: A chemical messenger released by neurons to stimulate other neurons, muscle, or gland cells.

Importantly, a lack of serotonin production in humans is associated with depression. Another group of researchers took gut bacteria from depressed humans and used the bacteria to colonize the guts of nondepressed mice. The investigators reported that the mice showed changes in their behavior that were characteristic of depression. Gnotobiotic mice that grew to adulthood exhibited deficits in social behavior.

In another series of behavioral experiments, investigators gave the mice four *Bifidobacterium* species known to be normal members of the human infant microbiome during early childhood development. The addition of these species "rescued" the mice; they grew up to have no behavioral deficits.

Now mice are not humans, so much more work needs to better understand how the gut microbiome can affect brain chemistry and how the microbes might influence emotions and stress. No one knows how these microbes, and the compounds they produce, affect the brain. It is not known how the gut–brain–microbiome axis controls the communication of these signals. Still, some researchers suggest that in the future it might be possible in human depression to improve mood control through a rebalancing of the gut bacterial community. Researchers are even thinking of how beneficial bacteria might help war veterans better cope with post-traumatic stress disorder. Again, until the gut microbiome is understood much better, these ideas are suggestive at best.

11.5 Microbes and Society: What Is a Human?

For years, our traditional view of human "self" was defined by our own bodies. Our 10 billion cells and 22,000 genes encoded by our genome build and control the tissues, organs, and organ systems that we define as a human being. We control or fail to control our social behavior and mood. Today, this apparent reality faces a major challenge because of the influence of the human microbiome with its 30 trillion cells and more than 2 million genes. Consequently, should a human be viewed as a single object (or "self") or as a dynamic and interactive community of human cells and microbial cells (**FIGURE 11.18**)? In fact, some scientists have referred to a human as a "superorganism." So, what is a human?

In 2018, three international professors published an article entitled "How the microbiome challenges our concept of self" (see Figure 11.18 for reference). They suggest that because microorganisms are a natural part of the human body, a new interpretation is needed for understanding what it means to be human. Social scientists have argued that unique human abilities, such as reasoning, language, and art, separate us from other animals. These "human qualities" cannot be ascribed to the laws of nature. Rather, these behaviors are concerned with society and the relationships among individuals within a society.

FIGURE 11.18 A New Perspective of the Human Self.
The traditional view represents humans as the interaction of nervous system (brain), immune system, and genome. The microbiome-era view portrays the individual human as interactions of brain, immune system, and genome with microorganisms of the human microbiome.

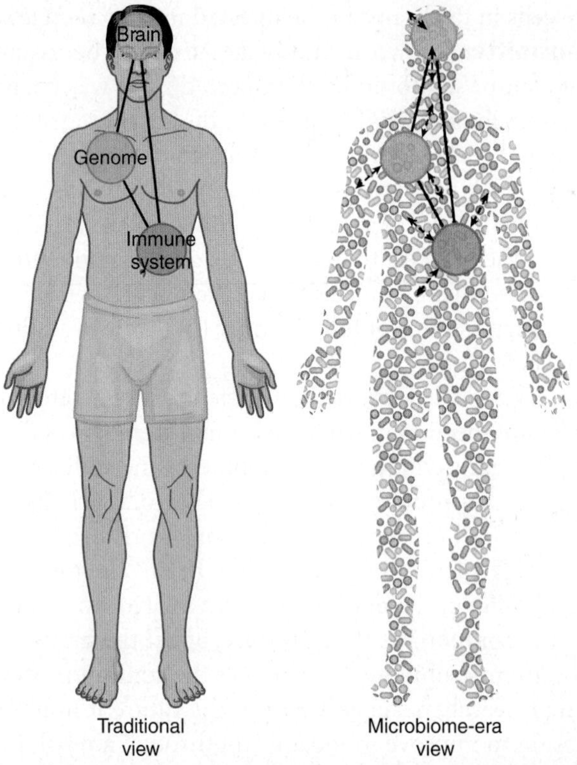

Data from Rees, T., Bosch, T., and Douglas, A. E. (2018). How the microbiome challenges our concept of self. *PLoS Biol* 16 (2): e2005358. https://doi.org/10.1371/journal.pbio.2005358.

The continual discoveries concerning the human microbiome confound the traditional distinction of being human. Microbes are greatly influencing many of our behaviors and our interactions in society.

In the article, the authors conclude by stating,

> *"The challenge is big, the opportunity even bigger: it is time, and perhaps past time, to rethink collaboratively [about] what it means to be a living human being at home in a microbial world, one on which we depend and with which we are inseparably interwoven. Microbiome science has the exciting—the important—potential to catalyze the breakdown of [outdated] barriers between the natural and the human sciences and enable a truly integrated understanding of what it means to be human, [because the] human is more than the human."*

A Final Thought

We are just at the tip of the iceberg in our understanding of the human microbiome, and there is so much more to learn and understand about it and its interactions with us. As more research is done and a better understanding of the crosstalk between the human body with its microbiomes comes forth, the new knowledge will provide innovative ways to keep us fit and conceivably produce a healthier society. Part II of the text leaves much of the human microbiome behind. It examines the applied side of microbiology and the numerous roles microbes play in very well-documented ways to benefit society.

Chapter Discussion Questions

What Was He Thinking?

From your reading of this chapter, identify and discuss five major points about the social activities of biofilms and microbiomes that the author was trying to get across to you.

Questions to Consider

1. Describe four advantages that bacterial cells in biofilms might have over free-living single bacterial cells.
2. Explain why most microbes around the globe, whether in soil, oceans, or associated with another organism, form biofilms.
3. How do microbes in biofilms use associative learning?
4. In a symbiosis, what types of "services" are provided by the host and by the symbionts?
5. After reading the material in this chapter about the gut–brain–microbiome axis, is there any truth to the proverbs "I have a gut feeling" or "My gut tells me . . ."?
6. The human body consists of 11 organ systems, including the circulatory, respiratory, digestive, excretory, nervous, and endocrine systems. Some microbiologists are suggesting that the human microbiome should be considered a 12th organ system of the body, or a "second brain." Based on what you have read in this chapter, would you support or reject this suggestion? Explain.
7. Hundreds of microbial species normally colonize the gut. Would you agree or disagree with the statement, "The microbes of a healthy person's gut microbiome are simply pathogens that are waiting for an opportunity to infect the body and cause disease."
8. A friend of yours is eating a cup of yogurt for lunch. She reads on the label that it is a "probiotic yogurt." Knowing you are taking a Microbes and Society course, she asks you, "What is a probiotic yogurt?" How would you answer her question?
9. Based on the material presented in this chapter, construct an argument for what you believe it means "to be human."

©NYCstocker/iStock/Thinkstock

PART II

Microbes and Human Affairs

Throughout much of Europe in the late sixteenth and seventeenth centuries, food was a valuable commodity, and the interaction of urban and agricultural sectors was an integral part of society. Perishable fruits and vegetables were expected to move quickly because damaged and spoiled produce posed economic risks. Fish markets operated only in the morning before the sun could rot the fish (see figure below). Damaged or spoiled food also was an opportunity for a moral lesson, as one person selling poor quality produce could make customers suspicious of the whole market.

Many fresh market scenes around the world have not changed much, though trade of perishable foods and the urgency to move them from farm to table have changed markets in many of the more developed countries. Still, mishandled food from a food vendor can spoil the lot and affect our lives.

In Part II of *Microbes and Society*, we look at applied microbiology. In Chapter 12, we explore how microbes both contaminate foods and lead to food spoilage, while in Chapter 13, we examine their beneficial roles in foods. Chapter 14 reveals how microbes make possible a stunning variety of products ranging from antibiotics to vitamins. Next, Chapter 15 describes the roles microbes perform on the farm in helping produce our meat and dairy products. Our look at the magic of microbes concludes in Chapter 16, which outlines how microbes continue to protect our environment while making possible life itself.

There is a darker side to some microbes. In the final three chapters, we discuss the disease-causing aspects of pathogens. Chapter 17 describes the disease process and the means by which our bodies develop resistance to disease. It also identifies ways we can help establish resistance using vaccines. Chapters 18 and 19 present a survey of several familiar and not-so-familiar infectious diseases.

Joachim Beuckelaer, 1568/Purchase, Lila Acheson Wallace Gift and Bequest of George Blumenthal, by exchange, 2015/Metropolitan Museum of Art.

CHAPTER 12

Microbes and Food: Food Preservation and Safety

Marco Polo and the Silk Road

We all know of Marco Polo's travels to China in the thirteenth century to obtain spices and explore new trade routes (see chapter opening image). What often is not mentioned is that the spices brought more than wealth and a tasty addition to food.

When humans were still hunters and gatherers, they often wrapped meat in the leaves of bushes, accidentally discovering that this process enhanced the taste of the meat. As time moved on, spices were used for medicinal purposes. Historical documents mention the use of spices for their health benefits as far back as ancient Egypt. These records on papyrus identify coriander, fennel, juniper, cumin, garlic, and thyme as health-promoting spices. Importantly, from about the eleventh to sixteenth centuries, spices also were used to mask unpleasant tastes and odors of rotten food, especially to cover the taste of spoiled meat. When leaves, seeds, or roots had a pleasant taste or agreeable odor, they grew in demand and gradually became a norm for that culture as a condiment.

Many of the world's most valuable spices came from China, India, and the islands of Indonesia (known then as the Spice Islands), and it was important to find the quickest way to bring these spices back to Venice and Europe. Consequently, Marco Polo's exploration of Asia in the late thirteenth century found such trade routes—the so-called Silk Road. Marco Polo mentioned spices frequently in his travel memoirs (*The Travels of Marco Polo*). His engaging stories described the flavor of sesame oil from Afghanistan and ginger and cassia from Peking. He wrote about wealthy people who ate meat pickled in salt and flavored with spices, while the poor ate foods prepared with garlic. In all these cases, besides flavoring the food, the spices he brought back were indispensable for improving the smell and taste of food that had been spoiled by microbial contamination.

CHAPTER 12 OPENER Marco Polo was one of many who brought spices from Asia to Europe. In addition to adding flavor to foods, the spices also hid the rotting smell of spoiled foods. From: *Livre des merveilles du monde (Book of the Wonders of the World)* by Marco Polo and Rustichello de Pisa, ca. 1350 (National Library of Sweden/World Digital Library).

FIGURE 12.1 Food Preservation. Many foods are subject to bacterial contamination that can lead to food spoilage and sometimes food poisoning.

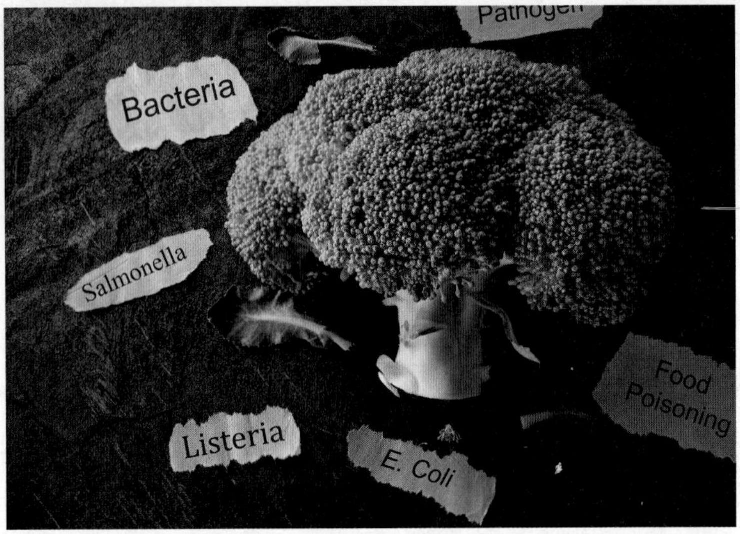

For centuries, methods to preserve foods have been important to all societies and civilizations. Today, food preservation and safety are important in every home, restaurant, and food service industry (**FIGURE 12.1**). In addition, society depends on very specific conditions and practices that preserve the quality of food to prevent spoilage and food decay. Therefore, the primary focus of this chapter is to examine the types of microbes that might contaminate and spoil foods and to understand how the food environment influences microbial growth. We then will review methods to prevent spoilage, a few of which have not changed since before the time of Marco Polo.

LOOKING AHEAD

After reading and completing this chapter, you will be able to:

12.1 Identify the intrinsic and extrinsic factors involved with food spoilage.
12.2 Assess the effects of food spoilage on foods (e.g., meats, fish, eggs, dairy, grains, and fruits and vegetables).
12.3 Discuss five methods used for food preservation.
12.4 Describe the challenges for food safety today.

12.1 Food Spoilage: Terms and Conditions

About 10,000 years ago, humans switched from being hunter-gatherers to agricultural societies. These societies now could produce more food than they could eat in a single meal. However, that brought challenges in storing excess food and preserving food from spoilage. Marco Polo's travels to China in the thirteenth century to obtain spices were essential for improving the smell and taste of spoiled food (**FIGURE 12.2**).

Every year nearly one-third of the world's food—1.4 billion tons—is lost through waste (**FIGURE 12.3**). More than 20% of this waste arises from **food spoilage**, a metabolic process causing foods to become undesirable or

FIGURE 12.2 Spices. A typical Asian spice stand in a large spice market in India.

FIGURE 12.3 Food Lost and Consumed. Percentages of food lost and consumed globally each year.

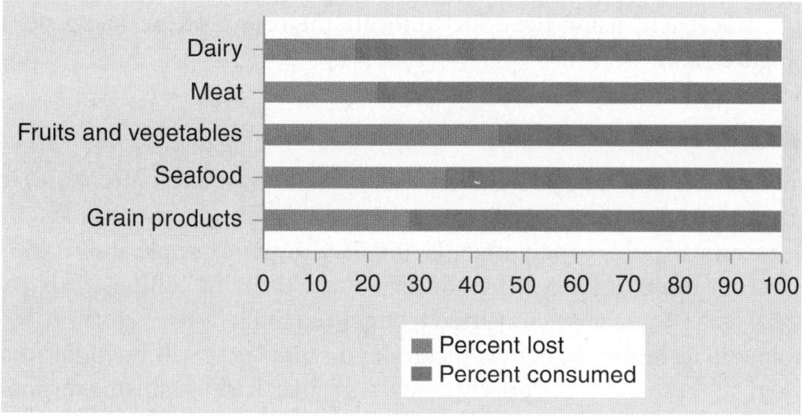

Data from Natural Resources Defense Council Issue Paper, August 2012. Source: Food and Agriculture Organization 2011.

unacceptable for human consumption. Spoilage is due to changes in sensory characteristics such as color, texture, and/or flavor caused by chemical or physical changes to the food product. It also can occur when microorganisms contaminate the food and produce enzymes that result in undesirable by-products in the food.

General Principles

The microbial content of foods usually has qualitative as well as quantitative aspects. The qualitative aspects refer to which microorganisms are present. The quantitative aspects refer to how many microbes are present. For example, in some food products, such as a can of vegetables, no microbes should be present; it must be **sterile**. By contrast, raw hamburger meat usually is considered acceptable if it contains up to 1,000 bacterial cells per gram of meat because cooking will kill the organisms. So, before cooking, a raw hamburger patty might have more than 200,000 bacterial cells. Because these bacterial cells can multiply in the meat and cause spoilage, meat processors must be vigilant to keep the number of bacterial cells as low as possible.

Sterile: Referring to the complete absence of all life, including viruses.

Several terms are used when judging microbial spoilage.

Label Date

Packaged foods usually have an **expiration date** (label date), indicated by "Best by …," "Use by …," or "Sell by …" (**FIGURE 12.4**). These designations represent the manufacturer's suggestion as to how long the product is at "peak quality" in terms of taste, smell, and/or freshness. It should not be interpreted as a safety date, and except for infant formula, the label date is not federally regulated. Manufacturers point out that eating a food product past its label date does not mean the food is spoiled or can cause illness. However, with enough time, spoilage microbes will multiply even if a food product is stored correctly.

Shelf Life

Shelf life refers to the time it takes a food product to decline to an unacceptable level. After this time, foods will exhibit a loss of quality attributes, including flavor, texture, color, and other sensory properties. Nutritional quality also might be affected during food deterioration. Thus, shelf life depends largely on the number of microbes present in a food product and the conditions under which it is stored. For packaged products, a **manufacturer code** is present on the product. If there is a contamination problem, investigators can do a "source traceback" using the code to locate the origin of the suspected food (see Figure 12.4).

Microbial Load

The numbers of microbes in a food product are known as its **microbial load**. Consumers generally expect the microbial load to be low in all foods, but in some cases, it is surprisingly high. In the hamburger example above, the load might seem quite high. A teaspoon of yogurt contains billions of harmless bacterial cells that have converted condensed milk into yogurt. Also, the microbial load in foods such as pickles and sauerkraut will be high because, like yogurt, these foods purposely contain the bacterial organisms responsible to produce the product.

Most dairy products, like milk, usually are pasteurized to remove any pathogens. Even so, a one-gallon milk carton could have over 3.7 million bacterial cells yet remain safe to consume.

FIGURE 12.4 Food Codes. Most foods on the market today carry a "best buy" date (top line) and a manufacturer code (bottom line).

Courtesy of Dr. Jeffrey Pommerville.

Sources of Microbial Contamination

Microbial contamination of many foods usually is impossible to avoid because contaminating microbes can enter foods from various sources.

- **Fresh Produce**. Fruits and vegetables can become contaminated when microbes in the air, soil, or water get on the skin or rind and then penetrate to the softer inner tissues when the skin is broken or cut. In addition, raw produce purchased at supermarkets might have been exposed to the contaminated hands of numerous shoppers. Many fresh fruits and vegetables are coming from abroad, as **A CLOSER LOOK 12.1** describes, providing another potential source for foodborne illnesses.
- **Livestock**. Meat and poultry products can become contaminated with animal feces. Livestock also can become cross-contaminated by contact with contaminated materials in the processing plant. If red meat, for example, is handled carelessly at the processing plant, bacterial organisms from the slaughtered animal's intestines can mix with the meat.
- **Shellfish**. Clams and oysters can become contaminated with pathogens or toxins. These shellfish concentrate microbes and toxins in their tissues by catching them in their filtering apparatus as they strain their food from contaminated water.
- **Rodents** *and Insects*. Rats and flies transport microbes on their feet and body parts as they move about among garbage, foods, and other unsanitary locations.

A CLOSER LOOK 12.1

Free Market Economy and Food Safety

Today, you can go into just about any supermarket and find fresh fruit that seems out of season. You can find fruits such as peaches and grapes in the middle of winter and perhaps winter squash in the peak of summer. Thanks to advances in agriculture and transportation, there has been a globalization of the food market. Peaches from Chile or raspberries from Mexico can be heading to U.S. markets within hours after being picked to ensure the produce does not spoil.

It's great to have fresh fruits and vegetables, but it brings along the risk of a foodborne illness, something we did not directly address in this chapter. So, let's briefly address this issue as part of food safety.

About 70% of many fresh fruits and vegetables we find in our local supermarket come from other countries. Sometimes these countries do not have the regulated and safe food practices (often identical to those to prevent food spoilage) found in the United States and other developed countries. Although outbreaks of foodborne illnesses are rare, they are occurring with greater regularity every year.

The United States Centers for Disease Control and Prevention (CDC) states that although foodborne infections have decreased by nearly 25% since 1996, "continued investments are essential to detect, investigate, and stop outbreaks promptly in order to protect our food supply."

Detecting outbreaks of foodborne illness is not easy because fresh produce is distributed across the nation and not to just one local community. In fact, physicians and local health departments might not see a local outbreak as part of a national event. Most local health departments are accustomed to responding to a foodborne illness outbreak at a social function or commercial establishment.

Globalization now means that local, state, and national food safety programs need to coordinate their surveillance methods and establish clear communication channels to properly recognize and quickly respond not only to food spoilage but also to foodborne illness outbreaks.

So, should you shy away from imported produce? Certainly not. In most cases, make sure there is no spoilage, and then properly wash fresh fruits and vegetables or cook the foods, which should eliminate most chances of contracting a foodborne illness.

Courtesy of Dr. Jeffrey Pommerville.

The strategy in the food industry is to control or minimize the contamination of foods. This demands good food management processes and a high level of sanitary practices. The rapid movement of the foods through processing plants and well-tested preservation methods also are key to keep foods safe for consumption by the public.

The Conditions for Spoilage

Food can be looked at as a culture medium for microorganisms. Therefore, the food's chemical and physical properties have a significant effect on what types of microbes it might harbor (**FIGURE 12.5**).

Intrinsic Factors

Conditions naturally present in foods that influence microbial growth are called **intrinsic factors**. These include:

- **Water Content**. One of the prerequisites for life is water. Therefore, to support microbial growth, food must be moist and have a minimum water content of 18% to 20%. Microbes do not grow in foods such as potato chips, dried pasta, nuts, rice, and flour because the water content is too low.

FIGURE 12.5 Food Spoilage. Several intrinsic and extrinsic factors determine whether foods are likely to spoil quickly or resist spoilage.

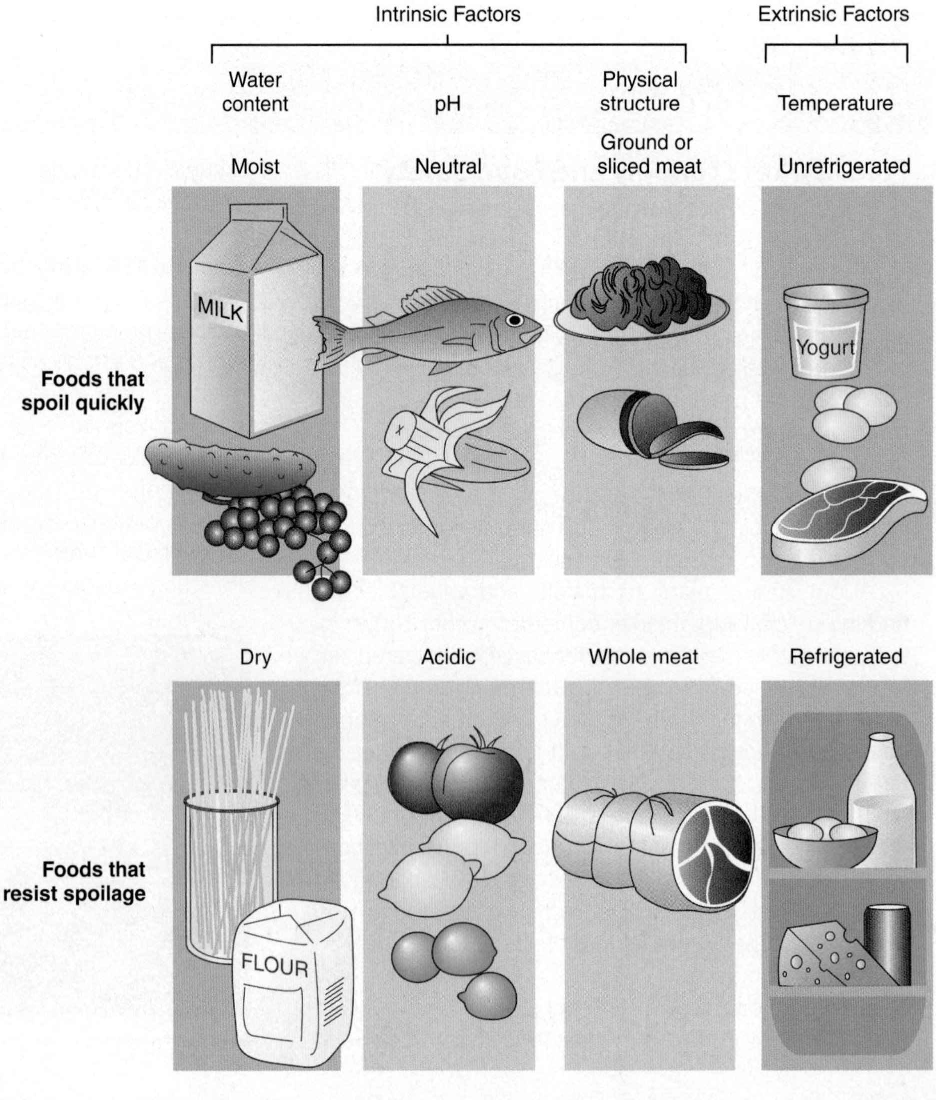

- **pH**. Most foods fall into the slightly acidic to neutral range on the pH scale, and numerous bacterial and mold species find these conditions ideal for growth. In foods with a pH of 5.0 or below, acid-loving molds are the predominant microbes. Consequently, citrus fruits generally escape bacterial spoilage but might be targets for mold growth and spoilage if damaged.
- **Physical Structure**. Whole meat products, such as steaks and roasts, are not likely to spoil quickly because microbes cannot easily penetrate beyond the surface of the meat. However, raw ground or sliced meat, such as uncooked hamburger or bacon, can deteriorate rapidly because microbes grow within the loosely packed structure of the moist meat as well as on the surface.
- **Chemical Composition**. A food's nutrient content is another intrinsic factor determining the type of spoilage possible. Fruits, for example, can support organisms metabolizing carbohydrates, whereas meats support microbes that decompose protein. Starch-digesting bacterial cells and molds might be found on raw potatoes and corn products.

Extrinsic Factors

Environmental conditions surrounding the food (food storage and packaging) make up the so-called **extrinsic factors** influencing microbial growth and spoilage (see Figure 12.5). The major factor here is temperature, which was described in more detail in the chapter on Growth and Metabolism.

- **Temperature**. How hot or cold raw foods or food products are can have a great influence on microbial growth and is a key factor for food safety. Refrigerator temperatures below 5°C (41°F) usually are too cold for the growth of most spoilage organisms, and freezer temperatures at or below −18°C (0°F) halt the growth of microbes (see Figure 12.12). However, the warm storage spaces of a ship or a humid, hot warehouse is an environment conducive to the growth of many spoilage organisms. It is common knowledge that contamination is more likely in cooked food at a warm temperature than cooked food kept at refrigerator temperature.

In summary, the food industry recognizes three groups of foods, loosely based on their intrinsic and extrinsic factors (**FIGURE 12.6**). **Highly perishable** foods are those that spoil rapidly. They include poultry, eggs, ground and sliced meats, most vegetables and fruits, and dairy products. Foods, such as shelled nuts, potatoes, and some apples, are considered **semiperishable**, meaning they spoil less quickly. **Nonperishable** foods often are stored in the kitchen pantry. Included in this group are cereals, rice, dried beans, dried pasta products, flour, and sugar.

FIGURE 12.6 The Perishability of Foods. Examples of highly perishable, semiperishable, and nonperishable foods. The physical and chemical properties of these foods are reliable indicators to their rate of perishability.

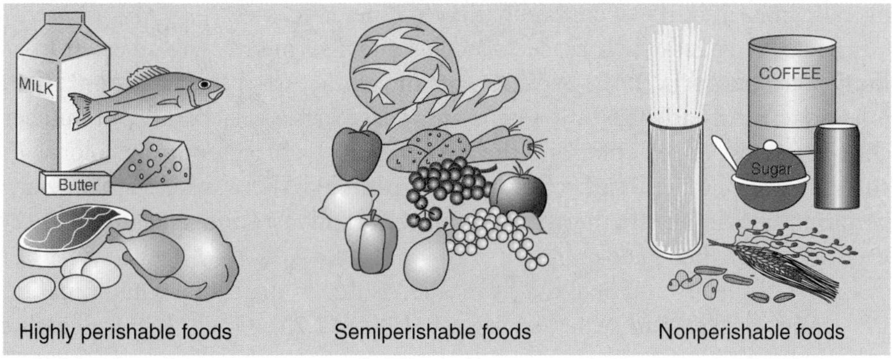

12.2 Microbes Causing Spoilage: Effects on Foods

Most of the foods in the human diet are rich in carbohydrates, fats, and proteins. Unfortunately, these are the same nutrients microbes find most favorable for growth. Therefore, food spoilage can be caused by a variety of microorganisms, including both bacterial and fungal (yeast and mold) species that come from the soil, water, or intestinal tracts of animals. They also might be transmitted through the air and water, and they can be carried by insects.

Microbial spoilage usually involves enzymes produced by the microbes. The resulting decay can be detected by a change in odor or color of the food, which comes from metabolites produced during microbial growth. Often there is a succession of different microbial populations that rise and fall as nutrients change during the deterioration process.

Let's examine the major food categories and uncover the challenges food processors and manufacturers face to inhibit the contamination and growth of spoilage microbes. Although many pathogens might also contaminate foods and cause disease, these organisms will be described in more detail in the chapters on Viral Diseases and Bacterial Diseases. In the descriptions below, a few pathogens will be mentioned only as a point of reference.

Fresh and Processed Meat, Poultry, and Seafood

All meats and meat products, as well as poultry and seafood, are subject to microbial spoilage.

Fresh Meat

Staphylococcus: staff-ih-loh-KOK-us

Micrococcus: my-kroh-KOK-us

Pseudomonas: sue-doh-MOH-nahs

Sanitation: The process of reducing the number of microbes to a safe level.

Lactobacillus: lack-toe-bah-SIL-lus

Leuconostoc: lou-koh-NOS-stock

Fermentation: In the food industry, any biological process that changes the properties of food, no matter if oxygen gas is present or not.

Putrefaction: Also called decay; the breakdown of proteins, which is often detected as sliminess.

Spoilage in meat and meat products is expected because of the nature of the food; beef are high in protein and fats. Although muscles of healthy animals are sterile and do not contain any microbes, when the animals are slaughtered, the extraordinarily high nutrient content in the fresh meat can support a large variety of bacteria (e.g., *Staphylococcus, Micrococcus. Pseudomonas*) as well as yeasts and molds—if contaminated with microbes from the hide, hair, or hooves. Thus, good **sanitation** practices are essential.

Many steps in meat processing provide multiple opportunities for microbial contamination. Chopping and grinding of meats can increase the microbial load as more surface area is exposed and the moisture content increases. In the processing of fresh meat, contamination also might be traced to dirty conveyor belts used to transport the meat. Improper temperature control while holding the meat or failure to distribute meat and meat products quickly are additional factors. Moreover, fecal matter from the animal's intestinal contents can contaminate the meat during slaughtering and cutting. Contaminated cutting blocks can be a source of contamination to other carcasses.

Bacterial genera such as *Lactobacillus* and *Leuconostoc* are referred to as **lactic acid bacteria** (**LAB**) because a major product from their **fermentation** is lactic acid. An early indication of meat spoilage in the meat counter or butcher shop is the loss of red color and the appearance of a brown or gray color with surface slime due to **putrefaction** by LAB. Although these species are not dangerous to human health, they can give an abnormal taste and smell to the meat even after it has been cooked.

Most of the ground beef that is recalled is due to microbial contamination in vacuum-packed and sealed packages (**FIGURE 12.7A**). Often the recall is due

FIGURE 12.7 Packaged Meats.
(A) Most recalls of ground meat have been from packages that were vacuum-wrapped and sealed.
(B) Processed meats contain preservatives and are subject to a different group of spoilage microbes than are wrapped ground meats.

to contamination with *Escherichia coli*. Although most strains are harmless, one strain, called *E. coli* O157:H7, has been associated with food poisoning outbreaks.

Escherichia coli: esh-er-EE-key-ah KOH-lee

Processed Meats

Processed meats, such as luncheon meats and frankfurters, are handled frequently and contain a variety of meat products (**FIGURE 12.7B**). Consequently, they can become contaminated with food spoilage microbes. These products might contain added salt, nitrites and/or nitrates, and other preservatives, which change the microbial population involved in spoilage. Fermented meats, such as sausages, rarely are contaminated because they contain a variety of organic chemicals toxic to spoilage microbes.

Cured meats, such as ham, bacon, and corned beef, often are treated with large doses of salt to draw water out of microbes and thereby kill them. However, LAB species tolerate the salt and ferment carbohydrates in the cured meat to lactic acid. This sours the meat and gives unpleasant odors and taste to the product. In addition, the bacterial cells produce slime and gas that cause the package to puff and form hydrogen peroxide. The gas changes the red meat pigments to green and gives the meat a spoiled look.

Some ready-to-eat foods, such as hot dogs and deli meats, can become contaminated with *Listeria monocytogenes* bacteria. Consumption of such meat might lead to a foodborne illness called listeriosis. This is a serious infection that primarily affects older adults, pregnant women, young children, and people with weakened immune systems. Additional caution is warranted because the pathogen can grow at refrigerator temperatures.

Listeria monocytogenes: lis-TEH-ree-ah mah-no-sigh-TAH-jeh-neez

Poultry and Eggs

Contamination in poultry often comes from the feathers, skin, and feet. Cross-contamination with the animals' feces can occur during transportation of birds to slaughter facilities and during processing.

Spoilage in poultry usually is restricted to the outer surfaces of the skin. It can be detected by off-odors, discoloration, and sliminess. Species of

Aeromonas: AIR-oh-mo-nass

Shewanella: shoo-ah-NELL-ah

Salmonella: sal-mon-EL-lah

Campylobacter: kam-pill-oh-BAK-ter

Proteus: PROH-tee-us

Serratia: ser-RAH-tee-ah

Pseudomonas, *Aeromonas*, and *Shewanella* are among the most common bacteria causing poultry spoilage.

Members of the genus *Salmonella* can cause diseases in poultry. If any of these strains is passed on to consumers via poultry products, infection can lead to salmonellosis. This foodborne disease is characterized by diarrhea, fever, vomiting, and abdominal cramps. Another common infection of poultry is caused by the bacterium *Campylobacter*. Eating contaminated raw or undercooked poultry might lead to campylobacteriosis. The symptoms are very similar to those of salmonellosis.

Chicken eggs normally are sterile when laid. However, the outer waxy membrane, as well as the shell and inner shell membrane, can be penetrated by some bacterial organisms. *Proteus* species cause black rot in eggs when they break down the amino acid cysteine and produce hydrogen sulfide gas. This results in black deposits and gives rotten eggs their horrid smell (**FIGURE 12.8**). Spoilage in eggs also occurs as the yolk develops a blood-red appearance from growth of red pigment-producing *Serratia*. In fact, the primary location of egg contamination is in the yolk rather than the white. The yolk is more nutritious, and the white has an inhospitable pH of approximately 9.0. Also, lysozyme in egg white is inhibitory to gram-positive bacterial species. Processed foods containing eggs, such as whole egg custard, mayonnaise, and eggnog, also might be sources of salmonellosis.

Seafood

Fish is a highly perishable food that is high in protein and amino acids. Because microbes usually contaminating fish are naturally adapted to the environment in which fish live, water temperature can have a great influence on the numbers and types of spoilage bacteria on the body of the fish. Very often, spoilage organisms concentrate in the gills of fish because these structures trap microbes as water is strained. Another source of contamination is the boxes used for storing and transporting fish. Cracks, pits, and splinters in the wood can trap microbes and spread them among the fish.

FIGURE 12.8 Rotten Eggs. While rare, a rotten egg, such as this one with black rot, might have been contaminated through a pinpoint hole in the shell that wasn't visible before being sent to the supermarket.

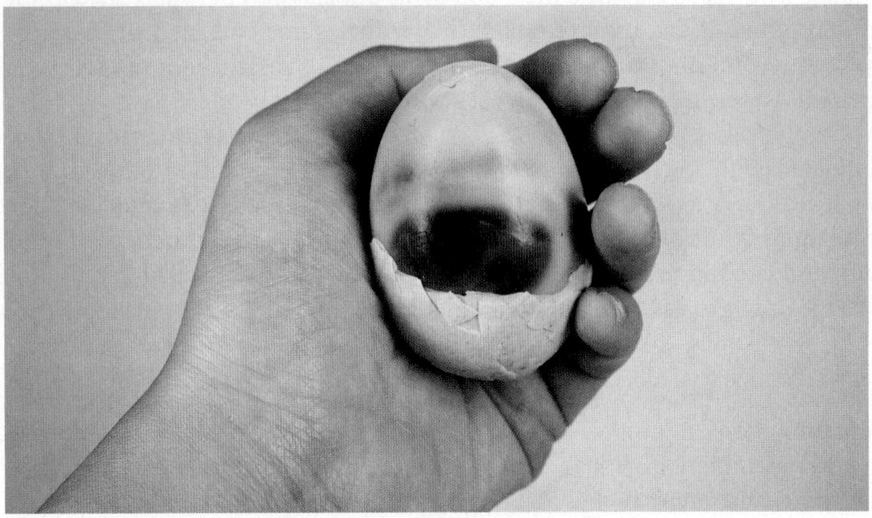

The microbial load is higher on fish caught in warm subtropical or tropical waters (typically *Bacillus* and *Micrococcus*) than on fish from colder waters (typically *Acinetobacter*, *Aeromonas*, *Pseudomonas*, and *Shewanella*). Therefore, refrigeration might not be as effective for long periods as it is for meat; freezing, salting, or drying fish is preferred.

Bacillus: bah-SIL-lus

Acinetobacter: a-sih-NEH-toe-bak-ter

The characteristic odor associated with spoiled fish is generally due to putrefaction. When fish are taken out of their natural habitat, they die quickly. If the fish are contaminated, several bacterial species degrade a compound in fish muscle tissue called trimethylamine oxide to trimethylamine. This chemical gives rotting fish its dreadful smell.

Shellfish, such as lobsters and crabs, are kept alive until they are heat processed. That means they will be less contaminated with spoilage microbes than shrimp that soon die after harvesting. Clams, oysters, and mussels obtain their food by filtering particles from the water. If the water is microbially contaminated, the shellfish can concentrate the microbes in their tissues. Typical reported illnesses include hepatitis A, typhoid fever, cholera, or other intestinal illnesses.

Milk and Dairy Products

Spoilage in dairy products can result from microbial growth beyond what microbes were originally present in the product. Spoilage is identified by off-flavors, off-odors, and changes in texture and appearance.

Milk

Milk is an extremely nutritious food. It is a solution of proteins, fats, and carbohydrates, with numerous vitamins and minerals. Milk has a pH of about 7.0 and is an excellent nutrient for humans and animals. Consequently, it also is a good growth medium for microbes.

About 87% of milk is water, 2.5% is protein, and about 5% is carbohydrates. The major carbohydrate is the milk sugar lactose, which is found naturally only in milk. It is digested by relatively few bacterial species, and these species usually are harmless. The last major component of milk is butterfat, a mixture of fats often churned into butter. When bacterial enzymes digest these fats into fatty acids, the milk or butter becomes rancid and develops a sour taste.

Milk spoilage can occur in the kitchen refrigerator or the supermarket dairy case. Contamination by species of *Pseudomonas* can come from its accidental introduction from the dairy farm's environment, milking and processing plant equipment, employees, or the air. *Pseudomonas* plays an important role in milk spoilage. The bacterial cells produce many enzymes that reduce both the quality and shelf life of processed milk. As a result, milk takes on off-flavors and has a rancid odor.

Milk is normally sterile in the udder of the cow, but contamination with LAB might occur as it enters the ducts leading from the udder. If not properly cooled (and pasteurized), LAB can grow to such large quantities that lactic and acetic acids accumulate, causing a souring of the milk. LAB also produce compounds that increase the viscosity of the milk, causing another defect in milk that produces a "ropy" texture.

Dairy Products

Dairy products are a different growth environment for spoilage microbes than fluid milk. This is because dairy products often have nutrients removed or concentrated, a lower pH, or a reduced water content. The presence of yeasts and molds can spoil yogurt, sour cream, and buttermilk because the higher acidity in these products inhibits many bacterial species. Cream might become rancid if bacterial species like *Pseudomonas* and *Enterobacter* contaminate the product and multiply.

Enterobacter: en-teh-roh -BACK-ter

Cheeses are less acidic than yogurt but have added salt and less water. If a wedge or block of cheese is left in the refrigerator too long, a mold might start growing and eventually spoil the product. Hard and semi-hard cheeses have a low moisture content (<50%) and a pH near 5.0, which limits or delays the start of growth of most molds. Soft cheeses, on the other hand, are less acidic (5.5–6.5) and have a higher moisture content (50%–80%). These cheeses might be spoiled more quickly. Common bacterial genera contaminating these cheeses include *Pseudomonas*, *Alcaligenes*, and *Flavobacterium*. Spoilage problems in cheese can arise from using a low-quality milk, unhygienic conditions in the processing plant, or simply mold spores (typically *Penicillium*) landing on the cheese after the package has been opened.

Flavobacterium: flay-voh-bak-TIER-ee-um

Alcaligenes: al-kah-LIH-jen-eez

Penicillium: pen-ih-SIL-lee-um

Fruits and Vegetables

Spoilage of fresh produce (fruits and vegetables) often involves changes in color, flavor, texture, or smell caused by microbe growth.

Fruits

Fruits refer to the seed-bearing organs of plants. Undamaged fruits normally have microbes on their surfaces. Consequently, once ripe, the cell walls of fruits weaken, and physical damage during harvesting might lead to breaks in the outer protective layers that the surface microbes can now exploit. Mold growth on citrus fruits, like lemons and oranges, is quite common (**FIGURE 12.9**). *Penicillium* and *Rhizopus* are frequently the spoilage organisms. Yeasts and some bacteria, including *Erwinia* and *Xanthomonas*, also can spoil some fruits, and these might be a particular problem for fresh-cut packaged fruits.

Rhizopus: rye-ZOH-puss

Erwinia: err-WIH-nee-ah

Xanthomonas: zan-tho-MO-nass

Fruit juices are relatively high in sugar and low in pH, which are favorable conditions for growth of yeasts, molds, and some acid-tolerant bacterial species. However, the lack of oxygen gas in most juices limits the growth of molds. LAB can spoil orange and tomato juices.

FIGURE 12.9 Moldy Fruit. A *Penicillium* species growing on an orange. Molds often prefer slightly acidic conditions for growth.

Courtesy of Dr. Jeffrey Pommerville.

Vegetables

Vegetables represent the non-fruiting parts of plants, including the roots, stems, and leaves. These plant parts have a neutral pH and a high water content, which are a source of nutrients for spoilage microbes. Vegetables are exposed to many soil microbes while growing, but many spoilage organisms, such as the LAB, are not common in the soil. Therefore, most microbial spoilage occurs after there has been mechanical and chilling damage to plant surfaces. Species of *Erwinia* cause most cases of spoilage and can be found associated with many types of vegetables.

The most common type of spoilage in vegetables is a softening of tissues in the skin called a **soft rot**. As the tissues are degraded, the whole vegetable might eventually disintegrate into a slimy mass. Starches and sugars are metabolized next, and unpleasant odors and flavors develop along with lactic acid and ethanol. Besides *Erwinia,* several *Pseudomonas* species and LAB are important spoilage microbes.

Molds belonging to several genera, including *Rhizopus, Alternaria,* and *Botrytis,* cause several vegetable rots that are identified by the color, texture, or acidic products of the rot.

Alternaria: all-ter-NARE-ee-ah

Botrytis: bow-TRI-tiss

Grains and Bakery Products

Cereal grains (e.g., corn, oats, rye, wheat, rice, barley) often are ground into flour or meal for pasta or bakery products. They also are processed and used in producing snacks and breakfast cereals. In these products, the water content is so low that most microbes cannot grow. Still, the statement found on these products, "Keep in a cool, dry place," is worth remembering.

Grains naturally are exposed to a variety of microbes during growth, harvesting, drying, and storage. Molds are the most common contaminants because of the low moisture levels in grains. Two examples of grain spoilage are described here.

The first contamination example is caused by the mold *Aspergillus flavus*. This fungus produces **aflatoxin**, a poisonous substance that accumulates in stored cereal grains (such as wheat) as well as peanuts and soybeans. The toxins can be consumed with grain products and meat from animals fed the contaminated grain. Scientists have implicated aflatoxins in liver and colon cancers in humans.

Aspergillus flavus: a-sper-JIL-lus FLAY-vus

The second example of mold grain spoilage involves *Claviceps purpurea*. Rye plants are particularly susceptible to this mold, but wheat and barley grains also might be affected (**FIGURE 12.10**). The toxins deposited by *C. purpurea* can cause ergot poisoning (**ergotism**) if the contaminated grain is consumed. Ergotism triggers neurological symptoms, including convulsions and hallucinations. In fact, the psychedelic drug lysergic acid diethylamide (LSD) was originally derived from the ergot toxin.

Claviceps purpurea: KLA-vi-seps purr-POO-ree-ah

In the production of bakery products, ingredients such as flour, eggs, and sugar are generally the sources of spoilage microbes. Although most microbes are killed during baking, *Bacillus* spores can survive in the interior of bread loaves baked at high oven temperatures. On cooling the bread, the spores can germinate, and the cells start growing. Some strains of *Bacillus* are notorious in causing a bread condition called "ropiness." This is where the bread has a soft, cheesy texture with long, stringy threads. This is due to the bacterial cells degrading starch and producing a slimy polysaccharide. Yeasts can cause spoilage of breads and fruitcakes, resulting in a chalky appearance on surfaces and producing off-odors.

Foods stored under the same conditions can spoil at different rates. Therefore, knowing the shelf life is important. **TABLE 12.1** lists the shelf life for a selected set of common foods.

FIGURE 12.10 Ergot. *Claviceps purpurea* can grow (the long dark structures) on cereal plants causing ergot. Consumption of grains or seeds contaminated with the fungus can cause ergotism.

TABLE 12.1 The Shelf Life for Several Food Products Found in a Home Refrigerator

Food Product	Typical Shelf Life
Milk	7 days
Yogurt	7–14 days
Fresh eggs	3–5 weeks
Ground beef	2 days
Hot dogs	2 weeks (unopened) 1 week (opened)
Bacon	1 week
Fresh chicken	2 days
Orange juice	7–10 days
Beer	6 weeks
Wine	1 week (uncorked)
Ketchup	5 months
Mayonnaise	3 months
Pickles	6 months
Peanut butter	4 months
Jelly	5 months

Data taken from FDA.

12.3 Food Preservation: Keeping Microbes Out

Centuries ago, humans battled the elements to keep a steady supply of food at hand. Sometimes, there was a short growing season; at other times, locusts descended on their crops; at still other times, they underestimated their needs and had to cope with scarcity. However, experience taught humans that they could prevail through difficult times by preserving foods, often by drying vegetables and salting meats and fish (**FIGURE 12.11**). Once Pasteur showed in the 1860s that food quality could be affected by microbes, others started to devise new food preservation methods to reduce the microbial population and maintain it at a low level until the food could be eaten. Modern preservation methods still have that objective.

Food preservation consists of the application of a variety of methods and procedures. It attempts to delay or prevent deterioration and spoilage of food products and extend their shelf life. The process also must assure consumers that a product is free of excessive spoilage and all pathogens. Though today's preservation methods are science based, advances have been counterbalanced by the increased complexity of food products and the great volumes of food to be preserved. Thus, the food preservation problems faced by early humans do not differ fundamentally from those confronting modern food technologists and the food industry today. As we will see in this section, some preservation methods are old standbys.

Physical Preservation Methods

The physical manipulation of foods can create conditions that are too extreme for microbial growth. Such conditions can inhibit microbial growth, kill microbial cells, or mechanically remove microbes from foods. These physical methods were discussed in more detail in the chapter on Growth and Metabolism, so only those aspects pertaining to food preservation will be described here.

Preservation by Heat

Heat is the most widely used method for eliminating microbes (**FIGURE 12.12**). In a moist heat environment, proteins lose their three-dimensional structure and biological activity. As their structural proteins and enzymes undergo this change, metabolism ceases, and the microbes die. That is why the Centers for

FIGURE 12.11 Salting of Fish. One food preservation method that has been used for centuries is to salt and dry fish. In this drawing, cod are being dried in Fiskerness, Greenland.

FIGURE 12.12 Important Temperature Considerations in Food Microbiology. Shown are several temperatures at which canning and sanitization procedures are carried out.

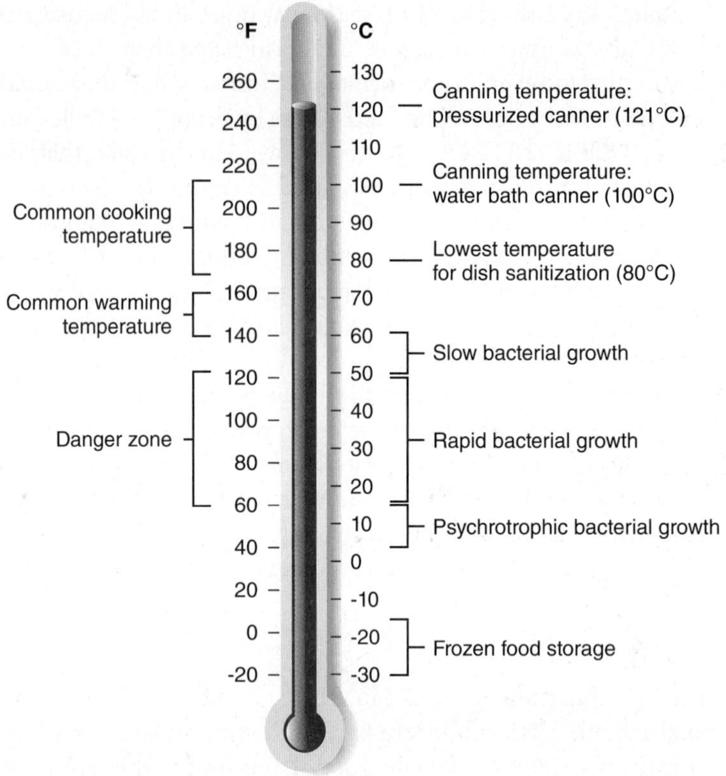

Disease Control and Prevention (CDC) suggest foods like raw hamburger meat be cooked to well done.

The process of **pasteurization** was developed by Louis Pasteur in the 1850s to eliminate bacterial cells in wines and keep the wine from turning to vinegar. His method of mild heating was applied to milk in Denmark about 1870 and was widely employed by 1895. Although the primary object of pasteurization is to eliminate all bacterial pathogens from milk, the heating process also eliminates up to 99% of the spoilage microbes (**FIGURE 12.13**). Nevertheless, some surviving microbes, such as *Streptococcus lactis*, can grow slowly in refrigerated milk. When their numbers reach about 20 million per milliliter, enough lactic acid has been produced to make the milk sour. The shelf life of the milk is based on an estimate of when souring is likely to occur.

Streptococcus lactis: strep-toe-KOK-us LAK-tiss

The most useful application of heat in food technology is **canning**, which is the **commercial sterilization** of food after packaging. Modern canning processes are complex. Machines wash, sort, and grade the food and then subject it to steam heat for 3 to 5 minutes. This last steam step, called **blanching**, destroys many enzymes in the food and prevents any further cellular metabolism. Next, the food is processed (peeled and cored, for example) and then put into tin cans, glass jars, or some other type of packaging. The air is evacuated from the container, which is then placed in a pressurized steam sterilizer similar to an autoclave at a temperature of 121°C or lower, depending on the food's pH, density, and heat penetration rate.

Commercial sterilization: The process that removes all pathogens from a food product.

All steps of the canning process must be carefully monitored to ensure contamination does not occur (**FIGURE 12.14**). Improper heating temperatures, a defective can, or an improper seal could result in contamination of the canned food product. At home, contamination is usually obvious because many spoiled canned foods have a putrid odor and produce gas that causes tin cans to bulge. Common contaminants include species within the genera *Escherichia*, *Serratia*,

FIGURE 12.13 A Milk Pasteurization Facility. Milk passes through pipes and is pumped into the large heating section of the milk heat exchanger where it is heat treated. The milk then is rapidly cooled to prevent any change in taste.

FIGURE 12.14 The Industrial Canning Process. Inspectors watch over the canning process to ensure sanitary conditions are maintained.

and *Enterobacter*. Growth of acid-producing microbes cannot be discerned from the can's shape as no gas is produced. However, such foods usually have a flat-sour taste from the acid produced.

Most contamination of canned food by pathogens is due to facultative or anaerobic bacteria such as *Clostridium tetani* that produces toxins that can cause botulism if the food product is ingested.

Clostridium tetani: kla-STRIH-dee-um TEH-tahn-ee

Preservation by Cold

Many of today's prepackaged foods are kept cold to maintain quality. In a refrigerator or freezer, the lower temperature reduces the rate of microbial enzyme activity and thus slows their rate of growth and reproduction, while extending the shelf life of food (see Figure 12.12). Although all the microbes are not killed, their numbers are kept low, and spoilage is minimized.

Modern refrigerators at 5°C (41°F) provide a suitable environment for preserving food without destroying the products, appearance, taste, or cellular integrity. However, **psychrotrophic** microbes survive in the cold, and given enough time, they will cause meat surfaces to turn green, eggs to rot, fruits to become moldy, and milk to sour.

Psychrotrophic: Referring to microbes that can survive or even thrive in a cold environment.

When food is placed in a freezer at −18°C (0°F), ice crystals form, which tear and shred microbes, killing substantial numbers of cells. However, many survive, and because the ice crystals are equally destructive to food cells, the microbes use the food nutrients to multiply quickly when the food thaws. Refrigerator thawing and cooking are therefore recommended for frozen foods. Moreover, food should not be refrozen; during the time it takes to thaw, nutrients are released from the food material and microbes will grow. If the thawed food is refrozen, the microbes can cause rapid microbial spoilage on re-thawing.

Deep freezing at −60°C (−76°F) results in smaller ice crystals, and although the physical damage to microorganisms is less severe, their biochemical activity is reduced considerably. Some food producers blanch their product (apply moist heat briefly) before deep freezing, a combination that reduces the number of microbes even further. A major drawback of freezing is freezer burn, which develops as food dries out from moisture evaporation. Another disadvantage is the considerable energy cost. Nevertheless, freezing has been a mainstay of preservation since Clarence Birdseye first offered frozen foods for sale in the 1920s. Approximately one-third of all preserved food in the United States is frozen.

Drying

Another centuries-old method of food preservation involves drying. In past centuries, people used the sun for drying, but modern technologists have developed sophisticated machinery for this purpose. For example, spray dryers expel a fine mist of liquid such as coffee into a barrel cylinder containing hot air. The water evaporates quickly, and the coffee powder falls to the bottom of the cylinder. Another machine for drying is the heated drum. Machines pour liquids such as soup onto the drum's surface, and the water evaporates rapidly, leaving dried soup to be scraped off. A third machine uses a belt heater that exposes liquids such as milk to a stream of hot air. The air evaporates any water and leaves dried milk solids.

During the past several decades, freeze-drying, or **lyophilization**, has emerged as a valuable preservation method. In this process, food is deep frozen. A vacuum pump then draws off the ice directly to its gaseous phase (water vapor) without passing through its liquid phase (**FIGURE 12.15**). The dry product is sealed in foil and easily reconstituted with water. Hikers and campers find considerable advantages in freeze-dried foods because of their light weight and durability; however, the ancient Incas of Peru were freeze-drying foods centuries earlier, as told in **A CLOSER LOOK 12.2**.

Other Preservation Methods

A few other preservation methods can be used with various types of foods. Some are centuries-old methods.

Natural Preserving Agents

Water in food allows microorganisms to grow and perhaps spoil the product. Using natural preserving agents such as salt and sugar, the water can be removed to prevent the growth of microbes. When living cells are immersed in large quantities of salt or sugar, water is drawn out of cells through their cell membranes and into the surrounding environment.

For centuries, "salting" has been used to preserve foods. For example, fish and raw meats, such as ham and bacon, can be preserved using sodium chloride (table salt). In these and other highly salted foods, microbes dehydrate, shrink, and die. Jams, jellies, fruits, maple syrup, honey, and similar products also are preserved this way using high concentrations of sugar, either added (as in jam) or naturally occurring (as in maple syrup and honey).

FIGURE 12.15 Freeze-Drying. Freeze-drying (lyophilization) is a four-step process. The final product is very light and retains most of its nutritional quality and taste.

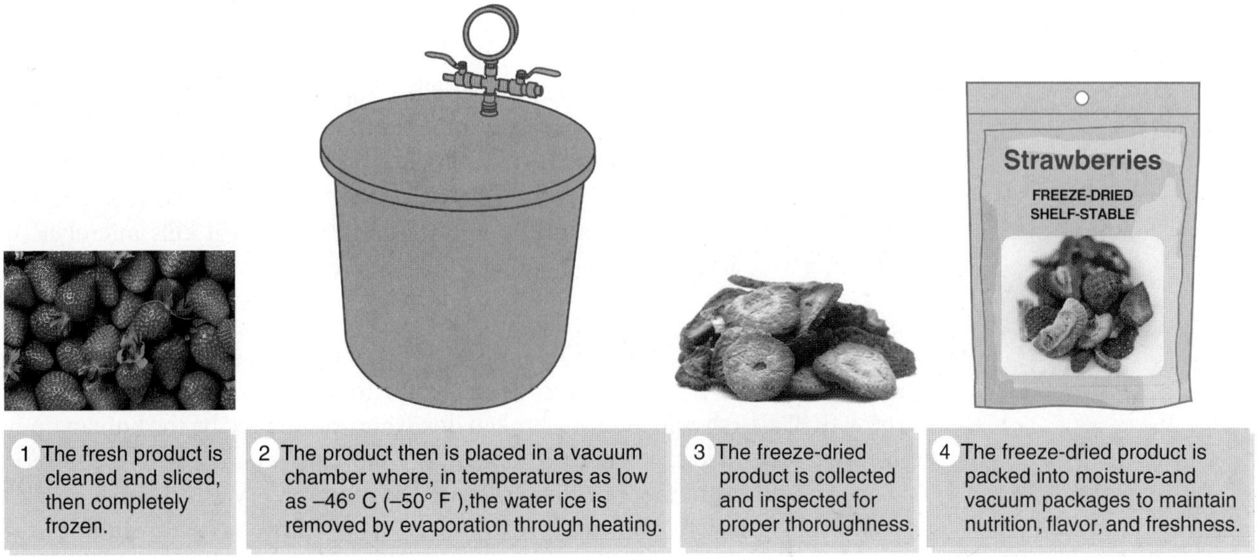
Strawberries
FREEZE-DRIED
SHELF-STABLE
1 The fresh product is cleaned and sliced, then completely frozen.
2 The product then is placed in a vacuum chamber where, in temperatures as low as –46° C (–50° F),the water ice is removed by evaporation through heating.
3 The freeze-dried product is collected and inspected for proper thoroughness.
4 The freeze-dried product is packed into moisture-and vacuum packages to maintain nutrition, flavor, and freshness.

https://www.produceforkids.com/skip-chips-add-crunch-lunchboxes-crispy-green/cg-freeze-dry-process/. Photos: (1) © GCapture/Shutterstock; (2) © P Maxwell Photography/Shutterstock; (3) © FeyginFoto/Shutterstock.

A CLOSER LOOK 12.2

It Started with "Stomped Potatoes"

Freeze-drying, or lyophilization, involves the removal of water from a substance without going through a liquid state. Such freeze-dried foods last longer than other preserved foods and are very light, which makes them perfect for everyone from backpackers to astronauts. However, the origins of freeze-dried foods go back to the ancient Incas of Peru.

The basic process of freeze-drying food was known to the pre-Columbian Incas who lived in the high altitudes of the Andes Mountains of South America. The Incas stored some of their food crops, including potatoes, on the mountain heights. First, most of the moisture was crushed out of the potatoes when the Incas stomped on them with their feet. Then, the cold mountain temperatures froze the stomped potatoes, and the remaining water inside slowly vaporized under the low air pressure of the high altitudes in the Andes.

Today, a similar procedure is still used in Peru for a dried potato product called *chuno*. It is a light powder that retains most of the nutritional properties of the original potatoes, but being freeze-dried, *chuno* can be stored for up to 4 years.

In the commercial market, the Nestlé Company was asked by Brazil in 1938 to help find a solution to save their coffee bean surpluses. Nestlé developed a process to convert the coffee surpluses into a freeze-dried powder that could be stored for long periods. The result was Nescafe coffee, which was first introduced in Switzerland. However, during World War II the process of freeze-drying developed into an industrial process. Blood plasma and penicillin were needed for the armed forces, and freeze-drying was found to be the best way to preserve and transport these materials.

Today, more than 400 commercial food products are preserved by freeze-drying. In the medical field, freeze-drying is the method of choice for extending and preserving the shelf life of enzymes, antibodies, vaccines, pharmaceuticals, blood fractions, and diagnostics. The benefit of all lyophilized products is that they rehydrate easily and quickly.

And to think—it all started with the Incas stomping on their potatoes.

FIGURE A A woman descendant of the Incas prepares potatoes for drying on a small farm high in the Andes Mountains.

© Joel Shawn/Shutterstock.

Irradiation

Irradiated: The process of exposing food and food packaging to radiation, such as gamma rays.

Though some of the public is apprehensive about **irradiated** foods, various forms of radiation have received approval by the U.S. Food and Drug Administration (FDA) for preserving foods. For instance, gamma rays are used to extend the shelf life of fruits, vegetables, fish, and poultry from several days to several weeks. This form of radiation also increases the distance fresh food can be transported and significantly extends the shelf life for food in the home. Health officials are quick to note that such irradiation of food does not cause food to become radioactive. As the radiation passes through the food product, it kills microbes by reacting with and destroying microbial DNA and other key organic compounds.

Chemical Preservatives

For a chemical preservative to be useful in foods, it must inhibit microbial growth. It also should be easily broken down and eliminated by the human body without side effects. These requirements are enforced by the FDA. The criteria have limited the number of chemicals used as food preservatives to a select few.

Organic acid: A small, carbon-containing compound with acidic properties.

A key group of chemical preservatives are **organic acids**, including sorbic acid, benzoic acid, and propionic acid. These acids are all natural, although today most are manufactured chemically. These chemicals damage microbial membranes and interfere with the uptake of essential nutrients such as amino acids. Sorbic acid is active against some mold and bacterial species and is added to syrups, cheeses, and baked goods. Benzoic acid protects beverages, jams, and jellies from mold growth, while propionic acid is added to baked goods and cheeses, where it also prevents the growth of molds. Other natural acids in foods add flavor while serving as preservatives. Examples are lactic acid in sauerkraut and yogurt and acetic acid in vinegar.

Sulfites

Sulfites, such as sulfur dioxide, are used as food additives to control food spoilage in fruits and vegetables, wines, sausages, and fresh shrimp. Used in either gas or liquid form, sulfur dioxide also retards color changes on dried fruit. Unfortunately, the FDA estimates that over 1% of the American population is sensitive to sulfites.

12.4 Maintaining Food Safety: The Challenges

Fueled by consumer awareness, the entire food industry has been placed under a food safety spotlight overseen by the FDA and the U.S. Department of Agriculture (USDA).

Among the most important food safety systems is the internationally recognized **Hazard Analysis and Critical Control Point** (**HACCP**). This is a set of science-based safety regulations enforced in the seafood, meat, and poultry industries (**FIGURE 12.16**). In an HACCP system, manufacturers identify individual processing sites, called **critical control points** (**CCPs**), where the safety of a food product could be affected. This includes sites that might be vulnerable to spoilage microorganisms as well as pathogens. The CCPs are supervised to ensure that any hazards associated with the operation are contained or, preferably, eliminated. When all possible hazards are controlled at the CCPs, the safety of the product can be assumed without further testing or inspection.

The standard regulations require food processors to monitor and control eight key sanitation areas, including the condition and cleanliness of utensils, gloves, outer garments, and other food contact surfaces; the prevention of cross-contamination from raw products and unsanitary objects to foods; and the control of employee health conditions that could result in food contamination. If the system and all the CCPs are kept up to standards, then the competition between humans and microbes will tip in our favor.

So, how are your food safety practices at home? **A CLOSER LOOK 12.3** provides the test.

FIGURE 12.16 HACCP System. A seven-step preventive approach to food safety to ensure the final product is safe for consumption.

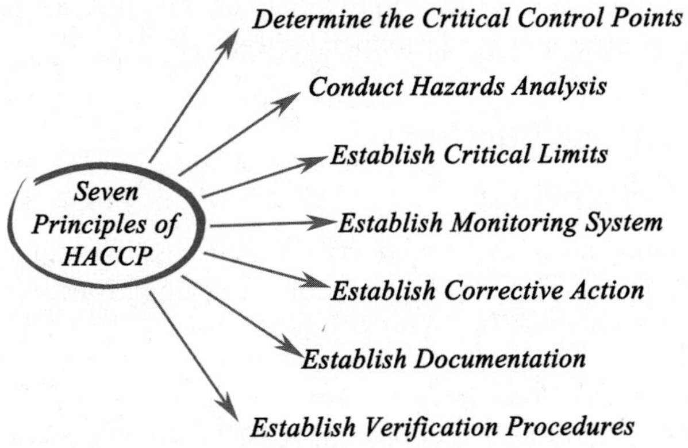

A CLOSER LOOK 12.3

Food Safety Quiz

How knowledgeable are you concerning food safety in your home or apartment? If you are, do you practice what you know? Take the following quiz (honestly) and see if you need to change safety methods to eliminate or keep the microbes under control.

1. Your refrigerator should be kept at what temperature?
   a. 0°C
   b. 4°C
   c. 10°C
2. Frozen fish, meat, and poultry products should be defrosted by ____________.
   a. setting them on the counter for several hours
   b. microwaving
   c. placing them in the refrigerator
3. After cutting raw fish, meat, or poultry on a cutting board, you can safely ____________.
   a. reuse the board as is
   b. wipe the board with a damp cloth and reuse it
   c. wash the board with soapy hot water, sanitize it with a mild bleach, and reuse it
4. After handling raw fish, meat, or poultry, how should you clean your hands?
   a. Wipe them on a towel
   b. Rinse them under hot, cold, or warm tap water.
   c. Wash them with soap and warm water.
5. How often should you sanitize your kitchen sink drain and garbage disposal?
   a. Once a week
   b. Every month
   c. Every few months
6. Leftover cooked food should be ____________.
   a. cooled to room temperature before being put in the refrigerator
   b. put in the refrigerator immediately after the food is served
   c. left at room temperature overnight or longer
7. How should you clean your kitchen counter surfaces that are exposed to raw foods?
   a. Use hot water and soap, then a bleach solution.
   b. Use hot water and soap.
   c. Use warm water.
8. Hamburgers should be cooked to ____________ for consumption.
   a. Medium-rare
   b. Medium
   c. Well-done
9. How often should you clean or replace your kitchen sponge?
   a. Daily
   b. Every few weeks
   c. Every few months
10. Normally dishes in the home should be cleaned ____________.
   a. by an automatic dishwasher and air-dried
   b. after several hours soaking and then washed with soap in the same water
   c. right away with hot water and soap in the sink and then air-dried

Best answers:

1. b; 2. c; 3. c; 4. c; 5. a; 6. b; 7. a; 8. c; 9. b; 10. a or c.

A Final Thought

Today, the FDA is responsible for ensuring the safety of about 80% of the food we eat. In addition, there are other non-food products that also fall within the FDA's authority. Every year the FDA recalls products for microbial contamination. The products include aqueous-based medicines, children's medications, several homeopathic drug products, cosmetics, and eye solutions. So, even though we examined food spoilage in this chapter, microbial contamination is a concern in many other nonfood products as well.

Chapter Discussion Questions

What Was He Thinking?

From your reading of this chapter, identify and discuss five major points about food preservation and safety that the author was trying to get across to you.

Questions to Consider

1. If a preservation method inhibited microbial growth, would the affected food product "last forever"? Explain.
2. In the last section on "Maintaining Food Safety: The Challenges," why is the word "challenges" used rather than "problems"?
3. Obtain a thermometer and measure the temperature of your home refrigerator and freezer (or a convenient one near you if you live on campus). Are these temperatures within the normal effective operating range?
4. From past experiences, describe some examples where you came across spoiled food? Was it at home, work, a restaurant, etc.? What type of food was spoiled? What did you do with the food?
5. Referring to Figure 12.3, identify reasons for food waste that are not due to food spoilage.
6. When looking at a food product's label, how do you know if the date stamped on the product represents the shelf life or the expiration date?
7. Foods from tropical nations tend to be very spicy, with lots of hot peppers, spices, garlic, and lemon juice. By contrast, foods from cooler countries tend to be much less spicy. Why do you think this pattern has evolved over the ages?
8. On Saturday, a man buys a steak and a pound of liver and places them in the refrigerator. On Monday, he must decide which to cook for dinner. Microbiologically, which is the better choice? Why?
9. Suppose you had the choice of purchasing "yogurt made with pasteurized milk" or "pasteurized yogurt." Which would you choose? Why? What are the "active cultures" in a cup of yogurt?
10. Look through your kitchen or that of a friend or family member. Identify examples of highly perishable, semi perishable, and nonperishable foods.
11. You look at a can of soup you bought at the grocery store. On the label, the nutrition facts indicate that a serving size of the soup contains 900 milligrams of salt. Because the can label says it contains two servings, there are 1.8 grams of salt in the can. Although these salt concentrations are high for human consumption, microbiologically, why has the soup manufacturer made the salt content so high?

CHAPTER 13

Microbes and Food: A Menu of Microbial Delights

Waiter! There Are Microbes in My Food!

"Maria, hurry up, or we're going to be late!" shouted Angela.

"Okay, okay! I am going as fast as I can," snapped Maria. "I don't understand what the big deal is with this dinner. Why are you so excited?"

"I'm excited because the dinner revolves around microbes and the foods we eat. I think all the courses will have microbes as a key ingredient," Angela explained.

"What! Are you nuts?" Maria yelled. "Am I going to say, 'Waiter! There are microbes in my food!'?"

"Very funny," Angela responded. "We learned in my intro bio class that since early human history, microbes have played a role in the food process. Wine, yogurt, and cheeses are all products of microbes, and they play critical roles in food production processes through the chemical reactions they perform. We learned that they engage in a process called fermentation…"

"Yes," Maria interjected. "I know yeasts are needed for wine and beer production. But these germs are in many of the foods we eat? Yuck! I thought most were bad and caused disease."

"You're right that some cause disease and decay," Angela explained. "But the clear majority of the planet's microorganisms are not harmful to us. In fact, they can be incredibly useful, producing alcohol, acids, and other molecules that add flavor, texture, and nutritional value to food. That is what this dinner is all about. How microbes are a fundamental part of the foods we eat."

Maria acknowledged, "As your roommate, I really do appreciate you asking me to this dinner. I'm glad it's paid for, too. I'm just not so sure about the ingredients."

"I understand, Maria," Angela said, "but often it's not the microbes themselves that are the food, but rather the substances like alcohols, gases, and acids

CHAPTER 13 OPENER Many food and beverage products we consume at the dinner table or restaurant and enjoy every day are either created by certain species of microbes, or they are part of the production process in making the food product.

they produce by fermentation that help give many foods their wonderful flavors and textures."

"Okay, I'm ready. Let's go. But this better be good!" Maria said.

Over the decades, many food-related microbes have received bad press: They have been linked to food spoilage and numerous foodborne illness outbreaks. One possible reason for this negative reputation is because people often are afraid of things they cannot see, and microbes certainly meet this criterion. But most are useful and even valuable.

This chapter emphasizes the beneficial roles for the food-related microbes and their necessity for producing and processing many of the items we eat and drink. We will use a fine restaurant as our venue to explore the relationships between microbes and foods, for it is here that the metabolic end products of microbes come to please our palates and heighten our senses.

LOOKING AHEAD

After reading and completing this chapter, you will be able to:

13.1 Define food fermentation.
13.2 Describe the process of wine fermentation.
13.3 Compare the roles of microbes in the fermentation processes for olives and cheese making.
13.4 Explain the role of yeast in bread making.
13.5 Outline the fermentation process for sausages and sauerkraut.
13.6 Discuss the process of brewing beer.
13.7 Identify the roles of microbes in coffee and chocolate production.

13.1 Microbes in Action: Fermentation

As we sample many of the microbial delights on the dinner table, a recurring theme will be fermentation, which is key to many of the flavors, textures, and composition of the food product. In the chapter on Growth and Metabolism, **fermentation** was defined as an enzyme-catalyzed process that occurred in the absence of oxygen gas. The metabolic pathways produced organic end products, like alcohol, carbon dioxide (CO_2) gas, and various acids. In **food fermentation**, the process often involves mixed populations of microbes producing a broad range of products resulting from the partial breakdown of carbohydrates and other large organic compounds in foods. These products add distinctive aromas and tastes to many foods, including those pictured in **FIGURE 13.1**. Moreover, they act as preservatives, making the foods safe to eat by holding in check any dangerous microbes that might be present.

In addition, eating fermented foods might help boost gut health because fermented foods are packed with **probiotics**. These are the "good bacteria" that can help encourage the growth of the microbes composing the gut microbiome, as described in the chapter on Microbial Crosstalk.

13.2 Beginning Our Meal: To Your Health!

To begin, we enter the restaurant and are seated at our table by our charming hostess, who extends the manager's greetings. Then, our attentive waiter approaches and inquires if we would like to start with a glass of wine.

FIGURE 13.1 An Array of Foods Produced by Microorganisms. Many foods, such as yogurt (left) and pickles (center) are the result of microbial fermentations.

A Glass of Wine

In ancient times, humans must have been awestruck by wine, for it was powerful enough to turn the mind, even though it began as mere grape juice. It was called *aqua vitae*, the "water of life" for much of history, because people marveled at the mystical abilities of this beverage and wondered how it came about. We might wonder, too, but for the moment, we will set aside the technicalities of fermentation. We must first decide if we prefer red wine or white wine. For making red wine, black grapes (red or blue tint) are used and the skins remain with the juice during the fermentation process. White wine can be fermented from black grapes or white grapes because the juice is colorless. For white wine, the skins would be removed before the fermentation process.

Once we have made our selection, the waiter leaves to fill our order. Now we can consider how alcoholic fermentation works.

Alcoholic Fermentation

If oxygen is present in their environment, many microbes—including yeasts—will live contentedly. If the oxygen is removed, however, yeast cells shift their metabolism to fermentation (**FIGURE 13.2**). Rather than shuttling pyruvate into **aerobic respiration**, yeasts undergo **alcoholic fermentation** that converts pyruvate into ethyl alcohol and CO_2 gas. This somewhat complex chemistry, which is the key to fermentation, was explored in depth in the chapter on Growth and Metabolism.

Aerobic respiration: A set of metabolic reactions and processes that take place in the cells of organisms to convert biochemical energy from nutrients into cellular energy in the form of ATP.

Before going much further, realize that few organisms other than yeasts can convert pyruvate to ethyl alcohol. Furthermore, this conversion is not particularly desirable for the yeast cells because they can gain more energy for their life processes by putting the pyruvate into aerobic respiration than by using it for fermentation. But the yeast cells have no choice when there is no oxygen in their environment; without using fermentation at those times, they would quickly die. For yeasts, fermentation therefore brings an evolutionary advantage in the struggle to survive—they can subsist by making a little bit of ATP

For winemakers, fermentation is economically advantageous as the basis for the wine industry—a $220 billion industry in the United States alone. In 2018, the United States ranked 55th in per-capita wine consumption (10 liters/2.6 gallons). At the top were several European nations where the per-capita consumption topped 56 liters (15 gallons) per year.

FIGURE 13.2 Metabolic Map of Aerobic and Fermentation Pathways for ATP Production. The production of ATP by microorganisms can be achieved through an aerobic respiration pathway or, in the absence of oxygen gas, some microbes, including many yeasts, can make a small amount of ATP through a fermentation pathway. The end products of that pathway are of human importance.

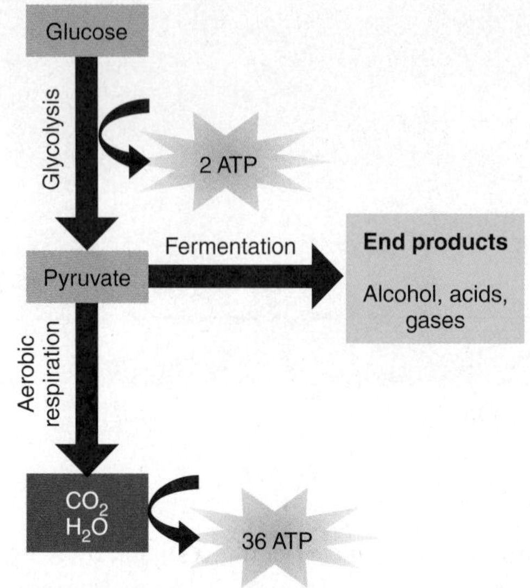

FIGURE 13.3 Wild Yeasts. Although cultured yeasts are usually used for wine making, when growing on the vine, grape skins are covered with wild yeast species as seen in the "coating" covering these black grapes.

Wild yeasts are plentiful wherever there are vineyards. The haze on grapes is a layer of wild yeasts (**FIGURE 13.3**). If those yeasts penetrate the skin and enter the soft flesh of the grape, they will ferment its carbohydrates and produce alcoholic grape juice. The same process occurs with other fruits or grains. The difference? Yeasts growing in fruit juice will produce a wine, while yeasts growing in grains like barley or rice will produce grain-flavored ethyl alcohol; that is, beer.

Making Wine

Vitis vinifera: VIH-tiss vih-NIF-ur-ah

Saccharomyces cerevisiae: sack-ah-roe-MY-seas seh-rih-VIS-ee-eye

Among grapes, the species *Vitis vinifera* is recognized as producing the highest-quality fruit for winemaking. The most common fermentation is performed by inoculating the grape juice with cultured yeasts, usually *Saccharomyces cerevisiae*, which also is used in bread making and brewing, as we will soon see (**FIGURE 13.4**). Because wild yeasts can produce undesirable qualities in a wine, many wineries kill these yeasts by adding sulfur dioxide ("sulfites") during the production process.

Characteristics of the grapes are determined by soil and climate conditions, such as temperature, humidity, and amount of water. Importantly, the role of microbes in the soil and on the grape vines is an important factor in producing the final unique flavor, aroma, and bouquet of a wine, as **A CLOSER LOOK 13.1** describes. Consequently, wine varies, and there are "vintage years" and "poor years."

FIGURE 13.4 ***Saccharomyces cerevisiae.*** A light microscope image of *S. cerevisiae*, the most common cultured yeast used to make wine. (Bar = 10 μm.)

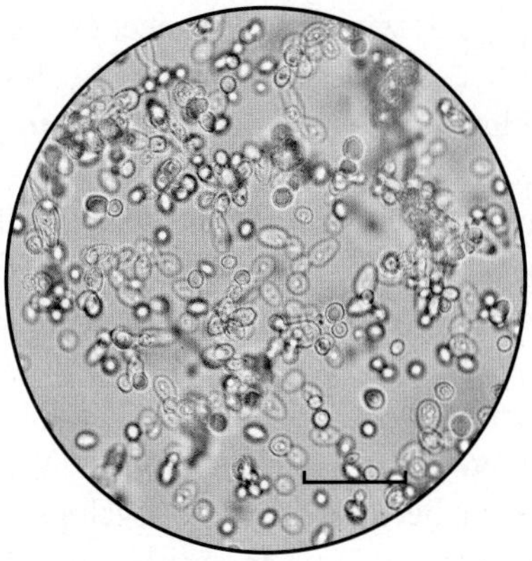

A CLOSER LOOK 13.1

A Microbial *Terroir*

The transformation of grape juice to wine depends on various microbial species, especially the yeasts used for the fermentation process. In addition, winemakers and grape growers have known for a long time that the surrounding environment, specifically the soil and climate—the so-called *terroir*—contribute to grapevine health. The *terroir* also provides various soil (organic) compounds that influence both the quality and "nose" (smell or aroma) of the wine (see **FIGURE A**). But is that all? Do microbes, besides yeasts, influence a wine's character?

In 2013, David Mills and colleagues at the University of California, Davis published a paper suggesting microbes on the grape surface and on grape stems—the grapevine microbiome—are unique to a plant's variety and geographic region (vinicultural zone). The team sequenced microbial genomes from 273 samples of grape "must" (juice from freshly crushed grapes) from many California grape-growing regions and the 2010 and 2012 vintages. Their results indicated that wine grapes from the same region, or same variety, have similar profiles for native bacterial and fungal species. Interestingly, for chardonnay musts, the bacterial and fungal profiles on the grapes and stems were specific to the wine region, whereas for cabernet sauvignon, only the fungal community was specific to a region. In addition, the bacterial and fungal profiles were similar when the two vintages were compared.

So, does climate play a role in the grape microbiome? The researchers say yes, as rainfall and temperature will influence the composition of the microbial community on the grapes and stems, and thus in the must.

The researchers believe their work represents a paradigm shift in understanding how factors beyond the grape variety itself influence the final characteristics of a wine. Their work identifies bacterial and fungal microbes as a completely new "factor," or set of factors, to be considered with both vineyard practices and wine production. As with the human body, understanding the relationships between the microbiomes and the host (humans—good health; grapes—great wines) will require much more work. Cheers and good health!

FIGURE A A California vineyard in the fall.

First, the grape juice—known as **must**—bubbles intensely and froths from the CO_2 produced during aerobic respiration, as oxygen gas is still present during this step. The CO_2 fills all the air spaces in the juice, and, so long as the juice remains still, the environment becomes oxygen-free, or **anaerobic**. The yeast cells now shift their metabolism from aerobic respiration to fermentation and start producing ethyl alcohol. However, there is a limit to this process. When the percentage of alcohol in the wine reaches about 18%, the population of yeast cells starts to die due to the toxic effect of the high alcohol content. Thus, no natural wines have an alcoholic content higher than about 16%.

The basic fermentation process is varied to obtain a broad variety of wine types. To produce a dry (still) wine, for instance, yeasts break down almost all the sugar in the grape juice. Should a sweet wine be desired, some sugar is left unfermented. Most table wines average about 12% to 15% alcohol, with red wines often having a higher alcohol content than white wines. Exceptions are the **fortified wines**, such as port, sherry, and Madeira. Brandy or other spirits are added to these wines to give an alcohol content approaching 22%. Because fortified wines are generally considered dessert wines, we will pass on them for now.

Although the production of alcohol by fermentation requires only a few days, the aging process might go on for weeks or months. Wooden casks have traditionally been used for aging because the wine develops its unique flavor, aroma, and bouquet with help from organic molecules in the wood (**FIGURE 13.5A**). Today, more and more wines are being aged in giant, stainless-steel tanks, and the desired "flavors" are imparted by adding chips, chunks, or even whole planks of wine-barrel wood suspended inside the tank (**FIGURE 13.5B**).

Before leaving wines, what about sparkling wines, like champagne? For these wines, a second fermentation takes place inside the bottle: Sugar cubes are added to the wine after the first fermentation, and the yeast is encouraged to continue fermenting the sugar within the bottle. CO_2 gas builds up and adds the sparkling bubbles to the wine. A thick bottle is needed for champagne because the gas pressure would cause ordinary glass to crack, and a wire cage is used to prevent the cork from popping out. By the way, a bottle of "cheap champagne" does not necessarily mean that quality is lacking. Some champagnes are "cheap" because they are mass produced in large vats rather than handled as individual bottles.

FIGURE 13.5 The Large-Scale Production of Wine.
(A) Many red wines and white wines, like chardonnay, are barrel-aged for months or years.

Courtesy of Dr. Jeffrey Pommerville.

(B) The juice, skins, and seeds go into large, temperature-controlled, stainless-steel fermentation tanks for primary fermentation.

Courtesy of Dr. Jeffrey Pommerville.

13.3 First Course: The Appetizers

As we anticipate our meal, our palates will first be stimulated by fermented olives and cheese brought to our table by our waiter.

Olives

Olives have traditionally represented the abundance of life, and in many cultures the olive branch is a symbol of peace. Unfortunately, the natural taste of olives is quite bitter.

Tradition and microbial fermentation resolved the bitterness problem long before the chemistry was understood. In regions of Western Europe, unripe olives were soaked in lye (sodium hydroxide) to neutralize their bitter taste, then washed and covered with brine. Next, the olives were sealed in casks, and bacteria normally present on the olive skins would ferment the carbohydrates. Some weeks later, when the fermentation was complete, the tasty Spanish, or "green," olive resulted (**FIGURE 13.6**).

In Greece, olives were eaten without the benefit of fermentation, and so they had to be preserved for later use. This was accomplished by allowing the olives to ripen on the tree, then picking them and exposing them to the air for weeks. During this time, chemical conversions of chemical compounds in the olive skins created black deposits, yielding Greek, or "black," olives. The Italians modified the process by placing the black olives in salt and encouraging the naturally occurring microbes in the olive skins to carry out fermentation. And certainly, lots of olives are processed into olive oil.

Cheese

Our Spanish, Greek, and Italian olives are accompanied by crackers and assorted cheeses. Cheese results when microbes interact with **casein**, the major protein in milk. In the dairy plant, the microbes produce enzymes that join with added enzymes to curdle the casein (**FIGURE 13.7**). These so-called **curds** then are separated as **unripened cheese**, such as cottage cheese and cream cheese. The remaining fluid is called **whey**.

To prepare different kinds of **ripened cheese**, the curds are washed, and salt is added to flavor the curds and prevent spoilage. Then, cultures of microbes are added to the curds. For example, if a fine Swiss cheese is to be made, at

FIGURE 13.6 Black and Green Olives. Black olives are ripened before fermentation, while unripe green olives are fermented for several weeks.

Courtesy of Dr. Jeffrey Pommerville.

FIGURE 13.7 Cheese Manufacture. The process of cheese manufacture begins with pasteurized milk and produces unripened and ripened cheeses of a wide variety. Swiss cheese, a hard cheese, and two mold-ripened soft cheeses, Camembert and Roquefort, depend on microbes to produce the final appearance, flavor, and texture to these cheeses.

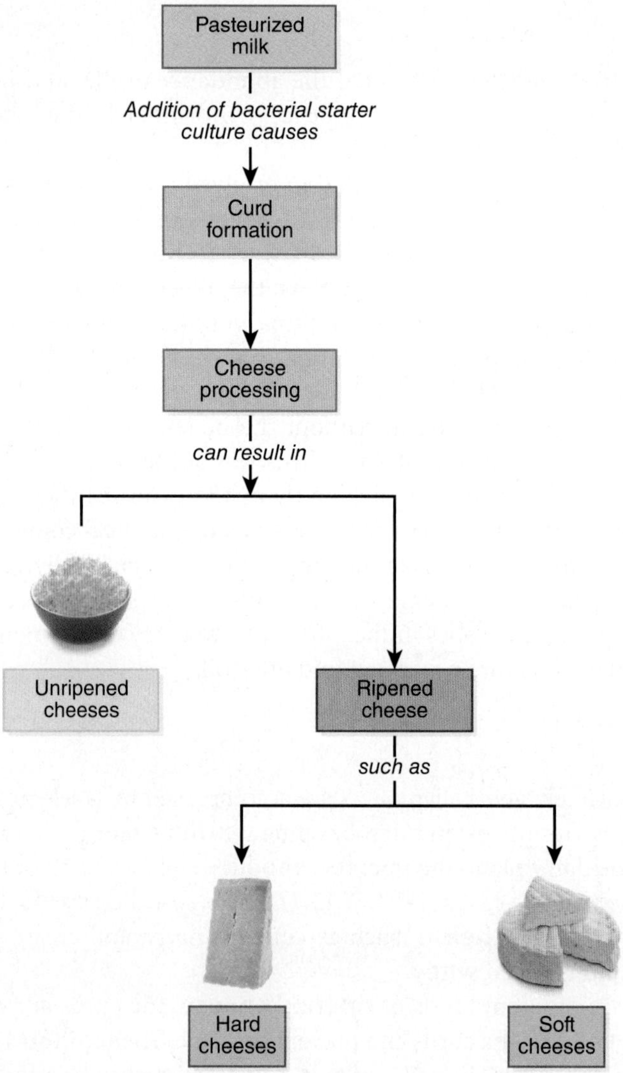

Lactobacillus: lack-toe-bah-SIL-lus

Propionibacterium: pro-pea-OHN-ee-bak-tier-ee-um

least two different bacterial species are added. During the aging process, these microbes bring about unique chemical changes. Species of *Lactobacillus* produce lactic acid to lend sourness. The acid also is consumed by species of *Propionibacterium*, which results in the production of more organic acids and CO_2. The organic compounds give Swiss cheese its distinctively nutty flavor, and the CO_2 gas accumulates as the holes, or eyes, in the cheese.

However, some of the dinner guests might instead select a mold-ripened, soft cheese for the appetizer. Among the choices in this category are Camembert and Roquefort cheeses (see Figure 13.7). Camembert, a soft cheese, is made by dipping milk curds into the fungus *Penicillium camemberti*. As the fungus grows on the outside of the curds, it digests the proteins and softens the cheese. To make Roquefort, a blue-green veined cheese, the curds are rolled with spores of the blue-green fungus *Penicillium roqueforti*. The fungus penetrates cracks in the curds and grows within them, thereby creating the distinctive blue-green veins in the cheese.

Penicillium camemberti: pen-ih-SIL-lee-um kam-am-BER-tee

Penicillium roqueforti: pen-ih-SIL-lee-um row-ko-FOR-tee

A CLOSER LOOK 13.2

The Microbial Ecology of Cheese

Globally, more than 20 million metric tons of cheese are produced annually, making cheese a major food product. Much of its delight comes from the wide-ranging variety of complex flavors, textures, and forms that different traditionally ripened cheeses possess. This especially is true when comparing these cheeses to the modern, mass-produced cheeses. Often the taste and texture of ripened cheeses come from the different molds and/or bacterial species added to the cheese during the production process.

How the various members of this complex cheese microbiome interact in the ripening process is being studied. Stilton cheese, for example, is an internally mold-ripened, semisoft blue cheese made from pasteurized cow's milk (see **FIGURE A**). The cheese-making process involves seeding the milk with *Lactococcus lactis*, and then the ripening is promoted by the mold *Penicillium roqueforti*. However, other bacterial species also are involved. What are these species and are they distributed randomly in the cheese, or are they located in specific areas?

In one study, samples of Stilton cheese were analyzed using various biochemical and DNA sequencing techniques to identify the microbial community structure. The analyses identified several bacterial species besides *L. lactis*. These species included *Enterococcus faecalis*, *Lactobacillus curvatus*, *Staphylococcus equorum*, *Leuconostoc mesenteroides*, and *Lactobacillus plantarum*. See Appendix A for organism pronunciations.

The studies found that in the core of the cheese, *L. lactis* made up about 70%, *L. mesenteroides* 15%, and *E. faecalis* made up the other 15% of the bacteria. *L. curvatus* was found in the margins of the veins, *S. equorum* on the vein surface, and *L. mesenteroides* in the central portion of the veins. The mold *P. roqueforti* was found in the veins, and *L. plantarum* and *L. curvatus* also were present under the crust of the cheese.

The investigators concluded that there is a differential distribution of species in the cheese, suggesting that there might be unique "ecological **niches**" in the cheese that the different microbial species prefer. Changes in the distribution of these microbes might be why different batches of cheeses, and often poor quality cheeses too, vary in their flavor and texture.

There is much more to understand about the microbiomes in cheese. However, evidence does support the view that different microorganisms in cheese play important roles in the development of the taste, smell, and texture, and possibly in the nutrient composition and shelf life. In addition, health experts believe eating cheese and its microbes might be helpful in maintaining good health by lowering the risk of heart disease and stroke.

FIGURE A Slice of blue Stilton cheese.

Stilton is another mold-ripened cheese in which the mold penetrates cracks within the curd, creating its distinctive veins. Microbiologists are finding that such cheeses have a unique microbial community (microbiome), as **A CLOSER LOOK 13.2** reports.

Niche: A term describing the relational position of a species or population in its environment.

13.4 The Salad Course: Of Vinegar and Bread

For the salad course, we have an assortment of mixed greens along with some other healthful and nutritious vegetables, such as tomatoes, red onions, and cucumbers. There is an oil and balsamic vinegar dressing (vinaigrette). Although microbes cannot claim to have produced the greens or the oil, they are essential for producing the vinegar.

Vinegar

Vinegar has traditionally been made by the souring of wine (the word "vinegar" is derived from the French word *vinaigre*, which means "sour wine"). Remember, one of Pasteur's first contributions to microbiology was discovering why

FIGURE 13.8 Balsamic Vinegar. Aging in wooden barrels speeds the acidification process, and the wood imparts characteristics flavors to the vinegar.

French wines were turning to vinegar. The sour wines were contaminated with acid-producing bacteria.

Vinegar can have different sources. For example, for apple cider vinegar, apple cider wine is fermented. Clear white vinegar begins as potato starch, which is fermented to potato wine, then converted to clear vinegar. This vinegar has no taste other than the natural sour taste of acetic acid. By contrast, balsamic vinegar acquires its sweet flavor from the wood barrels in which it is aged for many years; the barrels are made of balsam fir (**FIGURE 13.8**). The flavor of the wine vinegar also is a product of the original grape must as well as products of bacterial growth.

Acetobacter aceti: a-SEA-toh-bak-ter a-SET-ee

To turn the alcohol to vinegar, cultures of the bacterium *Acetobacter aceti* are added. The bacterial cells grow and multiply, and the bacterial enzymes convert the alcohol into acetic acid. Balsamic vinegar has an acetic acid content of about 3% to 5%.

Bread

Of course, we have not forgotten the bread basket. For this part of our meal, we once again turn to the yeasts, especially *S. cerevisiae*, in this case commonly called baker's yeast. This microbe is added to flour and water, the other two basic ingredients for making all the types of breads. In bread making, the yeast plays three roles. When yeast is added to flour and water (plus sugar, salt, and other ingredients at the whim of the bread maker), the fungus starts feeding on carbohydrates in the dough and produces substantial amounts of CO_2 gas. The gas expands the dough, and it rises seemingly by magic (**FIGURE 13.9**).

Gluten: A substance in cereal grains consisting of two proteins that add elastic texture to dough.

Enzymes also produced by the yeast help to strengthen and develop **gluten** in the dough and help give bread its spongy texture. In addition, yeast cells manufacture some ethyl alcohol in the dough during fermentation, but the high temperature of baking vaporizes the alcohol.

Yeast and flour can come together in an almost infinite range of variations, and the baker can lend some ingenuity to the process. Vienna bread, for example, is baked in a high-humidity oven and develops a flaky crust. Semolina flour is used to make semolina bread, while potato flour is used for potato breads and rolls. Bagels are boiled in water before baking; pizza dough is modified to give it high elasticity; and pumpernickel bread is produced from rye flour and yeast-fermented molasses. The yeasts are "busy bakers"!

FIGURE 13.9 Yeast and Dough Rising. The yeast *Saccharomyces cerevisiae* causes dough to rise while developing gluten in, and adding flavors to, the dough.

Microbes have contributed mightily to our dining experience thus far and will continue their influence as our culinary adventure unfolds.

13.5 The Main Course: Salmon, Sausages, and Sides

Assuming we have not overindulged during the previous courses, we now are ready for our main course. Our microbial menu contains two choices for the main course: teriyaki salmon or sausage and sauerkraut.

Teriyaki Salmon

The teriyaki salmon choice consists of fresh, wild-caught salmon marinated and then grilled in a teriyaki sauce. So, what's microbial here?

Scientifically speaking, the salmon is a product of microbial growth (although an indirect product). Salmon and some other so-called oily fish are excellent sources of omega-3s, which when consumed by humans benefit the heart, brain, and circulation. Because these fish need a healthy gut microbiome to regulate fat storage, the process of producing the omega-3s might be intimately dependent on their gut microbes.

For marinating and cooking the salmon, our chef is using teriyaki sauce, which consists of **soy sauce**, rice vinegar, and sugar. To make soy sauce, manufacturers begin with a starter mixture of soybeans and wheat bran that is inoculated with **koji**, the common name for the fungus *Aspergillus oryzae*. The mixture then is incubated at slightly over 30°C (85°F), which allows the fungus to break down complex proteins and carbohydrates into smaller molecules (**FIGURE 13.10**).

Aspergillus oryzae: a-sper-JIL-lus OH-rye-zeye

The mixture is blended with salt brine or coarse salt. The fungus continues to grow in the mixture and over the course of another year, the mixture continues to age as bacteria (*Lactobacillus* and *Bacillus* species) produce some lactic acid and *S. cerevisiae* yeasts add a small amount of alcohol. The liquid pressed from the mixture is soy sauce. It has a distinct meaty flavor called "umami," which in Japanese means "pleasant, savory taste." Because the fungus has contributed the predominant flavor and aroma to the soy sauce, our teriyaki salmon certainly has culinary roots tied to microbes.

Bacillus: bah-SIL-lus

FIGURE 13.10 The Production of Soy Sauce. Soybeans and wheat bran are inoculated with *Aspergillus oryzae* (koji) and allowed to ferment in pots at a soya sauce factory in Malaysia.

Sausages

The other main course consists of a variety of sausages accompanied by appropriate vegetables. Sausages generally consist of dry or semi-dry fermented meats. They include pepperoni from Italy, bratwurst from Germany, and kielbasa from Poland. To produce sausages, curing and seasoning agents are added to ground meat before the meat mixture is stuffed into casings and incubated at warm temperatures. Microbes in the product then multiply and produce a mixture of acids from the meat's carbohydrates, thereby giving the sausage its unique taste.

Sides Dishes

A side dish with the salmon is truffled potatoes. Like mushrooms, **truffles** are the reproductive body (spore-forming part) of the subterranean fungus. In the wild, truffles form a symbiotic relationship with the tree roots they grow on. This means the fungus helps the plant obtain water and minerals from the soil in exchange for sugars from the tree roots. The pungent, musky smell of truffles is thought to come from a combination of molecules given off by the truffles and the bacteria that live on the truffles. When prepared with truffle oil, roasted potatoes are given a powerful flavor punch.

For those who ordered sausage, there are some vegetable sides to go along with the sausage. One is "sour cabbage," or **sauerkraut**, as it is better known. Sauerkraut is a well-preserved and tasty form of cabbage that is an excellent source of vitamin C. Indeed, in the days of world exploration, the British often took sauerkraut on their ocean voyages to help prevent **scurvy** (as an alternative to the more expensive citrus fruits). On Captain James Cook's first world voyage in 1768, among the provisions were three tons of sauerkraut!

Scurvy: A disease resulting from insufficient vitamin C, the symptoms of which include spongy gums, loosening of the teeth, and bleeding into the skin and mucous membranes.

Modern researchers have verified the healthful benefits of sauerkraut. For example, scientists have noted that Polish women who had immigrated to the United States were less likely to develop breast cancer than non-Polish immigrant women. The reason? Polish foods, such as sauerkraut and fermented products of other cabbage family members (e.g., broccoli, cauliflower, and Brussels sprouts), contain compounds capable of blocking the activity of estrogen, which scientists say can be a possible stimulator of breast cancer. While the research is not complete, it is extremely interesting.

Species of *Leuconostoc* and *Lactobacillus* are members of the **lactic acid bacteria** (**LAB**) that are essential to the production of sauerkraut. These are

FIGURE 13.11 Pickled Vegetables. Many vegetables can be fermented and canned, including pickles (pickled cucumbers) and beets.

Courtesy of Dr. Jeffrey Pommerville.

both gram-positive bacterial genera naturally found in the leaves and tissues of a head of cabbage. Sauerkraut preparation begins by shredding the cabbage and adding about 3% salt. The salt ruptures the walls of the cabbage plant cells and releases their juices, while adding flavor. Then, the shredded and salted cabbage is packed tightly into a closed container to eliminate oxygen and stimulate fermentation. About a day later, the *Leuconostoc* species begins multiplying rapidly. The cells ferment the carbohydrates and produce lactic and acetic acids. After several days, the pH content of the cabbage is an acidic 3.5. Now, species of acid-tolerant *Lactobacillus* take over. They ferment the carbohydrates further and produce additional lactic acid to reduce the pH to about 2.0 and give the sauerkraut its tangy sourness.

A variety of pickled cucumbers and beets are available to accompany the sausages. For all types of dill, sour, and sweet pickles, the fermentation is similar, Manufacturers begin by placing cucumbers of any type or size in a high-salt solution, where the cucumbers change color from bright green to a duller olive-green. Then, the fermentation begins in an aging tank. The first bacterial genus to grow is *Enterobacter*. This gram-negative rod produces large amounts of CO_2 gas, which takes up the air space in the tank and establishes fermentation conditions. The next bacterial genera to proliferate are the LAB. They produce large amounts of acid to soften and sour the cucumbers. Yeasts also grow in the aging tank and establish many of the flavors associated with pickled cucumbers. Various herbs and spices are added to finish the process. Virtually any vegetable, including beets, can be substituted for cucumbers for an equally tasty result (**FIGURE 13.11**).

Enterobacter: en-teh-roh-BACK-ter

13.6 Washing It Down: A Refreshing Grain Beverage

The distinctive flavors of the main course are wonderfully complemented by the sparkle and tang of a glass of beer or sake. And, not surprisingly, microbes once more play a part in our meal.

Beer Making

Much of the chemistry of wine fermentation applies equally well to the production of beer. Beer making is thousands of years old. Records show that the ancient Egyptians, Greeks, and Romans understood the art of brewing beer. During the Middle Ages, the monasteries were the centers of beer making. By the thirteenth century, taverns and breweries were commonplace throughout the towns of Britain. Centuries would pass, however, before beer would make its appearance in cans. That auspicious event took place in the United States in 1935 in the city of Newton, New Jersey. The six-pack was the logical successor.

The word "beer" is derived from the Anglo-Saxon *baere*, which means "barley." This terminology developed because beer is traditionally a product of barley fermentation. The process begins by predigesting barley grains, a process called malting (**FIGURE 13.12**). During malting, the barley grains are steeped in water, and naturally occurring enzymes in the barley digest the starch into smaller carbohydrates, among them maltose (also known as malt sugar).

Next, the malt is ground with water in the process known as mashing, and the liquid portion, or **wort**, is removed. At this point, the brew master adds dried flowers (hops) of the hop vine *Humulus lupulus*. Hops give the wort its characteristic beer flavor, while adding color and stability. Then, the fluid is filtered, and a species of *Saccharomyces* is added.

Humulus lupulus: HU-mew-lus LU-pu-lus

Various species of yeasts are used to produce different types of beer. For example, *S. cerevisiae* gives a dark cloudiness to beer; it is called a "top yeast," because the yeast cells are carried to the top of the vat by the extensive CO_2 foam. The beer it produces is an English-type ale or stout. A different species, *S. carlsbergensis*, causes a slower fermentation and produces a lighter, clearer beer with less alcohol. This microbe is called a "bottom yeast," because there is less frothing, and the yeast cells settle to the bottom. Fermentation with this yeast is carried out at a cooler temperature (approximately 15°C/59°F) than is used for ale production (approximately 20°C/68°F). The beer produced by bottom yeast is Pilsner, also called lager. Almost three-quarters of the beer produced in the world is lager beer.

Saccharomyces carlsbergensis: sack-ah-roe-MY-seas ka-ruls-ber-GEN-sis

Initially, there is intense frothing as CO_2 is produced during aerobic respiration. Then, the frothing subsides, and the yeast shifts its metabolism to fermentation as alcohol production begins. The final alcoholic content of beer is approximately 4% to 5%.

After about a week in the fermentation tank, the "young beer" is aged for two weeks, after which it is transferred to another tank for secondary aging, also called **lagering**. This process might take an additional six months, during which time the beer develops its characteristic flavor and taste.

If the beer is to be canned, it is usually pasteurized at 60°C (140°F) for a period of 55 minutes to kill the yeasts. Alternatively, the beer can be filtered before canning to remove the yeasts; in this case, the beer is called "draft beer." If the beer is to be delivered directly to an alehouse, it is placed in casks and immediately chilled, making pasteurization unnecessary.

Sake Making

With the teriyaki salmon, a sake might be nice. Though many people consider sake a rice wine, it is more correctly a type of beer because it is fermented from rice (a grain rather than a fruit). To produce sake, steamed rice is mixed with the fungus *A. oryzae* and set aside. During the incubation period, enzymes from the mold break down the starches in the rice to simpler sugars. Fermentation by yeasts follows. The final product has the alcoholic concentration of a wine, about 13%, and is commonly drunk at room temperature or warmer.

But we cannot linger too long, for it is time for dessert.

FIGURE 13.12 A Generalized Process for Producing Beer. Barley grains are held in malting tanks while the seeds germinate to yield fermentable sugars. The digested grain, or malt, is then mashed in a mashing tank and the fluid portion, the wort, is removed. Hops are added to the wort in the next step, followed by yeast growth and alcohol production during fermentation. The young beer is aged in primary and secondary aging tanks. When it is ready for consumption, it is transferred to kegs, bottles, or cans.

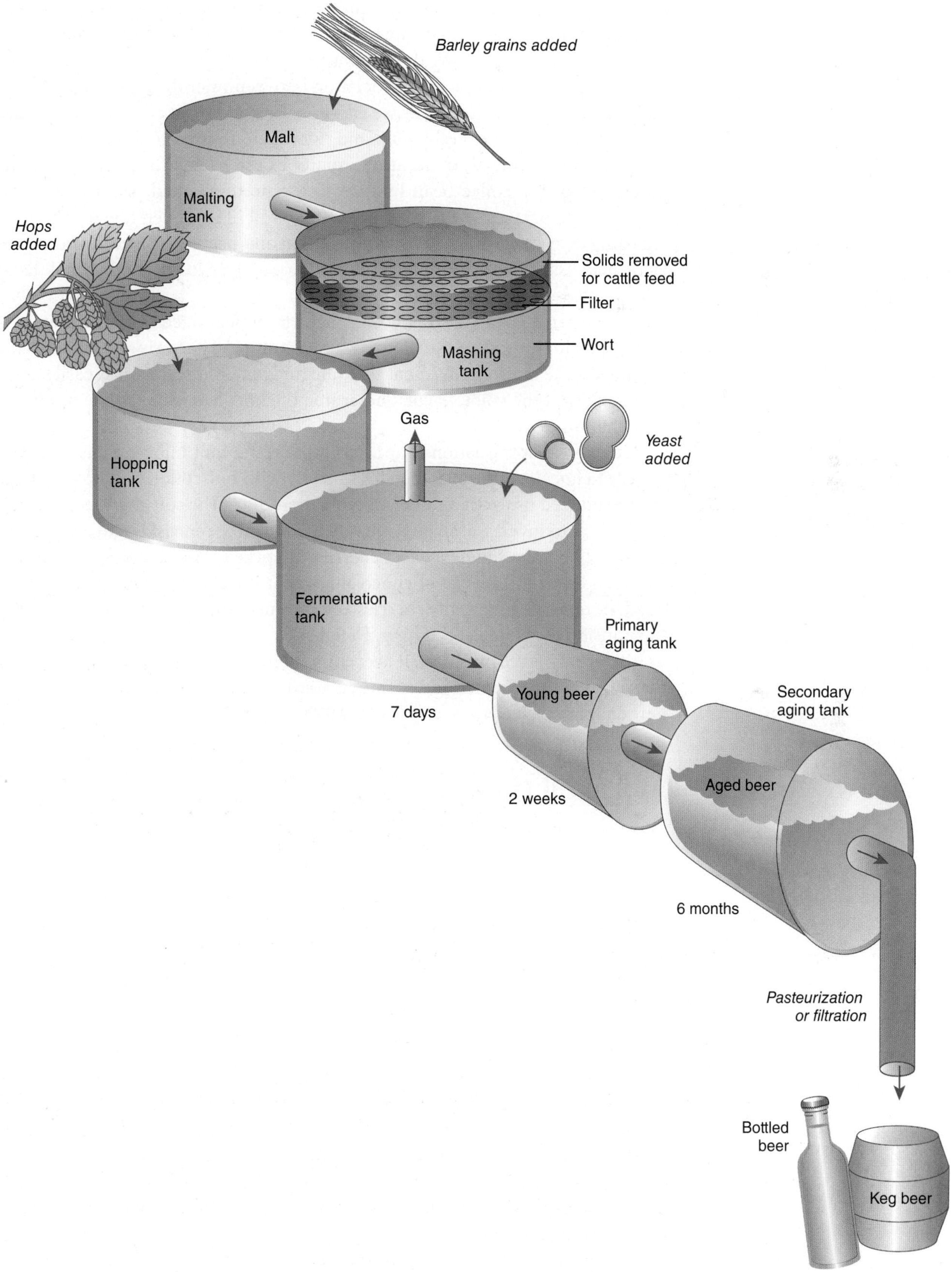

13.7 The Dessert Course: Coffee and Chocolates

Try as we might, we cannot really credit microbes for the luscious assortments of cakes, pastries, and cookies on the restaurant's dessert cart. However, some of the fruit fillings and toppings owe their flavors to slight degrees of fermentation, but this would be stretching the point a bit.

However, we can draw an association between microbes and the flavors of coffee and chocolate. Coffee, as we all know, is made from coffee beans. Coffee beans are covered by a fleshy pulp and microbes are used to ferment this pulp to assist in its removal. In addition, microbes are needed to help remove the outer skin of the coffee bean by digesting a protein called pectin that comprises a large portion of the bean skin. This protein is broken down by pectin-digesting enzymes produced by various fungi and bacteria. Bacteria producing lactic acid assist in pectin removal but do not appear to add to the final flavor of the coffee.

To prepare chocolate, cacao beans need to be separated from the pulp covering them in the pod. Manufacturers add a culture of bacteria and yeasts to the beans, and the microbes ferment the pulp and soften it so that it is easily shed. In fact, part of the color, flavor, and aroma of chocolate is due to the microbial fermentation process.

To express appreciation for our visit, the restaurant manager has sent over a tray of chocolate-covered cherries to enjoy with our coffee. Unknowingly, he has introduced us to yet another microbial product. The soft cherries owe their production to invertase, an enzyme obtained from *S. cerevisiae* and a species of *Bacillus*. For these candies, sweet cherries are pitted, mixed with the invertase, and then dipped into a vat of chocolate to coat the cherries. Over the next few days, the enzyme softens the cherries and produces the tasty liquor surrounding them inside the chocolate coating. Delightful!

Before we leave our meal, it is exciting to hear that chefs and cooks around the world are experimenting with new food fermentations to provide us with new culinary experiences, as **A CLOSER LOOK 13.3** explains.

A Final Thought

"Well, Maria. What did you think of the dinner?" asked Angela.

"I have to admit it was awesome," Maria proclaimed. "At first, I was wary of the microbes in the dinner. But I must say, I really enjoyed everything."

"It was pretty interesting how all the bacteria and molds and yeasts are used to give foods such great tastes," Angela commented. "Quite an epicurean feast!"

"I can actually say that I found fermentation fascinating," Maria remarked. "Not only does the process and the microbes produce strong flavors in foods, but they help preserve the food as well."

"Yeah. Hey, the next time we go grocery shopping, we need to look for more fermented foods," Angela suggested.

"Okay, but let's stay away from the truffle oil. It's awfully expensive," Maria suggested.

"Agreed!"

A CLOSER LOOK 13.3

Artisanal Food Microbiology

Carefully controlled fermentation processes are key to many well-known delicacies beyond the microbial products tasted at the dinner. Without microbes and fermentation, our culinary experiences would be sorely lacking.

The association of many foods with microbes and fermentation has a long history in many cultures and societies around the world. For example, here are a few other food products based on microbial fermentation:

- **Appam**. A pancake-like food from southern India that results from using fermented rice batter and coconut milk.
- **Gejang**. In Korean cuisine, a salted, fermented seafood, which is made by marinating fresh raw crabs and soy sauce.
- **Katsuobushi**. This food consists of smoked, dried, fermented tuna that is shaved and marinated with kelp to make *dashi*, one of the basic broths of Japanese cuisine.
- **Kefir**. A fermented milk product resulting from a mix of lactic acid bacteria (LAB) and yeast. Found around the world today, but its origins were in the Caucasus mountains of Western Asia.
- **Kombucha**. An acidic liquid or beverage made from fermenting sweetened tea with a mixed culture of yeasts and acetic acid bacteria. The origin of kombucha is unclear, though Russia, China, Japan, and Korea are all contenders. It, too, is found around the world today.
- **Kosho**. A specialty of the Kyushu region of Japan, the beverage is a fermented mixture of yuzu citrus peel, chili peppers, and salt.
- **Kvass**. A beverage in Baltic and Slavic countries based on beet- or rye-bread fermentation.
- **Miso**. A rich, sweet, tangy, salty, and umami-flavored seasoning from Japan that is produced by fermenting rice, barley, or soybean with salt and koji (see **FIGURE A**).
- **Pulque**. A milky-looking ancient drink of South-Central Mexico that is made from the fermented sap of the agave plant.
- **Tempeh**. An Indonesian soybean product resulting from the fermentation of soybeans with molds.
- **Worcestershire sauce**. A fermented liquid condiment first created in England that consists of a complex mixture of ingredients producing a flavor that is both savory and sweet.

With this short list as examples, cooks around the world are discovering (or rediscovering) the possibilities of using fermentation processes in new ways in the kitchen. Fermenting for flavor has not traditionally been something that chefs have been interested in exploring. Today, however, there is an emerging interest among cooks to explore the potential for microbes to create new flavors and dishes. Called "artisanal food microbiology," chefs are developing new ways to modify old fermentation techniques as new cooking tools to transform some of the foods we eat into new a taste experiences. It is giving chefs new methods for working in concert with microbes to explore new cuisines by creating new flavors.

Looking ahead, it should be an interesting and, hopefully, delicious time to see and taste what the next set of gastronomic delights might bring us.

FIGURE A Miso, a traditional Japanese seasoning.

Chapter Discussion Questions

What Was He Thinking?

From your reading of this chapter, identify and discuss five major points about food microbiology that the author was trying to get across to you.

Questions to Consider

1. One day, the students in a microbiology class presented the instructor with a basket of "microbial cheer" in recognition of her efforts on their behalf. From your knowledge of this and other chapters, can you guess some of the items that the basket contained?
2. Yeasts are sometimes known as the "schizophrenic microbes." What do you think that means?
3. Explain the difference between an unripened cheese and a ripened cheese. Give some examples based on the cheeses you eat.
4. Wine is part of many cultures worldwide, but one of the great mysteries of the human experience is how wine was discovered. Write a scenario depicting the circumstances under which wine (fermented grape juice) might have been first experienced by a human.
5. Suppose you decide to enter the pickling business. You intend to pickle tomatoes, peppers, and a host of other foods. How would you proceed with the science end of your new business?
6. Wine and beer both are the result of alcoholic fermentation. Although sake is called a "rice wine," why is it technically a beer?
7. This chapter has surveyed many foods of microbial origin, but the survey is far from complete. From your reading of other chapters and your general knowledge, what other fermented foods of microbial origin or involving microbes can you add to the list discussed in this chapter? For a long list, see *https://en.wikipedia.org/wiki/List_of_fermented_foods*

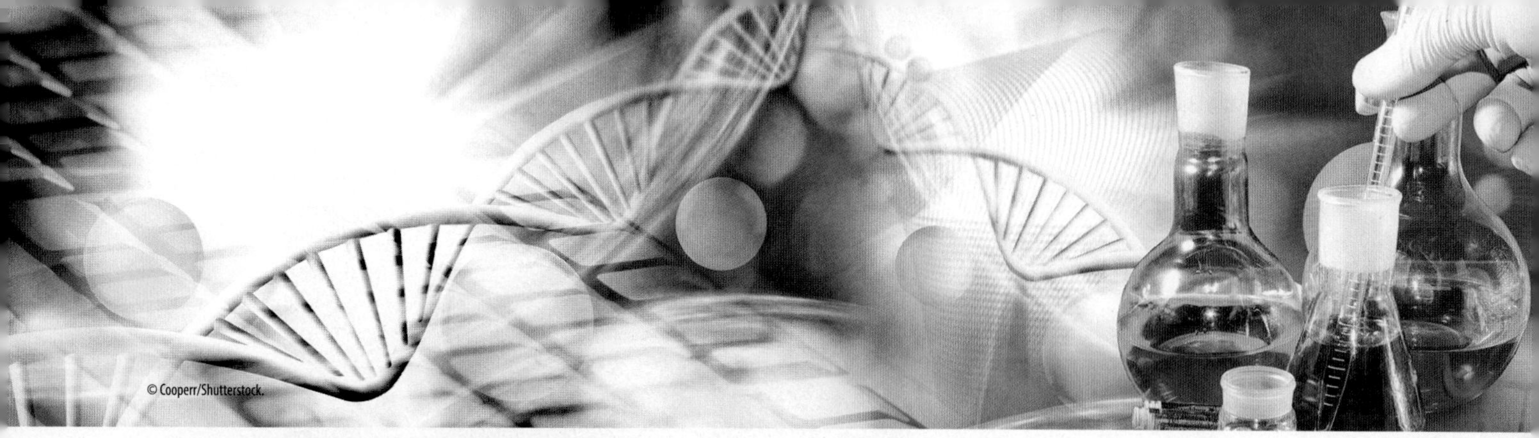

CHAPTER 14

Biotechnology and Industry: Putting Microbes to Work

The Spider's Silk Parlor

Will you walk into my parlour? said the spider to the fly.
Tis the prettiest little parlour that ever you did spy;
The way into my parlour is up a winding stair,
And I've a many curious things to show when you are there.

In this first line of Mary Howitt's *The Spider and the Fly* (1828), a crafty spider tries to trap an unsuspecting fly using seduction and flattery. Today, the clever spiders are attracting scientists for a different purpose—the spider's silk.

Spider silk is a protein fiber spun by spiders to make nests to protect their offspring. It also acts as a sticky net to catch and wrap up prey (see the poem above) (**FIGURE 14.1**). Spider silk is the strongest known natural fiber. Being stronger and tougher than steel, spider silk holds great promise for industrial and consumer applications. It could be used to make super thin, super strong surgical sutures. It might be used in clothing that would have a **tensile strength** greater than synthetic fibers like Kevlar that is fashioned into bulletproof vests.

Tensile strength: The maximum stress needed to break the fibers in a material.

Unfortunately, one cannot raise spiders like bacteria in concentrated colonies that would be needed by industry to produce large amounts of silk. So, what to do? Turn to microbes, of course.

In 2018, researchers at Washington University in St. Louis constructed a DNA sequence that coded for the spider silk protein. The genes were genetically engineered into bacteria. When the cells grew, the cells produced biosynthetic spider silk that matches its natural counterpart.

Unfortunately, the bacterial cells produce only small pieces of synthetic silk. They do not produce the long, continuous strands needed for commercial use.

CHAPTER 14 OPENER With the ability to synthesize and sequence DNA, microbial biotechnology and industrial microbiology have burst on the scene. These technologies apply scientific and engineering principles to the processing of materials by microorganisms to create useful therapeutics and commercial products for all aspects of society.

FIGURE 14.1 Spider Silk. A silky spider web.

Consequently, genetic engineers need to figure out how to modify the metabolic process. Once the "bacterial factories" can churn out long silk fibers, the process should be scalable and economical for practical use.

So, someday when you go to buy a pair of waterproof and strong hiking boots or a rain jacket, instead of a **Gore-Tex** label, you might find a label that reads "made with microbial silk."

Gore-Tex: Trademark for a waterproof, breathable fabric composed of Teflon.

Today, society depends on microbes to bring us numerous medical and industrial products. These microbes are the creation of **microbial biotechnology**. By altering the genome of microbes, the modified organisms manufacture substances they would not naturally produce (e.g., spider silk). Then, **industrial microbiology** steps in. It increases the quantity of these products to the mass scale needed commercially.

In this chapter, we will examine some of the products of microbial biotechnology and industrial microbiology (applied sciences) in the context of practical applications.

LOOKING AHEAD

After reading and completing this chapter, you will be able to:

14.1 Describe how bacterial cells can be genetically engineered to produce a specific product.

14.2 Identify several biotech enzymes, and describe their uses to alleviate human medical conditions and disorders.

14.3 Explain how DNA probes are used in diagnostic tests to identify an infectious pathogen.

14.4 Distinguish between primary metabolites and secondary metabolites.

14.5 Discuss the industrial role of microbes in the production of antibiotics, vitamins, enzymes, and biofuels.

14.1 Microbes and Biotechnology: Seeing the Promise

Every so often in science, a window opens and the theoretical becomes possible. Such a window opened in microbiology in the 1970s with the dawn of genetic engineering. As described in the chapter on Microbial Genetics, microbiologists and other scientists began to see impossible dreams become reality. The field of microbial biotechnology was born.

In the next section of this chapter, we will review how foreign genes are engineered into microbes. Then, we will consider some of the impossible dreams that have become reality from the genetic engineering of microorganisms.

Genetically Engineering Bacterial Cells

Genetic engineering is the process of altering the DNA in an organism's genome. Often it involves taking a gene from one organism and inserting it into the genome of another organism. Here is one example in which the biotech product helped alleviate a human disorder.

One of the essential hormones produced in the body is **human growth hormone (hGH)**. The hormone is produced by the **anterior pituitary** and it stimulates muscle and bone growth. Some children fail to produce enough hGH due to a damaged pituitary gland or **hypothalamus**. As a result, bone and muscle growth can slow, and the individual will experience growth failure and short stature (dwarfism). If diagnosed before puberty, the condition might be treatable with genetically engineered hGH (recombinant hGH; rhGH).

To produce rhGH, the hGH gene is isolated from human cells and placed (recombined) into a bacterial **plasmid** (**FIGURE 14.2**). The plasmid then is

Anterior pituitary: The gland, situated at the base of the brain, that secretes several hormones

Hypothalamus: The part of the brain that, among other functions, controls the pituitary gland.

Plasmid: Circular, independent molecules of DNA that carry nonessential genetic information in bacterial cells.

FIGURE 14.2 Genetic Engineering and Expression of Recombinant Human Growth Hormone (rhGH). After obtaining the gene for human growth hormone (hGH), the gene is inserted into a bacterial plasmid, which is then introduced into *Escherichia coli* cells. The cells produce the rhGH, which then can be isolated, purified, and marketed for treatment of short stature (dwarfism).

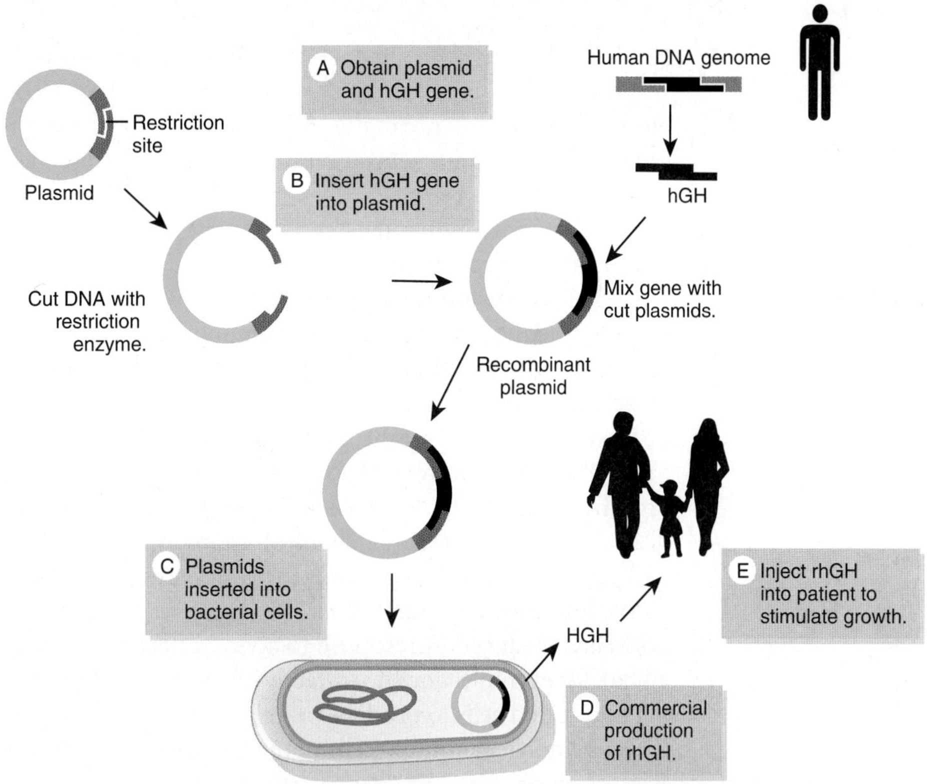

Escherichia coli: esh-er-EE-key-ah KOH-lee

transferred into *Escherichia coli* cells. In these cells, the hGH gene is transcribed and translated into the growth hormone protein. For medical use, treatments with rhGH (known commercially as Protropin) encourage growth spurts that permit children to reach a normal height for their age.

Some athletes and weightlifters believe rhGH can build muscle and improve physical performance. However, most clinical studies suggest rhGH will not increase physical performance, and the International Olympic Committee and the National Collegiate Athletic Association have banned the use of rhGH in sports competitions.

14.2 The Products of Microbial Biotechnology: Medical Therapeutics and Vaccines

Like rhGH, many of the therapeutic products of microbial biotechnology involve inserting a human gene into bacterial cells.

Therapeutics

One of the most economically profitable areas of microbial biotechnology, and the one with the highest pharmaceutical value, is the production of human hormones, blood products, and enzymes to alleviate human medical conditions and disorders (**TABLE 14.1**). Here are a few examples.

Insulin: A hormone made by the pancreas that helps keep blood sugar levels from getting too high or too low.

Subcutaneous: Referring to under the skin.

- **Type 1 Diabetes.** Also called insulin-dependent diabetes, this is a persistent condition in which the pancreas produces little or no **insulin**. The result is high blood glucose levels. The historic first step in using microbial biotechnology to treat the condition occurred in 1982. In this year, the United States Food and Drug Administration (FDA) approved the use of genetically engineered insulin (Humulin). With its approval, **subcutaneous** injections of the hormone allowed type 1 diabetics to better control their condition. Again, the process is explained in more detail in the chapter on Microbial Genetics.
- **Hemophilia A.** The most common inherited blood disorder in the United States is hemophilia A. It affects about 1 in 10,000 males whose blood doesn't clot normally. The reason is because these individuals cannot produce an essential blood clotting protein called clotting factor VIII.

  To help these individuals, recombinant factor VIII (rFVIII) now is produced through genetic engineering through a process like that described for rhGH and insulin. The availability of rFVIII has revolutionized the care and treatment of individuals with hemophilia A.

Ischemic: Refers to an inadequate blood supply in the body.

- **Ischemic Stroke.** If an individual has a blood clot that blocks a blood vessel, it will cut off blood flow to a part of the brain. Today, management of **ischemic** stroke might use recombinant tissue plasminogen activator (rtPA; called Activase) for treatment. This therapeutic product is a protein-digesting enzyme that stimulates other body enzymes to break down a blood clot (see Table 14.1). Many individuals owe their continued good health to this genetically engineered, microbe-derived protein.
- **Virus Infections. Interferons** (**IFNs**) are a group of signaling proteins normally produced in response to a viral infection. Interferon-alpha (IFN-alpha), for example, does not directly kill viruses. Rather, it signals surrounding uninfected cells to produce proteins that block viral genome replication. It also stimulates immune cells to the presence of the viral infection.

TABLE 14.1 A Sampling of Therapeutic Products of Microbial Biotechnology and Their Functions

Product	Function
Replacement Proteins and Hormones	
Factors VII, VIII, IX	Replace clotting factors missing in hemophiliacs
Growth hormone (rhGH)	Replaces missing hormone in people with short stature
Insulin	Treatment of insulin-dependent diabetes
Therapeutic Proteins, Hormones, and Enzymes	
Epidermal growth factor (hEGF)	Promotes wound healing
Granulocyte colony stimulating factor (hG-CSF)	Used to stimulate white blood cell production in cancer and AIDS patients
Interferon-alpha (IFN-alpha)	Used with other antiviral agent to fight viral infections and some cancers
Tissue plasminogen activator (TPA)	Dissolves blood clots; prevents blood clotting after heart attacks and strokes
DNase I	Treatment of cystic fibrosis

Naturally produced IFNs often act too slowly to arrest fast occurring viral infections. The molecules also might not provide a sustained response in some long-lasting infections. Therefore, recombinant IFN-alpha preparations have proved useful for some individuals suffering from viral diseases, such as hepatitis B, hepatitis C, and some forms of cancer.

- **Cystic Fibrosis (CF)**. Another serious genetic disorder is CF that affects more than 30,000 people in the United States and more than 70,000 worldwide. CF patients produce a thick, sticky mucus that clogs the respiratory passageways. Pathogens such as *Pseudomonas aeruginosa* find the sticky mucus to be a perfect home. As some bacterial cells die, they lyse and release their DNA into the mucus, making the material more viscous. Consequently, the mucus contributes to life-threatening respiratory tract obstructions and protects the living bacterial cells from host immune attack.

  One treatment option for CF patients is inhalation of recombinant human deoxyribonuclease I (rhDNase; called pulmozyme). This is an enzyme that selectively chops up DNA. Treatment with rhDNase greatly decreases the viscosity of the mucus in the lungs, which improves lung function. Now the *P. aeruginosa* cells are more susceptible to immune attack and antibiotic treatment might be more successful. Sales of this life-saving enzyme exceed $100 million annually.

  This short list of therapeutic products demonstrates that biopharmaceuticals are becoming first-line medicines. It also illustrates the potential for the field to develop additional protein therapeutics to benefit individuals and society.

Pseudomonas aeruginosa: sue-doh-MOH-nahs ah-rue-gih-NO-sah

Vaccines

Vaccines are made from microbial or viral substances (described in the chapter on Disease and Resistance). When injected into the body, these materials prepare the immune system to recognize those same microbial or viral agents. If the immune person then is exposed to one of these agents, the immune system will quickly attack the invader and eliminate it. Today, new vaccine formulations are being developed through biotech innovations. These products are more cost effective to manufacture and more easily transported than traditional vaccines. Importantly, these new vaccines give doctors more options for preventing illnesses such as hepatitis B, meningitis, the flu, and even cervical cancer. Vaccines are discussed in detail in the chapter on Disease and Resistance.

FIGURE 14.3 summarizes some of the therapeutic products of microbial biotechnology.

FIGURE 14.3 Developing New Products Through Microbial Biotechnology. Genetic engineering is a method for inserting foreign genes into bacterial cells and obtaining chemically useful products.

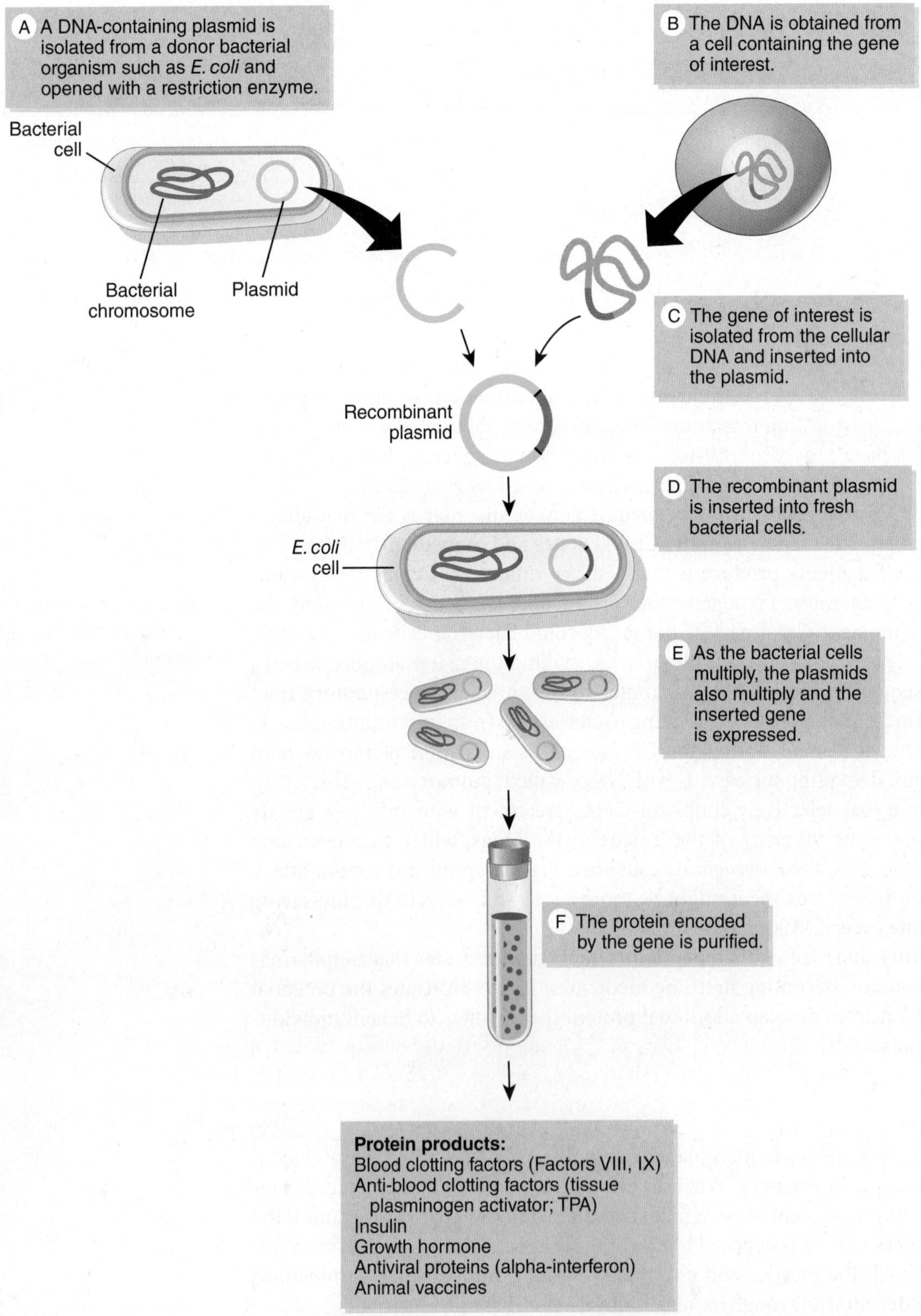

To finish off this portion of the chapter, let's explore a few of the diagnostic tools and tests for infectious diseases that are the result of microbial biotechnology.

14.3 The Products of Microbial Biotechnology: Diagnostic Tools and Tests

One of the more remarkable applications of microbial biotechnology is in the diagnostic laboratory when DNA samples are analyzed to identify unknown microbes. These methods allow the verification of numerous infectious and inherited diseases the presence of which previously could be predicted only by guesswork.

Diagnostic Tools

Two essential tools used in microbial technology are the polymerase chain reaction and the DNA probe.

Polymerase Chain Reaction

The **polymerase chain reaction** (**PCR**) is a method for producing billions of identical copies of a DNA sequence. The PCR process, which is carried out in a specially designed apparatus, has been described as a molecular equivalent of a photocopier (**FIGURE 14.4A**). Here is how the amplification process works.

To begin the process, a sample of DNA is collected from a pathogen or whatever target DNA is used. In the PCR machine, the double helix is unwound at a high temperature to yield single strands of DNA (**FIGURE 14.4B**). Then, a special, heat-resistant **DNA polymerase** enzyme called **Taq polymerase** is added. Now, the machine mixes in nucleotide molecules containing the four bases (adenine, thymine, guanine, and cytosine) together with a strand of primer DNA, which serves as a starting point for the copying process.

DNA polymerase: The enzyme that adds complementary nucleotides to a single strand of DNA.

By adding the nucleotides onto the primer DNA, Taq polymerase brings together the nucleotides to form a new strand of DNA that complements the single-stranded DNA. Next, the mixture is cooled, whereupon the new and old strands of DNA twist together to form double-stranded DNA molecules. At this point, the original number of DNA molecules has doubled. Then the process repeats itself. Under optimal conditions, a single DNA sequence can be amplified 1 billion times in less than 2 hours.

DNA Probe

One tool needed for a diagnostic test is a DNA probe. The **DNA probe** is a single-stranded DNA sequence that "hunts down" a complementary DNA (target) fragment in a diverse mixture of other DNA fragments.

To make a probe against a pathogen, a unique DNA sequence from the pathogen is produced and then subjected to PCR (see Figure 14.4). After a few hours, billions of identical fragments have been produced. They are separated into single strands. To identify the target fragment, a fluorescent or radioactive tag is attached to the DNA probe (**FIGURE 14.5A**).

Diagnostic Tests

DNA probe technology can be used for a range of new diagnostic tests. Let's look at one scenario.

FIGURE 14.4 The Polymerase Chain Reaction (PCR).
(A) A PCR machine.

(B) PCR is a laboratory technique for quickly amplifying a single DNA fragment a billion times.

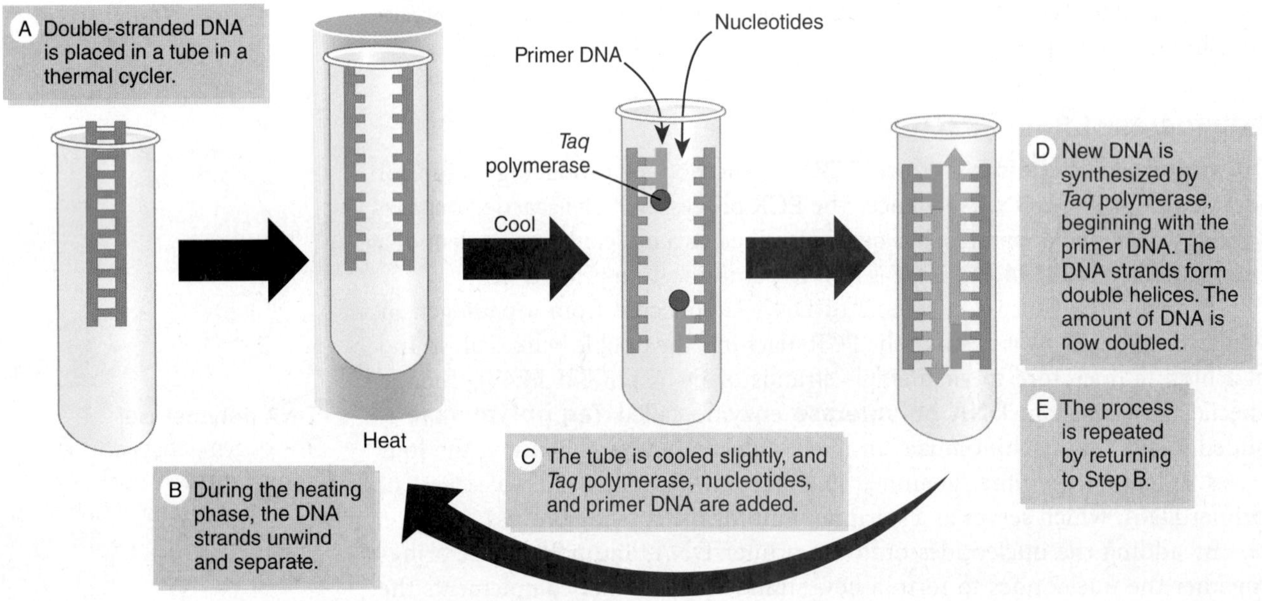

A Double-stranded DNA is placed in a tube in a thermal cycler.
Heat
B During the heating phase, the DNA strands unwind and separate.
Cool
Primer DNA
Nucleotides
Taq polymerase
C The tube is cooled slightly, and Taq polymerase, nucleotides, and primer DNA are added.
D New DNA is synthesized by Taq polymerase, beginning with the primer DNA. The DNA strands form double helices. The amount of DNA is now doubled.
E The process is repeated by returning to Step B.

Disease Diagnosis

Two individuals, patient 1 and patient 2, are concerned that they might have been infected with the human immunodeficiency virus (HIV). Because this virus is the causative agent of AIDS, both individuals want to be tested for the presence of the virus in their body.

As mentioned in the chapter on Viruses, HIV is a retrovirus. Part of its replication cycle involves the conversion of its single-stranded RNA into double-stranded DNA. This viral DNA segment, called a **provirus**, then is inserted into a chromosome in the infected cells.

One diagnostic test uses a DNA probe to detect the provirus in infected cells. DNA from a patient's tissues is isolated, fragmented into single-stranded pieces, and attached to a solid surface (**FIGURE 14.5B**). To perform the HIV test, a DNA probe capable of binding to the HIV provirus segment is mixed with the patient samples. If the target fragment (HIV provirus) is present, the fluorescent or radioactive tag reacts specifically with a target DNA fragment having complementary base pairs, much like a left hand matches up with its complementary right hand. The fluorescent or radioactive label provides a visual signal that binding has taken place. If there is no provirus present, there will be no binding, as the probe is washed away in the procedure. No signal will be detected. In other words, DNA probes assist in searching for the proverbial

FIGURE 14.5 DNA Probes.
(A) Construction of a DNA probe includes its labeling with an identifiable fluorescent or radioactive tag.
(B) Using an HIV probe, which patient is infected with the AIDS virus?

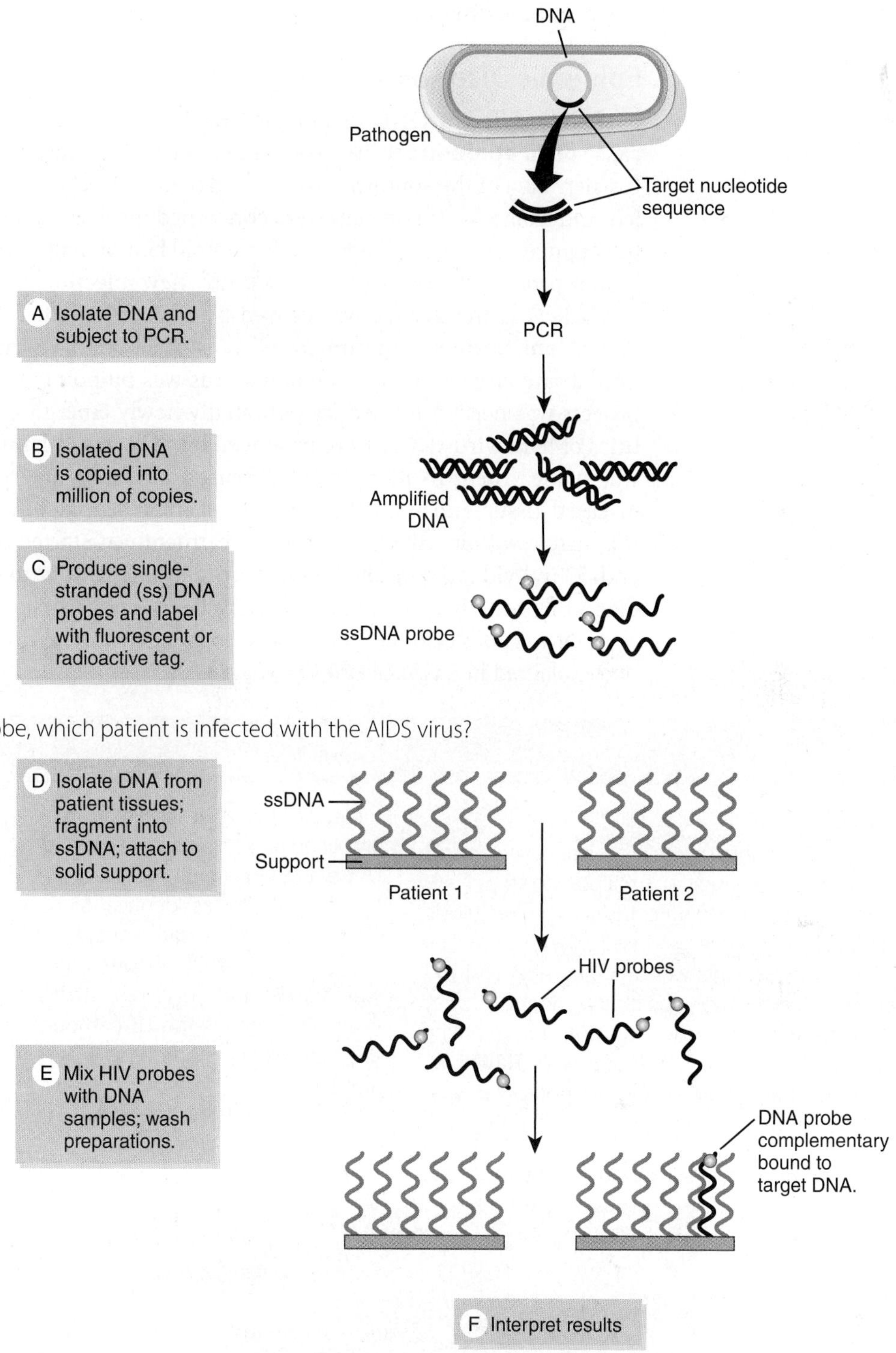

"needle in a haystack," where the needle is the target fragment of interest, and the haystack is the mass of other genetic material.

DNA diagnostic tests also are available for detecting many other infectious diseases. For example, Lyme disease is a blood disease due to a bacterial infection transmitted by ticks. The bacterium causing the disease is extremely difficult to grow in culture. However, thanks to microbial biotechnology, it is possible to secure blood from a suspected patient, amplify the bacterial pathogen DNA (if present in the blood), and then use a Lyme disease DNA probe to detect it.

These new microbial biotechnologies of PCR and DNA probes make the detection of a possible pathogen possible. It also reduces the guesswork often associated with the traditional methods of detection. In addition, these tests are rapid and allow for quicker treatment and might shorten the misery caused by some of these diseases.

Epidemic Diagnosis

Epidemic: The rapid and widespread occurrence of an infectious disease in many people in a given population.

Antibody: An immune system protein that neutralizes pathogens.

A slightly different DNA diagnostic procedure has been used to identify the cause of an **epidemic**. When a severe respiratory disease occurred in the Four Corners area of the southwestern United States (Arizona, Colorado, New Mexico, and Utah), health officials were concerned the disease was spreading among the population. Affected individuals suffered hemorrhaging, lung infection, and kidney problems. How could they identify new infections?

Believing the disease was caused by a virus, scientists laboriously mixed patient **antibodies** with numerous types of laboratory viruses, one at a time, until a rare virus known as the hantavirus was pinpointed. A faster diagnostic process was needed to identify potentially newly emerging cases. Large quantities of hantavirus DNA were produced by PCR, and a hantavirus DNA probe was made. Then, to identify the disease quickly in new patients, they took diseased tissue and used the probe to search for matching hantavirus DNA. If a match was identified, preventive treatment was started immediately. In the end, 53 individuals with the illness were identified and 32 died. Without a rapid diagnostic test, the epidemic might have been much worse.

DNA probes also have been used to solve historical mysteries, such as the one explained in **A CLOSER LOOK 14.1**.

A CLOSER LOOK 14.1

"Not Guilty"

Poor Christopher Columbus! Some historians have accused him of inadvertently bringing smallpox to the New World—and measles, whooping cough, and tuberculosis (TB)—and almost every other infectious disease. It makes it sound like the Santa Maria was a plague ship, and the New World was a sterile place, ripe for infection!

Well, not necessarily. While studying the remains of a mummified woman from Peru (see **FIGURE A**), biologists at the University of Minnesota noticed several lumps reminiscent of TB in the mummy's lung tissues. A molecular biologist extracted DNA from the lumps and amplified it so that there was enough DNA to study. Using a DNA probe for TB, the mummy's DNA turned out to have sequences identical to those of *Mycobacterium tuberculosis*, the causative agent of TB.

So, what does that prove? It turns out that the mummy was a thousand years old—that's right, one thousand years. That means the TB pathogen was already here in the New World hundreds of years before Columbus ever arrived.

So, Chris, you're off the hook—at least for TB.

FIGURE A A Peruvian mummy, similar to the one described.

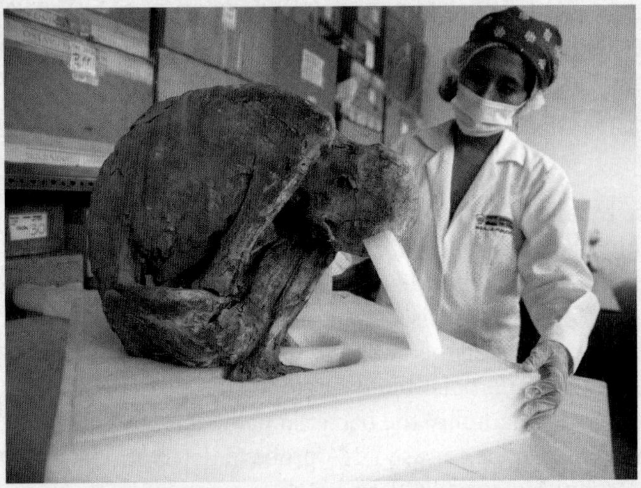

14.4 Microbes and Industry: Working Together

Long before microbial biotechnology came into existence, the use of microbes by human societies was commonplace. In fact, microbes were being used well before people realized they were taking advantage of these tiny organisms. For example, wine and beer making have been key activities of human culture and society for millennia. As these processes are covered in the chapter on Microbes and Food, let's examine some of the modern-day uses of microbes and see their relationships to industry.

The Industrial Microbes

There are numerous industrial and commercial products derived from microbes that are beneficial to humans. These end products or by-products (called **metabolites**) of metabolism fall into one of two groups (**FIGURE 14.6**).

- **Primary metabolites** are directly involved in the normal growth and reproduction of the microbe. They are formed during the **log phase** of growth. Pyruvate and the end products of the microbial fermentation pathways are examples of primary metabolites.
- **Secondary metabolites** form the bulk of products of interest to industrial microbiology. These metabolites often are a by-product of metabolism, and they are not essential for growth and reproduction. Look again at

Log phase: The period of growth of a population of microbes during which cell numbers increase exponentially.

FIGURE 14.6 Metabolite Synthesis. In culture, primary metabolites are mainly produced during the active log phase of growth. Secondary metabolites are produced near the end of the log phase and/or during stationary phase. See **Appendix A** for organism pronunciations.

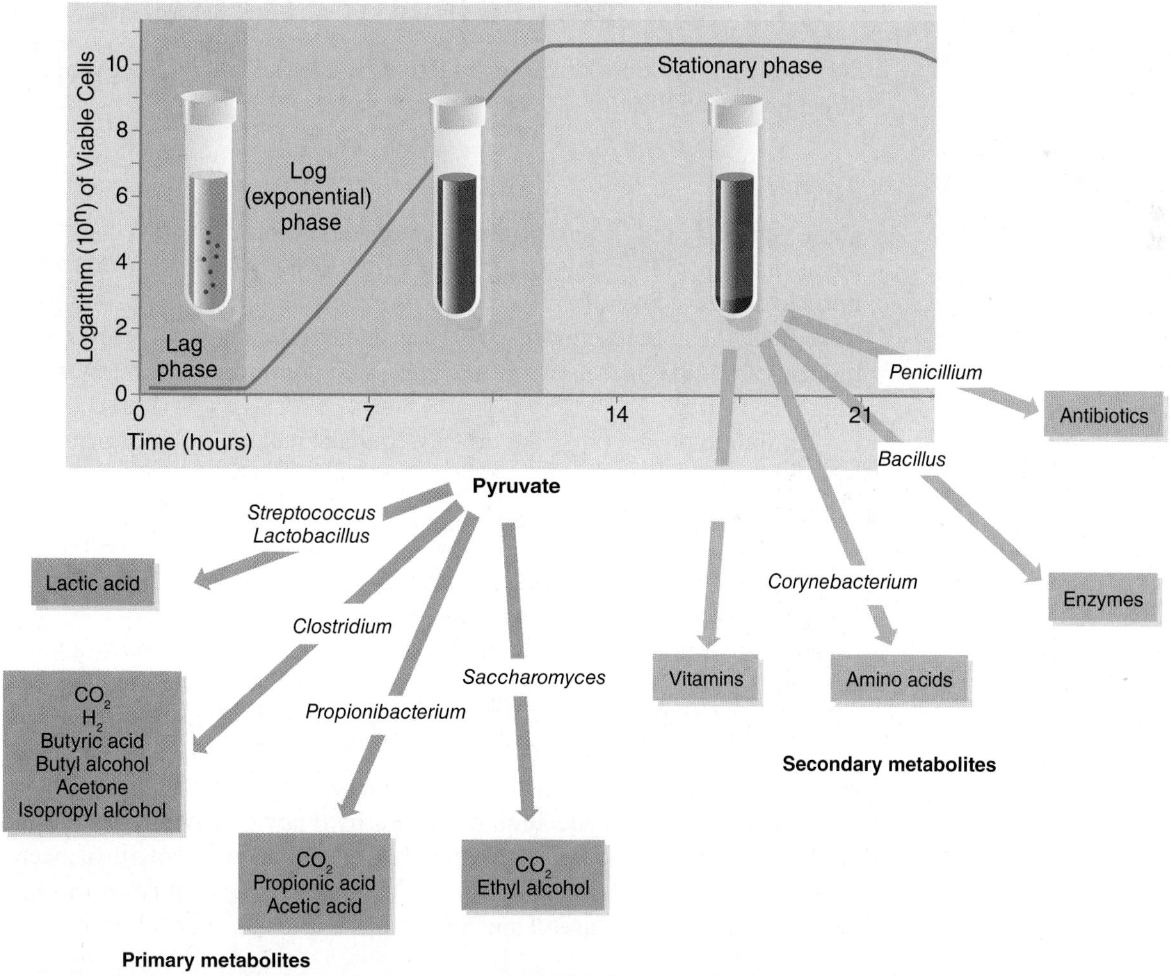

Stationary phase: The period of growth of a population of microbes during which the number of cells remains constant, even though some cells continue to divide, and others begin to die.

Figure 14.6. Notice that secondary metabolites usually are produced near or at the end of the microbe's log phase or in **stationary phase**. Most antibiotics, as well as some vitamins, amino acids, and enzymes of industrial interest, are examples of secondary metabolites.

Producing Metabolites and Growing Microbes in Mass

Tremendously large numbers of cells are needed to produce a significantly useful amount of metabolite at the industrial scale. On this large scale, industry often uses the term **industrial fermentation**. This refers to any procedure used to grow large masses of microbes, be it aerobic or anaerobic.

A major effort is needed to scale up microbial metabolic processes to a level capable of producing the required metabolite. Large, stainless steel tanks called **fermentors** or **bioreactors** frequently are used (**FIGURE 14.7A**). Technicians add the desired microbes to sterile growth medium in the fermentor where environmental conditions are computer monitored. Temperature, pH, oxygen, and nutrients—all the typical physical and chemical conditions that apply to growing microbes in a culture tube—apply here, just on a much larger scale. All these steps are needed to maintain the desired growth conditions and to ensure the microbes reach a state in which maximal metabolite is produced (**FIGURE 14.7B**).

After the maximal amount of metabolite has been produced, the liquid in the fermentor is filtered. The product then is extracted from the liquid and the purity determined. It then can be packaged and sent to the marketplace.

14.5 Microbes and Industry: The Products

Let's now survey some of the products of industrial microbiology that have direct human applications.

Antibiotics

Some bacteria and fungi produce chemical by-products of their metabolism that can kill or slow the growth of bacteria. More than 5,000 of these **antibiotics** have been described. However, only about 100 are useful for bacterial infections because most have toxic side effects, making them useless for human use. **TABLE 14.2** lists some of the more common antibiotics and their microbial source.

Antibiotic production, like other secondary metabolites, is carried out in large bioreactors that can hold 40,000 to 200,000 liters (10,500 to 52,000 gallons) of growth medium. As the carbon source in the medium becomes exhausted at stationary phase, the microbes start producing the antibiotic. After several weeks in the bioreactor, the cells are filtered out. The antibiotic then is extracted from the medium for further purification.

Of course, industrial production of antibiotics assumes you have a microbe capable of producing the antibiotic. The traditional screening method for discovering new antibiotics was rather straightforward. Microbes were isolated (usually from the soil) and cultivated in the laboratory at various temperatures and in a variety of growth media. Then, biochemists would attempt to harvest, purify, and identify the metabolites to identify if any of those by-products had antibiotic activity; that is, would they kill or inhibit the growth of bacterial cells? By these methods, novel compounds were discovered, but only rarely. Few proved to be medically useful and able to make it to commercial production.

FIGURE 14.7 The Industrial Fermentation Process.
(A) A pharmaceutical technician monitors a series of fermentors to ensure a maximum yield of metabolite.

(B) A general industrial process scheme for metabolite production and recovery in a fermentor.

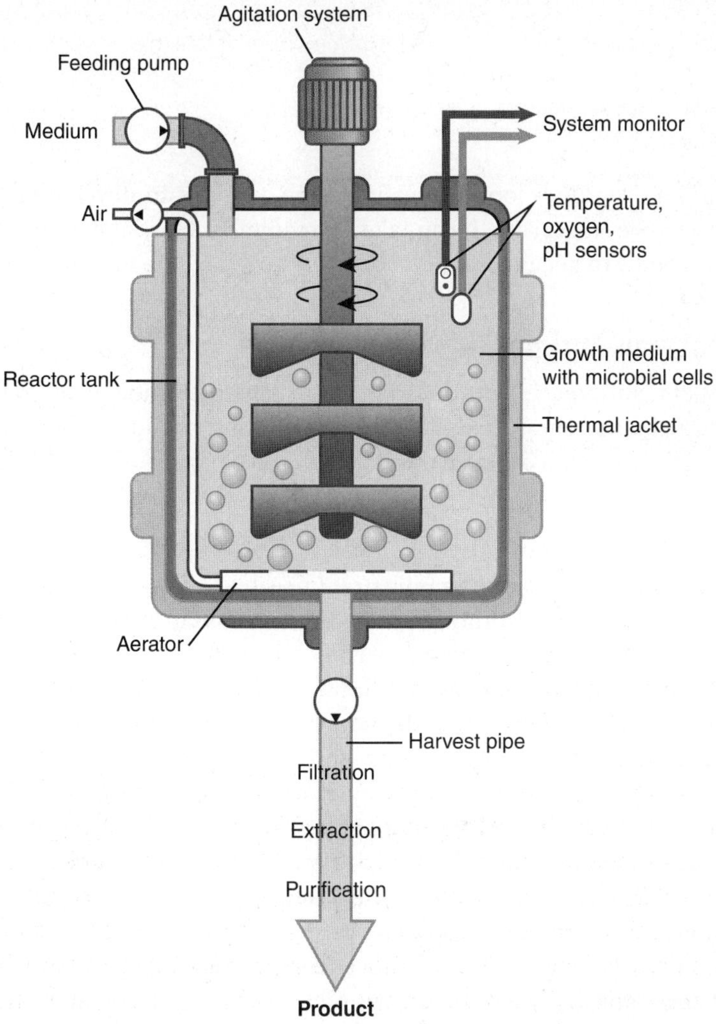

Modified from: https://en.wikipedia.org/wiki/Bioreactor.

TABLE 14.2 Some Commercially Produced Antibiotics (*)

Antibiotic	Source (microbe)
Bacitracin	*Bacillus licheniformis* (bacterium)
Carbapenem	*Streptomyces cattleya* (bacterium)
Cephalosporin	*Cephalosporium acremonium* (fungus)
Chloramphenicol	*Streptomyces venezuelae* (bacterium)
Erythromycin	*Streptomyces erythreus* (bacterium)
Gentamicin	*Micromonospora purpurea* (bacterium)
Penicillin	*Penicillium chrysogenum* (fungus)
Polymyxin B	*Bacillus polymyxa* (bacterium)
Streptomycin	*Streptomyces griseus* (bacterium)
Tetracycline	*Streptomyces rimosus* (bacterium)
Vancomycin	*Amycolatopsis orientalis* (bacterium)

* See Appendix A for organism pronunciations.

Even if a drug does look promising, time and cost in developing the new antibiotic can be staggering. It can take more than 10 years and up to $1 billion (not a misprint!) to get a truly useful drug approved by the FDA and to market.

Vitamins and Amino Acids

Vitamins and amino acids produced by industrial microbiology are referred to as **nutraceuticals**. They are used in the food industry and as nutritional supplements in the human diet and animal feeds.

Vitamins

The industrial production of vitamins ranks second only to antibiotics in total yearly commercial sales. Although most vitamins today are made by chemical processes, a few are easier to produce using microbes. Two of these are riboflavin (vitamin B_2) and cobalamin (vitamin B_{12}).

A deficiency of vitamin B_2 in the human diet can lead to cracking and reddening of the lips, inflammation of the mouth, sore throat, and an iron deficiency. The vitamin is found naturally in many foods, such as milk, cheese, and leafy vegetables. The filamentous mold *Ashbya gossypii* also produces vitamin B_2 as a by-product of metabolism. Through industrial fermentation, around 1 million kilograms (1,000 tons) of the vitamin is produced each year. It is added to (or used to fortify) white flour, breads, and breakfast cereals (**FIGURE 14.8**).

Ashbya gossypii: ASH-be-ah gos-SIP-ee-ee

Vitamin B_{12} plays a key role in brain and nervous system function. Like riboflavin, it cannot be made by the human body and must be taken in from foods or supplements (see Figure 14.8). If an individual has a B_{12} deficiency, it can cause fatigue, depression, memory loss, and loss of taste and smell. The deficiency

FIGURE 14.8 Vitamin Additives. Some vitamins, such as riboflavin (vitamin B_2) and vitamin B_{12} that are added to many foods we eat, are produced through industrial fermentation using microbes.

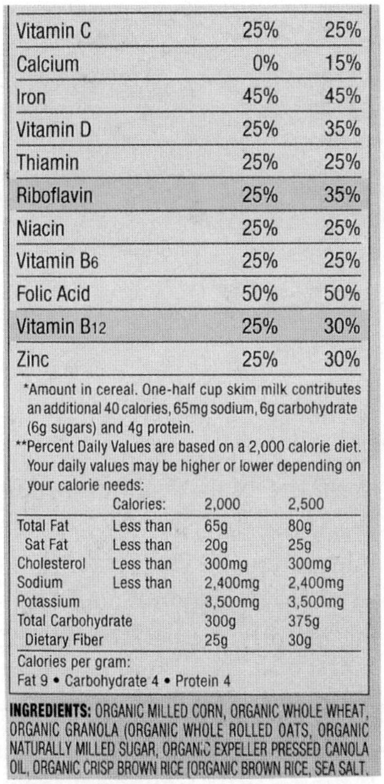

Vitamin C	25%	25%
Calcium	0%	15%
Iron	45%	45%
Vitamin D	25%	35%
Thiamin	25%	25%
Riboflavin	25%	35%
Niacin	25%	25%
Vitamin B6	25%	25%
Folic Acid	50%	50%
Vitamin B12	25%	30%
Zinc	25%	30%

*Amount in cereal. One-half cup skim milk contributes an additional 40 calories, 65mg sodium, 6g carbohydrate (6g sugars) and 4g protein.
**Percent Daily Values are based on a 2,000 calorie diet. Your daily values may be higher or lower depending on your calorie needs:

	Calories:	2,000	2,500
Total Fat	Less than	65g	80g
Sat Fat	Less than	20g	25g
Cholesterol	Less than	300mg	300mg
Sodium	Less than	2,400mg	2,400mg
Potassium		3,500mg	3,500mg
Total Carbohydrate		300g	375g
Dietary Fiber		25g	30g

Calories per gram:
Fat 9 • Carbohydrate 4 • Protein 4

INGREDIENTS: ORGANIC MILLED CORN, ORGANIC WHOLE WHEAT, ORGANIC GRANOLA (ORGANIC WHOLE ROLLED OATS, ORGANIC NATURALLY MILLED SUGAR, ORGANIC EXPELLER PRESSED CANOLA OIL, ORGANIC CRISP BROWN RICE [ORGANIC BROWN RICE, SEA SALT.

also leads to an inability to produce enough red blood cells, resulting in anemia. Therefore, many foods, such as breads and cereals, are fortified with B_{12}. The vitamin is produced by high-yield species of *Propionibacterium*, *Pseudomonas*, or *Streptomyces*, and the industrial fermentation of vitamin B_{12} is on the order of 10 million kilograms (10,000 tons) per year.

Propionibacterium: pro-pea-OHN-ee-bak-tier-ee-um

Streptomyces: strep-toe-MY-seas

Amino Acids

Amino acids are essential to the building of proteins in all organisms. Commercially, amino acids are important products of industrial microbiology. The amino acids are used extensively as additives in the food and animal husbandry industries and as nutritional supplements in the nutraceutical industry.

The most important commercial amino acid is glutamic acid, which is a product of an industrial fermentation process using *Corynebacterium glutamicum*. Annually, hundreds of millions of kilograms of glutamic acid are produced industrially using this bacterial species, and sales of the amino acid are in the billions of dollars. The most familiar use of glutamic acid is as a flavoring agent in many cuisines in the form of monosodium glutamate (MSG) (**FIGURE 14.9A**).

Corynebacterium glutamicum: KOH-ree-nee-back-tier-ee-um glu-TAH-meh-cum

Another widely used amino acid that is produced by microbes is lysine. Lysine, also a product of *C. glutamicum*, is used as a food additive, particularly in bread. Because this amino acid is not synthesized by the human body (it is an essential amino acid), it must be obtained in the diet. This nutritional requirement creates a strong consumer demand for lysine, making the amino acid a key export product of the United States.

Aspartic acid and phenylalanine are two other amino acids produced as microbial byproducts. In the beverage industry, they are used to produce the artificial sweetener **aspartame**. This nonnutritive sweetener is found in many diet soft drinks, chewing gum, and hundreds of other foods sold as low-calorie or sugar-free products (**FIGURE 14.9B**).

FIGURE 14.9 Chemical Additives.
(A) The flavoring agent monosodium glutamate is a product of microbial industrial fermentation.

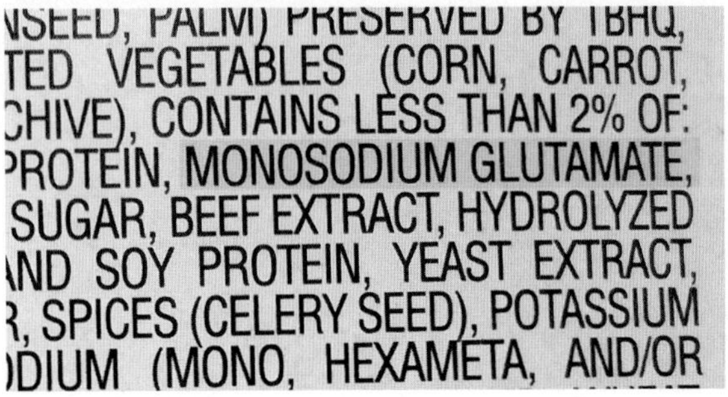

Courtesy of Dr. Jeffrey Pommerville.

(B) The artificial sweetener aspartame is another product of microbial industrial fermentation.

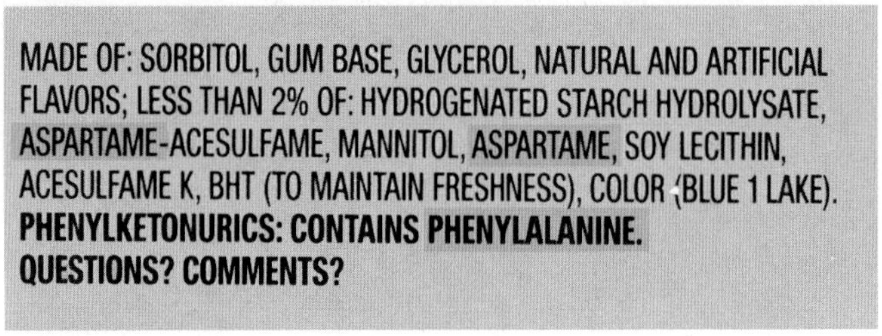

Courtesy of Dr. Jeffrey Pommerville.

Industrial Enzymes

Enzymes are well known as the biological compounds that make metabolic reactions possible in cells. Over the past few decades, microbiologists have identified and extracted many enzymes from microbes that have potential commercial value. As a result, microbial enzymes are found in many of the products used every day. Marketing experts predict that by 2020, the market value of industrially produced microbial enzymes will be over $6.2 billion.

Trichoderma: trick-oh-DER-mah

Bacillus subtilis: bah-SIL-lus SUH-til-iss

Aspergillus niger: a-sper-JIL-lus NYE-jer

Microorganisms produce many different enzymes, mostly in small amounts for the cell's own internal use. However, some microbial enzymes are excreted into the environment. These so-called **exoenzymes** are produced in much larger amounts. They digest large organic molecules (e.g., cellulose, proteins, and starches) into smaller ones that can be transported into the cell. Commercially, these enzymes have great value in the food, laundry, and health industries and in textile manufacturing. For example, an enzyme from the soil-dwelling fungus *Trichoderma* breaks down the cellulose in blue denim jeans, giving the fabric a soft, whitened, "stone-washed" appearance.

One of the most useful microbial enzymes is **amylase**. Several species of microbes, including *Bacillus subtilis* and *Aspergillus niger*, produce large amounts of amylase during their normal cycles of growth. Through industrial microbiology, these enzymes have been brought to market. Worldwide production of amylase is over 10 billion kilograms (11 million tons) per year.

Bakers add amylase to dough to promote the breakdown of starch to sugar, after which the sugar is used by yeasts. Amylase also is important to the digestion of starch for beer production. In household products, the enzyme is used in many spot removers to breakdown the starches in plant material that soils clothing. Amylases also are important in the non-alcoholic beverage industry. When amylase breaks down starch to glucose, the glucose can be converted by another enzyme to fructose, which is a much sweeter sugar than glucose. If the starting starch comes from corn, the final product is high-fructose corn syrup, which has been used as a sweetener in numerous juices and soft drinks (**FIGURE 14.10**).

FIGURE 14.10 Sugared Beverages. Many juices and soft drinks have added high fructose corn syrup, another microbial product of industrial microbiology.

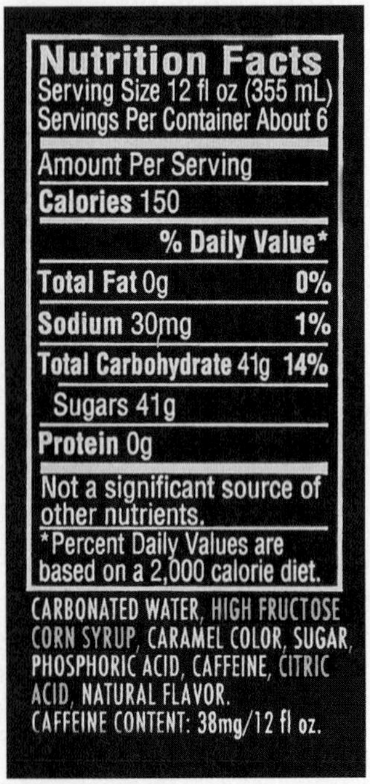

Courtesy of Dr. Jeffrey Pommerville.

Microorganisms are sources of **proteases**. These are enzymes that digest proteins into smaller fragments (peptides) and amino acids. Microbiologists can grow large quantities of protease-producing microbes in bioreactors. Most of the microbial proteases used in the detergent industry are alkaline proteases from the genus *Bacillus*. In laundry products, these added proteases are effective as spot removers for anything that contains protein, such as egg, blood, or milk.

In the baking industry, many bacterial and fungal proteases are used. The proteases encourage the breakdown of gluten proteins in flour, thereby increasing the nutritional value of the bread. They also are in the tenderizers sprinkled on meat before cooking. Here the enzymes break down the protein fibers and release the juices. In addition, some proteases are used in pizza dough to allow the dough to stretch and return to its original shape.

Rennin is an enzyme used to curdle the proteins in milk, an early step in the manufacture of cheese, which was described in the chapter on Microbes and Food. Historically obtained from the stomach lining of a calf, rennin now is produced by molds grown in bioreactors.

Many industrial processes occur most efficiently at higher temperatures. Until recently, those industry processes were limited in scope because most enzymes are destroyed at high temperatures. Then came the discovery of archaeal organisms and specifically the **hyperthermophiles**. These organisms prefer to grow at higher temperatures, typically around 80°C to 90°C (176°F to 194°F). To survive at these high temperatures, the archaeal cells contain thermostable enzymes, called **extremozymes**. Other extremozymes from other prokaryotes are resistant to acids, bases, and high salt. A common industrial use of these enzymes is in laundry products, which often are carried out at warm to hot temperatures. In research, the PCR technique mentioned earlier for amplifying specific DNA sequences depends on the use of Taq polymerase. This is an extremozyme that is tolerant to the higher temperature (72°C/162°F) used in the PCR process.

On the horizon, plastic-degrading enzymes are something that might help society deal with the huge amounts of plastic being produced—and spread—around the world, as **A CLOSER LOOK 14.2** describes.

Ocean gyre: A large system of circular ocean currents formed by global wind patterns and forces created by Earth's rotation.

A CLOSER LOOK 14.2

Plastic-Eating Microbes: Our Saviors?

Imagine this. Instead of throwing out your plastic water bottles, you spray them with a solution that will degrade the plastic before putting them in a "plastic degrading" bin. The time might soon come when this will be possible.

In 2018, scientists announced that undegradable plastics are accumulating in our oceans and on our beaches at an astounding rate. Billions of pounds of plastic have been found in swirling **ocean gyres** that make up about 40% of the world's ocean surfaces. The gyre in the North Pacific Ocean has been nicknamed the Great Pacific Garbage Patch (see **FIGURE A**). Such plastics pollution has a direct and deadly effect on wildlife, as thousands of sea turtles, seals, and other marine mammals are killed each year from ingesting plastic or getting entangled in it. At current rates, plastic probably will outweigh all the fish in the oceans by 2050. Plastic pollution has become a global crisis.

So, how can the massive amount of plastic already in the environment be reduced?

For decades, microbes have been driving the biotechnology industry by producing unique proteins and other products to help solve an array of problems in medicine, agriculture, and industry. Can microbes help solve the plastic problem? Possibly.

In 2016, Japanese researchers reported the discovery of a bacterium that has a big appetite for the common plastic polyethylene terephthalate (PET). PET is the most common polymer making plastics, used in everything from water bottles to clothing (polyester).

FIGURE A The Great Pacific Garbage Patch is located halfway between Hawaii and California. Scientists estimate there are more than 1.8 trillion pieces of plastic weighing more than 80,000 tons in the patch.

©Pro_Vector/Shutterstock.

The bacterium produces two enzymes capable of breaking down PET into digestible end products. More recently, other scientists have modified the enzyme to be even more efficient at breaking down PET and polyethylene furanoate (PEF), an alternative form of PET. However, it remains to be seen whether microbial biotechnology can ramp up enzyme activity to consume plastics completely and quicker than the rate at which plastics are currently being produced.

Lastly, for breaking down compounds in aromatic ("strong-smelling") vegetables, galactosidase from the fungus *A. niger* is used. This enzyme is the major ingredient in Beano, a commercial product used to relieve bloating and gas.

Organic Acids

Organic acid: A small, carbon-containing compound with acidic properties.

Citric acid is one of the most widely encountered **organic acids** found in consumable items such as soft drinks (see Figure 14.10). Hundreds of thousands of tons of citric acid are produced annually in the United States. It is used in such diverse products as soft drinks, candies, frozen fruits, and wines. Citric acid also is used to tan leather, to electroplate metals, and to activate slow-flowing oil wells. Most citric acid is produced by the normal metabolism of the fungus *A. niger*. The citric acid is naturally excreted making the fungus unwittingly a partner in an industrial process.

Lactobacillus: lack-toe-bah-SIL-lus

Still another microbial product is lactic acid. Several *Lactobacillus* species produce this acid from the whey portion of milk derived from cheese production. Lactic acid is used as a flavoring and preservative agent in many foods. It also is employed to finish fabrics, prepare hides for leather, and for bioplastics. It is approaching a $2 billion a year industry.

Biofuels

Gasoline and diesel are examples of fossil fuels produced from the breakdown of prehistoric biological matter by geological processes. Today, with fluctuating oil prices and a limited time before oil reserves run dry, alternate fuels are being explored. These contemporary fuels are called **biofuels**. They are composed of plant carbohydrates that contain potential energy in forms that can be released in ways similar to fossil fuels. The production of many of these biofuels depends on the actions of microbes.

The most common biofuel available today is ethyl alcohol (ethanol). As a major industrial process, over 60 billion liters (15 billion gallons) of ethanol are produced yearly worldwide. Most of this industrial fermentation depends on microbes. For example, in the United States, most ethanol is produced through yeast (*Saccharomyces*) fermentation of glucose obtained from cornstarch. For decades, Brazil has been using microbes to break down sugarcane into ethanol. Some cars there can run on pure ethanol rather than a blend of ethanol mixed with fossil fuels. Other major microbial-generated biofuels being tested and used include biodiesel made from vegetable oils and algal fuels, including alcohols and oils produced from eukaryotic green algae.

Saccharomyces: sack-ah-roe-MY-seas

Much of the gasoline in the United States is blended with ethanol to produce **gasohol**. Combustion of gasohol is a cleaner-burning fuel that produces lower amounts of carbon monoxide and nitrogen oxides than pure gasoline. Ethanol-rich fuels, such as E-85 (85% ethanol and 15% gasoline), reduce emissions of nitrogen oxides by nearly 90% (**FIGURE 14.11**). Therefore, these fuels represent a way to reduce important pollutants in the atmosphere, and to lower society's dependence on conventional sources of oil. Many major cities concerned about air pollution are retrofitting their public transportation systems, especially buses, to burn E-85.

Botryococcus braunii: bow-tree-oh-KOK-kus BRAWN-ee-ee

Our energy concerns today also have spurred research on "algaculture." For example, the green alga *Botryococcus braunii* excretes hydrocarbons having the consistency of crude oil (**FIGURE 14.12**). However, the major problem with algal petroleum is scale. The logistics of growing significant amounts of oil-producing algae to meet global oil demand are formidable.

FIGURE 14.11 Gasohol. Ethanol-rich fuels (gasohols) such as E-85 (85% ethanol and 15% gasoline) are cleaner-burning fuels than pure gasoline.

FIGURE 14.12 Algae as Source of Biofuels. This alga, *Botryococcus braunii*, produces petroleum, as seen by the secreted oil droplets along the edge of the colony. (Bar = 8 μm.)

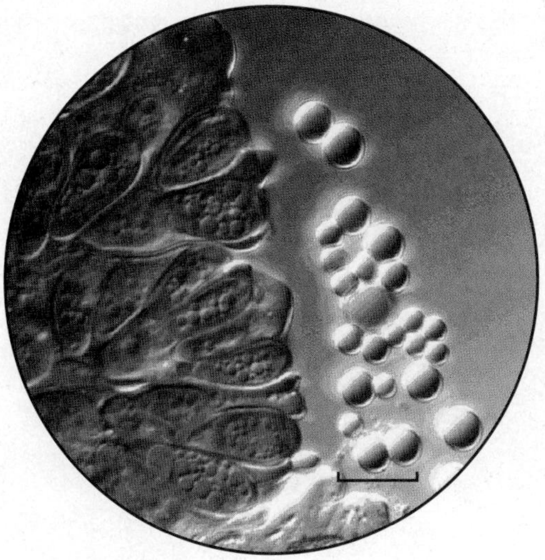

For the future, many believe a better way of making biofuels will be from grasses, such as switchgrass, which contains more cellulose (**FIGURE 14.13**). Cellulose is the tough material that makes up plants' cell walls, and most of the weight of a switchgrass plant is cellulose. If microbes can turn cellulose into biofuel, it could be more efficient than current biofuels and emit less carbon dioxide.

Engineering New Skills

The expanding tools available in microbial biotechnology and industrial microbiology are manipulating life processes for commercial gain. A blossoming area in microbial biotechnology today is **pathway engineering**. This new field attempts to alter and/or improve the metabolic capabilities inherent in microorganisms. In the example here, it is something as mundane as the dye for blue jeans.

FIGURE 14.13 Switchgrass. Switchgrass contains a high percentage of cellulose, a potential source for bioethanol production by microbes.

FIGURE 14.14 Indigo Dyes. Traditionally, indigo dyes, such as these in Thailand, were produced as the result of a fermentation process.

Blue jeans or denim is a $53 billion a year business, with an estimated 450 million pairs of jeans made every year just in the United States. Blue jeans are made from blue denim, and the blue in the denim comes from indigo dye that traditionally came from the fermentation of indigo plants (**FIGURE 14.14**). Industrially, today most indigo dye is made from coal or oil. However, the process produces potentially toxic by-products. To be more ecologically friendly, scientists have considered genetically altering *E. coli* to churn out the indigo pigment.

Biotech indigo can be made through pathway engineering by reconstructing bacterial metabolism (**FIGURE 14.15**). *E. coli* produces the amino acid

FIGURE 14.15 Pathway Engineering in Microbes.
Microbes, such as *Escherichia coli*, can have a metabolic pathway altered (engineered) to produce a nonmicrobial product, such as indigo. Inset: Engineered bacterial colonies producing indigo.

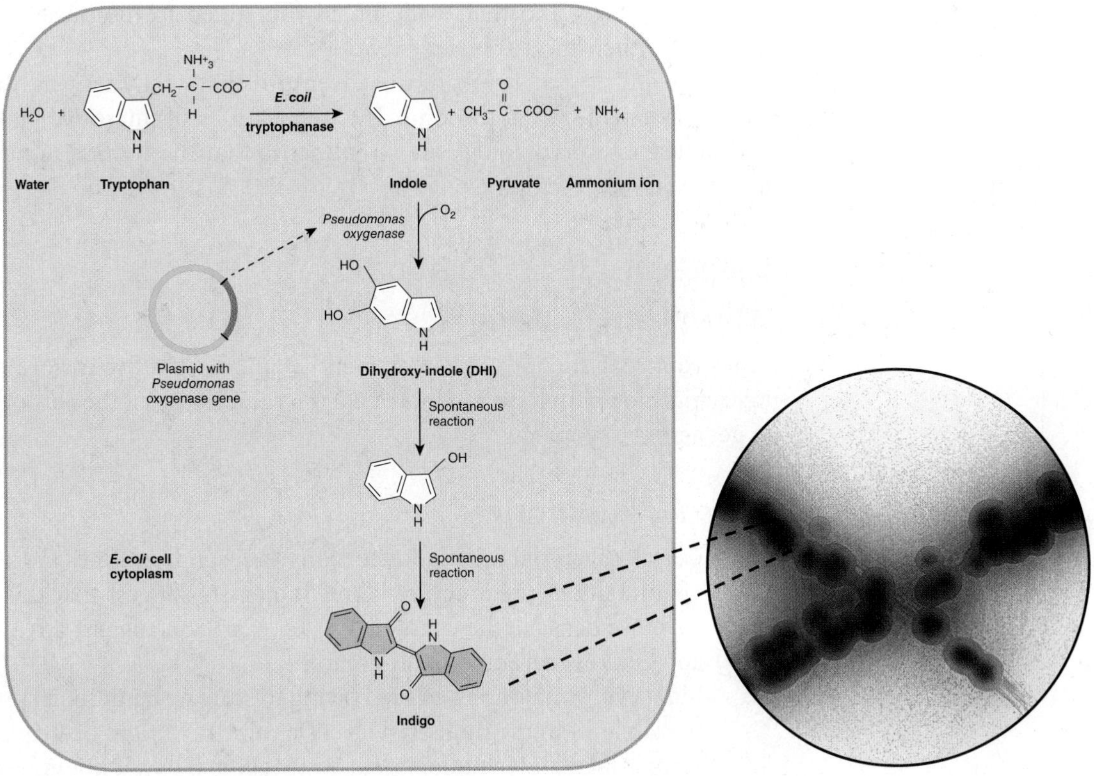

tryptophan, which the organism degrades with an enzyme it has naturally. The end products of the enzyme reaction include indole and pyruvate. Scientists also know that indole can be mixed with another enzyme that produces dihydroxy-indole (DHI), which spontaneously yields indigo in the presence of oxygen. So, the pathway engineering step was to take the gene from *Pseudomonas* that codes for DHI and engineer it into *E. coli* cells. When the engineered *E. coli* cells are cultured on agar, they soon turned blue due to the production of indigo (see Figure 14.15).

Today, approximately 16 million kilograms (18,000 tons) of indigo dye are made annually, and most of it is used to stain denim. The indigo dye made through microbial biotechnology is indistinguishable from the deep blue of the chemically made dye. So, if the cost of pathway engineering can be made more competitive with the chemical process, "biotech indigo" might be in your future pair of genes—jeans, that is!

A Final Thought

For many decades, humans stood by and watched the "game of life." They marveled at the wonders of nature; they searched out and cataloged the plants, animals, and microbes of the world; and they spent exhaustive hours trying to understand how these organisms fit into the scheme of things.

Then came the age of DNA science and biotechnology. Now many of the observers became manipulators as they learned how to change the character of an organism at its most fundamental level. They isolated its DNA, changed its metabolism, and inserted the new DNA into recipient cells to see what would happen. Bacterial and yeast cells began producing human hormones, therapeutic proteins, and enzymes. Also new vaccines and diagnostic tests were developed that were undreamed of a generation before. Microbes were at the center of biotechnology, and their widespread use added to the positive press they enjoyed for their industrial contributions.

We stand at the brink of an adventure that will carry us through the twenty-first century and beyond. The implications of microbial biotechnology and industrial microbiology are so immense that the human mind has yet to imagine all the possibilities.

Chapter Discussion Questions

What Was He Thinking?

From your reading of this chapter, identify and discuss five major points about microbial biotechnology and industrial microbiology that the author was trying to get across to you.

Questions to Consider

1. Certain bacterial cells produce many thousands of times more of specific vitamins than they require. Some biologists suggest that this makes little sense because the excess is wasted. Can you suggest a reason for this apparent overproduction in nature?
2. Identify the major obstacles in trying to scale up a microbial process from the lab level to the industrial level (Figure 14.7 might be helpful).
3. A microbial biotechnologist suggests that one day, it might be possible to engineer certain harmless bacterial cells to produce antibiotics and then to feed the cells to people who are ill with infectious disease. The bacterial cells would then serve as antibiotic producers within the body. Would you favor research of this type?
4. While studying for the exam that covers the material in this chapter, a friend asks you: "What is industrial microbiology?" How would you answer her question?
5. A bacterial species has been engineered so it is resistant to high levels of radiation. Can you think of any potential industrial uses for this species?
6. From the examples in this chapter, how many times in the last week have you had the opportunity to use or consume the industrial product of a microbe?
7. Although the products of microbial biotechnology have been of great benefit in numerous ways, they also have been abused. One example is the use of erythropoietin by athletes to increase their red blood cell counts and to give them an unfair advantage at competitive events. What other abuses of products of biotechnology can you think of?

© Rustic/Shutterstock.

CHAPTER 15

Microbes and Agriculture: No Microbes, No Hamburgers

The Man from Delft

He stood in front of the assembled farmers and made an outrageous proposal: "Don't plant your crops in the same field as last year," he said. "Leave the field alone and let the clover grow for the next year."

The year was 1887; the man was Martinus Willem Beijerinck; the country was The Netherlands. Because agricultural land was scarce in his small country, his proposal was outrageous.

Beijerinck: BY-yer-ink

"But Martinus. We cannot afford to let the field remain empty for a whole planting season. It will bankrupt us!" several farmers shouted.

Beijerinck was a local bacteriologist from Delft, Holland. While his colleagues in France and Germany were investigating the germ theory of disease and its implications Beijerinck was out in the agricultural fields. He noticed the land was very productive when it was freshly cleared of brush and trees and newly planted, but less productive after several years of use. Moreover, the fields yielded bountiful crops when the farmer let the field lie unplanted for a couple of seasons. And now he thought he had the answer to these mysteries.

"My dear friends. I understand your fear. But you need to let the fields regain needed nutrients removed by your crops," Beijerinck explained. "Plant elsewhere in the meantime and then when you do replant, your crop yield will be much greater."

Beijerinck was an agricultural expert, to be sure, but he also had a solid background in chemistry. He believed atmospheric nitrogen gas (N_2) was essential

CHAPTER 15 OPENER An idyllic farm setting. Cattle in a field of grass and clover. Farm silos in the background. Can you identify anywhere in this agricultural setting where microbes might be found? As we will learn in this chapter, the clover is a natural source of needed nitrogen that is converted into a useful form by microbes. The digestive system of cattle is inhabited by hundreds of species of microbes. The farm silos might be fermentation vats in which microbes convert cut crops or pasture grasses into cattle feed. Some farm crops might be protected from insect damage by microbial toxins. We are "down on the microbial farm!"

FIGURE 15.1 ***Rhizobium* Root Nodules.** Root nodules on clover roots.

for plant growth, and he thought bacteria in the soil were the link between atmospheric N_2 and the nitrogen needed for plant growth. He observed little lumps and bumps, called root nodules, on the roots of wild plants like clover that grew in untended fields (**FIGURE 15.1**). When he looked at these nodules with the microscope, he saw great hordes of bacterial cells. These nodules were absent from the roots of most agricultural crops.

So, his advice on that day in 1887 was direct and straightforward. By leaving the field alone for a spell, wild plants like clover would grow in the field and the bacterial populations in their root nodules would enrich the nitrogen content of the soil.

Here is one example of where microbes are important, if not indispensable, on the farm. In the root nodules of the clover, microbes are busy at work bringing usable nitrogen to plants and to animals when they eat the plants. As we will see in this chapter, these and other chemical transformations carried out naturally, or through biotechnology, would not occur without microbes. Microbes are doing what they do best, and they are indispensable for us.

LOOKING AHEAD

After reading and completing this chapter, you will be able to:

15.1 Describe the importance of nitrogen and discuss the role of microbes to ruminant digestion.

15.2 Describe how DNA is inserted into plants and assess the importance of bacterial insecticides.

15.1 Microbes on the Farm: If the Environment Is Suitable, They Will Perform

In soil and water, elements such as carbon, oxygen, and hydrogen are cycled and recycled by a broad variety of plants, animals, and, of course, microbes (as discussed in the chapter on Microbes and the Environment). However, the recycling of nitrogen is largely the responsibility of microbes alone.

Connecting the Nitrogen Dots

As early as Beijerinck's time, microbiologists knew the productivity of many agricultural crops was dependent on the conversion in the soil of nitrogen gas (N_2) to ammonia (NH_3), nitrate (NO_3^-), and organic nitrogen compounds. But why is nitrogen metabolism so important?

Nitrogen is an essential element of many organic molecules. This includes all the amino acids (and, consequently, all proteins) as well as all the nucleotides of nucleic acids (DNA and RNA). In all, between 9% and 15% of the organic matter (biomass) of a typical cell is composed of the element nitrogen. Ironically, even though 80% of the gas in Earth's atmosphere is N_2, neither plants nor animals can use the gas directly to synthesize the organic compounds they must have to survive. Instead, plants and animals depend on a limited number of bacterial species to bring nitrogen in a usable form into the cycle of life. This is one part of the so-called **nitrogen cycle**. Without these microbes, the entire nitrogen cycle would grind to a halt—and life on this planet would eventually disappear.

The nitrogen cycle can be broken into three parts: nitrogen fixation, nitrification, and denitification.

Nitrogen Fixation

The trapping of N_2 from the atmosphere is called **nitrogen fixation** (**FIGURE 15.2**). Two types of **nitrogen-fixing bacteria** have the enzymes needed to convert N_2 into NH_3.

- **Free-living Nitrogen Fixers.** The free-living nitrogen-fixers include species of *Azotobacter, Beijerinckia* (named for Beijerinck), and several genera of cyanobacteria, such as *Nostoc* and *Anabaena.* In the Arctic region, cyanobacteria are the most important nitrogen-fixers in the ecosystem.
- **Symbiotic Nitrogen Fixers.** The **symbiotic** nitrogen-fixers are species of *Rhizobium* that live in root nodules of **legume** plants such as soybeans, alfalfa, peas, beans, and clover (**FIGURE 15.3**). When the rhizobia enter the root hairs, they transform the nearby plant cells into a tumor-like nodule. Within the nodules, the bacterial cells assume distorted forms known as **bacteroids**. Bacteroids cannot live independently once they have entered the symbiotic relationship because they become dependent on sugars from the plant. In return, the bacteroids use their nitrogenase to fix N_2 for the plant's benefit.

So much organic nitrogen is captured that the net amount of nitrogen (and the net worth of the soil) increases considerably after a crop of legumes is harvested or when a crop such as clover or alfalfa is plowed under.

Azotobacter: a-ZOE-toe-bak-ter

Beijerinckia: bi-yeh-RINK-ee-ah

Nostoc: NOS-tock

Anabaena: an-nah-BEE-nah

Symbiotic: Referring to a close living relationship between two different organisms.

Rhizobium: rye-ZOH-bee-um

Legume: A plant that bears its seeds in pods.

FIGURE 15.2 Simplified Nitrogen Cycle. Special groups of bacteria are essential to the functioning of the nitrogen cycle.

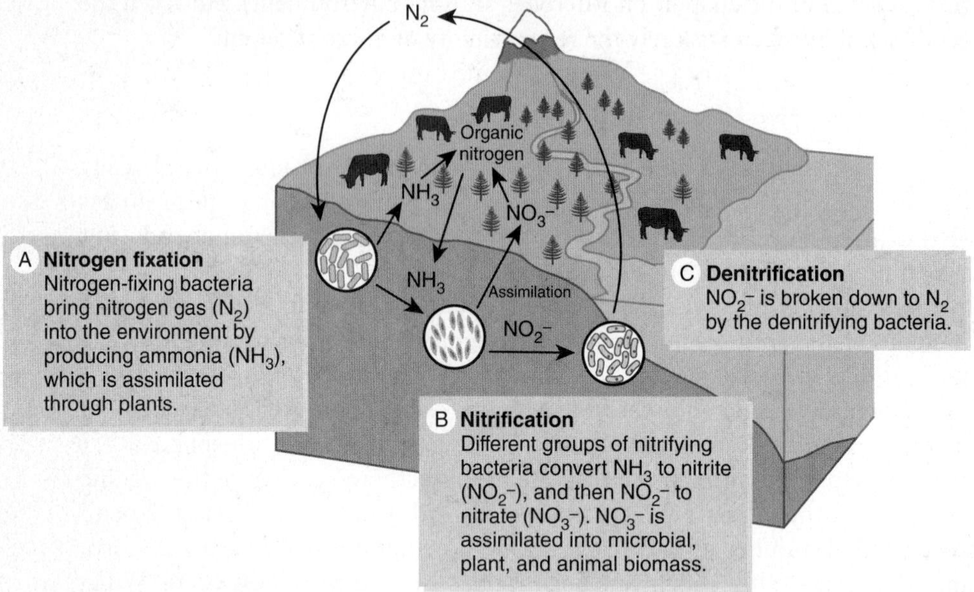

FIGURE 15.3 Cross-Section Through a Root Nodule. In this light microscope image, *Rhizobium* cells have infected a plant root hair and triggered the development of a root nodule, which, when mature, contains the bacteroids (red-stained bacilli). (Bar = 20 μm.)

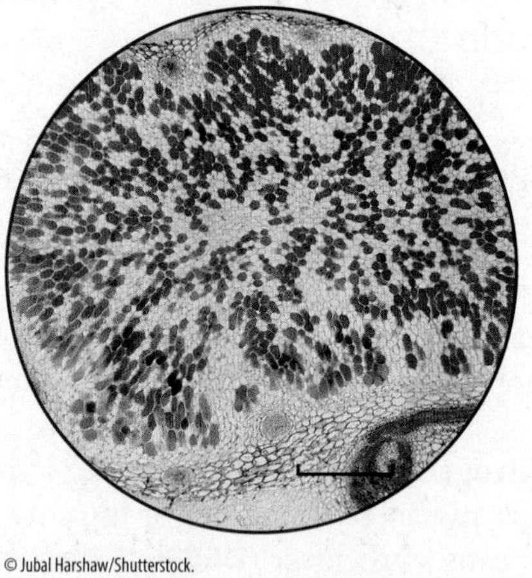

Nitrification and Denitrification

Nitrogen fixation is only one aspect of microbial involvement in the nitrogen cycle. Much of the ammonia deposited in the soil is converted to nitrate (NO_3) (see Figure 15.2). This **nitrification** process happens in two steps that require the so-called **nitrifying bacteria**. The NO_3 generated is used by plants (called **assimilation**) for making their organic nitrogen molecules, including proteins. Animals then get their organic nitrogen from eating the plants.

Not all the NO_3 gets assimilated. Some is converted to nitrite (NO_2^-). The **denitrifying bacteria** in the soil break down the NO_2 and produce N_2, which they release into the atmosphere (see Figure 15.2). This **denitrification** process removes nitrogen from the cycle of life. However, operating through nitrogen fixation, nitrogen-fixers again bring N_2 back into the cycle.

Engineering Nitrogen Fixation Genes

The importance of nitrogen fixation by symbiotic bacteria has not escaped the attention of microbial biotechnologists. Scientists foresee the day when they can use genetic engineering tools to enhance the nitrogen-fixing talents of *Rhizobium* to be even more efficient at nitrogen fixation. In the end, this would increase the efficiency of the bacteria-plant interaction.

Agricultural scientists also are looking for ways to engineer a *Rhizobium* species that would assume a symbiotic relationship with crop plants, such as wheat, rice, or corn. The association with nitrogen-fixers would greatly increase the yield from these plants. Such an advance could be a "societal jackpot," as it would go a long way toward solving world hunger problems.

Some microbial biotechnologists even look to the day when the genes for nitrogen fixation could be transferred directly from microbes to animals. Then, animals could extract their nitrogen directly from the atmosphere rather than depending on eating plants or other animals for their nitrogen. The prospect of an animal synthesizing its own amino acids from atmospheric nitrogen sounds like scientific fiction, but the idea fires the imagination of even the most innovative futurists.

Those Remarkable Ruminants

Ruminants are **herbivores**; that is, animals such as cattle, sheep, goats, and deer that survive on a diet of plants and grasses as the main component of their diet. The primary structural carbohydrate in these plants is **cellulose** that builds the plant cell walls. These wall polysaccharides are built from chains of energy-rich glucose molecules. Unfortunately, ruminants lack the enzyme cellulase needed to digest cellulose. In fact, no vertebrate animal, including humans, can produce cellulase or digest cellulose. In the cow, cellulase is provided by—you guessed it—microbes.

What happens is outlined in **FIGURE 15.4A**. The cow's large stomach is divided into several compartments, the first and biggest of which is called the **rumen**. The rumen has a constant temperature and is slightly acidic and anaerobic. This is the perfect environment for microbial fermentation.

Importantly, the rumen is loaded with numerous anaerobic microbes that produce cellulase. The microbial cellulase digests the cellulose into glucose units (**FIGURE 15.4B**). Then, the glucose molecules are fermented by microbes to simple organic acids such as propionic acid (propionate) and acetic acid (acetate). Some of the organic acids pass through the rumen wall and enter the cow's bloodstream, from where they are transported to its body cells. Here they are key energy sources fueling the cow's metabolism. Methane gas (CH_4) and carbon dioxide gas (CO_2) also are produced. These gases from the rumen are eliminated when the cow belches.

While they are performing their biochemical magic on cellulose, the microbes continue to grow and multiply at astoundingly high rates. The rumen is literally a fermentation tank where a single milliliter (20 drops) of rumen fluid might contain up to a trillion (10^{12}) microbes. As microbes continue to multiply furiously in the rumen, they soon overgrow the space and pass into the second part of the stomach, the reticulum (see Figure 15.4A). Here, the microbes are squeezed together with undigested plant material to form balls of cud material.

FIGURE 15.4 Microbes and the Cow Rumen.
(A) A schematic diagram of the rumen and other components of the digestive tract of the cow. Arrows show the pathway of food; dashed lines represent the pathway of regurgitated food when the cow chews its cud. Note the large size of the rumen relative to the other compartments of the stomach.

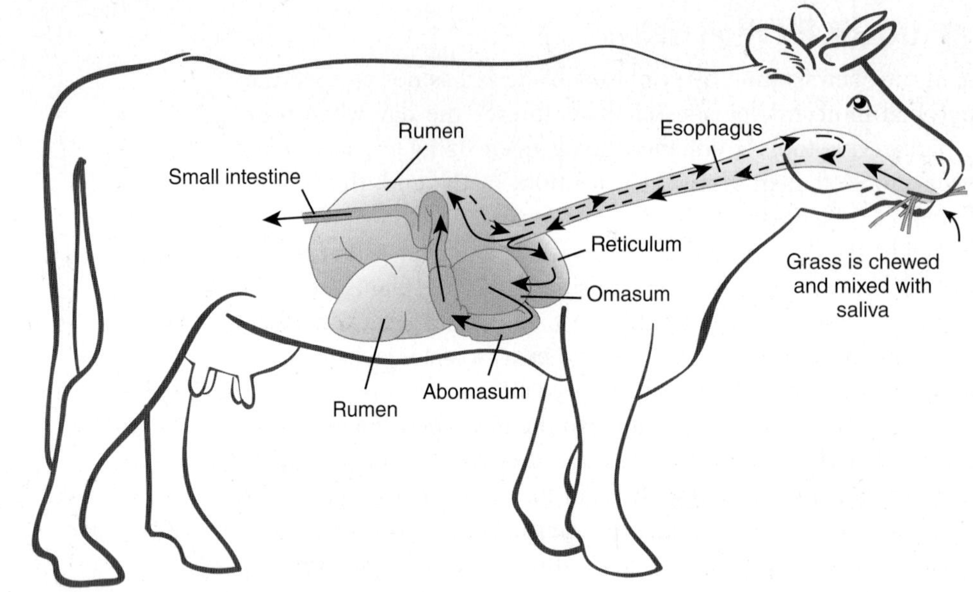

(B) Some of the microbial chemical reactions taking place in the rumen.

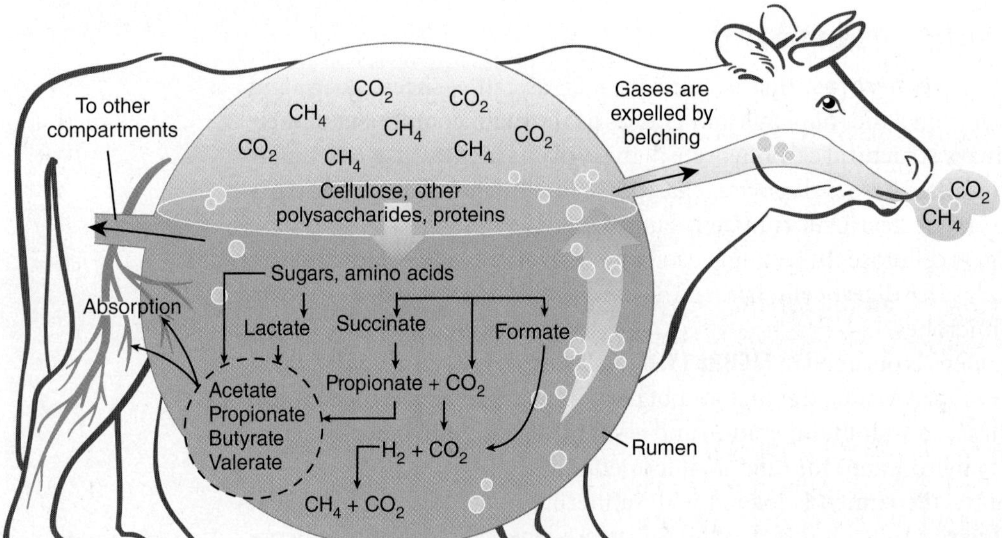

The cow regurgitates the cud, mixes it with copious amounts of saliva, and chews it to crush the fibers. Then, the cud is reswallowed and returned to the rumen, where it is redigested by more cellulose-digesting microbes. Next, it passes back into the reticulum for regurgitation and rechewing. Finally, the plant-microbe mass passes into the omasum and the abomasum (together, the "true" stomach of the cow).

In addition to the organic acids mentioned above, the microbes synthesize protein for the cow, using normally toxic urea as a nitrogen source. From this point on, the physiology of the cow is like that of other vertebrates (including humans). As nutrients are passed into the cow's bloodstream, they are used to build muscles and organs (e.g., potential steaks and hamburgers).

FIGURE 15.5 Silage. Silage is fermented plant materials fed to cattle for feed in winter or the dry season.

Non-ruminant grass-eaters, such as horses, rabbits, and guinea pigs, have a side pocket in their gut called the **cecum**. In this organ, a rich microbial population performs the same task as the microbes in the cow rumen. However, instead of regurgitating the cud, certain "cecum animals" such as rabbits instinctively eat their fecal pellets, which is called **coprophagy**. This gives the food a second pass through their intestine. It also provides the animal with the opportunity to obtain the vitamins produced by the bacterial cells in the cecum.

Before we leave our ruminant friends, there is one more example where microbes impact the lives of these animals. Cows, goats, and other barnyard animals are fond of **silage**, a product of microbial fermentation. In tall, cylindrical silos, the farmer loads grasses, grains, legumes, and other plant materials. Under tightly packed, anaerobic conditions, numerous bacterial species break down the plant material by carrying out fermentation. In about three weeks, the silage can be fed to cattle (**FIGURE 15.5**).

On the hazardous side, the buildup of microbial fermentation gases in the silo can be deadly, as **A CLOSER LOOK 15.1** explains.

At the Dairy Plant

Besides meat from cattle, the milk from cows is rich in various proteins and carbohydrates, primarily the disaccharide lactose ("milk sugar"). Moreover, milk has a variety of fats, vitamins, and minerals.

Unpasteurized milk rapidly undergoes a natural souring if allowed to stand at room temperature for some time. The souring is caused by bacterial fermentation of the lactose and the production of lactic acid. If excess lactic acid is present, the protein will curdle. Dairy farmers have put this biochemical process to work to produce a variety of fermented milk products (**FIGURE 15.6**). Here are a few common examples.

- **Buttermilk.** Buttermilk is produced by *Lactobacillus bulgaricus* and *Leuconostoc citrovorum*. At the dairy plant, these bacterial species are added to skim milk. The *L. bulgaricus* produces lactic acid from the lactose and sours the milk. *L. citrovorum* synthesizes polysaccharides to make the milk slightly thick.
- **Sour Cream.** Sour cream is produced in the same way as buttermilk, except cream is used instead of skim milk. The acid in sour cream retards microbial spoilage.

Lactobacillus bulgaricus: lack-toe-bah-SIL-lus bull-GAIR-ee-kus

Leuconostoc citrovorum: lou-koh-NOS-tock sit-roh-VOR-um

A CLOSER LOOK 15.1

Toxic Atmospheres

Your reading in this text hopefully has made you aware of the variety of products produced by microbial metabolism, and the speed with which it occurs. Here is one particularly striking example.

Silage is an important way for farmers to feed cows and sheep during times when the pasture isn't productive, such as winter or the dry season. For many farms, silos are used for storing silage (see **FIGURE A**). When a silo is filled in late summer and early fall, within 48 hours the plant materials start to undergo fermentation. The process is complete in about two weeks. Although silage itself poses no danger, microbial fermentation produces toxic "silo gases" that can be a hazard to agricultural workers.

Normally, the gases produced remain at low concentrations if the hatches and vents of a silo are open. Should silo gases build up, they include carbon monoxide (CO) and nitric oxide (NO). The NO can react with oxygen (O_2) in the air and form nitrogen dioxide (NO_2), which is toxic and can cause permanent lung damage if inhaled. Other harmful gases arising from fermentation are methane (CH_4), which is flammable or explosive, ammonia (NH_3), and hydrogen sulfide (H_2S).

Should an agricultural worker enter a silo with high concentrations of toxic silo gases, the worker might go into respiratory distress, collapse, and die. Unfortunately, it is possible to work in the presence of the toxic gases for some time without ever feeling major discomfort. In fact, victims of silo gases have been known to die many hours later, sometimes in their sleep, from **pulmonary edema**.

Therefore, the process of filling and maintaining a silo requires several safety precautions. As mentioned, the vents and hatches should be open to allow the gases to escape. If one must enter the silo, respirators should be worn, and ventilator fans must be operational. Also, farmers can lower the risk of exposure by waiting one month after filling before entering a filled silo.

It is a very sad situation if an individual is harmed or killed by silo gases, but it clearly demonstrates the speed with which microbes in mass numbers can produce end products of metabolism.

FIGURE A A grain silo.

Pulmonary edema: A buildup of fluid in the lungs.

Streptococcus lactis: strep-toe-KOK-us LAK-tiss

Lactobacillus acidophilus: lack-toe-bah-SIL-lus ah-sid-OFF-ill-us

- **Yogurt.** A very common fermented milk product is yogurt. Among the principal bacterial species used in yogurt production are *Streptococcus thermophilus*, *Lactobacillus bifidus*, *S. lactis*, and *L. acidophilus*. Milk is first concentrated by adding dried milk protein. The bacterial cells (the "active cultures") are added by mixing in a sample of previously prepared "mother" yogurt. At a high temperature (about 60°C/166°F), the streptococci are the first to ferment the lactose. Then, the lactobacilli take over the fermentation and produce more acid. This gives yogurt its characteristic texture and consistency. Although yogurt is really a "spoiled milk" product, it is

FIGURE 15.6 Products of Fermentation. Many dairy products found in the supermarket involve microbes and a fermentation process.

Courtesy of Dr. Jeffrey Pommerville.

considered a healthful addition and a **probiotic** in the human diet. The chapter on Microbes and Food addresses the pluses and minuses of dietary probiotics.

Probiotic: A product containing live microorganisms intended to improve or help restore the gut microbiome.

- **Butter.** Butter can be considered a microbial product, as microbes contribute to its production. One popular method for preparing butter begins by adding cultures of streptococci and *Leuconostoc* species to pasteurized sweet cream. The streptococci sour the cream slightly by producing lactic acid. The *Leuconostoc* species synthesize a substance called diacetyl, which gives butter its characteristic aroma and taste. Once the reactions have been completed, the slightly sour milk is churned to aggregate the fat globules into butter. Diacetyl also is an ingredient in butter-flavored popcorn and provides the buttery smell in margarine, candy, and baked goods. It occurs naturally in fermented drinks like beer, and it gives some chardonnay wines their buttery taste.
- **Other Products.** Many types of fermented dairy products are produced in different parts of the world. They vary according to the source of the milk, temperature of incubation, and species of microbes used. For example, an increasingly popular fermentation product is kefir. It is produced using lactobacilli, streptococci, and the yeast *Saccharomyces kefir*. The fermentation gases (CO_2) give the product a unique effervescent quality.

Saccharomyces kefir: sack-ah-roe-MY-seas KEY-fur

The chapter of Microbes and Food: A Menu of Microbial Delights discusses other fermentation products.

15.2 Biotechnology on the Farm: The Coming of the Transgenic Plant

The techniques of genetic engineering were discussed in detail in the chapter on Microbial Genetics. Those aspects of the field that affect human health were described in the chapter on Biotechnology and Industry. However, microbial biotechnology also has found ways to help agriculture. The technology has been used to dramatically increase crop yields and substantially decrease plant disease. It also has been employed for novel agricultural purposes.

DNA into Plant Cells

One of the breakthroughs in agricultural biotechnology was the ability to cultivate a whole plant starting from a single plant cell. That means such a cell can be genetically modified by adding the gene or genes of interest that will produce a more robust plant. Let's see how this is done to make soybean plants tolerant to herbicides.

Individual soybean plant cells are cultivated in a growth medium of carefully balanced plant nutrients and hormones. Initially, each cell will develop into a random mass of cells called a **callus**. Some days or weeks later, the beginnings of roots, stems, and leaves appear (**FIGURE 15.7A**). Shortly thereafter, the plantlet is transferred to a container until ready for planting outdoors.

Almost all the world's soybeans grown today have been genetically modified to be herbicide tolerant (**FIGURE 15.7B**). Such plants are called **transgenic** plants because they contain a foreign gene deliberately inserted into its genome. All such organisms having been genetically engineered in this way are referred to as **genetically modified organisms (GMOs)**.

FIGURE 15.7 Genetically Modified Soybeans.
(A) A soybean cell that has been genetically modified can be grown in a special plant culture, similar to what is shown here.

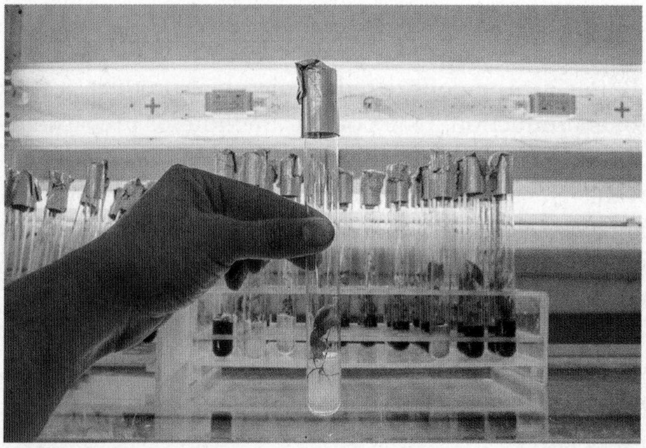

(B) Once the plantlet has developed into a true plant, it can be planted in the field.

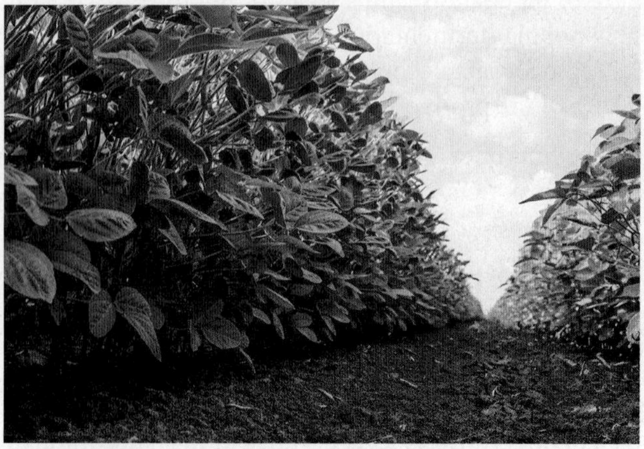

Inserting foreign genes, such as herbicide tolerance, into plant cells requires the proper transfer tools. A valuable tool for gene delivery is the **Ti plasmid** (short for tumor-inducing plasmid) obtained from the bacterium *Rhizobium radiobacter*. This microbe causes common plant tumors and a disease called crown gall (**FIGURE 15.8A**). Crown gall develops when the bacterium releases its Ti plasmid in the plant cell. The plasmid then inserts into a chromosome, and its genes disrupt plant development. A plant tumor is the result.

Rhizobium radiobacter: rye-ZOH-bee-um ray-de-oh-BACK-ter

For produce transgenic plants, Ti plasmids are isolated from bacterial cells. The tumor-inducing gene is removed, and the gene or genes (e.g., herbicide tolerance) of interest added (**FIGURE 15.8B**). Then, the plasmids are inserted into individual plant cells, where the plasmids transport the foreign genes into the plant cell chromosome.

Beside herbicide-tolerant soybean plants, plant biotechnologists have designed, developed, and carried out several successful improvements to plants using the Ti plasmid system. These enhancements include resistance to insecticides and improvement of product quality.

Bacterial and Viral Insecticides

Bacterial products and viruses are being used in unique ways to protect crop plants from insect pests.

Bacterial Insecticides

To be useful as an insecticide, a microbial toxin should act only on the targeted pest and should act rapidly. It should be stable in the environment and easily

FIGURE 15.8 *Rhizobium radiobacter* and the Ti Plasmid.
(A) *R. radiobacter* induces tumors in plants and causes a disease called crown gall. A lump of tumor tissue forms at the infection site, as the photograph shows.

Courtesy of Lian Bruno.

(B) The vehicle for gene transfer is a Ti plasmid.

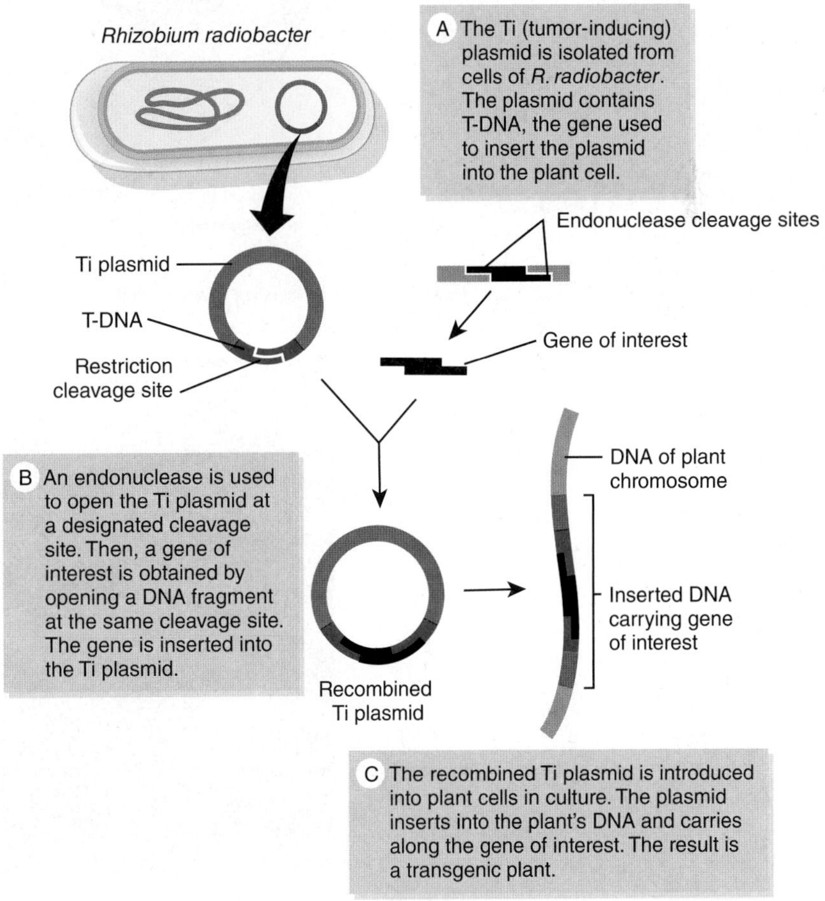
Rhizobium radiobacter
A The Ti (tumor-inducing) plasmid is isolated from cells of R. radiobacter. The plasmid contains T-DNA, the gene used to insert the plasmid into the plant cell.
Endonuclease cleavage sites
Ti plasmid
T-DNA
Restriction cleavage site
Gene of interest
DNA of plant chromosome
B An endonuclease is used to open the Ti plasmid at a designated cleavage site. Then, a gene of interest is obtained by opening a DNA fragment at the same cleavage site. The gene is inserted into the Ti plasmid.
Inserted DNA carrying gene of interest
Recombined Ti plasmid
C The recombined Ti plasmid is introduced into plant cells in culture. The plasmid inserts into the plant's DNA and carries along the gene of interest. The result is a transgenic plant.

dispensed, as well as inexpensive to produce. Finding a microbe that produces such a toxin to fit these criteria has been an ongoing challenge for microbial biotechnologists.

Bacillus thuringiensis: bah-SIL-lus thur-in-je-EN-sis

Caterpillar: The larval form of butterflies, moths, and related insects.

Bacillus thuringiensis (commonly called *Bt*) is a common spore-forming, soil bacterium (**FIGURE 15.9A**). During the period of spore formation, the bacterium produces crystalline proteins that are toxic to insects (**FIGURE 15.9B**). When these **Bt toxins** are deposited on leaves, they are ingested by insect **caterpillars**. In the caterpillar gut, the proteins are converted to their toxic form. When this happens, the toxins break down the cells of the gut wall, causing cell death and eventually death of the caterpillar.

FIGURE 15.9 *Bacillus thuringiensis* and Its Insecticide.
(A) A light microscope image of gram- stained *B. thuringiensis* cells. The cells produce a crystalline toxin that is poisonous to insect caterpillars. (Bar = 20 μm.)

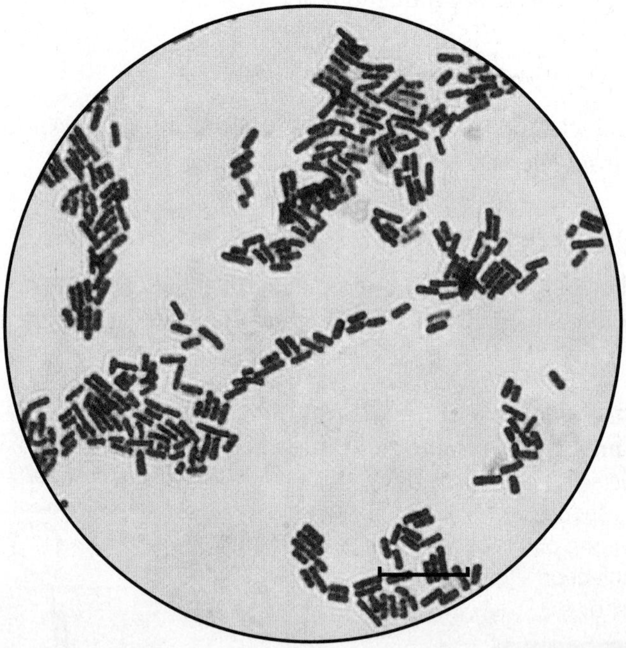

Courtesy of Dr. Jeffrey Pommerville.

(B) When applied as a powder to plants, Bt toxin can kill insect caterpillars that are eating the plant leaves.

© Floki/Shutterstock.

Bt is harmless to plants and mammals, including humans. As a commercial insecticide, the bacterial cells are harvested at the onset of spore formation and the toxins extracted. They then are dried to make a dusting powder applied to crops. The product is used against tomato hornworms, corn borers, and other agricultural insect pests.

More recently, microbial biotechnologists have inserted the gene for the Bt toxin directly into plant cells to give them protection against insect pests. To accomplish this, they have identified and cloned the gene and spliced it into the Ti plasmid (see Figure 15.8B). Then, they used the Ti plasmid to carry the toxin gene into plant cells. When an insect eats the plant, it consumes the toxin and is soon killed. This approach is advantageous because only the insect attacking the plant is subjected to the toxin.

Other GMOs are used widely in agriculture. Bt maize (corn) has revolutionized pest control, and many farmers have benefited by being able to use much less insecticide. Bt cotton also has been planted on millions of acres in the United States as well as many other countries. Caution has been warranted because the successful control of insects by Bt maize and Bt cotton has some scientists concerned that overuse of the transgenic crops could produce pests resistant to Bt toxins. In fact, there have been reports of resistance to Bt toxin emerging in some caterpillars.

Another useful species is *Bacillus sphaericus*. This bacterial species also produces a toxin that targets at least two species of mosquito larvae. To increase the efficiency of the toxin, researchers inserted the gene that encodes the toxin into the bacterium *Asticcacaulis excentricus*. Using *A. excentricus* as a gene carrier has several advantages: It is easy to grow in large quantity; it tolerates sunlight better than *B. sphaericus*; and it floats in water, where mosquitoes feed. Scientists have reported insecticidal activity by *A. excentricus* against mosquitoes that transmit such microbial diseases as malaria and viral encephalitis.

Bacillus sphaericus: bah-SIL-lus SFEH-rih-kus

Asticcacaulis excentricus: as-tik-ka-CAW-liss ek-SEN-trih-kus

The fear that insect pests might develop resistance to the bacterial toxins has forced scientists to look for alternative pesticides and insecticides. One potential is the bacterium *Photorhabdus luminescens*. Like the Bt toxin, *P. luminescens*' toxin, known as Pht, attacks the gut lining of insect larvae. However, the spectrum of activity of Pht is wider than that of Bt, and it can kill numerous species of caterpillars as well as cockroaches. Normally, *P. luminescens* lives in the intestines of soilborne worms called nematodes. The worms invade tissues of insects in the soil, and the toxins produced by *P. luminescens* kill the insects. The toxin-encoding Pht genes have been isolated, and efforts are under way to introduce them into plant cells.

Photorhabdus luminescens: fo-tow-RAB-dus lu-mih-NES-senz

Due to the mode of action of the bacterial toxins on insect guts, some individuals are concerned about potential health impacts on human ingestion of plants expressing these toxins. Over two decades of review, the United States Environmental Protection Agency (EPA) and numerous other scientific studies report that Bt toxin and transgenic Bt-crops are not harmful to humans. Other individuals disagree with the safety of Bt-crops. The pros and cons continue to be debated.

Viral Insecticides

Many viruses exist that specifically infect agricultural pests. The use of these viruses as potential insecticides is being developed. Commercially, microbial biotechnologists can harvest virus-infected insects. The infected insects are ground up into a powder. The powder containing the viruses then is used to dust the crop plants. Several agricultural pests, including the cotton bollworm, the cabbage looper, and the alfalfa caterpillar have been successfully controlled with these virus powders.

Researchers also have developed a viral insecticide using a toxin found in the venom of scorpions. The toxin, which paralyzes moth larvae, is encoded by a gene isolated from the cells of the scorpion. To make the viral insecticide, this gene is attached to a **baculovirus**, a type of DNA virus that targets insect tissues. After spraying the genetically modified baculoviruses onto the plant foliage, caterpillars consume the viruses while eating the foliage. In the caterpillar, the virus infects the cells and produces the toxin, which then kills the caterpillars. Baculovirus-based insecticides are currently being used worldwide.

Herbicide-Tolerant Crops

Herbicides are weed-killing chemicals used to clear the land of all plant growth before a field is sown. However, some weed seeds usually survive and remain in the soil among the crop seeds. Consequently, weeds and crops grow together, and the weeds often rob the crop plants of vital nutrients while crowding them out. It is generally agreed that it would be useful to create **herbicide-tolerant (HT)** crop plants. Then, herbicides could be sprayed on the field during the growing season to kill any weeds (**FIGURE 15.10**).

One commonly used herbicide is Roundup. The active ingredient, **glyphosate**, inhibits the activity of enzymes that synthesize essential amino acids in plant chloroplasts. By a coincidence of nature, *Escherichia coli* cells possess a gene encoding an enzyme to inhibit glyphosate activity. Biotechnologists have isolated this gene and inserted it into the Ti plasmid. Then, they used the plasmid to deliver the gene to the cells of tobacco and soybean plants. Here, the gene encodes the *E. coli* enzyme and makes the plants more tolerant to glyphosate than any surrounding weeds. Thus, when glyphosate is sprayed on the field, the weeds die, but the crop plants live.

Escherichia coli: esh-er-EE-key-ah KOH-lee

Today, several HT crops are being grown globally (**FIGURE 15.11**). These include soybeans, cotton, and corn. Some experts estimate that HT crops, and the derived genetically modified (GM) foods, could help alleviate malnutrition affecting more than 850 million people globally.

Pharm Animals

Besides transgenic plants, many animals are inoculated with hormones produced by gene-altered microbes. Animals treated with these hormones and other pharmaceutical products have acquired the catchy name "pharm animals."

Among the first pharm animals was the dairy cow. As early as 1983, the gene for bovine growth hormone (BGH) was isolated and inserted into *E. coli* cells. Soon the cells were producing the hormone in high yield. Injected into beef and dairy cattle, the recombinant BGH (known as recombinant bovine somatotropin or rBST) promotes growth of bone and muscle. It also increases milk production by as much as 25%. Although the United States Food and Drug Administration (FDA) has declared that milk produced by hormone-treated cows is safe to drink, some people are concerned about rBST and the perceived potential risk of developing cancer. Although the scientific evidence does not support this claim, many dairies have returned to marketing non-rBST milk.

FIGURE 15.10 Effect of Herbicide Spraying.

(A) This crop field was not treated with herbicide to kill weeds.

© Pamas/Shutterstock.

(B) This field was treated with an herbicide to which the corn plants are tolerant.

© Mailsonpignata/Shutterstock.

FIGURE 15.11 Most Prevalent GMOs in the Global Market. Food crops make up four of the top five GMO crops.

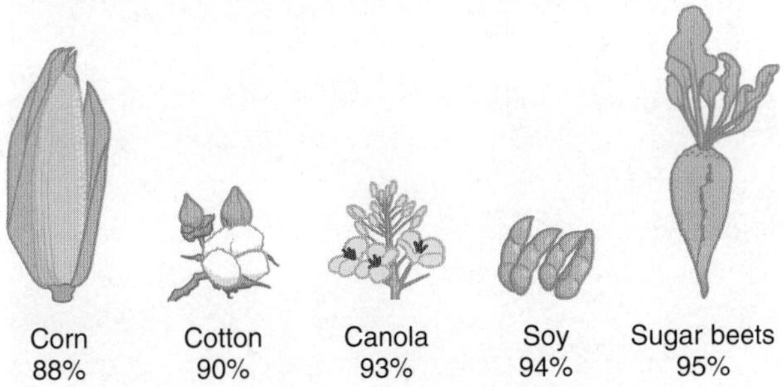

Data from South Carolina University. (2016). The GMO Facts and Controversy. Retrieved from http://www.southcarolinaliberty.com/the-gmo-facts-and-controversy/.

Another interesting example of pharm animals are sheep. Shearing sheep can be an arduous task. To ease wool collection from sheep, biotechnologists have engineered bacterial cells to produce a hormone called sheep epidermal growth factor. When the hormone is injected into the sheep, it weakens the follicles, allowing the fleece to come off the sheep in a single, unmutilated sheet.

Other Imaginative Examples of Plant Biotechnology

The processes of winemaking, breadmaking, and food fermentation can be traced back to antiquity, but there are some imaginative uses of biotechnology in contemporary food production. One of the first **genetically modified** (**GM**) **foods** was the Flavr Savr tomato. It was engineered to slow its ripening process, while still retaining flavor and color (**FIGURE 15.12**). Unfortunately, this tomato never made it to the market because of numerous technical problems (among them a skeptical public). However, the technology used in its development illustrates this innovative thinking. Here is one last example.

Edible Vaccines

Some fruits and vegetables are being studied as potential vaccine sources. Basically, the idea is that some crop plants, such as bananas, tomatoes, and

FIGURE 15.12 Genetically Modified Foods. One of the first commercially grown GM foods was the Flavr Savr tomato.

FIGURE 15.13 "Edible Vaccines." Several fruits and vegetables have been proposed for the development of vaccines that one could eat to obtain immunity to specific infectious diseases.

potatoes, could be genetically engineered to produce vaccines in their edible parts (**FIGURE 15.13**). The GMO plants then could be consumed when inoculations were needed. Whimsically called "edible vaccines," they are desirable because they are inexpensive to store. Also, they are readily accepted in developing nations, where vaccines are needed most. Use of the vaccines would not necessitate exposing people to potentially contaminated syringes. Moreover, the vaccines would be safer than vaccines containing whole organisms or their parts.

Unfortunately, brilliant ideas often run into reality. Such has been the case for edible vaccines, where technical, social, and governmental hurdles have impeded edible vaccine development. Still, interest in plants as biosynthetic factories for human and livestock pharmaceuticals is ongoing down on the "pharm."

A Final Thought

Although still in its infancy, agricultural biotechnology has breathtaking possibilities for improving human health and nutrition. GM foods stand at the forefront of this revolution. But no revolution is without controversy, and the critics of GM foods are numerous and loud. The American government's position is that genetically engineered crops are safe, they resist pests and disease better, and, as such, more food can be produced to feed starving nations.

The European Union (EU) has banned all imports into Europe of GM food products. The EU believes there might be a risk that GM foods could harm the public and have negative environmental consequences. They and others point to the potential for allergic reactions in consumers, the possibility of gene transfers in the environment to create "superweeds," and the death of unintended animal victims of insecticides in nature.

Supporters of GM foods respond by saying no food is 100% safe, whether it is genetically modified or not. In fact, most foodborne illnesses today do not come from GM foods. To date, the reports of adverse reactions to a GM food are few and therefore the risks are outweighed by the benefits. They point to a burgeoning world population that must be fed—an estimated 10 billion by 2050—and suggest biotechnology can dramatically improve productivity where food shortages arise from pest damage and plant disease. They point to the steady decline in the world's arable land, a decline that will accelerate precipitously in the years ahead. They believe biotechnology can raise crop yields in developing countries by over 25%.

But, as with any technology, caution should be taken. GM foods need to be subjected to strenuous testing before they are widely used in agriculture.

Chapter Discussion Questions

What Was He Thinking?

From your reading of this chapter, identify and discuss five major points about microbes and agriculture that the author was trying to get across to you.

Questions to Consider

1. This chapter mentions some of the innovative and imaginative work being done in agricultural biotechnology. Suppose you were the head of a biotech lab and had the choice of pursuing any project you wished. What would that project be? Why would you pursue it? What would be the chances of your success?
2. It is intriguing to stop and wonder how microbes came to be where they are. How, for example, did microbes find their way into the rumen of cattle and evolve to become as important as they now are? And what if they had found their way inside some other type of animal? Suppose they had found their way into the human intestine? What answers might you give to these questions?
3. Italians are fond of a dish called pasta e fagioli [?]. The dish consists of pasta with beans, chick peas, or another legume. Italians know that if they eat enough pasta e fagioli, they don't have to eat quite so much meat, which is often in limited supply. Microbiologically speaking, why does this idea make sense? (Hint: Think nitrogen fixation and legumes).
4. Scientists have noted that one of the side effects of the use of Bt toxin has been a reduction in the population of Monarch butterflies. Why do you think this might have happened, and what might you as a concerned citizen do about it?
5. *War of the Worlds* is a classic novel by H. G. Wells made into several movies. The story details the invasion of Earth by aliens from the planet Mars.

All the might and resources of earthlings are exhausted as the aliens make their way through American cities destroying everything in their path. When all appears lost, the aliens suddenly die, and their space vehicles crash to Earth. After a pregnant pause, the narrator solemnly explains, "In the end, the invaders were exposed to 'germs' in the Earth's atmosphere and died." Although *War of the Worlds* is fiction, a case nevertheless can be made for microbes contributing mightily to the well-being of society. In what ways has this chapter supported their case?

6. Futurists tell us that a great variety of genetically modified (GM) foods will soon be appearing in the supermarkets of the world. Explain what GM foods are, and assemble a list of the perceived pros and cons of GM foods.

© Photos.com.

CHAPTER 16

Microbes and the Environment: No Microbes, No Life

"All for One and One for All"

In 1844, Alexander Dumas published his novel *The Three Musketeers*. It's a story that revolves around D'Artagnan and his three inseparable friends (the musketeers) who all lived life by the motto: "All for one and one for all, united we stand, divided we fall" (**FIGURE 16.1**). Here's an intriguing comparison: Suppose a bacterium was not a single-celled creature, but rather, a unit of an inseparable, united global population of bacteria. All these bacteria are linked metabolically, and all depend on one another for the common good.

During the 1960s, the renowned scientist and writer R. Buckminster Fuller (1895–1983) described our planet as "Spaceship Earth." Fuller envisioned Earth as a self-sustaining entity whirling through space, completely isolated from everything but sunlight. He hoped to convey the notion that some of Earth's resources, such as oil and gas, are nonrenewable, but most others are renewable—if they are recycled.

An essential element of Fuller's vision was that Spaceship Earth's life support depends on life itself. The living organisms aboard this great planet produce its oxygen, cleanse its air, adjust its gases, transfer its energy, and recycle its waste products—all with great efficiency. And high on the list for maintaining life support on Spaceship Earth are the bacteria.

Most bacteria live in mixed communities whose metabolisms complement one another. Such an association can be viewed as a sort of multicellular superorganism. Importantly, the communities of bacteria appear to improvise in

CHAPTER 16 OPENER Microbes live just about anywhere you want to look on planet Earth. From miles down beneath the surface to miles up in the atmosphere, microbes can be found. They live in Earth's coldest environments in Antarctica to areas around Earth's boiling hot volcanic fields. They thrive on all types of Earth's nutrients, including oil, toxic wastes, and rock. Every time you step on the soil, you step on billions of microbes. But in all these and other Earth environments, microbial metabolism is returning nutrients from dead organisms and nonliving substances to forms used by plants and animals. In turn, plants and animals feed the world. From microbes to humans, planet Earth represents a superorganism.

FIGURE 16.1 The Musketeers. Statue of d'Artagnan and the Three Musketeers.

nature and adapt to changes in the environment. Essentially, the bacteria are "All for one and one for all." If they do not remain united, Spaceship Earth's life support is in trouble, and "divided we fall." As Fuller once said

> *"We are not going to be able to operate our Spaceship Earth successfully nor for much longer unless we see it as a whole spaceship and our fate as common. It has to be everybody or nobody."*

In this chapter, we will see the important roles microbes play in successfully operating Spaceship Earth. We will discover that many important resources of Spaceship Earth could not be recycled without microbial intervention. In fact, without microbes, the plants and animals of Earth would have depleted its resources countless eons ago, and the great experiment of life would have failed miserably. However, life did evolve, which is testament to the power and adaptation of microbes. They have managed to fill every conceivable environment on Earth. They have evolved so they can sustain themselves on anything Spaceship Earth has to offer. Consequently, microbes have adapted to participate in the intricate web of metabolic activities that permits Spaceship Earth to continue its long journey through the universe.

LOOKING AHEAD

After reading and completing this chapter, you will be able to:

16.1 Identify the essential roles of microbes in maintaining the biogeochemical cycles.
16.2 Describe the three steps of the waste treatment process.
16.3 Discuss the steps of the water treatment process.

16.1 The Cycles of Nature: What Goes Around, Comes Around

Take a pinch of rich soil, and hold it in your hand. You are holding an estimated billion microbes representing over 10,000 different microbial species. If you were to try cultivating them in the laboratory, only a few species would grow. The remaining thousands of species remained unknown until scientists developed the right combination of nutrients and environmental conditions to cultivate them, or they devised the molecular techniques to identify these species. Indeed, because microbes are invisible, their importance in the environment might go unsuspected. Yet, without microbes, life could not continue.

Spaceship Earth

Microbes provide the basic underpinnings to the cycles of elements on Earth that are needed to build biological molecules. Like all other organisms, microbes do not operate alone. Rather, they act as part of an **ecosystem**, which consists of all plants, animals, and microbes interacting with the nonliving components within a defined space. Ecosystems like a lake might be small compared to other enormous ecosystems like the oceans. Nevertheless, whatever its size, an ecosystem is a type of "superorganism" having the ability to respond to and modify its environment. How these organisms interact with one another is an important part of ecosystem dynamics.

Ecosystem Dynamics

The physical space or location where a species lives is called its **habitat**. An ecosystem can contain different habitats, such as freshwater lakes, marine environments, and soils. **Soil**, for example, is a complex mixture of mineral particles, along with organic remains from decaying plants and animals. Within this habitat, bacterial and fungal species occupy an important **niche**. They are the preeminent recyclers of minerals and decaying matter.

Niche: A term describing the relational position of a species or population in its environment.

Two major features of an ecosystem are energy transfer and resource recycling (**FIGURE 16.2**).

- **Energy Transfer**. Energy enters an ecosystem as sunlight. A portion of that light energy is absorbed by photosynthetic organisms (cyanobacteria, algae, and green plants). They convert the light energy into chemical energy in the form of organic compounds. These photosynthetic organisms therefore are known as **producers**. Some of the chemical energy in the organic compounds then gets transferred to the **consumers** (animals) when they eat the producers. These "feeding ranks" are referred to as **trophic levels**.

FIGURE 16.2 The Pathway of Energy in an Ecosystem. This diagram illustrates how light energy (orange arrows) is transferred through and lost from an ecosystem. The chemical elements recycle within the closed system (blue arrows).

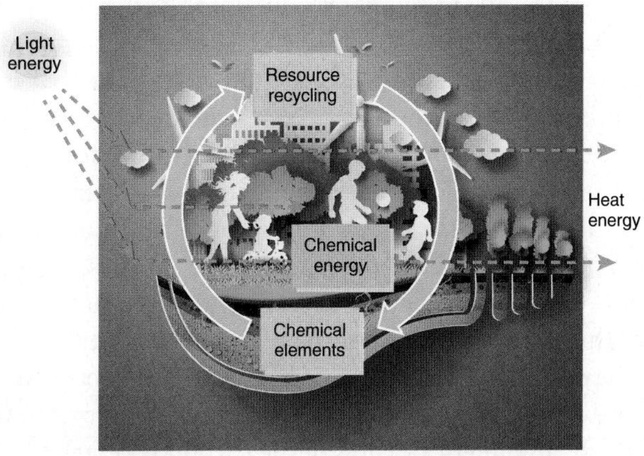

FIGURE 16.3 A Simple Energy Pyramid. Approximately 10% of the energy at one trophic level is transferred as useable energy to the next level. Depending on the ecosystem, there can be more or fewer trophic levels.

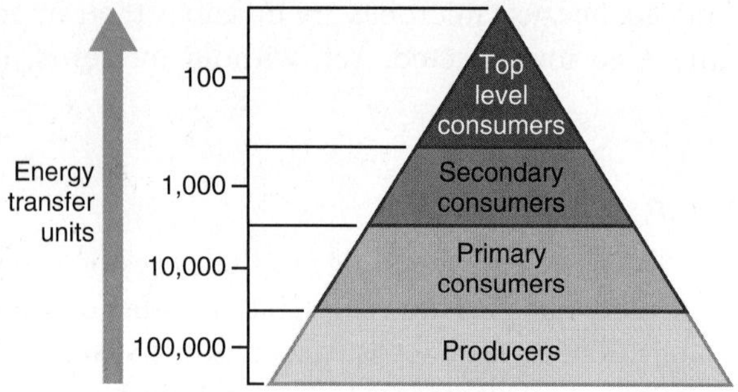

The transfer of energy between trophic levels in an ecosystem is represented as an **energy pyramid** (**FIGURE 16.3**). Importantly, when organic compounds (resources) are passed from one trophic level to the next, only about 10% of the chemical energy is used by the next level. For example, a **primary consumer**, such as a rabbit, only incorporates 10% of the energy from the plants it ate. In turn, only about 10% of the chemical energy in primary consumers gets incorporated by **secondary consumers (carnivores)** when they consume the primary consumers. Consequently, about 90% of the energy in an ecosystem is lost as heat, used up during **cellular respiration**, or excreted as indigestible wastes in animal feces. Importantly, energy cannot be recycled, so lost energy is constantly being replenished by new light energy from the Sun.

Cellular respiration: The process by which cells harvest the energy stored in food to generate cellular energy (ATP).

- **Resource Recycling**. Chemical resources are an essential supply of chemical elements that organisms need for growth and reproduction. The bacteria and fungi that represent the **decomposers** break down dead plant and animal matter (see Figure 16.2). This ensures a continual recycling of raw materials and chemical elements needed to build organic compounds. Consequently, bacteria and fungi play a vital role in recycling nutrients within an ecosystem.

This resource recycling is the basic process underlying what are called **biogeochemical cycles**. These cycles illustrate how chemical elements are reused through a series of naturally occurring biological, geological, and chemical processes. In the next section, the roles that microbes play in a few of these cycles will be made clear.

The Carbon Cycle

Carbon (C) is one of the most important elements because it is the backbone of all organic compounds in living organisms. This includes all carbohydrates, lipids, nucleic acids, and proteins. Like all chemical resources, there is a finite amount of carbon on Earth, and it is the **carbon cycle** that is critical to the recycling of carbon between the atmosphere, land, and ocean (**FIGURE 16.4**).

Photosynthetic organisms assimilate carbon in the form of CO_2 from the atmosphere. They use the CO_2 and sunlight energy to form a variety of carbon-rich carbohydrates. The vast jungles of the world, the grassy plains of the temperate zones, and all the algae and cyanobacteria in the seas display the results of this process. These producers, in turn, might be consumed by animals, fish, and humans. These consumers use some of the carbohydrates as energy sources and convert the remainder of the transferred chemical energy to make cell parts. Although some carbon is released by respiration and returns to the atmosphere as carbon dioxide, the major portion of the carbon is returned to the soil when an organism dies.

FIGURE 16.4 Simplified Carbon Cycle. Photosynthesis represents the major process by which carbon dioxide gas (CO_2) is incorporated into organic matter, and respiration accounts for its return to the atmosphere. Microorganisms are crucial to all decay in soil and ocean environments.

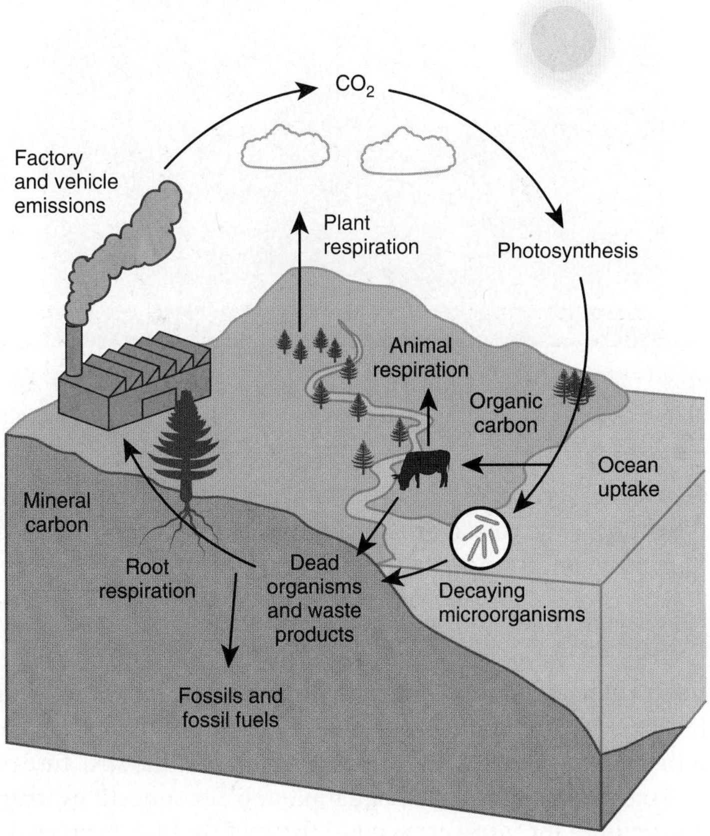

In the soil, the decomposers work in their countless billions to consume dead plant and animal matter. In addition, along with the animals and plants, they release CO_2 into the atmosphere as a result of cell respiration. Importantly, much of the organic matter remains unused, in part, because the microbes are killed by the toxic products of decay. This leftover organic matter then combines with mineral particles to form **humus**, a dark-colored material that retains air and water and is excellent for plant growth.

Our planet would be a garbage dump of accumulating animal waste, dead plants, and organic debris without the decomposers. Fortunately, in soil and water they break down this debris and, in the process, contribute the greatest share of organic matter to the environment. Microbes accomplish a similar feat under controlled conditions. Many microbes decompose manure and other natural waste materials into **compost**, which can be used as a fertilizer for the farm and home garden (**FIGURE 16.5**).

Scientists believe coal, being a fossil fuel, was formed 360 to 290 million years ago when plants in marshes and lakes became buried. Under high temperatures and pressure, the microbial decay process was interrupted, and the vegetation was transformed into peat and then to coal. Petroleum in the sea beds might have originated in the same way when the organic matter of the phytoplankton and zooplankton in brackish waters sank to the bottom and was incorporated in clay sediments that became sedimentary rocks (see Figure 16.4). This so-called oil shale, again under high pressure and temperature, had the oil (i.e., petroleum) squeezed out into the porous rock.

FIGURE 16.5 A Home Compost Pile. Bacteria and fungi account for most of the decomposition of organic matter that takes place in a compost pile.

Lastly, when fossil fuels are burned (combustion), CO_2 is released into the atmosphere (see Figure 16.4). Unfortunately, factory and vehicle combustion are upsetting the balance of CO_2 in the atmosphere. This increase is one of the significant factors concerning global warming.

The Nitrogen Cycle

The **nitrogen cycle** is described at length in the chapter on Microbes and Agriculture. In summary, the cyclic movement of nitrogen is of vital importance to life on Earth because this element is an essential part of the structure of nucleic acids and proteins. Without the essential microbial connections, there would be little opportunity for nitrogen to enter the cycle of life, nor would there be any way for nitrogen to be combined into nucleic acids and proteins. There would be few amino acids, few proteins, few enzymes, few structural materials, and few of anything built around nitrogen. In the end, all life would eventually disappear.

The Phosphorus Cycle

Phosphorus (P) also has a place in the chemistry of living things. It is an important component in nucleic acids and phospholipids in membranes. It also is part of the all-important energy molecule adenosine triphosphate (ATP).

One portion of the **phosphorus cycle** begins when phosphorus enters the sea in the form of phosphates (PO_4) in solution (**FIGURE 16.6**). Microbes come into play right at the outset. Unicellular algae and other microbes take up the PO_4 as they multiply in the waters. These producers manufacture the all-important nucleic acids and other phosphorus-rich compounds. The phosphorus compounds are concentrated as the algae are consumed by other microbes, which then are eaten by shellfish and fin fish. These consumers feed upon one another and use the phosphorus for organic molecules and such body parts as bones and shells. The death of these animals returns the phosphorus to the sea.

Another portion of the cycle involves phosphorus weathering from rocks. In addition, phosphorus used in mineral fertilizers generates soluble PO_4 in the soil. Field crops incorporate the phosphorus, and the crops are eaten by animals. In time, animal waste and the action by decomposers release soluble PO_4. Some of it eventually returns to the sea, where some solidifies to rock. In this form, it would be unavailable for eons of time. Fortunately, microbes live

FIGURE 16.6 The Phosphorus Cycle. Phosphorus enters the cycle as phosphate (PO_4) from various points. Decomposers make soluble phosphate available to plants in the soil and water.

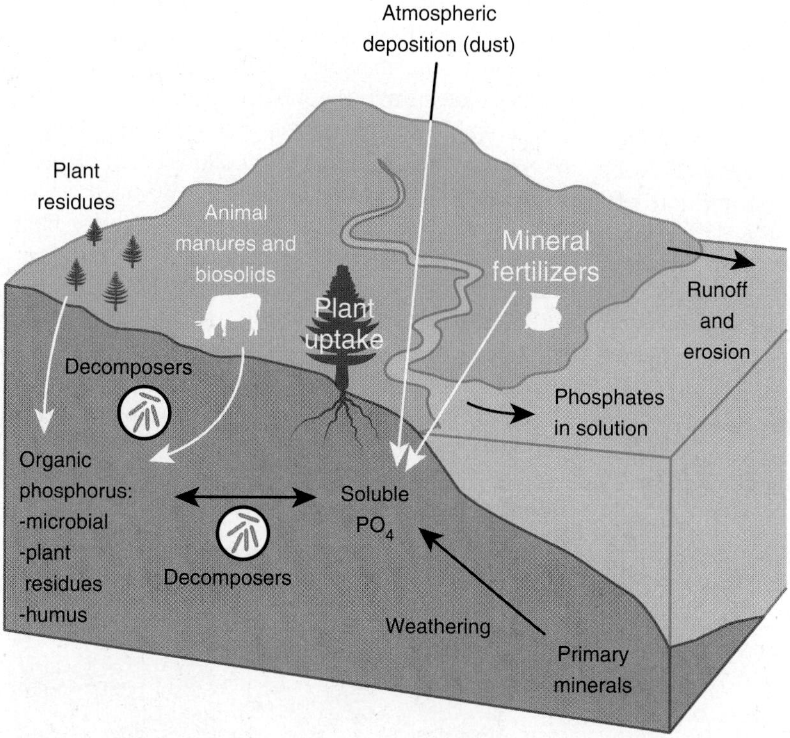

in the sea. As above, the unicellular algae and other microbes trap much of the phosphates for reuse in the cycle of life. They keep the "wheel of life" turning.

Microbes occupy a prominent position in the elemental cycles of life. To be sure, they keep organic matter fresh and vital as they sustain Spaceship Earth. Moreover, the work of microbes does not end here. Microbes also have a key place in preserving Earth's environment. They are the waste digesters, water purifiers, and soil cleansers par excellence. In these respects, they help society solve its age-old problem of how to keep the environment a fit place to live. We rely on microbes even more than we might realize, as the next section illustrates.

16.2 Preserving the Environment: Sanitation and Waste Removal

There is an adage that says, "The solution to pollution is dilution." This somewhat simplified view recognizes the fact that industrial, agricultural, and human waste can be added to bodies of water and the water will disperse the waste and remain pure. For a small village situated along a quick-flowing river, the adage might be true. However, in today's industrialized society, the intervention of microbes is necessary to keep water clean.

Sanitation is the process of reducing the number of microbes to a safe level. It came into full flower in the mid-1800s as a relatively modern phenomenon. Before that time, living conditions in some Western European and American cities were almost indescribably grim. Garbage and dead animals littered the streets, and human feces and sewage stagnated in open sewers. Many rivers were used for washing, drinking, and excreting. All told, filth was rampant. Conditions like these fueled the Great Sanitary Movement of the mid-1800s, as described in **A CLOSER LOOK 16.1**.

A CLOSER LOOK 16.1

The Great Sanitary Movement

In the early 1800s, the steam engine and its product, the Industrial Revolution, brought crowds of rural inhabitants to European cities. To accommodate the rush of people, row houses and apartment blocks were erected. Owners of already crowded houses took in more tenants. Not surprisingly, crowding brought easier transmission of infectious diseases. Death records from the time show alarming numbers of people were dying from typhoid fever, cholera, influenza, dysentery, and other diseases.

As the death rates rose, a few activists spoke up for reform. Among them was an English lawyer and journalist named Edwin Chadwick (see **FIGURE A**). Chadwick subscribed to the then-novel idea that humans could eliminate many diseases by doing away with filth. In 1842, well before the germ theory of disease was established, Chadwick published a landmark report indicating that poverty-stricken laborers suffered a far higher incidence of disease than those from middle or upper classes. He attributed the difference to the horrible living conditions of workers and declared that most of their diseases were preventable. His report established the basis for the Great Sanitary Movement.

Chadwick was not a doctor, but his ideas captured the imagination of both scientists and social reformers. He proposed that sewers be constructed using smooth ceramic pipes, and enough water be flushed through the system to carry waste to a distant depository. In order to work, the system required the installation of new water and sewer pipes, the development of powerful pumps to bring water into homes, and the elimination of older sewage systems. The cost would be formidable.

Chadwick's vision eventually came to reality, but it might have taken decades longer were it not for an outbreak of cholera. In 1849, cholera broke out in London and terrified so many people that public opinion began to sway in favor of Chadwick's proposal. Another epidemic occurred in 1853, during which an English physician named John Snow proved that water was involved in transmission of the disease. In both outbreaks, the disease reached the affluent as well as the poor, and more than half of the sick perished. Construction of the sewer system began shortly thereafter.

As the last quarter of the 1800s unfolded, Europe's sanitary movement needed a "smoking gun" to hammer home its point. That incident came in 1892 when a devastating epidemic of cholera erupted in Hamburg, Germany. For the most part, Hamburg drew its water directly from the polluted Elbe River. Just west of Hamburg lay Altona, a city where the German government had previously installed a water filtration plant. Altona remained free of cholera. The contrast was sharpened further by a street that divided Hamburg and Altona. On the Hamburg side of the street, multiple cases of cholera broke out; across the street, none occurred. The sanitarians could not have imagined a more clear-cut demonstration of the importance of sewage treatment and water purification.

FIGURE A Edwin Chadwick (1800–1890).

Types of Waste Systems

Treatment systems for human waste all operate under the same basic principle. Water first is separated from the waste. Then, the solid matter is broken down by microorganisms into simple, nontoxic compounds for return to the soil and water. These systems range from the primitive outhouse, which is nothing more than a hole in the ground, to the sophisticated sewage-treatment facilities used by many large cities.

Cesspools

In some localities, especially rural areas, human waste is emptied into underground **cesspools** (**FIGURE 16.7A**). These are concrete cylindrical rings with

FIGURE 16.7 Waste Treatment Systems.
(A) Cesspools are underground containers for the temporary storage of liquid and solid waste.

© 88studio/Shutterstock.

(B) In a septic tank system, the water flows into a leach field and the sediment in the tank is periodically removed.

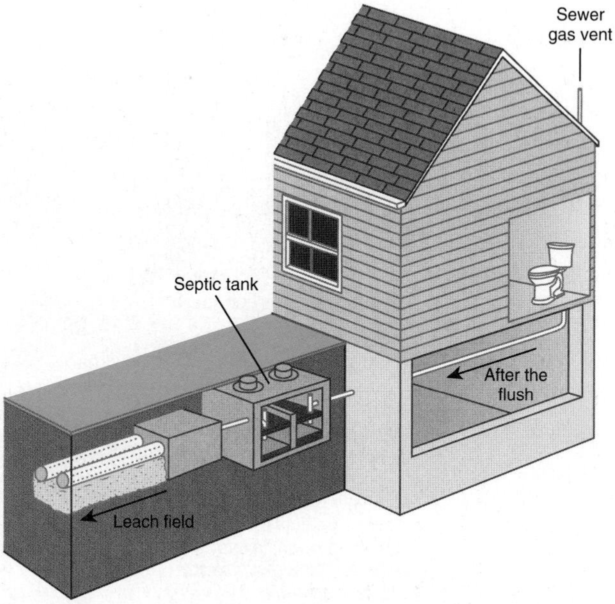

(C) Oxidation lagoons are large, shallow ponds where waste water is treated with microbes.

© Karnwadee/Shutterstock.

(D) A municipal sewage treatment plant separates water from the waste. The solid matter then is broken down by microbes into simple compounds that are returned to the soil and water.

© Mariusz Szczygiel/Shutterstock.

pores in the wall. Water passes into the soil through these pores. Pores on the bottom accumulate the solid waste. Microorganisms in the waste, especially anaerobic bacterial species, digest the solid matter into soluble products that move through the pores into the soil.

Septic Tanks

Some homes have a **septic tank**, an enclosed concrete box for collecting waste from the house (**FIGURE 16.7B**). Solid organic matter accumulates on the bottom of the tank, while water rises to the outlet pipe and flows to a distribution box. The water then flows into perforated pipes that empty into the surrounding soil called the leach field. Because digested organic matter is not absorbed into the ground, the septic tank must be pumped out regularly.

Oxidation Lagoons

Some small towns collect sewage into large ponds called **oxidation lagoons** (**FIGURE 16.7C**). The sewage is left undisturbed in the lagoon for up to 3 months.

FIGURE 16.8 A Diagram of the Sewage Treatment Process. Sewage-treatment facilities use physical, chemical, and biological processes to treat and remove all water contaminants.

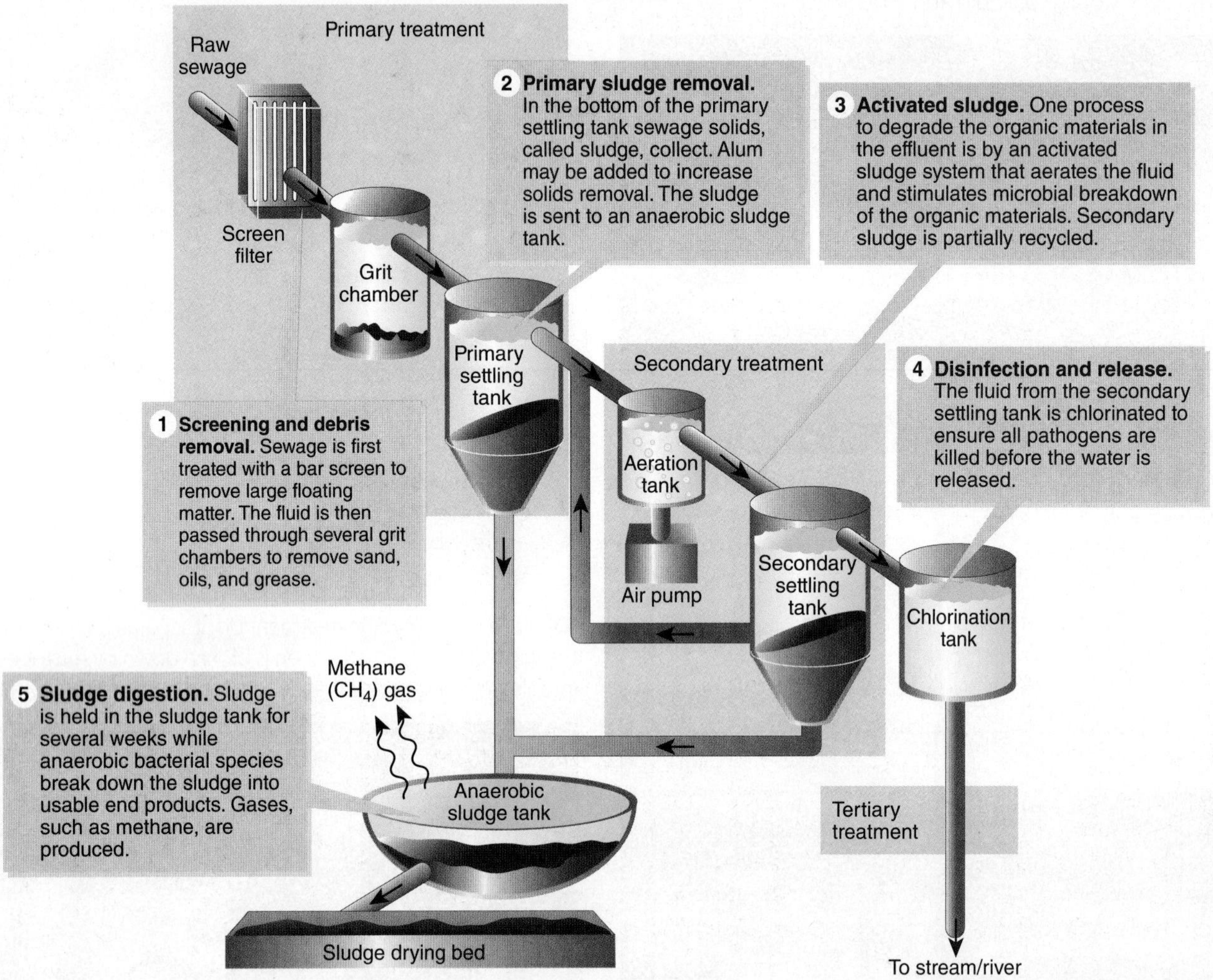

During this time, aerobic bacterial species digest the organic matter in the water as anaerobic organisms break down the settled material. Under controlled conditions, all the waste can be converted to simpler compounds, such as carbonates, nitrates, and phosphates. At the cycle's conclusion, the bacterial cells die naturally. As the water clarifies, it can be emptied into a nearby river or stream.

Large municipalities rely on a mechanized sewage-treatment facility to handle domestic wastewater, which contains massive amounts of waste and garbage (**FIGURE 16.7D**).

Sewage Treatment

Sewage treatment is the process of removing household sewage plus some industrial waste from municipal wastewater. The process involves primary, secondary, and sometimes tertiary treatment (**FIGURE 16.8**).

Primary Waste Treatment

Sedimentation: In waste and water treatment, it refers to the tendency for particles in the liquid to settle out of the fluid by gravity and come to rest at the bottom of the container.

For **primary waste treatment**, a screen is used to remove grit and large, insoluble waste. Then, the raw sewage is piped into huge open **sedimentation** (primary settling) tanks for organic waste removal. This waste, called **sludge**, is passed into anaerobic sludge tanks for further treatment. Flocculating materials, such as alum, might be added to the raw sewage to drag microorganisms and debris to the bottom.

Secondary Waste Treatment

The next step is **secondary waste treatment**. It uses aerobic decomposers to degrade the biological content of the fluid (effluent) coming from the primary treatment. The products are carbon dioxide gas (CO_2) and water (H_2O). Two common secondary sewage treatment processes are the activated sludge system and trickling filters. They differ primarily in the way oxygen is supplied to the microorganisms and the rate at which organisms metabolize the organic matter.

- In the activated sludge system, the bacterial species *Zoogloea ramigera* in the aeration tank degrades the organic matter in the primary effluent. Following 3 to 8 hours in the system, the active microorganisms (**activated sludge**) are separated from the liquid by sedimentation. The clarified liquid (secondary effluent) is further processed. A portion of the activated sludge is recycled back to the aeration tank to maintain a constant concentration of microbes. The remainder is removed and sent to the anaerobic sludge tank.
- The "trickling filter" (also called a "biofilter") system includes a basin or tower consisting of a bed of stones or gravel, molded plastic, or other porous substances that maintain a high surface-to-volume ratio (see circular structures in Figure 16.8). Microorganisms become attached to the bed and form a biofilm. The primary effluent containing the organic matter is sprayed over the bed. As the organic matter in the water percolates through the biofilm, the organic matter is rapidly metabolized by the aerobic microbes. Oxygen is normally supplied to the film by the natural flow of air.

Zoogloea ramigera: ZO-oh-glee-ah rah-mih-JER-ah

The secondary effluent from secondary treatment processes might still contain a few pathogens. Therefore, the water is disinfected, usually by **chlorination**, before it is released into a stream, river, or ocean.

Finally, the primary and secondary sludge in the sludge tank are digested by anaerobic microbes. Methane-producing archaeal species predominate and give off methane (CH_4) and CO_2 gases. The methane gas might be used to provide the fuel to heat the sludge tank. The digested sludge is dried and used as a soil amendment or placed in landfills.

Tertiary Waste Treatment

Sometimes, further treatment of wastewater is needed as not all the water pollutants have been removed. Such **tertiary waste treatment** systems are costly. However, often they are essential for the removal of pesticides, fertilizers, and phosphates, which can pose problems if dumped into a lake or stream.

Biofilms and Bioremediation

Using bacterial species and other microbes to break down waste in sewage treatment has been very effective. Consequently, scientists have put microorganisms to work in a similar manner to "clean up" environmental chemical spills and contamination. The goal is to limit the human footprint in the environment.

For decades, microbiologists have come to realize the importance of bacterial species living in biofilms. A **biofilm** is an immobilized population of bacterial species (or other microorganisms) living in a matrix of tangled polysaccharide fibers adhering to a surface. Examples of environments where biofilms are naturally found include the surfaces of human teeth (dental plaque), dense growths of algae in a stream, and, as we just saw with sewage treatment, on stones of a trickling filter. In the chapter on Microbial Crosstalk, biofilms are described in natural settings.

Bioremediation is the process of using microbes to neutralize or remove toxic wastes or other synthetic products of industry from water or soil. The

FIGURE 16.9 Bioremediation of an Oil Spill. During the cleanup of the *Exxon Valdez* oil spill, microbes were stimulated to help degrade the oil.

intention is to return these areas to their natural state. To do this, microbial biofilms can be used to stimulate or augment the degradation of the contaminating substance. Here are a few examples.

Biostimulation

Biostimulation is when scientists and engineers attempt to modify the environment by adding nutrients to stimulate the growth of resident bacteria. Among the first attempts to use bacterial biofilms for bioremediation was after a major oil spill from the tanker *Exxon Valdez* in 1987 (**FIGURE 16.9**). Previous studies showed that where oil is spilled, there are oil-degrading bacterial species already present. Scientists and engineers only had to encourage their growth. Thus, after the oil spill occurred along the Alaskan coastline, the oil-soaked water was "fertilized" with nitrogen sources (e.g., urea), phosphorus compounds, and other mineral nutrients. It was hoped that these measures would stimulate biofilm development. If so, these areas could be cleared of oil significantly faster than non-remediated areas. The biostimulation process worked. In fact, the oil degraded five times faster when microbes were enlisted in the cleanup.

Bioaugmentation

The process of **bioaugmentation** refers to adding specific bacteria to an environment. As these organisms grow into a biofilm, they would speed up bioremediation.

Polychlorinated biphenyls (PCBs) are man-made insulating materials that were used widely in industrial and electrical machinery. Often, PCBs were released into soil and water, where they remained. Commercial production of PCBs ended in 1977 when evidence was presented that these compounds can accumulate in the food chain. Accumulation of PCBs in the body can cause cancer in animals and might cause cancer in humans. Today, about 10% of the PCBs produced remain in the environment where they continue to contaminate rivers, lakes, and soil.

Microbes have come to the rescue. Scientists have discovered a few naturally occurring bacterial species capable of producing enzymes that degrade

PCBs. If these species can be grown in large quantities, they could greatly accelerate the cleanup of areas containing these contaminants.

Trichloroethylene (TCE) once was a common cleaning agent and solvent. Although TCE causes liver damage and nervous system dysfunction and is a carcinogen, at the time, scientists did not realize TCE would diffuse through the soil and contaminate underground wells and water reservoirs (aquifers). To combat the problem and degrade the TCE, scientists began using bioaugmentation to exploit the ability of bacterial strains in biofilms to detoxify TCE. As the bacterial cells grow into a biofilm, they destroy the TCE.

Today, microbes are being used for their ability to degrade flame retardants, chemical warfare agents, radioactive materials, and numerous other waste products of industry. Again, the aim is to reduce the human footprint on the environment by using microbes to reduce or eliminate industrial contaminants.

Which brings us to one more human-produced problem—water pollution.

16.3 Preserving the Environment: Water Pollution and Purification

Dirty water is the world's biggest health risk, threatening both the quality of life and public health for billions of people. According to a 2019 report from the World Health Organization (WHO) and the United Nations Children's Fund (UNICEF), 30% of people worldwide (2.2 billion) lack access to safe, readily available water at home (**FIGURE 16.10**). Almost 60% (4.2 billion) lack safely managed sanitation. Consequently, each year millions of people (almost half are children under 5 years of age) die from waterborne diseases contracted from unsafe water.

Drinking or **potable** water (*pot* = "drink") refers to water that is safe to drink. In the United States, a typical family of four uses up to 400 gallons of potable water each day. Even though most Americans have safe water, water pollution is still a major issue. **Water pollution** is the contamination of lakes,

FIGURE 16.10 Populations Having Access to Basic Drinking Water Services. Parts of Central and South America, and some countries in Africa and Southeast Asia lack access to clean water. Countries in white provided insufficient data.

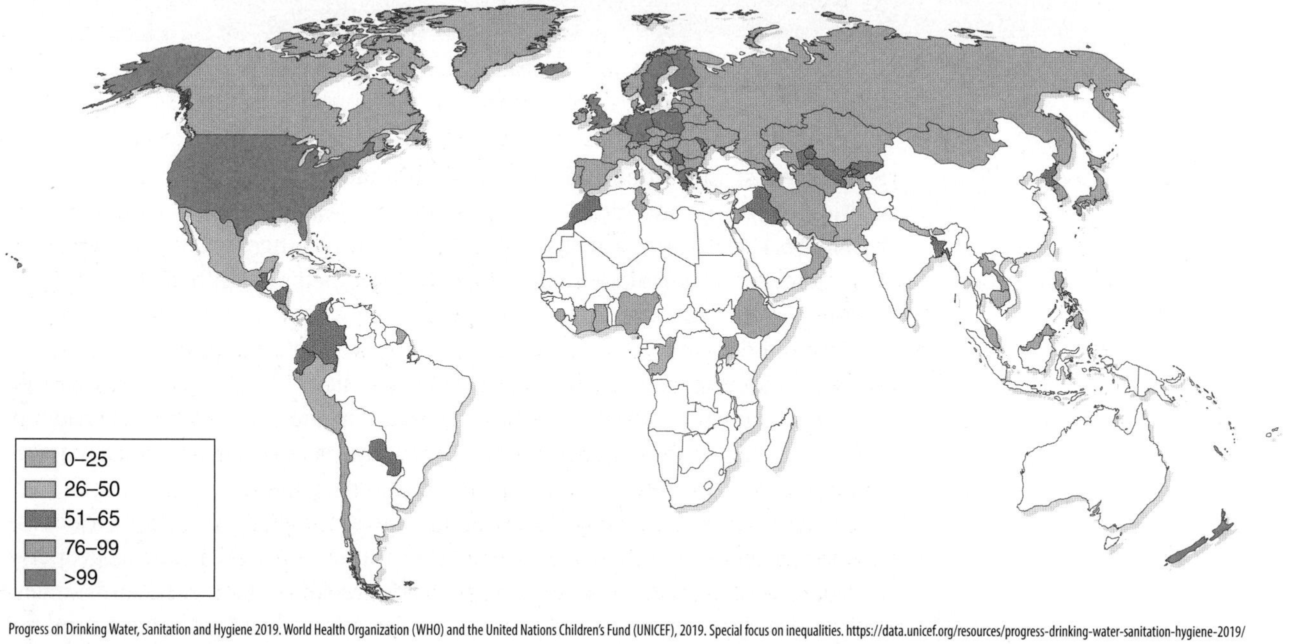

Progress on Drinking Water, Sanitation and Hygiene 2019. World Health Organization (WHO) and the United Nations Children's Fund (UNICEF), 2019. Special focus on inequalities. https://data.unicef.org/resources/progress-drinking-water-sanitation-hygiene-2019/

FIGURE 16.11 The Death of a River. The introduction of nutrients can lead to an algal bloom from which other microorganisms derive nutrients. This metabolism can quickly deplete the oxygen in the water, leading to the death of larger organisms, such as fish.

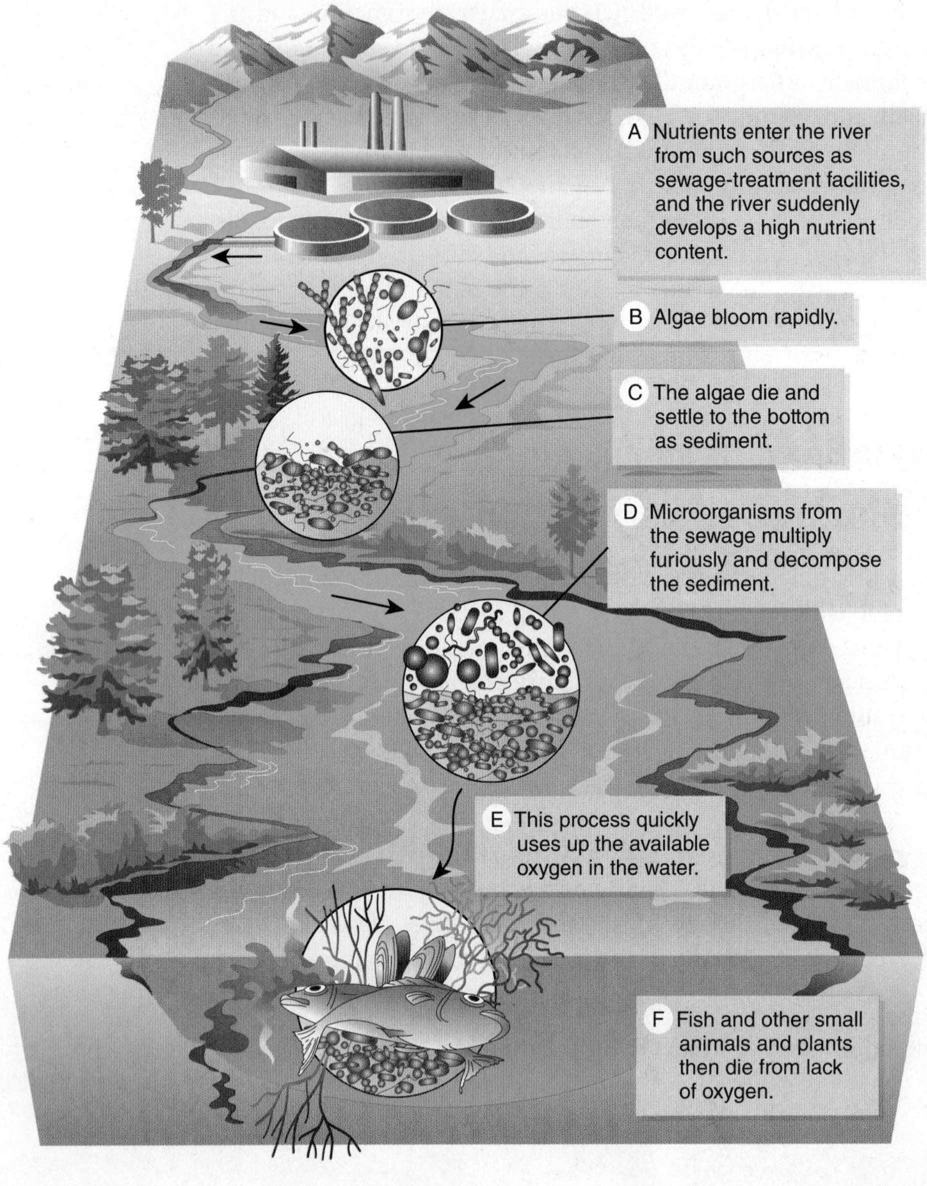
A Nutrients enter the river from such sources as sewage-treatment facilities, and the river suddenly develops a high nutrient content.
B Algae bloom rapidly.
C The algae die and settle to the bottom as sediment.
D Microorganisms from the sewage multiply furiously and decompose the sediment.
E This process quickly uses up the available oxygen in the water.
F Fish and other small animals and plants then die from lack of oxygen.

rivers, oceans, and groundwater through human activities. It is caused by the addition of sewage and industrial wastes discharged into nearby water supplies (**FIGURE 16.11**). Although in the United States and other developed countries these practices are regulated, pollutants still find their way into water supplies.

Many pollutants are toxic substances that destroy marine and freshwater species. The decomposition of aquatic life sparks bacterial growth, which uses up the available oxygen gas (O_2) in the water. As the O_2 drops, fish, small arthropods, and plants die. Then, anaerobic bacterial species thrive in the sediments containing the dead material. They also produce gases, such as hydrogen sulfide, that give the water a stench reminiscent of rotten eggs.

Water also can be a vehicle for the transport of pathogenic microbes causing human illness. This can be the result of humans who fail to deal properly with their sewage and waste. It also can be the result of natural circumstances. Heavy rains, for example, wash pathogens from the soil and into rivers used for drinking water. However, whether from human or natural sources, pollution must be detected. Therefore, public health departments and water companies continually test water to identify any potential health hazards.

Microbe Detection Methods

Typhoid fever, cholera, and hepatitis A are a few of the diseases that can be transmitted in water. Therefore, one responsibility of public health agencies is to make sure water supplies do not become contaminated with human intestinal pathogens.

It is impossible to test water for all potential intestinal pathogens. Therefore, specific **indicator organisms** are used as signs of contamination. Among the most frequently used is *Escherichia coli*, because it is found in the intestines of virtually all humans. If *E. coli* is identified, the likelihood is high that the water has been contaminated with intestinal microbes, including possible pathogens. Several tests have been devised to detect *E. coli* contamination.

Escherichia coli: esh-er-EE-key-ah KOH-lee

The **membrane filter technique** is a common laboratory test. A water technologist collects a 100 milliliter (mL) sample of water (**FIGURE 16.12A**). Back in the water-testing lab, the water sample is passed through a cellulose-based membrane filter to trap any bacterial cells. The filter then is transferred to a plate of growth medium and incubated. Any bacterial cells trapped on the filter will grow and form visible colonies on the surface of the filter (**FIGURE 16.12B**). It is assumed that each trapped bacterial cell will grow into a colony. The technologist then counts the colonies and thereby determines the original number of bacterial cells in the sample of water.

These more traditional detection methods can take several days to complete. Therefore, today the techniques of biotechnology can shorten this time considerably.

A sample of polluted water is filtered. Any bacterial cells trapped on the filter are broken open to release their DNA. The DNA is amplified by the **polymerase chain reaction** (**PCR**), and a **DNA probe** specific for *E. coli* DNA is added. Both tools are described in the chapter on Biotechnology and Industry. In this test, the technologist attempts to identify *E. coli* DNA, rather than *E. coli* cells. The process is extremely sensitive. It has been estimated, for example, that a single *E. coli* cell can be detected in a 100-mL sample of water. In addition, a wealth of other microbes, including many pathogenic species, can

DNA probe: A known segment of single-strand DNA that is complementary to a desired DNA sequence of a bacterial species.

FIGURE 16.12 Water Analysis.
(A) A water sample is collected for analysis.

(B) In the water-testing lab, potential indicator organisms are identified by the membrane filtration technique.

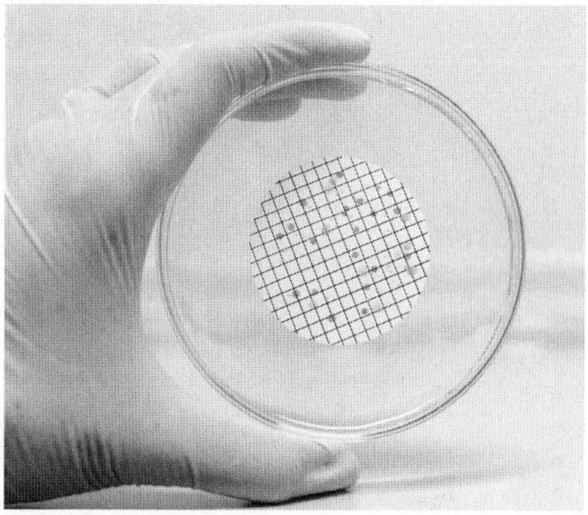

FIGURE 16.13 Steps in the Purification of Municipal Water Supplies. Water purification involves sedimentation, filtration, and chlorination.

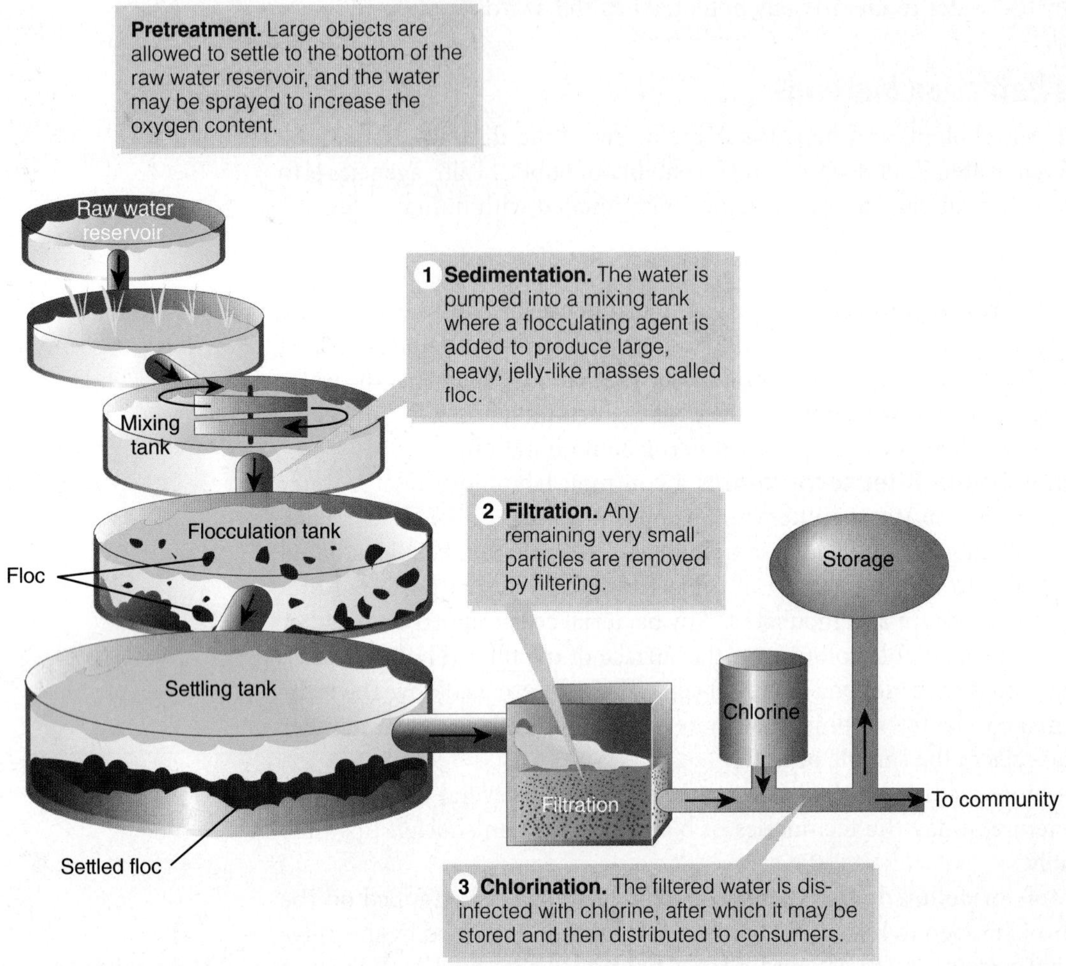

be directly identified by DNA probe analysis, thereby eliminating the search for indicator organisms.

Ensuring the safety of water supplies for consumers is a high priority of the public health system. Rapid determination of a potential disease health risk permits public health officials to implement measures to benefit the most people. High on the list of these measures is the treatment of water to interrupt the spread of disease.

Water Treatment

It is rare to locate a water source that does not need treatment before consumption. The general rule is that water must be treated to remove potentially harmful microbes and to improve its clarity, odor, and taste.

Three basic steps are included in the preparation of water for drinking: sedimentation, filtration, and chlorination (**FIGURE 16.13**). As we go through the steps, realize the treatment process does not necessarily produce sterile water. Rather, it yields clean, potable water that is free of pathogens.

Sedimentation

The first step of water purification is called **sedimentation**. This process uses large reservoirs (settling tanks) to remove leaves, particles of sand and gravel,

A CLOSER LOOK 16.2

Purifying Water with the "Miracle Tree"

In developed nations, water purification usually uses chemical powders such as aluminum sulfate or iron sulfate for the flocculation process. Many developing nations do not have such chemicals available, nor can they readily afford to purchase them. Could something else be used for the flocculation step in water purification? Yes! It's the "miracle tree." Scientifically known as *Moringa oleifera,* this tropical tree was given the name "miracle tree" because of the many uses the tree has—including an environmentally friendly way to purify water.

M. oleifera survives in arid areas and produces elongated seedpods (see **FIGURE A**). The local people use the leaves and roots for food, the wood is used in building, and some parts of the tree are used in traditional medicines. High-grade oil for lubrication and cosmetics is produced from the seeds. However, the seeds also have another use—purifying water. In those parts of Africa and India where the trees grow, people grind up the seeds and add the powder to cloudy water to precipitate all the solid particles. Clean drinking water results.

Ian Marison is a research director at École Polytechnique Fédérale de Lausanne in Switzerland. He and his colleagues have examined the ground-up seed residue and have isolated a charged peptide. When they add this peptide to cloudy water, within 2 minutes the water goes from cloudy to clear.

As the Swiss group studied the peptide in more detail, they discovered it has bactericidal properties. It can kill some drug-resistant strains of *Staphylococcus, Streptococcus*, and *Legionella*, as well as some nonbacterial waterborne pathogens. In fact, Marison's group discovered they could tweak the peptide's structure and increase its antimicrobial effect.

This is certainly good news for the developing nations where the trees mainly grow. However, if commercially produced, the plant flocculant also would be useful to developed nations because the traditional chemical flocculants often have safety and environmental concerns associated with them.

As Marison says, "It's a biological, biodegradable, sustainable resource. *Moringa* grows where there is very little water; it grows very, very fast and it costs almost nothing." It truly is a "miracle tree."

FIGURE A A Moringa tree (*Moringa oleifera*) bearing seedpods.

and other materials from the soil. Then, chemicals such as aluminum sulfate (alum) are dropped as a powder onto the water. The alum causes small particles in the water to stick together and form jelly-like masses of congealed material called **flocs**. In the flocculation tank, these masses fall through the water and cling to organic particles and microorganisms, dragging a major portion to the bottom sediment. **A CLOSER LOOK 16.2** describes a new, environmentally safe flocculant for many developing parts of the world.

Filtration

The second step in water treatment is **filtration**. Most filtration steps use a layer of sand and gravel to trap microorganisms. A slow sand filter, containing fine particles of sand a meter deep, is efficient for smaller-scale operations. Within the sand, a biofilm of bacterial, fungal, and protist cells acts as a supplementary filter. A slow sand filter can purify over 3 million gallons of water per acre per day. To clean the filter, the top layer is removed and replaced with fresh sand.

Some water treatment plants use a membrane filter system, where the water flows through submerged hollow fibers with tiny pores. The water molecules can pass through the pores, but the larger contaminants stay within the fibers and are filtered out. The filtered water then is pumped on to the next stage of treatment.

Both forms of filtration eliminate more than 99% of the microbes and other very small particles from water. Often granular activated carbon, a black, sand-like material, is used to remove natural organic matter. This step reduces bad taste and odors in the water.

Chlorination

The final step of water purification is **chlorination**. Treated water is disinfected with chlorine gas, which reacts with any organic matter in water. It is important, therefore, to continue adding chlorine until a residue is present. Under these conditions, most remaining microbes die within 30 minutes. In the case of home swimming pools, or in a public swimming pool, the chlorine level is maintained at a level to ensure any fecal microbes have been killed.

Some communities and homeowners also soften water by removing magnesium, calcium, and other salts. Softened water mixes more easily with soap, and soap curds do not form. Water also might be fluoridated to help prevent tooth decay. Scientists believe fluoride strengthens tooth enamel and makes the enamel more resistant to the acid produced by anaerobic bacterial species commonly found in the mouth.

In some instances, it is necessary to treat water for drinking on the spot. For example, raw sewage might find its way into water supplies and contaminate the water. Moreover, during drought conditions, sediment from the bottom of reservoirs might be stirred up, bringing bottom-dwelling microbes into the water and making it hazardous to health. Under conditions like these, consumers are advised to boil their water for a few minutes before drinking. This treatment kills all microbes except for bacterial spores. However, with few exceptions, these spores do not represent a hazard to health. The water is safe to drink, but it is not sterile.

To disinfect natural stream water, campers, backpackers, and hikers are advised to use commercially available chlorine or iodine tablets. Many filtration systems also are available to eliminate any pathogens. If these cannot be obtained, the Centers for Disease Control and Prevention (CDC) recommends a half-teaspoon of household chlorine bleach in 2 gallons of water, with 30 minutes contact time before consumption.

A Final Thought

In late 2018, the UN Intergovernmental Panel on Climate Change (IPCC) released a report saying that global warming must be kept to a 1.5°C increase. If not, within 12 years there will be an increase in the risks of drought, floods, extreme heat and poverty for hundreds of millions of people. As described in this chapter, microbes are key factors in many environmental processes. Therefore, with climate change, what positive and/or negative effects could microbes have in the environment?

Marine microbes inhabit the largest ecosystem on Earth. In the oceans, they are the most abundant biological factors, and they play a crucial role in global biogeochemical cycles. For example, phytoplankton form the basis of the ocean's food web and produce 50% of the world's oxygen. How will climate change affect these dominant microbes? Changes to the ocean's microbes could have a significant impact on all types of marine life throughout the food web.

Changes to the ocean's microbes also might have an impact on global weather. Microbes generate atmospheric compounds that help form clouds. Would climate change result in more clouds? This might cool air temperatures as more heat from the sun would be reflected.

Drought conditions arising from climate change might alter soil at the microbial level. Extreme weather conditions can change vegetation composition and soil moisture, which in turn impact underlying microbial communities. Therefore, climate change might have widespread impact on biogeochemical cycles, which would impact the wider ecosystem.

Because climate change is a complex and incompletely understood phenomenon, it is difficult to make predictions on what might happen. At this point, it is critical that more research is done to investigate how climate change might change the microbial communities and how the microbial communities might change the environment. The chapters on Diseases in Humans will examine how climate change might affect the spread of infectious disease.

Chapter Discussion Questions

What Was He Thinking?

From your reading of this chapter, identify and discuss five major points about microbes and the environment that the author was trying to get across to you.

Questions to Consider

1. The English poet John Donne once wrote: "No man is an island, entire of itself." This statement applies not only to humans, but to all living things in the natural world. From this chapter, mention are some roles microbes play in the interrelationships among living things?
2. Approximately 50% of the O_2 gas produced each day comes from cyanobacterial photosynthesis. Hypothesize what might happen if all the cyanobacteria suddenly disappeared.
3. In the 1970s, a popular bumper sticker read: "Have you thanked a green plant today?" The reference was to photosynthesis taking place in plants. Suppose you saw this bumper sticker: "Have you thanked a microbe today?" What might the owner of the car have in mind?
4. Explain how sewage treatment is just another example of bioremediation.
5. The late syndicated columnist Erma Bombeck wrote a humorous book entitled *The Grass Is Always Greener Over the Septic Tank* (an adaptation of the expression "The grass is always greener on the other side of the fence"). Indeed, the grass often is greener over the septic tank. Why is this so? How can you locate your home's cesspool or septic tank in the days following a winter snowfall?
6. Bioremediation holds the key to solving numerous types of environmental problems in the future. Do you know of any examples in your area or state where bioremediation is being used?
7. When sewers were constructed in New York City in the early 1900s, engineers decided to join storm sewers carrying water from the streets together with sanitary sewers bringing waste from the homes.

The result was one gigantic sewer system. In retrospect, was this a good idea? Why?

8. The author of a biology textbook writes: "Because the microorganisms are not observed as easily as the plants and animals, we tend to forget about them, or to think only of the harmful ones…and thus overlook the others, many of which are indispensable to our continued existence." How do the carbon, nitrogen, and phosphorus cycles support this outlook?
9. Refer back to R. Buckminster Fuller's quote about Spaceship Earth in the chapter opening. After reading about microbes in this chapter, how insightful was his remark? Give examples in your answer.

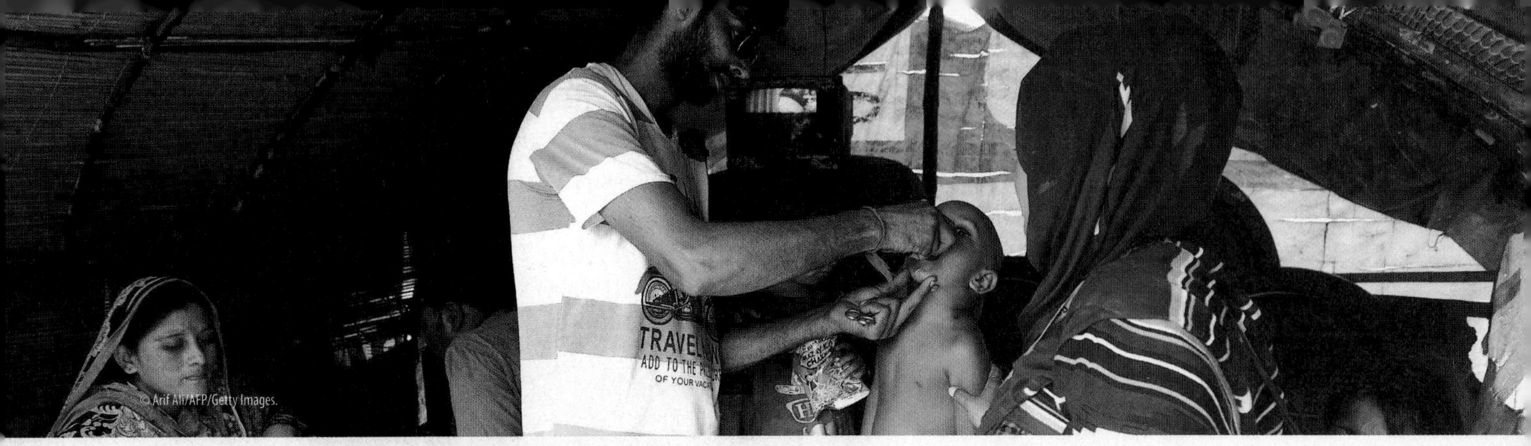

CHAPTER 17

Disease and Resistance: The Wars Within

The Plague of Athens

The Peloponnesian War (431–404 BCE) was a fight for power between Athens (and its empire) and Sparta (with its allies). Sparta had assembled large armies that were nearly unbeatable. In 430 BCE, the huge Spartan threat forced the Athenians to retreat behind the city walls of Athens. The retreat meant throngs of people from the countryside took refuge in the city. Due to the density of people and poor hygiene, Athens became a breeding ground for disease. The result was a devastating epidemic, today called "The Plague of Athens" (**FIGURE 17.1**).

In his *History of the Peloponnesian War*, the contemporary Greek historian Thucydides gave an eyewitness account of the coming catastrophe.

> *The bodies of dying men lay one upon another, and half-dead creatures reeled about in the streets. The catastrophe became so overwhelming that men cared nothing for any rule of religion or law.*

Today, we still are not sure what this "plague" was. Several bacterial and viral diseases have been suggested.

As the epidemic spread, Thucydides made an extraordinary observation about resistance to the disease. He wrote:

> *Yet it was with those who had recovered from the disease that the sick and dying found most compassion. These knew that it was from experience and now had no fear for themselves, for the same man was never attacked twice, never of the least fatality.*

CHAPTER 17 OPENER Vaccines provide resistance against infection, such as this oral vaccine against polio being given in Pakistan in 2018. Today, through a global vaccination effort, polio is close to eradication.

FIGURE 17.1 The Plague of Athens. In 430 BCE, a disastrous plague hit Athens, Greece, taking the lives of up to 100,000 people.

In other words, if individuals survived the disease, as Thucydides did, they would never again develop the illness. They were immune.

By the end of the epidemic, historians believe 75,000 to 100,000 people sheltered within Athens' walls died of the disease. The sight of the burning funeral pyres of Athens caused the Spartan army to temporarily withdraw for fear of the disease. But they would be back.

In this chapter, we explore the infectious disease process and the factors contributing to the establishment of disease. We then will survey the immune defenses by which the body develops resistance to infectious diseases. We will end the chapter by looking at the types of vaccines and the importance of the vaccination process in helping protect individuals and populations from contracting an infectious disease.

LOOKING AHEAD

After reading and completing this chapter, you will be able to:

17.1 Describe the direct and indirect methods of disease transmission.
17.2 Distinguish exotoxins from endotoxins.
17.3 Discuss the human body's physical defenses and innate immunity responses.
17.4 Contrast the roles of cell-mediated immunity and antibody-mediated immunity.
17.5 Compare the molecular makeup of whole-agent and genetic-engineered vaccines.

17.1 Concepts of Infectious Disease: Individuals and Populations

In the late 1960s and early 1970s, many believed infectious diseases had been conquered. It was thought that the use of antibiotics and vaccines would make the threat of infectious disease of little consequence. However, antibiotic resistance and new emerging viral diseases have thwarted such optimism. In 2018, approximately 57 million humans died worldwide. Of these, more than 25% (15 million) died from infectious diseases, making them the second leading cause of death behind cardiovascular disease (**FIGURE 17.2**). Globally, infectious diseases are the leading cause of death in children under 5 years of age.

Infection and Disease Within Individuals

Infection refers to the entry, establishment, and multiplication of a pathogen in the **host**. A host whose resistance is strong remains healthy, and the pathogens are either driven from, or assume a temporary relationship with, the host. By contrast, if the infection leads to tissue or organ damage, disease develops. The term **disease** therefore refers to any change from the general state of good health. Importantly, infection and disease are not the same; a person might be infected without suffering a disease.

Host: A cell or organism in which a microbe or virus can live, feed, and reproduce (replicate).

Whether a disease is mild or severe depends on the pathogen's ability to do harm to a susceptible host. An organism that consistently causes disease, such as the typhoid bacillus, is said to exhibit a high degree of **virulence**.

Virulence: The severity or harmfulness of a disease.

FIGURE 17.2 Infectious Disease Deaths Worldwide. This pie chart depicts the leading causes of infectious diseases and the number of worldwide deaths as reported by the World Health Organization. Tropical diseases: African sleeping sickness, Chagas disease, schistosomiasis, leishmaniasis, filariasis, and onchocerciasis. Childhood diseases: diphtheria, measles, pertussis, polio, and tetanus.

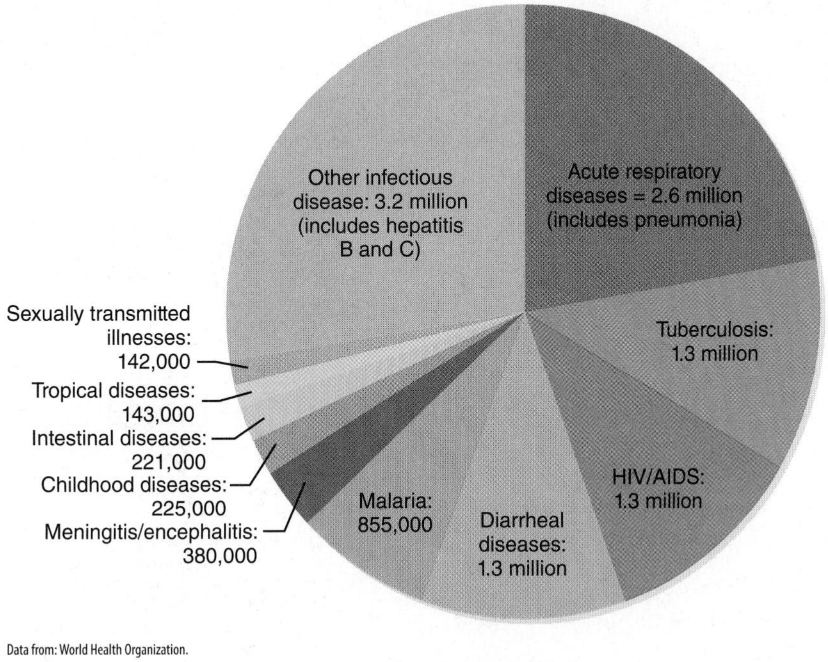

Data from: World Health Organization.

Candida albicans: KAN-did-ah AL-bih-kanz

By comparison, a pathogen that sometimes causes disease, such as the yeast *Candida albicans*, is of moderate virulence. Certain organisms, described as **avirulent**, are not regarded as disease agents. The lactobacilli and streptococci normally found in yogurt are examples.

Infection and Disease Within Populations

Disease **epidemiology** involves the study of diseases and their spread in whole populations. In these cases, infectious diseases are described according to the level at which they occur in a population (**FIGURE 17.3**). An **endemic** disease, for example, persists at a low level in a certain geographic area. By comparison, an epidemic disease (or an **epidemic**) breaks out in explosive proportions within a population. This should be contrasted with an **outbreak**, which is a more contained epidemic. The abnormally high number of measles cases in the United States in 2018–2019 was classified as an outbreak. Should it spread widely throughout the country, it would be called an epidemic. A pandemic disease (a **pandemic**) occurs worldwide. Two newsworthy examples are the current AIDS pandemic and the 2009 swine flu pandemic.

Transmission of Infectious Diseases

The microbial agents of disease can be transmitted in various ways. One way is by contact transmission, which can be direct or indirect (**FIGURE 17.4**).

Direct-Contact Transmission

Person-to-person transmission of a pathogen is an example of direct contact. Here, physical contact occurs between a person who is infected, or has the disease, and one or more susceptible individuals. Activities such as shaking hands with or kissing an infected individual are examples where an infectious agent can be transmitted. A bite or scratch from an infected animal

FIGURE 17.3 Types of Infectious Diseases. Infectious diseases can be categorized as endemic, epidemic, or pandemic. The dashed arrows indicate the spread of the disease.

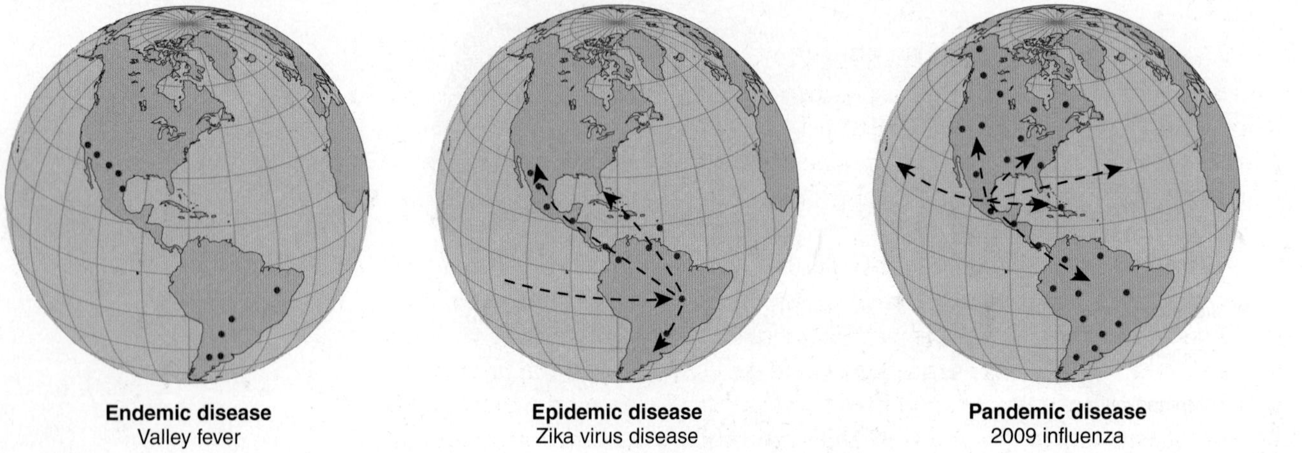

FIGURE 17.4 Methods of Transmitting Disease Through Contact. Infectious diseases can be transmitted by direct methods and indirect methods.

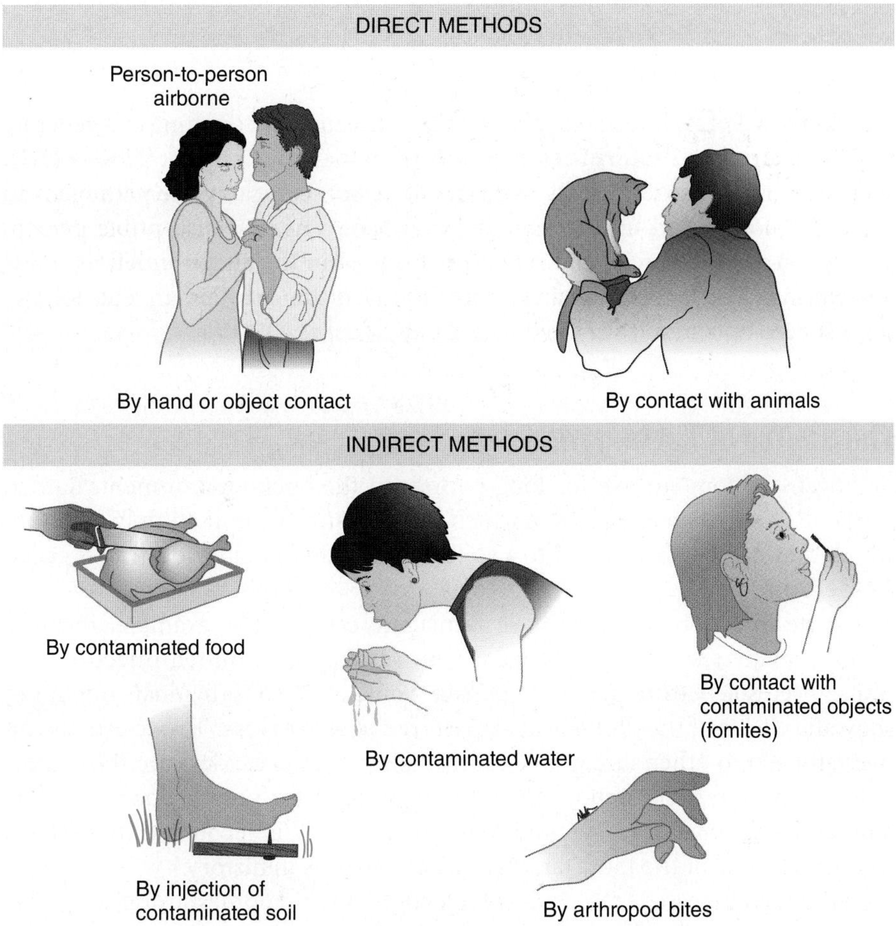

also involves direct-contact transmission. Sexual intercourse is another direct-contact mechanism for sexually transmitted infections.

Indirect-Contact Transmission

Among the indirect ways of disease transmission are the consumption of contaminated food or water. Foods might be contaminated during processing or handling. They also might be contaminated through contact with a diseased animal.

Indirect contact also can occur through touching inanimate objects called **fomites**. For instance, doorknobs or drinking glasses might be contaminated with flu or cold viruses that are transmitted to another person when that person touches the object. This shows the importance of hand washing. Also, contaminated syringes and needles can transport the viruses that cause hepatitis B and AIDS.

Arthropods (e.g., insects) are responsible for another indirect method of transmission. In some cases, the arthropod represents a **mechanical vector** of disease because it transports pathogens mechanically by contact with its legs and other body parts. Houseflies would be examples of mechanical vectors. In other cases, the arthropod itself is infected and represents a **biological vector**. The malaria protist and West Nile virus, for instance, infect mosquitoes.

The infected mosquito can inject the pathogen into a human during its next bite (blood meal).

Airborne Transmission

Respiratory droplet: A small drop of moisture expelled from the nose or throat through sneezing or coughing.

The simple act of an infected person talking, sneezing, or coughing can generate infectious airborne **respiratory droplets** from the nose or throat (**FIGURE 17.5**). The larger respiratory droplets from the ill person can carry the pathogens to the eyes, mouth, and upper respiratory tract of a nearby susceptible person. Smaller airborne particles can remain suspended in the air for indefinite periods, during which time they can be spread some distance by air currents following a sneeze or cough from the infected individual.

The Source of Pathogens

Pathogens seldom survive for long periods in the open environment. Rather, they must have a suitable place in which they can survive and multiply and from which they can be transmitted to a suitable host. This suitable location is called a **reservoir**.

Some infectious diseases have human reservoirs. For example, measles, mumps, many respiratory pathogens, and sexually transmitted infections are transmitted solely from person to person. Some infected individuals might not show the effects of the illness and are referred to as **carriers**. They can transmit the pathogen to other susceptible individuals. Chronic carriers are those individuals who can harbor and transmit the pathogen for months, perhaps years after an initial infection. Typhoid Mary, whose story is recounted in **A CLOSER LOOK 17.1**, is one of the most famous chronic carriers in history.

Animals also can be reservoirs of infection. A rabid domestic dog can transmit the rabies virus to humans through a bite. Such a disease transmitted from animals to humans is called a **zoonosis (pl. zoonoses).** Besides rabies, the Black Death (plague), West Nile disease, and malaria are zoonoses. In these examples, the reservoir for the plague bacillus is rats, and the reservoir for the West Nile virus disease virus and the malaria parasite are mosquitoes.

FIGURE 17.5 Airborne Disease Transmission. Sneezing or coughing represents a method for airborne transmission of pathogens.

A CLOSER LOOK 17.1

Typhoid Mary

By 1906, typhoid fever was claiming about 25,000 lives annually in the United States. During the summer of that year, a puzzling outbreak occurred in the town of Oyster Bay on Long Island, New York; one girl died, and five others contracted typhoid fever. Eager to find the cause, public health officials hired George Soper, a well-known sanitary engineer from the New York City Health Department. Soper's suspicions centered on Mary Mallon, the seemingly healthy family cook. She had disappeared three weeks after the disease surfaced. Soper was familiar with Robert Koch's theory that infections like typhoid fever could be spread by people who harbor the organisms. Quietly, he began to search for the woman who would become known as Typhoid Mary.

Soper's investigations led him back over the decade during which Mary cooked for several households. Twenty-eight cases of typhoid fever occurred in those households, and each time, Mary left soon after the outbreak.

Soper tracked Mary through a series of leads from domestic agencies and finally came face-to-face with her in March 1907. She had assumed a false name and was now working for another family in which typhoid had broken out. Soper explained his theory that she was a carrier and pleaded with her to be tested for typhoid bacilli. When she refused to cooperate, the police forcibly brought her to a city hospital on an island in the East River off the Bronx shore. Tests showed her stools swarmed with typhoid organisms. Fearing her life was in danger, Mary adamantly refused the gall-bladder operation to eliminate the organisms (the causative agent, *Salmonella typhi*, often colonizes the gallbladder). As news of her imprisonment spread, Mary became a celebrity. Soon public sentiment led to a health department policy deploring the isolation of carriers. She was released in 1910.

But Mary's saga had not ended. In 1915, she turned up again at New York City's Sloane Hospital working as a cook under another new name. Eight people had recently died of typhoid fever, most of them doctors and nurses. Mary was taken back to the island, this time in handcuffs. Still she refused to have her gallbladder removed and vowed never to change her profession. Doctors placed her in isolation in a hospital room while trying to decide what to do (see **FIGURE A**). The weeks wore on.

Eventually Mary became less stubborn and assumed a permanent residence in a cottage on the island. She gradually accepted her fate and began to help with routine hospital work. However, she had to eat in solitude and was allowed few visitors. Mary Mallon died in 1938 at the age of 70 from the effects of a stroke. She was buried without fanfare in a local cemetery.

FIGURE A Typhoid Mary in a hospital bed.

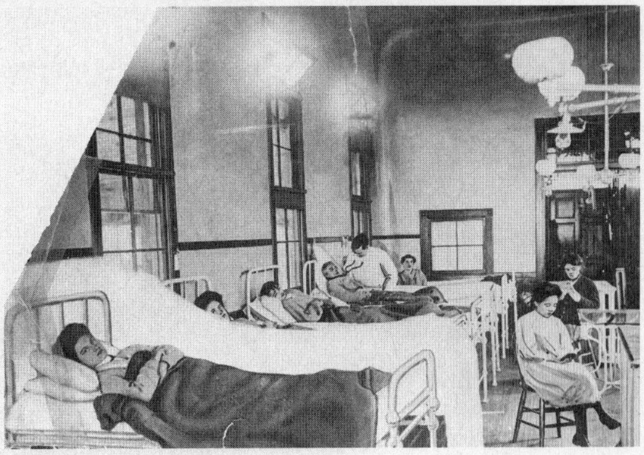

Soil and water can be reservoirs for some infectious agents. For example, the fungal pathogen that causes valley fever survives and multiplies in the soil. The reservoir for the cholera bacterium often can be part of the normal microbial population of estuaries.

FIGURE 17.6 summarizes the three elements (host, transmission, and source) forming the three factors contributing to infectious disease.

The Course of a Disease

In most instances, there is a recognizable pattern in the progress of the disease following the entry of the pathogen into the host. Often these periods are identified by **signs**, which represent evidence of disease detected by an observer (e.g., a physician). A low-grade fever or bacterial cells in the blood would be examples of signs. Disease also can be noted by **symptoms**, which represent

FIGURE 17.6 Infectious Disease Elements. Assuming there is a susceptible host, infection requires both a source of the pathogen and a method for transmission from source to host.

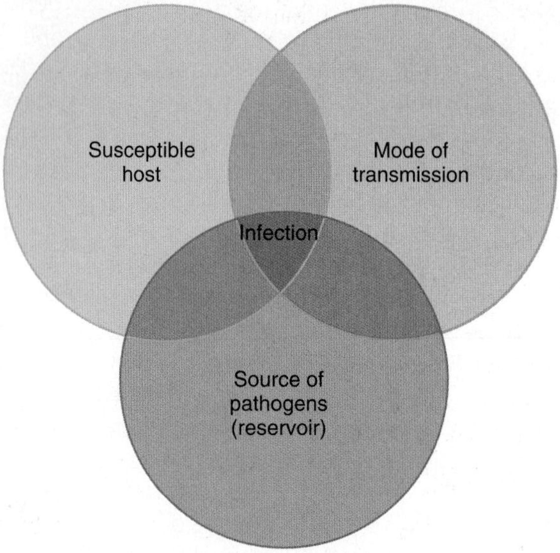

FIGURE 17.7 The Course of an Infectious Disease. Most infectious diseases go through a five-stage process. The length of each stage ("Time" on the horizontal axis) would depend on the specific pathogen and the host response.

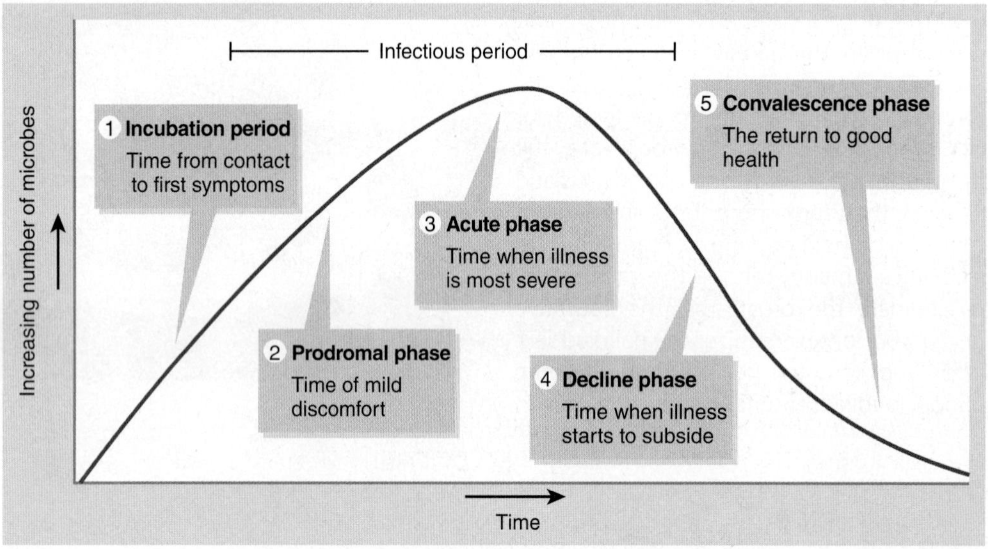

changes in body function sensed by the patient. A sore throat and a headache are examples of symptoms.

Diseases often are characterized by a specific collection of signs and symptoms called a **syndrome**. The syndrome caused by HIV, acquired immunodeficiency syndrome (AIDS), is an example where over time the individual exhibits a typical set of signs and symptoms following infection and disease progression. Thus, a doctor's **diagnosis** often is based on a patient's signs and symptoms.

Diagnosis: The identification of a disease or illness.

Disease progression is distinguished by five stages (**FIGURE 17.7**).

Incubation Period

The episode of disease begins with an **incubation period**. This represents the time elapsing between the entry of the pathogen into the host and the appearance of the first symptoms. For example, an incubation period might be as short as 2 to 4 days for the flu, 1 to 2 weeks for measles, or 3 to 6 years for leprosy. Such factors as the number of organisms, their generation time, and level of host resistance determine the incubation period's length.

Prodromal Phase

The next phase in a disease's progression is a time of mild signs or symptoms, called the **prodromal phase**. For many diseases, this period is characterized by indistinct and general symptoms such as headache and muscle aches. In this phase, the affected individual does not know for sure what are causing the symptoms.

Acute Period

In the third stage of a disease, called the **acute period** or **climax**, the signs and symptoms are of greatest intensity. For the flu, patients suffer high fever, chills, a headache, cough, body and joint aches, and loss of appetite. This collection of signs and symptoms sometimes is referred to as a flu-like syndrome.

Decline and Convalescence Periods

Provided the immune system eventually gets the upper hand, the signs and symptoms of the disease begin to subside during the **decline period**. For the flu, sweating can be common as the body releases excessive amounts of heat.

The sequence concludes after the body passes through a **convalescence period**. During this time, the body's systems eventually return to normal.

17.2 The Establishment of Disease: Overcoming the Odds

The establishment of disease is a complex series of interactions between pathogen and host. The progression of interactions is described next.

Pathogen Entry and Invasion

A pathogen must possess unusual abilities if it is to overcome host defenses and cause disease. Before it can express these abilities, the pathogen must enter the host in high enough numbers to establish a population. Then, it must invade the tissues and grow at that location. Let's look at a simplified scenario that can lead to disease (**FIGURE 17.8**).

Portal of Entry

The site where the pathogen enters the host is called the **portal of entry**. There are several potential entry routes.

- **Respiratory Portal.** Inhalation brings airborne pathogens into the respiratory system.
- **Gastrointestinal Portal.** Ingestion brings pathogen-contaminated food or water containing fecal material into the digestive system. This is often referred to as the **fecal-oral route**.
- **Sexual Transmission Portal.** Blood, semen, vaginal, and other bodily fluids transmit pathogens through sexual contact.
- **Non-Oral Portals.** Piercing the skin through cuts, animal/insect bites, wounds, or injections transmit pathogens from fomites or insects into the host.

FIGURE 17.8 The Flow of Events for a Disease. A portal of entry and an infectious dose are required to initiate infection and disease.

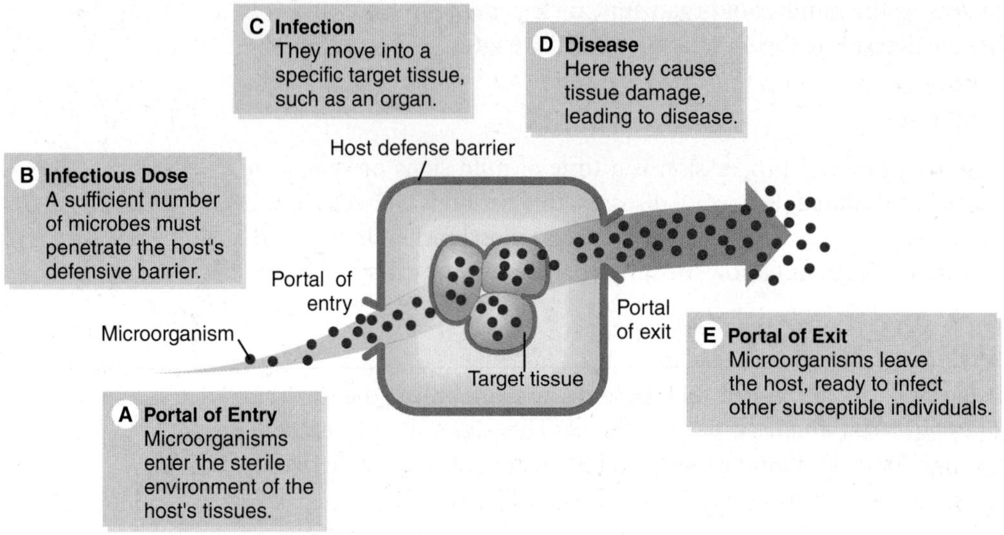

Staphylococcus aureus: staff-ih-loh-KOK-us OH-ree-us

Let's assume that an individual has ingested via the fecal-oral route a food item contaminated with ***Staphylococcus aureus***.

Infectious Dose

Having reached the appropriate portal of entry, the ability of a pathogen to establish an infection depends on the number of cells or viruses that are transmitted. This is referred to as the **infectious dose** (see Figure 17.8). For example, the consumption of a few thousand typhoid bacilli will probably lead to typhoid fever in a susceptible individual. By contrast, millions of *S. aureus* bacteria must be ingested for an infection to be established in an individual. One explanation for the difference is the high resistance of typhoid bacilli to the acidic conditions in the stomach. Most staphylococci perish under these conditions. Also, it might be safe to eat fish taken from water that contains hepatitis A viruses, but eating raw clams from the same water can be dangerous. Clams filter water to obtain nutrients. In doing so, they concentrate any hepatitis A viruses to a concentration representing an infectious dose.

Pathogen Infection and Disease

Specific pathogens possess a wide array of structures and molecules that enhance their infective and disease-causing abilities. These so-called **virulence factors** are coded by genes that would be absent in nonpathogenic microbes.

Surface Virulence Factors

After a pathogen has successfully entered the host, it must establish and maintain itself. Two structural virulence factors that help bacterial pathogens establish an infection and avoid the host's immune defenses are the glycocalyx and pili. Both structures were described in the chapter on the Prokaryotic World. For example, *S. aureus* cells have a polysaccharide glycocalyx that allows them to attach to the surface of the digestive tract. This surface structure also helps protect them from immune system attack.

Enzyme Virulence Factors

For some pathogens, only adhesion is necessary for infection and disease development. For other pathogens, enzymes are important early on to establish infection. These enzymes help the pathogen in penetrating deeper into a tissue of the body to cause disease. For example, *S. aureus* produces an enzyme that digests a polysaccharide that binds host cells together in a tissue. This activity forms gaps between the cells, allowing entry and spread to other parts of the body. A large group of enzymes also assist other pathogens in spreading. One major result of these enzymes is damage to the affected tissues that can lead to disease.

Toxin Virulence Factors

Toxins are microbial poisons affecting the establishment and course of disease. These virulence factors are of two types.

- **Exotoxins.** These toxins are produced and released by some bacterial pathogens. They are protein molecules that act locally or diffuse to their site of activity in the body. *S. aureus* produces several exotoxins. Some affect the digestive tract, while others affect the skin or blood cells.
- **Endotoxins.** These toxins are part of the cell wall of gram-negative bacterial species. They are released on disintegration of the cell. Endotoxins show their presence by certain signs and symptoms, including increased body temperature, substantial body weakness and aches, and general **malaise**. Damage to the circulatory system and **shock** also might occur.

Malaise: An overall feeling of discomfort, illness, or lack of well-being.

Shock: A state of physiological collapse, marked by a weak pulse, coldness, sweating, and irregular breathing due to too low a blood flow to the brain.

Pathogen Exit

To spread an infection, a pathogen must escape the host through some suitable **portal of exit** (see Figure 17.8). For food poisoning caused by *S. aureus*, the portal of exit would be through the feces as a result of diarrhea. Other common mechanisms for exit of other pathogens are coughing and sneezing, which easily spread nasal secretions, saliva, and sputum as respiratory droplets. Exit also could be by blood transfusion or an insect taking a blood meal.

Human pathogens have an arsenal of structural and chemical weapons to give the invaders a competitive advantage over the host. If the host is to survive an episode of infectious disease, the body's immune system must come into play. How the immune system responds is explored in the next two sections of this chapter.

17.3 Nonspecific Resistance to Infection: Natural-Born Immunity

Immunity refers to the ability of the body to resist infections. Think of the human body as protected by a three-layered "immunological umbrella" that can shield itself from the torrent of potential microbial pathogens to which it is continually exposed (**FIGURE 17.9**). If the umbrella remains intact, the body stays safe. However, if "holes or tears" develop, then the body runs a higher risk of an infection. If breaks in the first layer develop, a second tier of resistance is encountered. This so-called innate immunity attempts to kill off the pathogens that have entered. Cells of innate immunity also send out chemical signals to a third layer of resistance called adaptive immunity. Should the pathogen break through innate defenses, this third layer has acquired the ability to defend against the specific invading pathogen.

Let's examine each level of the immunological umbrella. To simplify the process, we will assume pathogens are bacteria or viruses.

FIGURE 17.9 The Relationship Between Host Resistance and Disease. Host resistance can be likened to a three-layered "microbiological umbrella" that forms a barrier or defense against pathogen or toxin invasion.

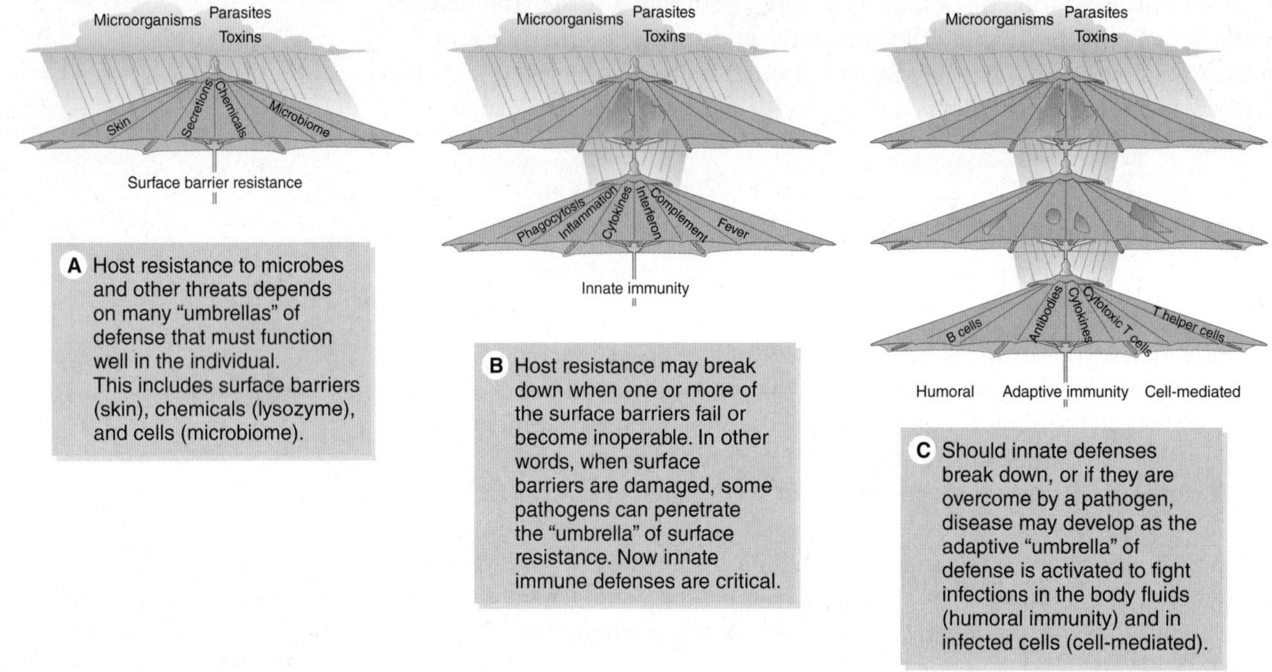
Microorganisms
Parasites
Toxins
Skin
Secretions
Chemicals
Microbiome
Surface barrier resistance
A Host resistance to microbes and other threats depends on many "umbrellas" of defense that must function well in the individual. This includes surface barriers (skin), chemicals (lysozyme), and cells (microbiome).
Microorganisms
Parasites
Toxins
Phagocytosis
Inflammation
Cytokines
Interferon
Complement
Fever
Innate immunity
B Host resistance may break down when one or more of the surface barriers fail or become inoperable. In other words, when surface barriers are damaged, some pathogens can penetrate the "umbrella" of surface resistance. Now innate immune defenses are critical.
Microorganisms
Parasites
Toxins
B cells
Antibodies
Cytokines
Cytotoxic T cells
T helper cells
Humoral
Adaptive immunity
Cell-mediated
C Should innate defenses break down, or if they are overcome by a pathogen, disease may develop as the adaptive "umbrella" of defense is activated to fight infections in the body fluids (humoral immunity) and in infected cells (cell-mediated).

Surface Barrier Resistance

The surface barriers against infection consist of the skin, mucous membranes, and human and microbial cells.

The Skin

The intact skin provides an important nonspecific defense against all microbial invaders. The skin itself not only provides mechanical protection, but its cells are constantly shed, and with them go any attached microbes. In addition, the skin is a poor source of nutrients for microbes, and the low water content of the skin makes the surface a veritable desert. The skin's oil glands add additional defenses by secreting antimicrobial substances and lowering the skin pH. Therefore, unless injury to the skin occurs through a cut or abrasion, a wound, or an insect bite, infections of and through the skin are rare.

Secretions

The surface of the respiratory, gastrointestinal, and urogenital tracts is lined with a **mucous membrane**. The cells composing the mucous membranes secrete **mucus**, which is a viscous substance that can trap microbes. For example, to help eliminate inhaled cold and flu viruses, the amount of mucus increases substantially when we are sick, as **A CLOSER LOOK 17.2** points out.

Lysozyme: An enzyme found in tears and saliva that hydrolyzes the peptidoglycan of gram-positive bacterial cell walls.

Other antimicrobial substances, such as **lysozyme**, are secreted in tears, saliva, and perspiration. In the stomach, the extremely low pH of the gastric juices kills most pathogens before they can enter the intestines.

The Cells

The nonspecific defense barriers also include the normal human microbiome. These nonpathogenic microbes form a biofilm preventing colonization by

A CLOSER LOOK 17.2

Going with the Flow

Every time we get a cold, the flu, or a seasonal allergy, we often end up with the sniffles or a truly raging runny nose. When this happens, it is simply your body's response to what it believes is an infection—or a hypersensitivity to cold temperatures or spicy food. As a defensive barrier against infection, increased mucus flow is the best way to wash respiratory pathogens out of the airways.

You always are producing—and swallowing—mucus. Most people are not aware of their mucus production until their body revs up mucus secretion in response to a cold or flu virus, or allergens. So how much mucus is produced? In a healthy individual, glands in the nose and sinuses are continually producing clear and thin mucus—often more than 200 milliliters (about one cup) each day! An individual is not aware of this production because the mucus flows down the throat and is swallowed. Now, if that individual comes down with a cold or the flu, the nasal passages often become congested, forcing the mucus to flow out through the nostrils of the nose. This requires clearing by blowing the nose or (to put it nicely) expectorating from the throat. With a serious cold or flu, the mucus can become thicker and gooier, and it can have a yellow or green color. The revved-up mucus flow in such cases can amount to about 200 milliliters every hour; if you blow your nose 20 times an hour, each blow could amount to anywhere from 2 to 10 milliliters of mucus. If you have watery eyes as well, then that tear liquid can enter the nasal passages and combine with the mucus, producing an even larger "flow per blow."

So, although a runny nose is usually just an annoyance, make sure you drink plenty of water to make up for the liquid lost as mucus from that runny nose.

pathogens. As described in the chapter on Microbial Crosstalk, as a biofilm, the human microbiome outcompetes pathogens for nutrients and attachment sites on the skin and mucous membranes. In addition, some special types of immune cells exist under the skin and mucous membrane surfaces. These cells recognize invading pathogens and attempt to eliminate them.

Innate Immunity

The skin and mucous membranes form challenging barriers to infectious agents. However, with the appropriate virulence factors they can be penetrated. Consequently, if pathogens manage to cross the skin or mucous membranes, another level of immune defense is ready to act. This defense is called **innate immunity** because everyone is naturally born with this immune response. These responses to infection include phagocytosis, inflammation, fever, and interferon production.

Innate: Referring to something that is present in an individual since birth.

Phagocytosis

Phagocytosis (literally, "cell eating") is the process by which certain white bloods cells respond to an infection by attacking and ingesting (eating) the invaders. This reaction to infection is a form of nonspecific resistance because the response is identical no matter what pathogen infects the body.

Groups of white blood cells, called **phagocytes**, carry out the process. Among the most important phagocytes are the **macrophages** and **neutrophils** found in many body tissues and the blood. When a phagocyte encounters a microbe, the cell binds the pathogen to its cell surface using pseudopods (**FIGURE 17.10**). The cell then encloses the microbe with a portion of its

FIGURE 17.10 The Mechanism of Phagocytosis.

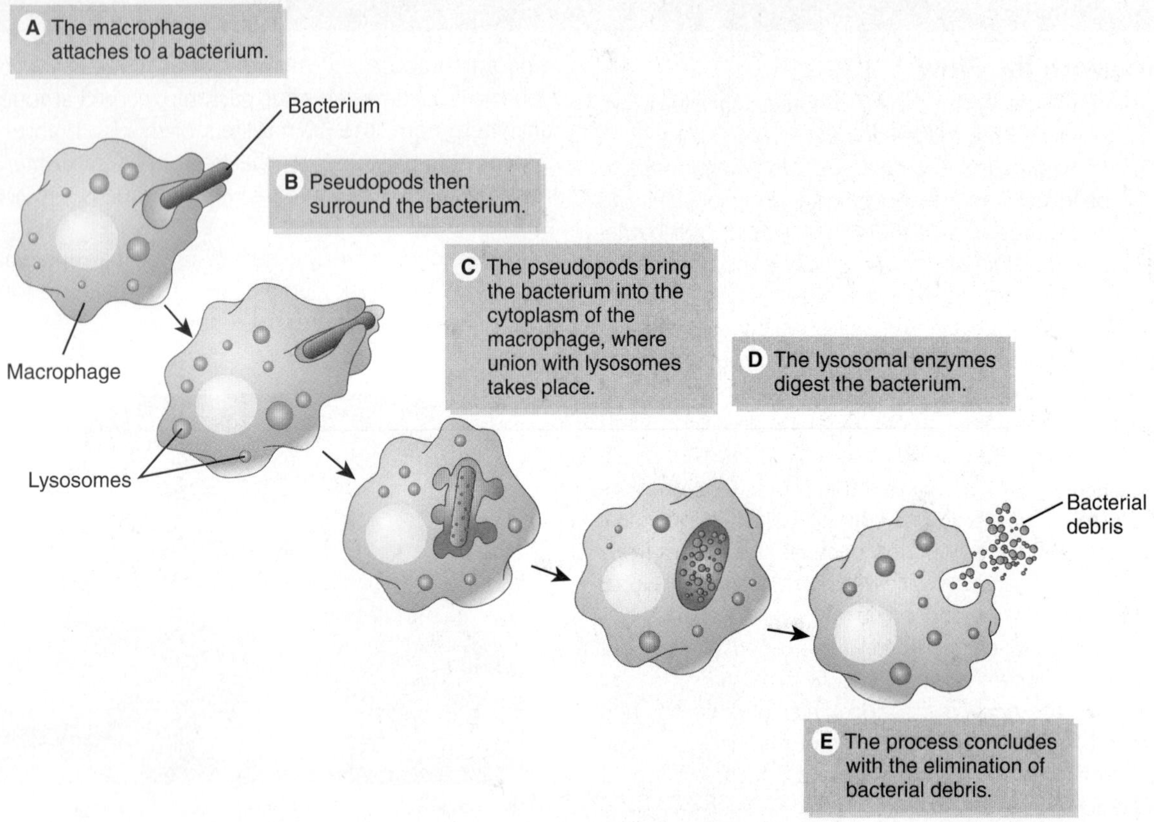

cell membrane to form an internal compartment containing the microbe. Now, **lysosomes**, which contain a large variety of digestive enzymes, fuse with the internal compartment. The digestive enzymes destroy and digest the enclosed microbe. Any indigestible debris is eliminated.

Inflammation

Inflammation is a defensive response that occurs when an injury or traumatic event occurs in human tissues. The trauma sets into motion an innate, nonspecific process to limit the extent of the injury (**FIGURE 17.11**).

For an infection originating from a cut or skin puncture, the damaged tissues send out chemical signals that initiate the inflammatory response. The response is identified by its characteristic signs: redness, warmth, swelling, and pain. The redness and warmth are due to the dilation (expansion) of the blood vessels that are bringing an increased flow of blood and phagocytes to the infection site. Swelling comes from the accumulation of fluid, the pressure of which on nerve endings accounts for the pain. These signs and responses culminate in bringing phagocytes quickly to the site of the infection where they phagocytize the invading pathogens. In addition, it sets in motion the events leading to repair of the damaged tissue.

Fever

Fever is an abnormally high body temperature that often accompanies inflammation. It is triggered by fever-producing substances called **pyrogens**. These substances, which include immune proteins produced by macrophages and

FIGURE 17.11 The Process of Inflammation in Response to Infection. A series of nonspecific steps composes the inflammatory response to trauma caused by an infection, such as being pierced by a plant thorn harboring bacterial cells (see inset). Histamine causes blood vessel dilation, which increases blood flow and defense cell influx. Neutrophils and macrophages release chemicals to attract additional defense cells. The arriving phagocytes engulf the bacterial cells.

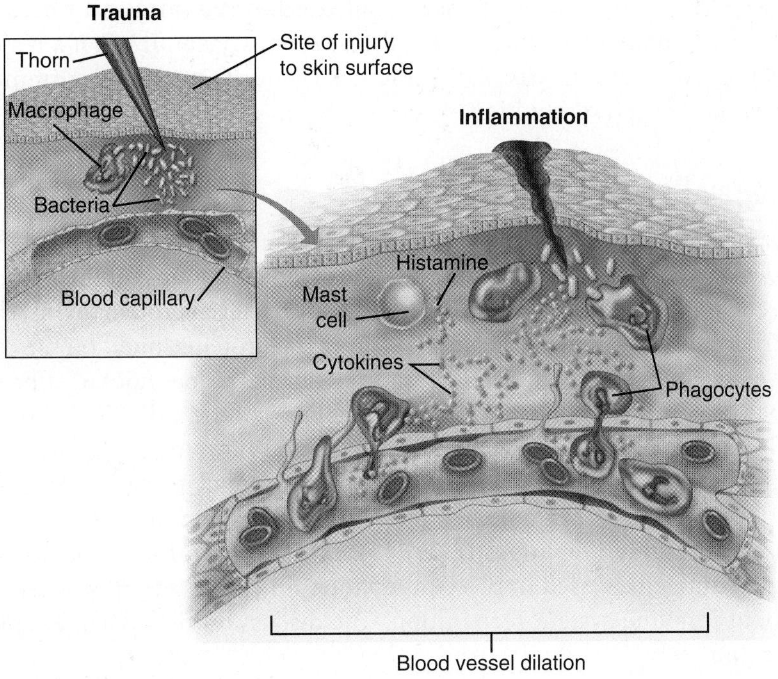

pathogen fragments in the blood. Once in the blood, these pyrogens are carried to the brain where they affect the anterior **hypothalamus**. The result is an increase in body temperature above the normal core (oral) temperature of 37°C (98.6°F).

Hypothalamus: The hypothalamus is an area of the brain producing hormones that direct a multitude of important functions, including body temperature, hunger, sleep, and thirst.

A low to moderate fever (up to 38.3°C/101°F) is a natural defensive response to infection. By elevating body temperature, the rapid growth of pathogens is slowed, and bacterial toxins are inactivated. Fever also heightens phagocytosis and encourages rapid tissue repair. However, if the fever is prolonged, or develops into a high fever (above 40°C/104°F), damage to the host tissues could occur. Without medical attention, convulsions and death might follow.

The coordinated innate immune response of phagocytosis, inflammation, and fever usually occur no matter what pathogen is causing the infection. However, when a viral infection occurs, an additional innate defense is triggered.

Interferon

When a virus infects a cell, the innate immune defenses kick into action. However, the infected cell also produces and secretes the protein **interferon**. Interferon is a warning signal to adjacent uninfected cells. These nearby cells respond to the signal by producing antiviral proteins inside the cytoplasm. These proteins now are at the ready if the virus infects the alerted cells. Then, if infection does occur, the antiviral proteins attempt to block viral nucleic acid replication. Obviously, this innate defense mechanism is not 100% effective because we still get colds, the flu, and other viral diseases. However, if interferon was not produced, most of us would fall ill to viral infections much more often.

17.4 Specific Resistance to Infection: Adaptive Immunity

Specific resistance refers to those immune system reactions that occur when a specific pathogen has managed to get past the host surface barriers and innate defenses. This third layer of defense is called **adaptive immunity** because it adjusts its response after experiencing the infectious agent. It is not a response one is born with or can pass on to the next generation. Adaptive immunity starts with pathogen recognition.

Pathogen Recognition

To respond to a specific pathogen, adaptive immunity first must identify the intruder. It does this by recognizing unique chemical groups on the pathogen. These chemical groups, called **antigens**, include bacterial protein toxins, chemical structures found on flagella and pili, and viral proteins found on the capsid or envelope. Importantly, adaptive immunity normally does not target proteins or other chemical structures that are a normal part of the individual's body.

The whole antigen itself does not trigger the immune response. Rather, stimulation is accomplished by identifying just small parts of the antigen called **epitopes**. For example, an antigen like the bacterial flagellum might have several different epitopes (**FIGURE 17.12**). However, for simplicity, often the term "antigen" is used in place of "epitope" in conversation or when writing about immune system recognition. The discussion below will adopt this convention.

T Cells and B Cells

Besides the phagocytes discussed with innate immunity, the cornerstones of adaptive immunity are a set of white blood cells known as **lymphocytes**. Lymphocytes are small cells, about 10 to 20 μm in diameter, each with a large nucleus (**FIGURE 17.13**). Two types of lymphocytes can be distinguished based on developmental history, cellular function, and unique biochemical properties. The two types are **T lymphocytes (T cells)** and **B lymphocytes (B cells)**. They both arise from lymphoid **stem cells** in the bone marrow. However, they each have a different role to play during an infection.

Stem cell: A nonspecialized cell from which all other cells with specialized functions are generated.

T Cells

Some of the lymphoid stem cells leave the bone marrow and mature into T cells in the **thymus** (T for thymus) (**FIGURE 17.14A**). Here, the cells are modified by the addition of surface receptor proteins. Once they have acquired their proper

Thymus: An organ located in the upper chest cavity that is involved in development of cells of the immune system.

FIGURE 17.12 Antigen and Epitopes. Antigens are major parts of microbes that contain small chemical groups called epitopes.

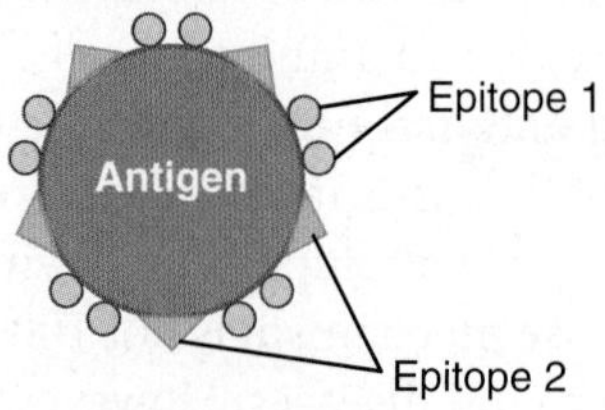

FIGURE 17.13 A T Lymphocyte. Light microscope image of a T lymphocyte (lower center) showing the typical large cell nucleus. (Bar = 8 μm.)

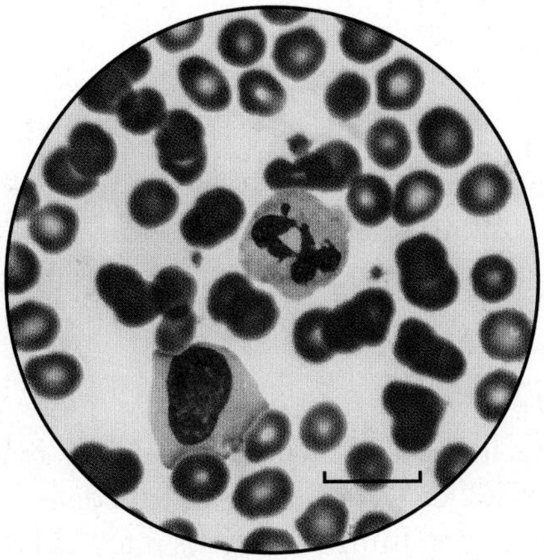

FIGURE 17.14 The Fate of Lymphoid Stem Cells. (A) Lymphocytes arise in the bone marrow and mature in the bone marrow (B cells) or thymus (T cells). **(B)** The T cells are involved in cell-mediated responses while **(C)** the B cells are involved with the antibody-mediated responses.

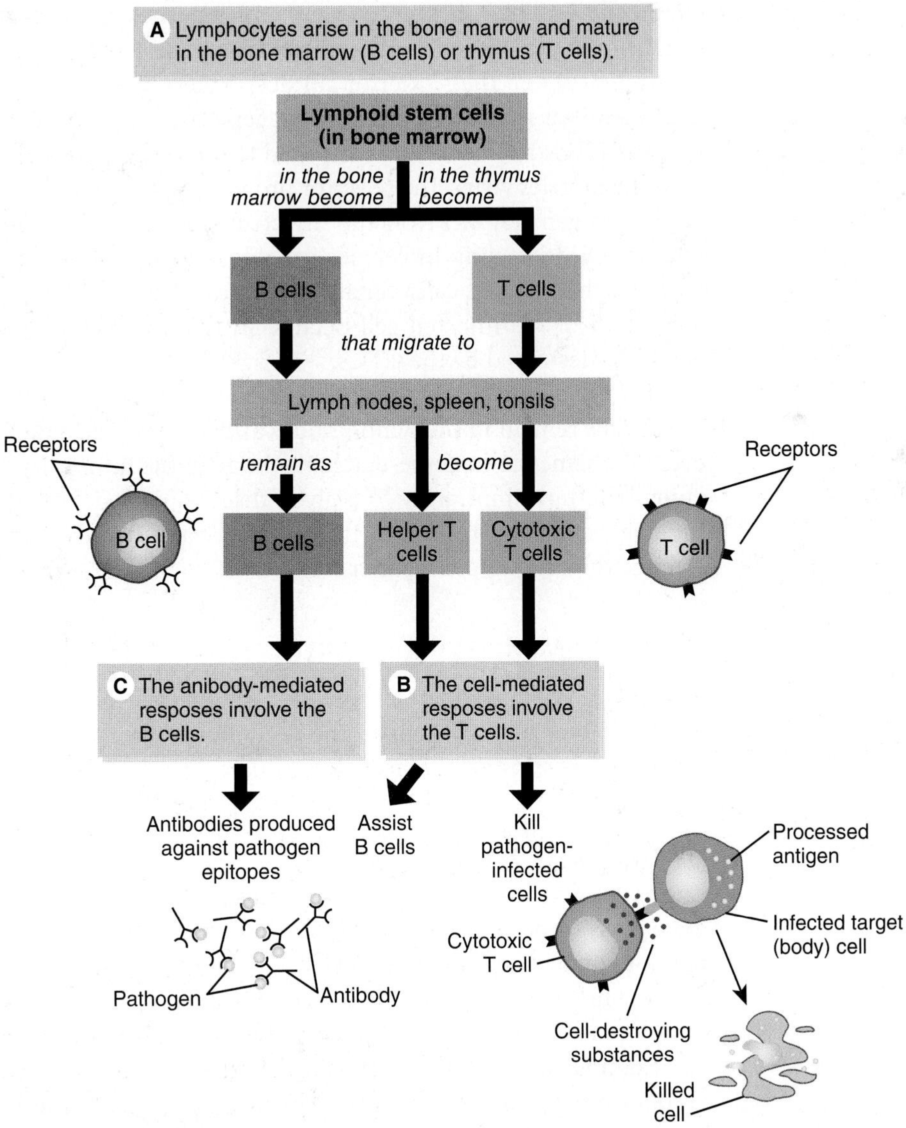

receptors, the T cells leave the thymus and migrate to the lymph nodes, spleen, and tonsils. At these locations, they mature as either "helper T cells" or "cytotoxic T cells."

B Cells

The B cells acquire a different "education." They stay in the bone marrow (B for bone), where they also acquire surface receptors. From there, the cells move to the lymph nodes, spleen, and tonsils where they take up residence with the T cells.

Cell-Mediated Response

The portion of adaptive immunity involving T cells is referred to as the **cell-mediated response** (**FIGURE 17.14B**). The cell-mediated response originates with the entry of antigen-containing pathogens into the body.

Death by Cytotoxic T Cell

Many pathogens, including certain bacterial species and viruses, infect body cells. These infected cells need to be destroyed, and that is the main job of the **cytotoxic T cells (CTCs)**. For example, when measles virus enters and infects cells, the infected cells destroy some of the viruses. They then present viral antigen fragments on their exterior surface. This combination is essentially a "red flag" identifying the cell as an infected cell.

The CTCs now leave the lymphoid tissue and go in search of cells infected with the measles virus. The receptor proteins of these CTCs recognize and bind to the antigen fragment on the infected cells. Binding triggers the CTCs to release cell-destroying molecules capable of poking holes through the membrane of the infected cells. Small molecules, fluids, and cell structures escape, and death of the infected cell occurs. In this manner, all such infected cells would be destroyed by the CTCs.

The cell-mediated response also involves the formation of **memory T cells**. These cells remain in the lymph nodes and can provide a rapid response in the event the same pathogen re-enters the body in the future. This is one reason we enjoy long-term immunity to a given disease (such as the measles) after having contracted and recovered from that disease. Remember what Thucydides remarked in the chapter opener.

Antibody-Mediated Response

While the cell-mediated response is killing infected cells, there might be large numbers of pathogens in the body fluids like the blood. The **antibody-mediated response** attempts to eliminate these pathogens (**FIGURE 17.14C**).

Mopping Up with Antibodies

The antibody-mediated response depends on the activity of **antibodies**, which are a class of proteins circulating in the blood. These proteins can react with epitopes on bacterial toxins as well as on whole pathogens.

The major cell type in this response is the B cell. When an antigen enters a lymph node, there is a specific B cell that recognizes and binds to the antigen. For example, if the pathogen is the measles virus, there are specific B cells that recognize measles virus epitopes. For other pathogens, there are other B cells that recognize those pathogens. In other words, there is a different B cell for every conceivable antigen (pathogen) epitope.

FIGURE 17.15 Antibody-Mediated Response. The antibody-mediated response results in the production of specific antibodies that bind to epitopes on the antigen.

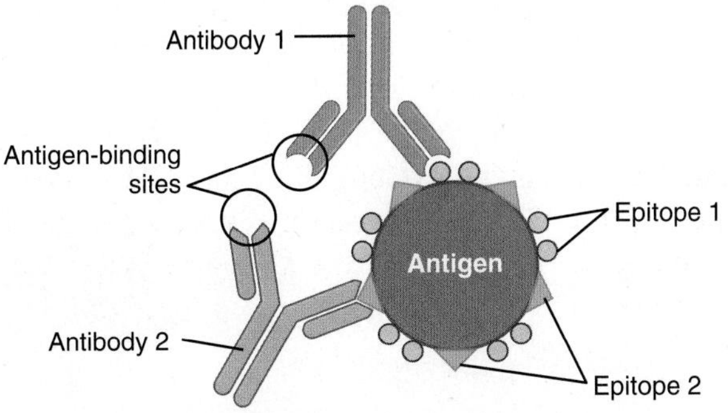

Besides binding to the antigen, B cells often require assistance from **helper T cells (HTCs)** to trigger antibody production. Binding between the two cells stimulates the B cell to divide into a large population of identical B cells. These B cells now mature into one of two types of cells. Some become **memory B cells** and, like memory T cells, will be ready for a future exposure to the same pathogen. Most B cells mature into **plasma cells**. These are large cells that produce antibodies to the specific pathogen infecting the body. In our example, the plasma cells would produce antibodies that would recognize and bind to epitopes on the measles virus (**FIGURE 17.15**).

The binding of antibody to the pathogen "coats" the antigen and acts as a "marker" that is recognized by phagocytes. The phagocytes, such as macrophages, recognize the antibodies coating the antigen and bind the antigen-antibody complex. The cells then phagocytize and destroy (mop up) the entire complex.

Antibody Structure

The basic antibody molecule consists of four polypeptide chains. There are two larger, identical **heavy (H) chains** and two smaller, identical **light (L) chains** (**FIGURE 17.16A**). These chains are joined together by chemical linkages to form a Y-shaped structure.

Each polypeptide chain, light or heavy, has both a constant and variable region. It is the amino acids of the variable region that allows the antibody to recognize a specific epitope on an antigen. Thus, the variable regions of a light and heavy chain combine to form a highly specific, three-dimensional structure somewhat analogous to the active site of an enzyme. This portion of the antibody molecule is called the **antigen binding site**. Because its "arms" are identical, there are two antigen binding sites, so it can combine with two identical epitopes. The tail region, composed of the two heavy chains, acts as the marker that is recognized by phagocytes.

Antibody Classes

Antibodies make up a group of proteins called **immunoglobulins (Igs)**. Five classes of antibodies have been identified and are designated IgG, IgM, IgA, IgE, and IgD.

- **IgG**. The classical "gamma globulin" is called IgG. This class of antibody comprises about 80% of the total antibody content in normal **serum**, the fluid portion of the blood. The IgG antibodies can be detected about 24 to 48 hours after B cell binding to the antigen. The IgG class provides long-term resistance to disease. It also is the maternal antibody that crosses the placenta and renders immunity to the fetus and newborn, which cannot fully produce antibodies until about 6 months of age.

FIGURE 17.16 Structure and Classes of Antibodies.
(A) The basic structure of an IgG antibody consists of two light and two heavy polypeptide chains. The variable domains in each light and heavy chain form a pocket called the antigen-binding site.
(B) Besides the IgG class, there are four other Ig classes. Note that the IgM class has 10 arms and 10 identical antigen binding sites and IgA has four arms and four identical antigen binding sites.

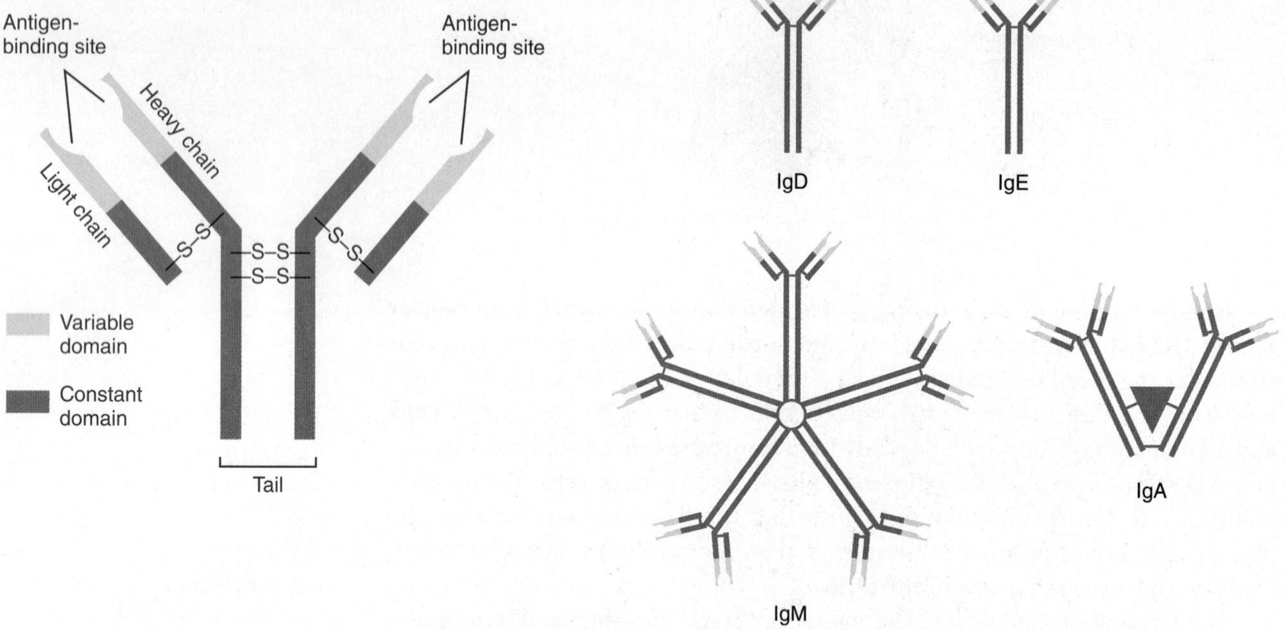

The remaining antibody classes are composed of one or more of the Y-shaped immunoglobulin units (**FIGURE 17.16B**).

- **IgM.** The first class of antibody to appear in the circulatory system after B cell stimulation is IgM. It is the largest antibody molecule and has 10 identical antigen binding sites. Because of its size, most of it remains in circulation and accounts for 5% to 10% of the antibody in the serum.
- **IgA.** The IgA class secreted and accumulates in body secretions. This antibody class is composed of two Y-shaped Ig units and four identical antigen binding sites. The antibody provides resistance in the respiratory and gastrointestinal tracts, where it helps block attachment of pathogens to the tissues. It also is produced in tears and saliva and in the colostrum, the first milk secreted by a nursing mother. When consumed by a child during nursing, the secretory IgA provides resistance to potential gastrointestinal pathogens.
- **IgE.** The IgE class plays a major role in allergic reactions by sensitizing cells to certain antigens.
- **IgD.** This class of antibody is the cell surface receptor that B cells use to bind to antigen.

Before concluding this section, it might be worth considering whether your immune system is influenced by "thinking good thoughts." **A CLOSER LOOK 17.3** considers this idea.

17.5 Vaccines: Stimulating Immune System Defenses

The first time the adaptive immune response encounters an antigen it takes about 10 to 14 days or longer for the system to become active. Therefore, when a pathogen infects the body for the very first time, it will take up to two weeks or more before substantial antibodies and CTCs are present to combat the

A CLOSER LOOK 17.3

Can Thinking "Well" Keep You Healthy?

Does a person's optimism affect the health of their immune system? A group of psychologists at the University of Kentucky and the University of Louisville wanted to find out. They followed university law students for 6 months to see if their expectations about their future careers (optimistic or not) was reflected in their immune responses.

The results of the study showed an optimistic disposition made no difference in their immune responses. However, as each student had highs and lows in law school, their immune response showed a similar up and down response. In optimistic times, their immune system was quicker to respond to an immunological challenge. In pessimistic times, their immune system was slower to respond to a similar challenge.

Because of studies like these, a new field called "psychoneuroimmunology" has emerged. One of the discoveries in this field is that there is a strong correlation between a patient's mental attitude and the progress of disease. Many studies report that an aggressive determination to conquer a life-threatening disease can increase the life span of those afflicted. For example, in one study, patients were provided with relaxation techniques, as well as mental imagery suggesting disease organisms were being crushed by the body's immune defenses. When these patients' immune system responses were examined, the immune responses appeared to accelerate the mobilization of the immune defenses.

Scientists at Ohio State University have looked at immune system responses to stress by examining what effects yoga might have on the system (see **FIGURE A**). The scientists found that after a stress experience, women who practiced yoga had low levels of immune-suppressing chemicals in their blood. Those women experiencing stress but not practicing yoga had higher levels of the chemicals. The levels of these immune-suppressing chemicals are important. Higher levels might contribute to the onset of such conditions as rheumatoid arthritis, inflammatory bowel disease, osteoporosis, multiple sclerosis, and some types of cancer. It is this intense stress that the "thinking well" movement attempts to address. Few reputable practitioners believe that behavioral therapies like yoga are the cure-all. However, the psychological devastation associated with many diseases, such as AIDS, cannot be denied. Very often, a person learning he or she is HIV positive goes into severe depression, which can adversely affect the immune system. Perhaps by relieving the psychological trauma, the remaining body defenses could handle the virus better.

As with any emerging treatment method, there are numerous opponents of behavioral therapies. Some opponents argue that naive patients might abandon conventional therapy. Another argument suggests therapists might cause enormous guilt to develop in patients whose will to live cannot overcome their failing health. Proponents counter with a growing body of evidence suggesting patients with strong commitments and a willingness to face challenges—signs of psychological hardiness—have relatively greater numbers of T cells than passive, nonexpressive patients.

To date, no study can conclude mood or personality has a life-prolonging effect on immunity. Still, doctors and patients are encouraged by the possibility of using one's mind to stay healthy.

FIGURE A A yoga class.

infection (**FIGURE 17.17A**). Using our measles example, the infected individual will suffer the symptoms of measles before the adaptive immune system can respond. Luckily, there is another way to generate these immune responses in a much shorter time. This other way is by **vaccination** (**FIGURE 17.17B**).

A **vaccine** is composed of an altered pathogen or part of a pathogen. Once vaccinated, the vaccine components will stimulate an individual's immune system to develop adaptive immunity. Importantly, these altered pathogen components usually do not trigger the disease. The person vaccinated does not become ill.

Vaccines simply strengthen and "educate" the body's adaptive immune system for potential future exposures to pathogens. Suppose an individual receives the measles vaccine. In the body, the vaccine components trigger an adaptive

FIGURE 17.17 Two Ways to Acquire Active Immunity.

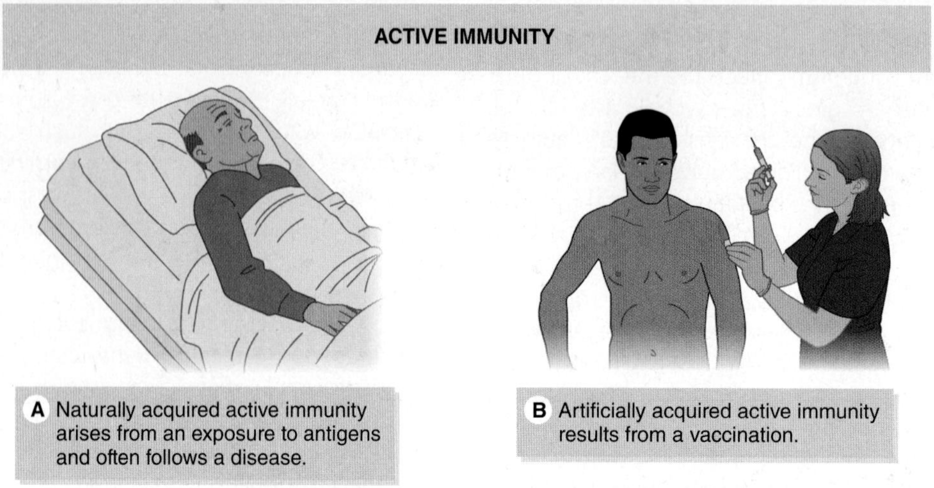

ACTIVE IMMUNITY
A Naturally acquired active immunity arises from an exposure to antigens and often follows a disease.
B Artificially acquired active immunity results from a vaccination.

TABLE 17.1 Major Bacterial and Viral Vaccines Currently in Use, United States

Whole-Agent Vaccines	Genetically Engineered Vaccines
Live, Attenuated Vaccines ▪ Chickenpox (varicella) ▪ Measles, mumps, rubella (MMR vaccine) ▪ Rotavirus ▪ Yellow fever **Inactivated Vaccines** ▪ Flu (shot only) ▪ Hepatitis A ▪ Polio (shot only) ▪ Rabies **Toxoid Vaccines** ▪ Diphtheria ▪ Tetanus	**Subunit Vaccines** ▪ Hepatitis B ▪ Hib (*Haemophilus influenzae* type b) disease ▪ HPV (human papillomavirus) ▪ Meningococcal disease ▪ Pneumococcal disease ▪ Shingles (zoster) ▪ Whooping cough [part of the diphtheria-tetanus-pertussis (DTaP) combined vaccine]

immune response as described above. As part of that response, long-lived B memory cells will be produced. If that person is exposed to the natural measles virus months or years later, the B memory cells act within just a few days to produce plasma cells and antibodies. With a vaccination, the infection is eliminated before it can make the individual sick.

The terms "vaccination" and "immunization" often are used interchangeably. However, the terms are not quite the same. Vaccination is the process of administering a vaccine that produces immunity in the body. **Immunization** is the process by which an individual becomes protected from a pathogen.

Let's now examine the different types of vaccines. The major viral and bacterial vaccines currently in use in the United States are summarized in **TABLE 17.1**.

Whole-Agent Vaccines

Many vaccines consist of the whole bacterium, virus, or bacterial toxin as the antigen. However, they are prepared in different ways.

Live, Attenuated Vaccines

Some pathogens can be weakened such that they should not cause disease. Such **attenuated vaccines** are said to contain "live" pathogens because the bacteria or viruses still can multiply or replicate. However, they do so at an extremely slow pace. These vaccines are the closest to the natural pathogens, and therefore, once vaccinated they generate a strong and long-lasting immune response. With one or two doses of the vaccine, the person often develops life long immunity.

The downside of a few attenuated vaccines is that on very rare occasions, the agent might revert to its virulent form and cause the very disease the vaccine was meant to prevent. The oral (Sabin) polio vaccine is an example of one such vaccine. It is no longer used for polio immunizations in most countries. In fact, there are no attenuated bacterial vaccines routinely used in the United States today (see Table 17.1).

Inactivated Vaccines

Another strategy for preparing vaccines is to use whole bacterial cells or viruses that have been chemically or heat killed. However, the chemicals or heat used for these **inactivated vaccines** alters the microbe's structure and shape. As a result, the vaccine produces a weaker immune response and the vaccinated individual might need several doses (booster shots) to maintain immunity.

Inactivated vaccines are safer than some attenuated vaccines because the pathogen in the inactivated vaccine cannot multiply or replicate. Therefore, the vaccine cannot cause disease in a vaccinated individual. Inactivated vaccines currently in use are identified in Table 17.1.

Toxoid Vaccines

For protection against bacterial diseases caused by exotoxins like diphtheria and tetanus, the bacterial toxin is inactivated chemically. Like some inactivated vaccines, these **toxoid vaccines** contain exotoxins with a chemically modified shape. For this reason, the vaccinated person might need booster shots to regain strong protective immunity.

Genetically Engineered Vaccines

The genetically engineered vaccines available today sometimes are called second generation vaccines because they are the product of more modern production processes. These vaccines contain only a genetically engineered subunit or fragment of the bacterial cell or virus. With only a small piece of the pathogen, a vaccinated individual cannot contract the disease from the vaccine. Scientists in industry and universities continue to design and develop new types of genetically engineered vaccines against bacterial and viral pathogens.

Subunit Vaccines

Unlike the whole-agent vaccines, the strategy for a **subunit vaccine** is to have the vaccine contain only those pieces (subunits) of the pathogen that stimulate a strong immune response. These subunits can be a bacterial polysaccharide, a bacterial protein, or a **capsid** protein of a virus (see Table 17.1). Subunit vaccines represent the safest vaccines that are available.

Capsid: The protein coat surrounding the virus genome.

Vaccine Need and Safety

Today, some individuals and parents have questions about vaccines. They are especially concerned about the need for vaccines in today's society and nervous as to the potential safety of vaccines.

The Need for Vaccination

Most childhood diseases like diphtheria and measles are rarely (but see below) seen in the United States. So, why vaccinate? The answer to that question is straightforward: It is because of vaccines that these diseases are rare or nonexistent today. Historically, with the introduction of each vaccine, the number of reported cases has dropped swiftly (**TABLE 17.2**). Vaccines are extremely effective. Unfortunately, many people have forgotten, or never experienced, how terrible some vaccine-preventable diseases can be.

In other parts of the world, many of these childhood diseases are still prevalent. Should an unvaccinated person be exposed to an infectious individual, the unvaccinated person might become ill. If that person then contacts other susceptible individuals or pockets of unvaccinated individuals, the pathogen can spread and an outbreak is likely. Therefore, it is important that as many people as possible get vaccinated. A situation in which most of a population is immune to an infectious disease (through vaccination and/or prior illness) to make its spread from person to person unlikely is called **herd (community) immunity** (**FIGURE 17.18**).

If herd immunity is high, even individuals not vaccinated, such as newborns and those with chronic illnesses, are indirectly protected. This is because the infectious agent has little opportunity to spread within the community. According to health experts and disease epidemiologists, when greater than 85% of the population (around 95% for some diseases like measles) is vaccinated, the spread of the disease is stopped. Although the rest of the "herd" or population remains susceptible, there are so many vaccinated people in the "herd" that it is unlikely an infected person could easily spread the disease.

The 2018-2019 measles outbreak in the United States (highest number of measles cases in 25 years) and the surge in measles cases globally are particularly instructive. Most people who developed measles were unvaccinated. Although measles was eliminated (absence of endemic cases and disease transmission for greater than 12 months) in the U.S. in 2000, foreign travelers with measles can bring the virus into the country. Also unvaccinated Americans traveling abroad can get infected with the virus and bring the illness back to the U.S. An outbreak can occur if the infected individual visits a community where herd immunity is

TABLE 17.2 Decline in Disease Cases, United States

Disease	Reported Number of Cases		
	Pre-Vaccine Era [year vaccine approved]	Current (2017)	% Decrease
Diphtheria	>200,000 (1923)	0	100
Measles	>530,000 (1971)	120	>99
Mumps	>162,000 (1967)	6,109	>96
Pertussis (whooping cough)	>200,000 (1930, 1991, 2005)	18,975	>91
Polio	>58,000 (1955)	0	100
Rubella	>47,000 (1969, 1979)	7	>99
Smallpox	>29,000 (1931)	0	100

FIGURE 17.18 Herd Immunity.
(A) If no individuals in a community are vaccinated, it is likely that an outbreak or epidemic could occur.

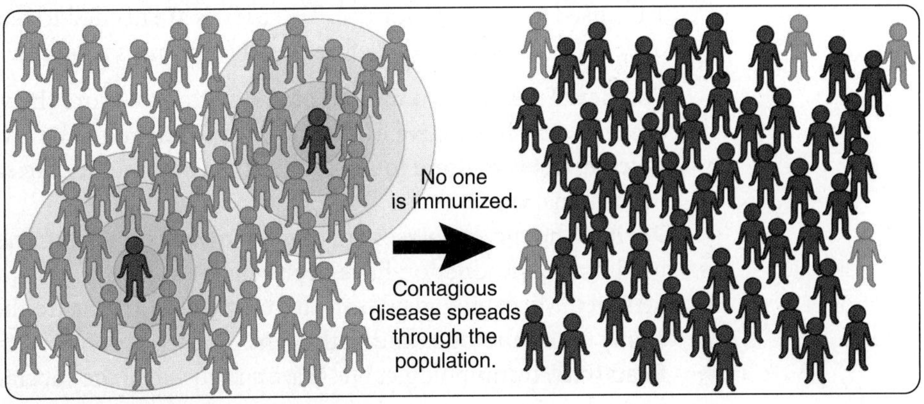

(B) If some individuals in the community are vaccinated, it still is not enough to confer herd immunity.

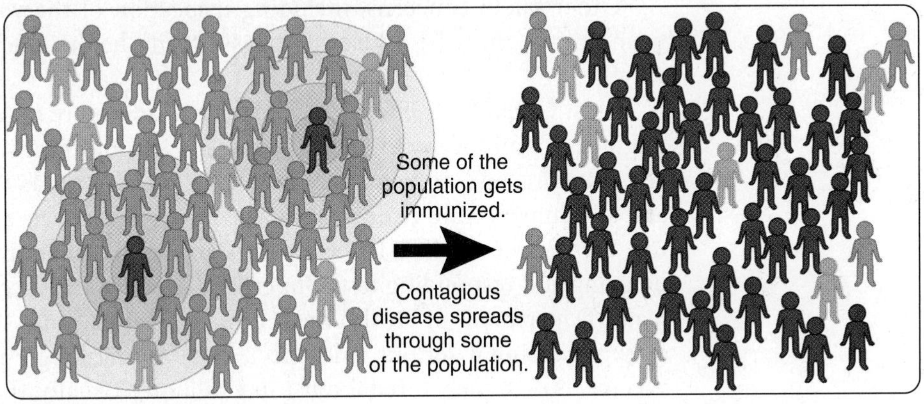

(C) If a large percentage of a population is vaccinated, outbreaks of disease can be prevented.

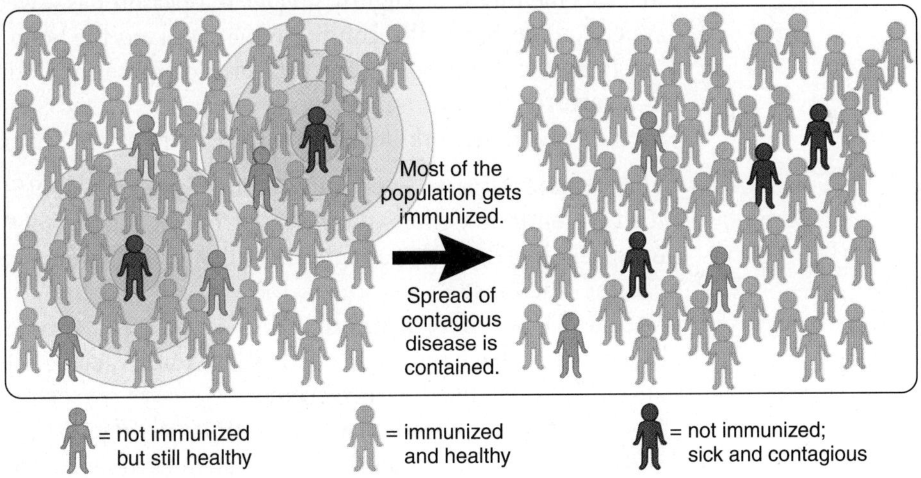

below the protective threshold (95% for measles); that is, where there are unvaccinated individuals or small groups of unvaccinated people. Thus, herd immunity is essential because below the threshold, the measles virus can "find" susceptible individuals and spread among other vulnerable members in the community. Measles is discussed in greater depth in the chapter on Virus Diseases of Humans.

Vaccine Safety

Vaccines are given to millions of healthy people around the world every day. Consequently, vaccines must be held to very high safety standards. The US Food and Drug Administration (FDA) requires vaccine manufacturers to follow extensive safety procedures when producing vaccines used in the United

States to ensure they are safe and effective. After years of lab and animal tests have been completed, a promising vaccine must undergo clinical trials before being licensed by the FDA for use. These trials are needed to make sure the vaccine meets certain standards before the FDA licenses a vaccine.

- **Potency.** Tests with volunteers are done to make sure the vaccine works as expected and is effective. If the vaccine is designed to be given with another vaccine, tests are done to make sure the two vaccines are safe and effective in combination.
- **Purity.** Chemical analyses are done to make certain extraneous ingredients used during production have been removed.
- **Sterility.** Microbiological tests are carried out to ensure the vaccine is free of any other microbes and pathogens.
- **Reactivity.** Immunological tests are run to verify the vaccine does not cause an adverse immune reaction.

Other safety concerns, including inspection of the vaccine manufacturing plant, must be addressed satisfactorily before the FDA licenses a vaccine.

Most vaccine side effects in children and adults are mild. A vaccination might cause minor swelling and redness at the injection site, a mild fever, and tiredness and body aches. These effects pass in a day or two. Parents and all adults need to realize that these minor side effects are an indication that the vaccine worked; the symptoms are due to the immune system recognizing and responding to the microbial agent in the vaccine. The United States Department of Health and Human Services states that the chances of experiencing a serious side effect, such as an allergic reaction from a vaccine, is extremely rare (about 1 in a million). Usually, these cases can be treated with standard medical care. However, the chances of a bad reaction to a vaccine might be slightly greater if a person has a weakened immune system or another health condition that weakens the body.

Seizure: A sudden attack of illness, especially a stroke or abnormal brain activity, such as convulsions or spasms.

Even rarer is the possibility of experiencing a serious "adverse event" such as a **seizure**. To "catch" any adverse events, the United States has one of the most advanced networks for monitoring vaccine safety. The networks include:

- **The Vaccine Adverse Events Reporting System (VAERS).** Established by the FDA and the Centers for Disease Control and Prevention (CDC), anyone, including doctors, patients, individuals, and parents can report adverse vaccine reactions.
- **Vaccine Data Safety Datalink (VSD).** The CDC and several healthcare organizations have set up this datalink to track vaccine safety and monitor any side effects linked to the vaccine.
- **Post-Licensure Rapid Immunization Safety Monitoring System (PRISM).** The FDA established this system to monitor products, including vaccines, after they are licensed for use.
- **Clinical Immunization Safety Assessment Project (CISA).** The CDC and vaccine safety experts carry out clinical vaccine safety research at the request of providers to evaluate possible vaccine safety side effects.

Autism spectrum disorder: A developmental disorder that affects communication and behavior generally occurring within the first two years of life.

Recently, the vaccine of most concern regarding side effects and childhood development is the measles-mumps-rubella (MMR) vaccine. In 1998, a medical paper (now discredited) was published suggesting that there might be a link between the MMR vaccine and **autism spectrum disorder** in children. The idea was put forward that the vaccine might affect a child's immune system and cause neurological damage. To make a long story short, as of 2019, more than a dozen studies carried out by independent scientists and researchers around the world have unanimously found no evidence to support the autism claim. The MMR vaccine does not lead to increased rates of autism in children.

Through mandatory immunization, diseases like polio, smallpox, and chickenpox have been eradicated or almost eradicated. Although a few of the millions of people vaccinated each year might suffer a serious consequence from vaccination, the risks of contracting a disease (especially in infants and children) from not being vaccinated are thousands of times greater than the risks associated with any vaccine. In fact, the chances of developing a seizure from the MMR vaccine are less likely than developing a seizure from an actual measles infection. In addition, the licensed vaccines are always being reexamined for ways to improve their safety and effectiveness.

A Final Thought

It might have occurred to you that an episode of disease is much like a war. First, the invading microbes must penetrate the natural barriers of the body. Then, they must evade the phagocytes constantly patrolling the body's circulation and tissues. Finally, they must elude the antibodies or T cells the body sends out to combat them. How victorious the body is in this battle will determine whether the individual survives the disease.

To be sure, antibiotics and other antimicrobial drugs help in those cases where diseases pose life-threatening situations. In addition, sanitation practices, insect control, care in the preparation of food, and other public health measures prevent microbes from reaching the body in the first place. However, in the final analysis, the immune system represents the bottom line in protection against disease, and, as history has shown, it usually works very well. The late Lewis Thomas, an American physician, poet, essayist, and researcher, said it best in his book, *The Lives of a Cell*:

> *"A microbe that catches a human is in considerably more danger than a human who catches a microbe."*

Chapter Discussion Questions

What Was He Thinking?

From your reading of this chapter, identify and discuss five major points about the immune system and vaccines that the author was trying to get across to you.

Questions to Consider

1. In his classic book *The Mirage of Health* (1959), the French scientist René Dubos develops the idea that health is a balance of physiological processes, a balance that considers such things as nutrition and living conditions. (Such a view opposes the more short-sighted approach of locating an infectious agent and developing a cure.) From your experience, describe several other items or events you would add to Dubos' list of "balancing agents."
2. The transparent windows placed over salad bars are commonly called "sneeze guards" because they help prevent respiratory droplets from reaching the salad items. What other suggestions might you make to prevent disease transmission via a salad bar?

3. An environmental microbiologist has created a controversy by maintaining that a plume of water is aerosolized when a toilet is flushed, and that the plume carries bacteria to other items in the bathroom, such as toothbrushes. Assuming this is true, what might be two good practices to follow in the bathroom?
4. Suppose you are a microbe hunter assigned to make a list of the ten worst "hot zones" in your home. The title of your top-ten list will be "Germs, Germs Everywhere." What places will make your list, and why?
5. The ancestors of modern humans lived in a sparsely settled world where infectious diseases were probably very rare. Suppose one of those individuals was magically thrust into our present-day world. How do you suppose he or she would fare in relation to infectious disease? What is the immunological basis for your answer?
6. In the book and 1966 classic movie, *Fantastic Voyage,* a group of scientists is miniaturized in a submarine (the Proteus) and sent into the human body to dissolve a blood clot. The odyssey begins when the miniature submarine carrying the scientists is injected into the bloodstream. What perils, immunologically speaking, would you encounter if you were to take such an adventure?
7. It is estimated that when at least 85% of the individuals in a population have been immunized against a disease, the chances of an epidemic occurring are very slight. The population is said to exhibit "herd immunity." In fact, members of the population (or herd) unknowingly can transfer the immunizing agent to other members of a population and eventually immunize the entire population. What are some ways by which the immunizing agent can be transferred?
8. When children are born in Great Britain, they are assigned a doctor. Two weeks later, a social services worker visits the home, enrolls the child on a national computer registry for immunization, and explains immunization to the parents. When a child is due for an immunization, a notice is automatically sent to the home, and if the child is not brought to the doctor, the nurse goes to the home to learn why. Do you believe a method like this can work (or should be used) in the United States to achieve uniform national immunization?
9. If a parent or any adult is concerned about vaccine safety, what vaccine monitoring systems are in place in the United States to oversee vaccine safety and possible vaccine side effects?

© Universal History Archive/Universal Images Group/Getty Images.

CHAPTER 18

Viral Diseases of Humans: AIDS to Zika

The Flu

Every year there are seasonal influenza (flu) outbreaks around the world. Sometimes the outbreaks are mild. At other times, they become epidemic and spread quickly through a large, susceptible population. Occasionally, a new type of influenza virus emerges that can infect many individuals and spread easily person to person. The result of the worldwide spread is a pandemic. The largest influenza pandemic occurred in 1918 to 1919. Called the "Spanish flu," it is estimated that a fifth of the world's population was infected by a very virulent form of the flu virus. Within that one-year period, an estimated 20 million to 50 million people died.

Influenza continues to be an ongoing problem in the twenty-first century. Just look back to 2009. In April of that year, the United States Centers for Disease Control and Prevention (CDC) identified two children in California who were infected with a new strain of influenza virus. This strain was traced back to the major flu outbreak in Mexico that originated in pigs. By the end of April, cases of this new flu strain, originally called "swine flu," were being reported in cities around the world. It appeared that a potentially dangerous flu pandemic was emerging.

The number of cases of swine flu continued to increase in the spring and early summer. Fortunately, most infected individuals exhibited only mild symptoms. The CDC estimated that between 150,000 and 500,000 people died worldwide which is similar to the numbers of deaths due to the seasonal flu. In the United States, there were an estimated 61 million cases and 275,000 hospitalizations.

CHAPTER 18 OPENER 2018 marked the 100th anniversary of the so-called Spanish flu pandemic that swept the globe in 1918–1919. It was one of the deadliest disease outbreaks in recorded history. Here, nurses with the Red Cross Motor Corps are on duty. They are wearing masks to try to prevent the spread of the flu virus. Still, about 675,000 people died in the United States, mostly from complications resulting from the "Spanish flu."

FIGURE 18.1 2017–2018 Seasonal Flu—United States. The burden of the 2017–2018 flu season was unusually heavy.

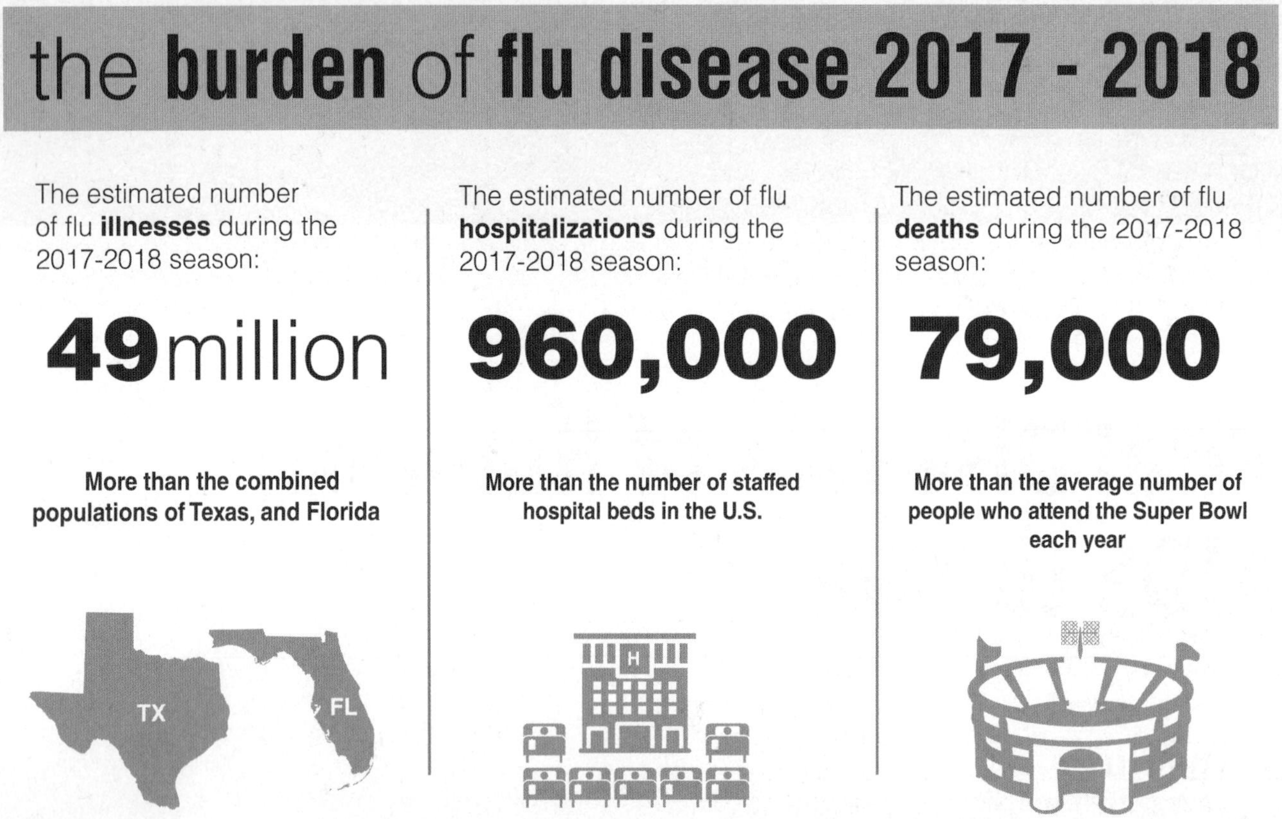
the burden of flu disease 2017 - 2018
The estimated number of flu illnesses during the 2017-2018 season:
49 million
More than the combined populations of Texas, and Florida
TX
FL
The estimated number of flu hospitalizations during the 2017-2018 season:
960,000
More than the number of staffed hospital beds in the U.S.
H
The estimated number of flu deaths during the 2017-2018 season:
79,000
More than the average number of people who attend the Super Bowl each year

DATA: Influenza Division program impact report 2017-2018, https://www.cdc.gov/flu/about/burden/index.html

Courtesy of CDC.

There were about 12,500 deaths reported, which was less than the 19,000 that usually die every year from the seasonal flu. This time, the pandemic did not develop into a monster Spanish flu-like pandemic.

There were at least three major take-home lessons from these "flu events." First, in this age of globalization, new viral threats can come from anywhere and spread with the speed of jet travel. Second, we need to be on "virological alert" worldwide to identify new virus strains before they cause major health issues. Third, people need to be vaccinated. For the 2017–2018 flu season, the CDC reports that about half of the U.S. population was vaccinated (**FIGURE 18.1**).

Influenza is just one of the present-day viral diseases we will discuss in this chapter on viral disease in humans. Let's get started.

LOOKING AHEAD

After reading and completing this chapter, you will be able to:

18.1 Identify and describe several viral diseases of the skin.
18.2 Draw the structure of a type A flu virus and compare flu symptoms with common cold symptoms.
18.3 Identify three viral diseases of the nervous system and explain why they are so dangerous.
18.4 List and describe several diseases of the blood, liver, and gastrointestinal tract.
18.5 Contrast an HIV infection versus AIDS.

18.1 Viral Diseases of the Skin: From Mild to Deadly

The viral diseases of the skin compose a diverse collection of human maladies. Some infections, such as herpes simplex, remain epidemic in our time. Many childhood diseases, such as measles, mumps, and chickenpox, usually are under control through effective vaccination programs. All these so-called "**dermotropic** diseases" normally are transmitted by contact with an infected individual, and at least some of the symptoms appear on the skin tissues. Unfortunately, there are relatively few antiviral drugs for these diseases, so prevention programs remain the major course of action.

Dermotropic: Referring to an illness that is attracted to, is localized in, or has entered by way of the skin.

Cold Sores and Genital Herpes

Two of the most prevalent human herpes viruses are herpes simplex virus type 1, or HSV-1, and HSV-2. Humans are the only host for both viruses.

Cold Sores

HSV-1 generally infects the facial area around the lips and is the cause of most **cold sores**, also called fever blisters (**FIGURE 18.2**). The first occurrence of cold sores can arise from kissing or sharing eating utensils or towels with a person having active lesions on the lips. Therefore, it is important to avoid contact with someone who has active blisters and to not share items that can spread the virus. With active blisters, frequent handwashing is important. In some rare cases, HSV-1 can spread to the brain or eyes. In the brain, virus infection can cause herpes **encephalitis**, and in the eyes cause herpes **keratitis**.

A person infected with HSV-1 remains infected for life. This is because the virus can lie dormant in sensory nerve cells of the face and emerge sometime later as another active cold sore infection on the lips. Often, recurrence is triggered by some form of stress, such as fatigue, sunburn on the lips, menstruation in women, or anything that lowers the body's resistance to infection. That is why the malady is called "cold sores" or "fever blisters"; reactivation often comes from the body's stress of having a cold or a fever. Although HSV-1 infections

Encephalitis: An inflammation of the brain.

Keratitis: An inflammation of the cornea, which is the dome-shaped tissue in the front of the eye.

FIGURE 18.2 Herpes Simplex Viruses Can Cause Cold Sores. Cold sores (fever blisters) erupt as tender, itchy sores that eventually burst, drain, and form scabs. Contact with the sores accounts for spread of the virus.

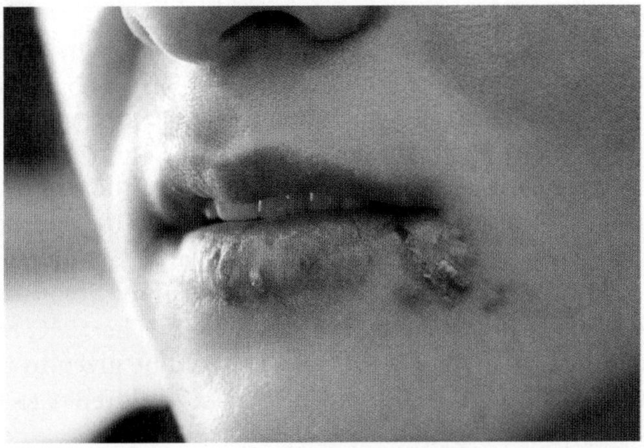

cannot be cured, there are a few antiviral pills or creams that accelerate the healing process and lessen the severity of recurrent episodes.

Genital Herpes

HSV-2 usually is responsible for causing **genital herpes**, a **sexually transmitted infection** (**STI**). However, oral sex can spread HSV-1 to the genitals and HSV-2 to the lips. Genital herpes affects between 10 and 20 million Americans yearly, the majority in people aged 14 to 49 years. Most people are unaware of the infection because there can be few, if any, symptoms. When symptoms do occur, painful blisters on or around the genital area, rectum, or mouth might erupt, crust over, and disappear, usually within about 3 weeks. Again, a lifelong infection occurs because the virus can remain dormant in the neurons around the spine. Therefore, genital herpes might reappear, although it often is less severe than the first episode.

In a pregnant woman, herpes simplex viruses sometimes pass to the fetus via the placenta. Infecting the placenta, the virus causes **neonatal herpes**, which can produce neurological problems and/or mental impairment in the newborn. Miscarriage and premature birth also can occur as a result of infection. Therefore, it is critical that a pregnant woman tell her doctor if she has ever experienced any symptoms of, been exposed to, or been diagnosed with genital herpes. If so, she might be given antiviral medication to reduce the risk of another episode. Obstetricians recommend women with active genital herpes consider giving birth by cesarean section, which lessens the chance of virus spread to the newborn.

Chickenpox and Shingles

In the centuries when pox diseases regularly swept across Europe and other parts of the world, people had to contend with the Great Pox (syphilis), the smallpox, the cowpox, and the chickenpox. Today, chickenpox and its associated disease, shingles, still cause significant illness.

Chickenpox

Chickenpox, or varicella disease, is highly communicable to people who have never had the disease or have not been vaccinated. The infection is caused by the varicella-zoster virus (VZV), a DNA virus that localizes in the skin and in nerves close to the skin surface. Infection leads to the formation of raised pink bumps that develop into a telltale rash composed of small, teardrop-shaped, fluid-filled blisters (**FIGURE 18.3A**). The blisters develop over 3 or 4 days in a succession of crops before breaking open to yield highly infectious, virus-loaded fluid. The blisters eventually form crusts and scabs, and they heal over several days. The chickenpox infection usually lasts about 5 to 10 days. The drug acyclovir lessens the symptoms of chickenpox and hastens recovery.

Chickenpox used to be very common in the United States. According to the CDC, in the early 1990s, there were about 4 million reported cases of chickenpox, 11,000 hospitalizations, and more than 100 deaths each year. Today, two chickenpox vaccines are available. Varivax contains only attenuated VZV, while ProQuad contains a combination of **attenuated** (weakened) measles, mumps, rubella viruses, and VZV. The CDC recommends two doses of either vaccine. The first dose should be given to infants at 12 to 15 months old and the second dose given to children 4 to 6 years old.

FIGURE 18.3 The Lesions of Chickenpox and Shingles.
(A) A typical case of chickenpox. The lesions might be seen in various stages, with some in the early stage of development and others in the crust stage.

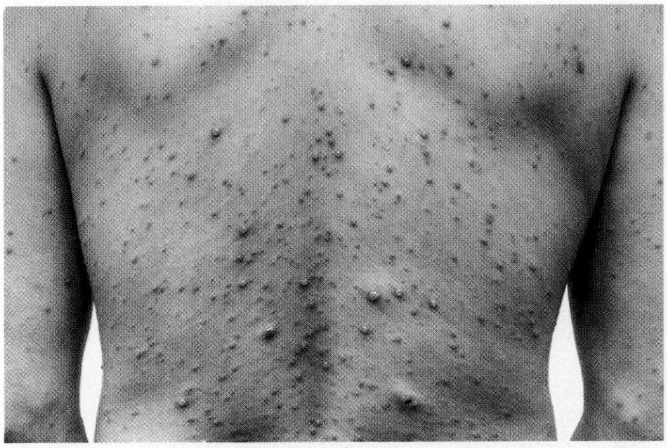

(B) Dermal distribution of shingles lesions on the skin of the body trunk. The lesions contain less fluid than in chickenpox and occur in patches as red, blister-like areas.

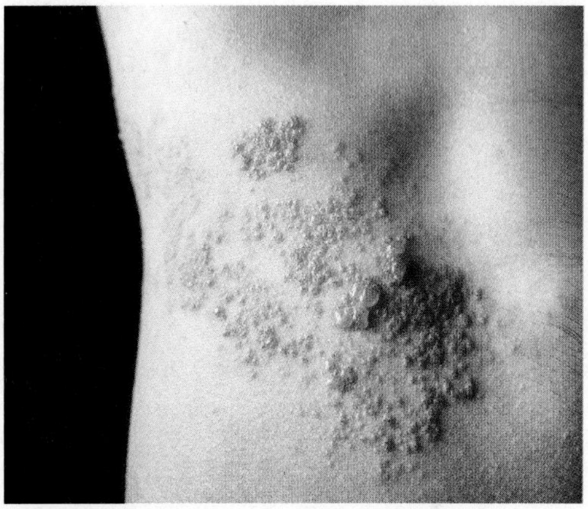

Immunization with the vaccines is responsible for the dramatic decrease in the occurrence of chickenpox in the United States. The CDC reported only 38 cases in 2017.

Shingles

Shingles, or herpes zoster, usually is an adult disease also caused by VZV. Like the herpes simplex viruses, VZV remains dormant in nerve tissue near the spinal cord in an adult who had chickenpox as a child. Years later, the viruses might reactivate and multiply in nerve tissue. They then travel down the nerves to the skin, where they can cause an excruciatingly painful rash often encircling either the left or right side of the torso (**FIGURE 18.3B**). Many sufferers also experience headaches as well as facial paralysis and sharp "icepick" pains. The condition can occur repeatedly, and its occurrence is linked to emotional and physical stress, a suppressed immune system, or aging. If a person who has never had chickenpox encounters a person with shingles, the susceptible person, if infected, will get chickenpox.

Antiviral medications can lessen the pain and speed recovery. Two shingles vaccines are available for adults over 50 years of age. Zostavax has been in use since 2006. A recombinant zoster vaccine called Shingrix has been in use since 2017 and is now the preferred vaccine.

Other Dermotropic Diseases

A few other viral infections also affect the skin. Some are common childhood diseases, such as measles and mumps. Another disease, smallpox, was once a worldwide scourge.

Measles

Measles, also called rubeola, is a highly contagious disease. It is an infection of the respiratory tract and usually is transmitted by **respiratory droplets** coughed out by an infected person during the early stages of the disease. Its

FIGURE 18.4 The Measles Rash. A child with measles, showing the typical rash on face and torso.

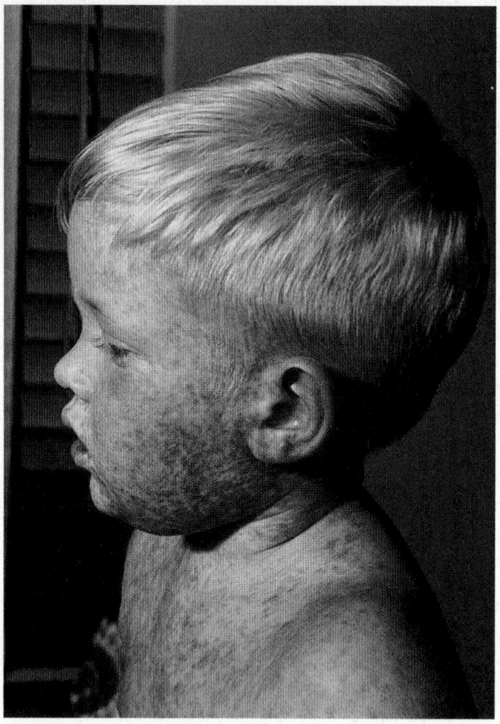

Courtesy of CDC.

recognizable symptoms appear on the skin, and so it is considered a skin disease.

Measles symptoms commonly include a hacking cough, sneezing, runny nose, eye redness, sensitivity to light, and a high fever. The characteristic red rash soon appears, beginning as pink-red, pimple-like spots. The rash breaks out at the hairline, covers the face, and spreads to the trunk and extremities (**FIGURE 18.4**). Within a week, the rash turns brown and fades.

Measles infections and deaths are still common in many parts of the world. The World Health Organization (WHO) reported more than 20 million cases and 110,000 deaths in 2018. Most of these cases and deaths were from complications of pneumonia and encephalitis in unvaccinated children under 5 years of age. In addition, in mid 2019, the United States and several other countries are experiencing record numbers of reported measles cases, the majority in people who were not vaccinated (**FIGURE 18.5**). These cases are arising from infected individuals who are coming to the U.S. and unvaccinated American travelers who got infected while traveling abroad. In both cases, these individuals have transmitted the measles virus to individuals or small groups of people who have not been vaccinated. Thus, failure to immunize people in the countries receiving infectious visitors is what is fueling the measles resurgence today.

This is shocking because measles is a vaccine-preventable disease. The CDC states that two doses of the measles, mumps, rubella (MMR) vaccine (the first at 12 to 15 months; the second at 4 to 6 years) are about 97% effective at preventing measles in children who receive the vaccine. With the introduction of the MMR vaccine in 1971, measles deaths have fallen. Globally,

FIGURE 18.5 Annual Number of Reported Measles Cases, United States. In recent years, there has been an increase in the number of reported cases, reaching a 25 year high in mid 2019.

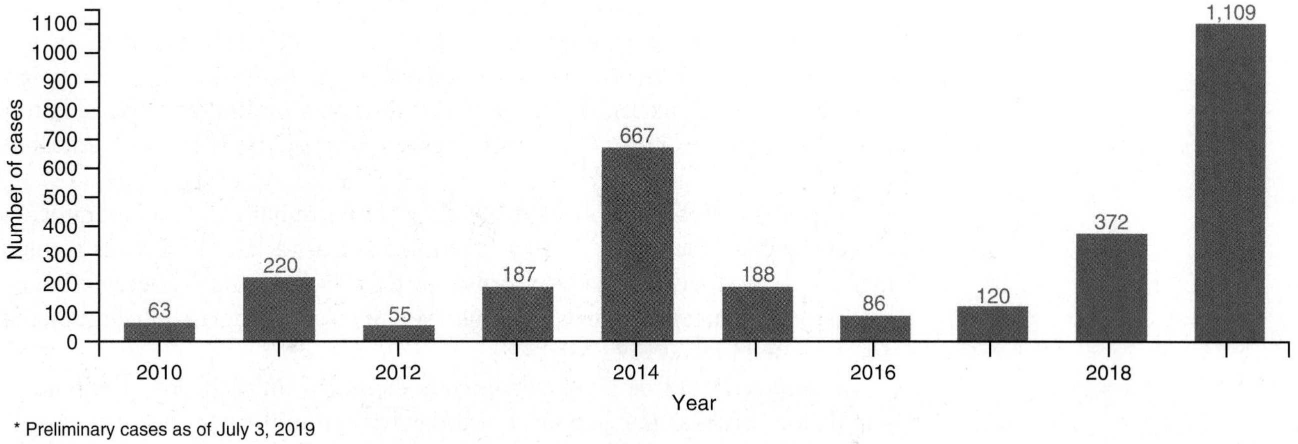

Courtesy of CDC.

measles deaths have decreased by 80% between 2000 and 2017. That translates into a saving of more than 21 million lives due to vaccination.

Immunization of children in the United States is not mandatory. However, the large measles outbreaks are causing health and government officials to reconsider exemptions that are not medical in nature (Figure 18.5). Nevertheless, compliance is a thorny social issue because a very small minority of parents object to some or all immunizations based on religious or philosophical beliefs. In addition, some people incorrectly believe vaccines have serious and dangerous side effects. In fact, such complications are extremely rare, as the vaccine information in the chapter on Disease and Resistance describes.

Mumps

When a person has **mumps**, ducts leading from the **parotid glands** become obstructed, which retards the flow of saliva and causes the characteristic swelling under one or both ears. The skin overlying the glands becomes taut and shiny, and patients experience pain when the glands are touched.

Parotid gland: A salivary gland present on either side of the mouth and in front of both ears.

Mumps is spread by respiratory droplets in coughs and sneezes. Transmission to an adult male can lead to an infection of the testicles and cause a lowering of the sperm count. This condition, however, rarely leads to fertility problems. Still, it can be painful, so to prevent the infection, susceptible adult males are told to avoid children with mumps.

The number of mumps outbreaks in the United States has risen sharply in recent years. These outbreaks are occurring even though most of the infected individuals have received the recommended two-dose series of the mumps vaccine. The reason for this increase is unclear. It might be because there are more people who are unvaccinated, or the immune protection provided by the mumps vaccine is weakening. To stop the growing numbers of mumps in the United States, public health officials need to stress the importance of timely childhood MMR vaccination.

Smallpox

Throughout history, **smallpox** was a horrible, contagious disease that could sometimes be fatal. Those individuals surviving the disease often had pitted scars called pockmarks. The earliest symptoms include fever, body aches, and a pink-red rash that develops into large, fluid-filled blisters (**FIGURE 18.6**). Eventually, the blisters break open and emit pus.

At present, there is no smallpox disease or wild smallpox virus in the world. Starting in 1966, the WHO coordinated a global vaccination campaign to eradicate smallpox and by late 1977, healthcare workers reported isolation of the last smallpox case. The disease was certified by the WHO as eradicated in 1980.

Smallpox viruses (also called variola viruses) remain in two laboratories: one at the CDC facility in Atlanta, the other in Russia. The WHO has recommended these stocks of smallpox viruses be destroyed, especially because scientists have sequenced the virus' genome. Not all scientists agree. The arguments are presented in **A CLOSER LOOK 18.1**.

Smallpox would be a very dangerous disease if unleashed on the human population. Because few people alive today have immunity to the virus, smallpox or any bioterrorism weapon, could cause a calamity of unprecedented proportions. **A CLOSER LOOK 18.2** looks at the topic of bioterrorism.

18.2 Viral Diseases of the Respiratory Tract: Flus and Colds

Several viral diseases occur within the human respiratory tract. The most common are influenza and the common cold that annually affect millions of people around the world.

Influenza

Influenza (the flu) is an acute, contagious disease that is transmitted in respiratory droplets. The influenza virus is quite unusual. It is composed of eight separate RNA segments (**FIGURE 18.7**). The viral envelope contains a group of protein projections called spikes. Most of the spikes are composed of the enzyme **hemagglutinin** (**H spikes**) that assist the entry of the virus into its host cell. The rest of the spikes contain **neuraminidase** (**N spikes**) that help the virus to exit the host cell. All the H spikes and N spikes on any one virus particle are the same.

Chemical changes, which are due to mutation or gene reassortment, occur periodically in these two spike enzymes, thereby yielding new strains of flu viruses. These changes have practical consequences. The antibodies an individual produced during last year's seasonal influenza will either partially or completely fail to recognize this year's strain because of the change to the spikes. That means the individual will suffer another bout of disease if exposed and not vaccinated. Consequently, every year there are seasonal flu outbreaks.

The type A flu virus can lead to outbreaks and epidemics, and occasionally to pandemics, as described in the chapter opener. The type B virus is less widespread but can cause more mild seasonal outbreaks. The identification of a specific influenza type A virus is based on the H and N spikes present on the virus. Influenza A is divided into subtypes based on the 17 potential unique H spike proteins, numbered H1 to H17. Identification also is based on the 10 potential

FIGURE 18.6 Smallpox.

(A) Smallpox lesions are raised, fluid-filled vesicles on the skin surface.

(B) False-color transmission electron microscope image of smallpox viruses. (Bar = 200 nm.)

Courtesy of Jean Roy/CDC.

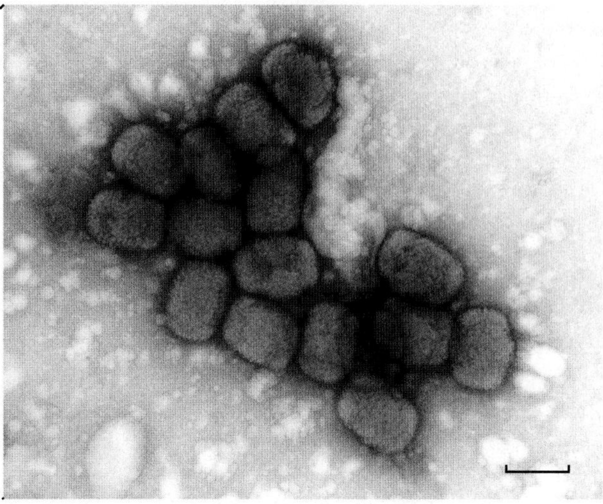

Courtesy of Dr. Fred Murphy/CDC.

A CLOSER LOOK 18.1

Should We, or Shouldn't We?

One of the liveliest global debates in disease microbiology is whether the last remaining stocks of smallpox viruses in Russia and the United States should be destroyed. Here are some of the arguments.

For Destruction:

- People are no longer vaccinated, so if the virus should escape the laboratory, a deadly epidemic could ensue.
- The DNA of the virus has been sequenced and many fragments are available for performing research experiments; therefore, retaining the whole virus is no longer necessary.
- The elimination of the remaining stocks of laboratory virus will eradicate the disease and complete the project.
- No epidemic resulting from the theft or accidental release of the virus can occur if the remaining stocks are destroyed.
- If the United States and Russia destroy their smallpox stocks, it will send a message that biological warfare will not be tolerated.

Against Destruction:

- Future studies of the virus are impossible without the whole virus. Indeed, certain sequences of the viral genome defy deciphering by current laboratory means. Insights into how the virus causes disease and affects the human immune system cannot be studied without having the genome and whole virus. The virus research might identify better therapeutic options that can be applied to other infectious diseases.
- Mutated viruses could cause smallpox-like diseases, so continued research on smallpox is necessary in order to be prepared.
- No one knows where all the smallpox stocks are located. Smallpox virus stocks might be secretly retained in other labs around the world for bioterrorism purposes, so destroying the stocks might create a vulnerability in protecting the public. Smallpox viruses also might remain active in buried corpses.
- Destroying the virus impairs the scientists' right to perform research, and the motivation for destruction is political, not scientific.
- Today, it is possible to create the smallpox virus from scratch. So, why bother to destroy it?
- Because the smallpox virus might have evolved from camelpox, who is to say that such evolution could not happen again from camelpox?

Discussion Point

Now it's your turn. Can you add any insights to either list? Which argument do you prefer?

Note: In 2011, the World Health Assembly of the World Health Organization (WHO) met to again consider the evidence for retention or destruction of the smallpox stocks. The stalemate continues in 2018.

A CLOSER LOOK 18.2

Bioterrorism: What's It All About?

The anthrax attacks that occurred on the east coast of the United States in October 2001 confirmed what many health and governmental experts had been saying for over 10 years—the concern is not *if* bioterrorism would occur, but *when* and *where*. **Bioterrorism** represents the intentional or threatened use of primarily microorganisms or their toxins to cause fear in, or inflict death or disease upon, a large population for political, religious, or ideological reasons.

Is Bioterrorism Something New?

Bioterrorism has a long history, beginning with infectious agents that were used for biowarfare. In the United States, during the aftermath of the French and Indian Wars (1754–1763), British forces, under the guise of goodwill, gave smallpox-laden blankets to rebellious tribes sympathetic to the French. The disease decimated the Native Americans, who had never been exposed to it before and had no immunity. Between 1937 and 1945, the Japanese established Unit 731 to carry out experiments designed to test the lethality on Chinese soldiers and civilians of several microbiological weapons as biowarfare agents. In all, some 10,000 "subjects" died of bubonic plague, cholera, anthrax, and other diseases. In 1973, after years of their own research on biological weapons, the United States, the Soviet Union, and more than 100 other nations signed the Biological and Toxin Weapons Convention, which prohibited nations from developing, deploying, or stockpiling biological weapons. Unfortunately, the treaty provided no way to monitor compliance. As a result, in the 1980s the Soviet Union developed and stockpiled many microbiological agents, including the smallpox virus and anthrax and plague bacteria.

In the United States, several biocrimes have been committed. **Biocrimes** are the intentional introduction of biological agents into food or water, or by injection, to harm or kill groups of individuals. The most well-known biocrime occurred in Oregon in 1984 when the Rajneeshee religious cult, in an effort to influence local elections, intentionally contaminated salad bars of several restaurants with the bacterium *Salmonella*. The unsuccessful plan sickened over 750 citizens and hospitalized 40.

What Microorganisms Are Considered Bioterror Agents?

A considerable number of human pathogens and toxins have potential as microbiological weapons. These so-called "Tier 1 agents" include bacterial organisms, bacterial toxins, and viruses. The seriousness of the agent depends on the severity of the disease it causes (virulence) and the ease with which it can be disseminated. Tier 1 agents can be spread by aerosol contact, such as anthrax and smallpox, or added to food or water supplies, such as the botulinum toxin (see **TABLE A**).

Why Use Microorganisms?

Perhaps as many as 12 nations have the capability of producing bioweapons from microorganisms. Such microbiological weapons offer clear advantages to these nations and terrorist organizations in general. Perhaps most important, biological weapons represent "The Poor Nation's Equalizer." Microbiological weapons are cheap to produce, compared to chemical and nuclear weapons, and they provide those nations with a deterrent every bit as dangerous and deadly as the nuclear weapons. In addition, microorganisms can be deadly in minute amounts to a defenseless (nonimmune) population. They are odorless, colorless, and tasteless, and unlike conventional and nuclear weapons, microbiological weapons do not damage infrastructure, yet they can contaminate such areas for extended periods. Without rapid medical treatment, most of the select agents can produce high numbers of casualties that would overwhelm medical facilities. Lastly, the threatened use of microbiological agents creates panic and anxiety, which often are at the heart of terrorism.

How Might Microbiological Weapons Be Used?

All known microbiological agents (except smallpox) represent organisms naturally found in the environment. However, most of the select agents must be "weaponized"; that is, they must be modified into a form that is deliverable, stable, and has increased infectivity and/or lethality. Nearly all the microbiological agents in tier 1 are infective as an inhaled aerosol.

Dissemination of biological agents by conventional means would be a difficult task. Aerosol transmission, the most likely form for dissemination, exposes microbiological weapons to environmental conditions to which they are usually very sensitive. Excessive heat and ultraviolet light would limit the potency and persistence of the agent in the environment. The possibility also exists that some nations could develop more lethal bioweapons through genetic engineering and biotechnology.

Conclusions

Ken Alibek, a scientist and defector from the Soviet bioweapons program, has suggested the best biodefense is to concentrate on developing appropriate medical defenses that will minimize the impact of bioterrorism agents. If these agents are ineffective, they will cease to be a threat. To that end, vaccination perhaps offers the best defense. The United States has stated it has stockpiled enough smallpox vaccine to vaccinate the entire population if a smallpox "event" occurred.

In the end, we cannot control the events that occur in the world, but by understanding bioterrorism, we can control how we should react to those events—should they occur in the future.

TABLE A Some Tier 1 Agents and Perceived Risk of Use *

Type of Agent	Disease (Microbe Species or Virus Name)	Perceived Risk
Bacterial	Anthrax (*Bacillus anthracis*) Plague (*Yersinia pestis*) Tularemia (*Francisella tularensis*)	High Moderate Moderate
Viral	Smallpox (Variola virus) Hemorrhagic fevers (Ebola virus)	Moderate Low
Toxin	Botulinum toxin (*Clostridium botulinum*)	Moderate

* See **Appendix A** for organism pronunciations.

unique N spike proteins, numbered N1 to N10. The current seasonal flu subtypes are A(H1N1) and A(H3N2).

The onset of influenza is abrupt, with sudden chills, fatigue, headache, and pain that is most pronounced in the chest, back, and legs. Over a 24-hour period, the body develops a fever and a severe cough. Despite these severe symptoms, influenza is normally short-lived and has a favorable prognosis. However, secondary complications might occur if bacterial pathogens, such as staphylococci, invade the damaged respiratory tissue. These bacterial cells might cause a form of pneumonia. In fact, most individuals that "die from the flu" actually die from a pneumonia infection.

Salmonella: sal-mon-EL-lah

There are antiviral drugs to treat influenza. The brand names are Tamiflu (generic name oseltamivir) that comes in pill form and Relenza (generic name zanamivir) that is an inhaled powder. Both inhibit the functioning of the neuraminidase enzyme. However, they do not cure, but rather, if taken early in the infection, diminish the symptoms and shorten slightly the time a person is sick.

Everyone who is at least 6 months of age should get a flu vaccination every fall to protect themselves and others from the seasonal flu for that year. The vaccine prepared for the "flu shot" uses inactivated influenza viruses. It contains either two type A and one type B virus or two A and two B that represent the major flu types that were identified for the impending flu season.

FIGURE 18.7 The Influenza A Virus. This diagram of the influenza A virus shows its eight segments of RNA and the envelope with hemagglutinin and neuraminidase spikes protruding.

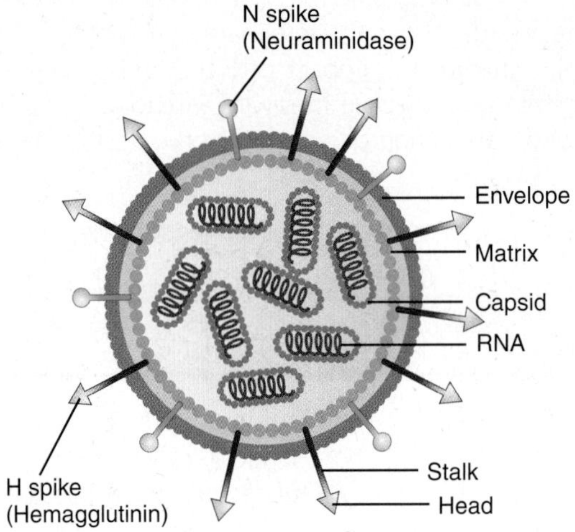

Importantly, the vaccine does not contain any "live" influenza viruses. Some people have the impression that they can get the flu from the vaccine. From the shot-in-the-arm vaccine, that is impossible. However, one can still get the flu in the following ways.

- **Unsuccessful Immune Stimulation.** In some cases, the flu vaccine simply does not "take." For some reason, the recipient's immune system did not respond to the vaccine and did not generate immunity.
- **Presence of Other Subtypes.** In any flu season, there are other subtypes of the flu virus circulating. So, a person could still get the flu if he or she was infected with one of these other nonvaccine strains.
- **Development of Resistance.** It takes about 10 to 14 days to develop antibodies and resistance to the seasonal flu after receiving the vaccine. Therefore, during this 10- to 14-day window period, a person still is susceptible to contracting the flu because the body has not had time to develop immune resistance.

There also is a second type of flu vaccine, one that is given as an inhaled nasal spray. The vaccine contains attenuated flu viruses (weakened), but it is still extremely unlikely the recipient would get the flu from the vaccine. It is available only for healthy people 2 through 49 years of age who are not pregnant.

The Common Cold

Upper respiratory infections called **head colds** are caused by the rhinoviruses. These viruses make up a group of over 100 different RNA viruses that take their name from the Greek *rhinos*, meaning "nose" and referring to the infection site. Adults typically suffer two or three colds and children up to six colds a year, usually in the fall and spring (**FIGURE 18.8**).

A head cold involves the typical symptoms of headache, chills, and a dry, scratchy throat. A "runny nose" and obstructed nasal passages are the dominant symptoms. A cough might occur, and a fever often is absent or slight. The illness

FIGURE 18.8 The Seasonal Variation of Viral Respiratory Diseases. This chart shows the seasons associated with various viral diseases of the respiratory tract (and their annual percentage). Enteroviruses cause diseases of the gastrointestinal tract as well as respiratory disorders and are usually acquired from the environment.

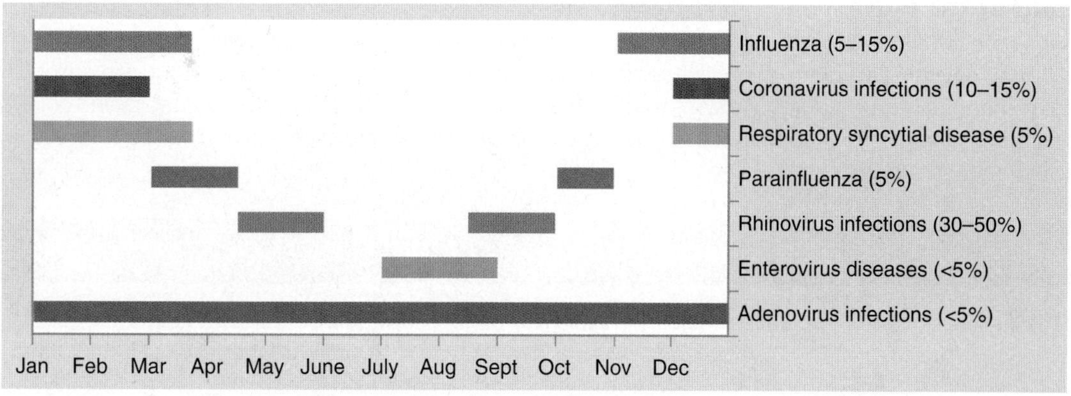

usually lasts 7 to 10 days. Antihistamines can be used to relieve cold symptoms. Because so many different viruses are involved in producing cold-like symptoms, the prospect for developing a vaccine for head colds is not promising.

So, what are the symptomatic differences between the flu and a common cold? **A CLOSER LOOK 18.3** compares the two illnesses.

Hantavirus Pulmonary Syndrome

In the summer of 1993, a brief outbreak occurred in the southwestern United States. It was named the "Four Corners disease" for the place where four states (Arizona, New Mexico, Colorado, and Utah) come together. The disease is a rapidly developing, flu-like illness, the symptoms of which include hemorrhaging blood and respiratory failure. The disease has been named **hantavirus pulmonary syndrome** (**HPS**) because the viruses responsible for it are a group of viruses called hantaviruses.

CDC investigators identified airborne viral particles from the dried urine and feces of rodents (especially deer mice) as the responsible agents. People become infected primarily by breathing air aerosols containing hantaviruses that are shed in infected rodent urine and droppings. A person suffering from HPS cannot spread the disease to other people. Treatment options are limited, as there are no vaccines or effective antiviral drugs available. Therefore, the best protection against HPS is to avoid rodents and their habitats.

18.3 Viral Diseases of the Nervous System: Some Potential Life-Threatening Consequences

The human nervous system can suffer substantial damage if viruses replicate in its tissues. Rabies, a highly fatal and well-known disease, is characteristic of these diseases.

A CLOSER LOOK 18.3

Is It a Cold or the Flu?

Do you know the differences in symptoms between a common cold and the flu? Both are respiratory illnesses but are caused by different viruses. Yet both often have some similar symptoms. In general, the flu has a sudden onset (3–6 hours) while a cold comes on more gradually. Flu symptoms are more severe than cold symptoms (see table) and colds generally do not progress to more serious complications, such as pneumonia and bacterial infections, nor do they usually require hospitalization.

Symptoms	Common Cold	Flu
Fever	Uncommon	Common; 38°C to 39°C (100°F to 102°F) [occasionally higher, especially in young children (40°C/104°F)]; lasts 3 to 4 days
Headache	Uncommon	Common
Chills	Uncommon	Fairly common
General aches and pains	Mild	Common and often severe
Fatigue and weakness	Sometimes	Usual and can last up to 2 to 3 weeks
Extreme exhaustion	Uncommon	Usual and occurs at the beginning of the illness
Stuffy nose	Common	Sometimes
Sneezing	Common	Sometimes
Sore throat	Common	Sometimes
Cough	Mild to moderate hacking, wet (mucus-producing) cough	Dry, cough that can become severe

Data from National Institute of Allergy and Infectious Diseases website www.niaid.nih.gov/topics/Flu/Pages/coldOrFlu.aspx

Rabies

Rabies is notable for having the highest mortality rate of any human disease, once the symptoms have fully materialized. Few people in history have recovered from rabies.

Early signs are abnormal sensations such as tingling, burning, or coldness at the site of the bite. Then, increased muscle tension develops, and the individual becomes alert and aggressive. Soon, there is paralysis, especially in the swallowing muscles, and saliva drips from the mouth. Brain deterioration, together with an inability to swallow, increases the violent reaction to the sight, sound, or thought of water. The disease therefore has been called **hydrophobia**—literally, “fear of water.” Death follows from respiratory paralysis.

Rabies can occur in most warm-blooded animals, including dogs, cats, horses, rats, skunks, and bats. The virus enters the tissue through a skin wound contaminated with the saliva, urine, blood, or other fluid from an infected animal. The incubation period can vary from days to years, depending on the amount of virus entering the tissue and the wound’s proximity to the central nervous system.

Once symptoms set in, there is little chance of survival. However, a person who is bitten or has had contact with a rabid animal should be given a post-exposure immunization with the rabies vaccine as soon as possible. The procedure consists of four injections in the deltoid muscle of the upper arm. The first dose is given right away, and the additional doses are given on the 3rd, 7th, and14th days. Because of the long incubation period of the disease, there hopefully is time for the immunization to produce protective immunity. The affected individual also should receive a shot called rabies immune globulin at the same time as the first vaccine dose. The immune globulin contains antibodies to inhibit the rabies virus from infecting more cells.

The number of human cases of rabies in the United States is usually less than five per year, due in large measure to immunizations of domestic animals. Today, some 92% of reported rabies cases to the CDC occur in wild animals (raccoons, skunks, bats, foxes, and rodents).

Polio

Polio is caused by the polio viruses, which are very small, RNA-containing viruses. They usually enter the body in contaminated water or food. About 75% of infected people never develop any symptoms, while about 20% have minor symptoms such as fever, sore throat, upset stomach, or flu-like symptoms. They do not develop any paralysis or other serious complications.

Regrettably, up to 5% develop serious paralytic symptoms due to nervous system damage caused by the viruses. If inflammation or swelling of the **meninges** occurs, paralysis of the arms, legs, and body trunk can result. In the most severe form of polio, the viruses infect the **medulla** of the brain. Swallowing is difficult, and paralysis develops in the tongue, facial muscles, and neck. Paralysis of the **diaphragm** muscle causes labored breathing and might lead to death.

Meninges: The three membranes that surround the brain and spinal cord.

Medulla: The lower half of the vertebrate brain and continuous with the spinal cord.

Diaphragm: A sheet of internal skeletal muscle that extends across the bottom of the rib cage.

There are two polio vaccines. The Salk, or injected, vaccine contains inactivated viruses, while the Sabin, or oral, vaccine contains attenuated polio viruses. The Salk vaccine is currently the vaccine of choice in the United States and most countries.

As a result of widespread polio vaccination programs, polio was eliminated from the United States in 1979. In the early 1950s, before polio vaccines were available, polio outbreaks caused more than 15,000 cases of paralytic polio each year. Following introduction of vaccines, the number of polio cases fell rapidly to less than 100 in the 1960s and fewer than 10 in the 1970s.

In 1988, the Global Polio Eradication Initiative set out to eradicate polio around the world. Prior to the initiative, polio paralyzed more than 1,000 children worldwide every day. Since then, more than 2.5 billion children have been immunized against polio and the number of reported polio cases has decreased by 99%. There now are only 3 countries (Afghanistan, Nigeria, and Pakistan) that have not yet stopped polio transmission. Hopefully, worldwide eradication through continued vaccination efforts will soon bring an end to this potentially crippling human disease.

West Nile Virus Infection

West Nile virus (WNV) infection, as the name suggests, is caused by the West Nile virus. It is transmitted by mosquitoes and first appeared in the United States in 1999. Since then, it has spread across the continental United States and Canada (**FIGURE 18.9**).

Most people get infected with WNV through the bite of an infected mosquito that itself became infected when it took a blood meal from an infected

FIGURE 18.9 West Nile Virus Disease Cases, United States. For 2018, many disease cases occurred in California and the upper Midwest.

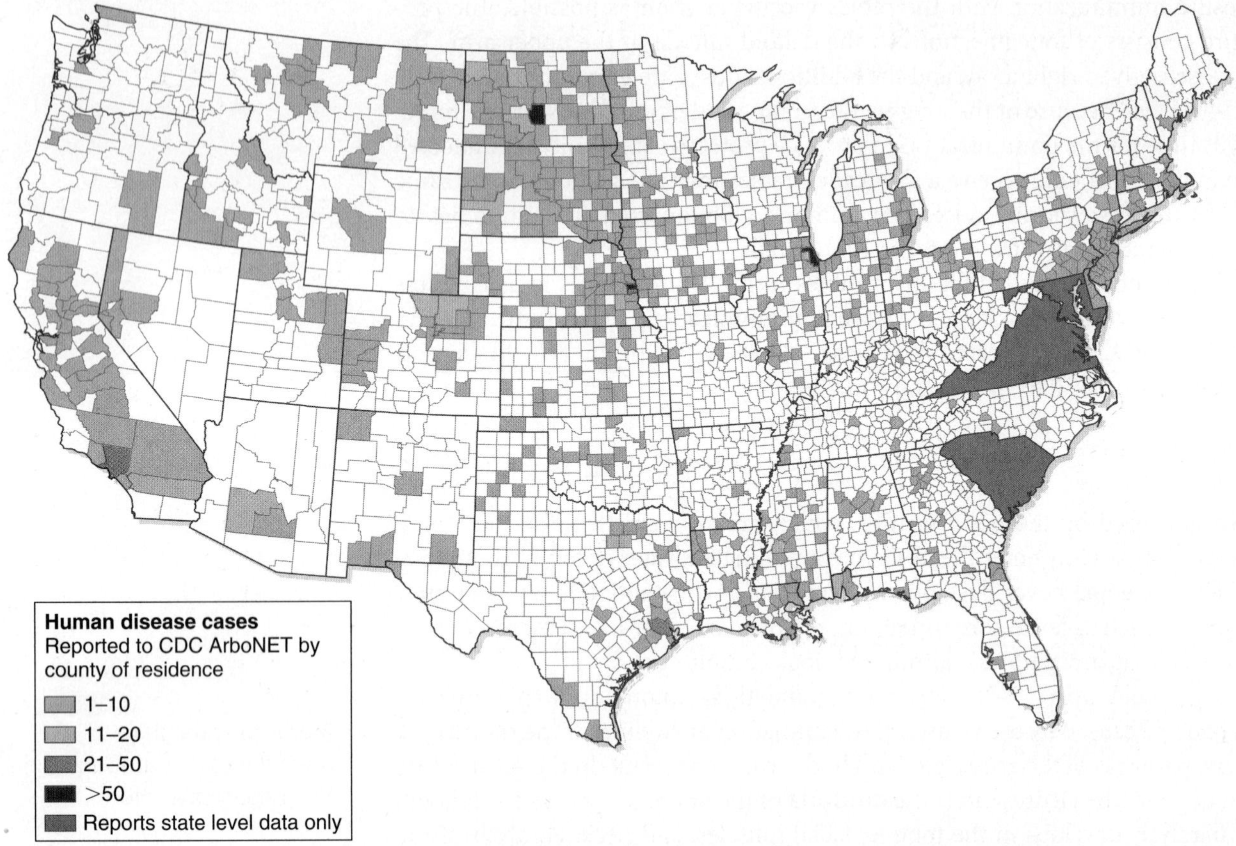

Courtesy of CDC.

bird. Infected mosquitoes can spread the virus to humans and other animals (**FIGURE 18.10**).

The risk of infection is highest for people who work outside or participate in outdoor activities as they are at risk of being bit by an infected mosquito. Most people who get infected show no symptoms. However, about 20% of infected individuals develop a non-neuroinvasive illness. It is characterized by fever, headache, body aches and joint pains, vomiting, diarrhea, and a rash. Most of these individuals fully recover after an extended period of weakness and fatigue.

Around 1% of infected individuals develop a neuroinvasive illness. This might involve infection and inflammation of the brain (encephalitis) or of the protective membranes (meninges) covering the brain and spinal cord (**meningitis**). Symptoms of a neuroinvasive WNV infection include a very high fever, a severe headache, disorientation, and possibly a series of convulsions before lapsing into a coma. About 10% of those patients developing neurological symptoms might die.

There are no medications or antiviral drugs to treat the disease, nor are there any vaccines to prevent virus infection. In more severe cases, patients often are hospitalized to receive supportive treatment, such as intravenous fluids, pain medication, and nursing care.

Zika Virus Infection

One of the most recent emerging virus diseases is **Zika virus (ZIKV) infection**. In 2016, ZIKV rapidly spread across Brazil and to more than 50 other

FIGURE 18.10 The Transmission of the West Nile Virus. Shown here is the generalized pattern of West Nile virus transmission among various animals, including humans. An incidental infection involves a host that can become infected but is not required for the survival of the pathogen; in this case, survival requires mosquitoes and birds.

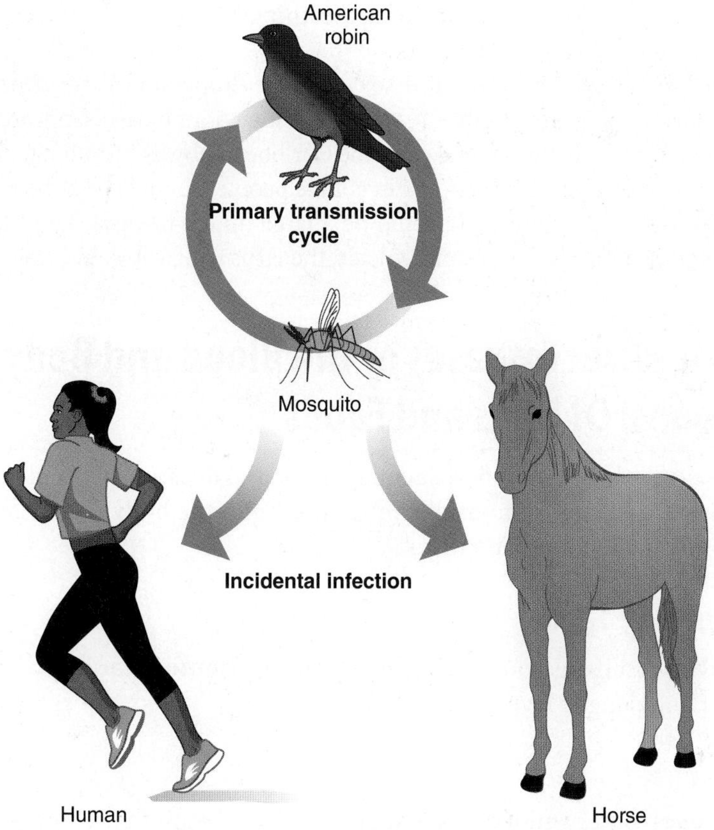

countries and territories in the western hemisphere. There were more than 800,000 reported cases of ZIKV infection. From earlier infections in Africa and the South Pacific, ZIKV was thought to cause only mild to moderate illness, and no serious complications or deaths. However, reports of a dramatic increase in the birth of infants with **microcephaly** during the 2016 epidemic led the WHO to declare Zika-associated microcephaly a global health emergency.

Microcephaly: An infant with an abnormally small head and brain size.

Zika virus is spread by some species of mosquitoes. Human infections have been identified in amniotic fluid and the placenta, and the virus can spread from mother to fetus during the early **perinatal** period. Sexual transmission also has been documented. Health experts report that 80% of individuals infected with the virus will not develop any symptoms. Of the 20% that do have symptoms, a rash, joint pain, and conjunctivitis (red eyes) can appear 2 to 7 days after the mosquito bite. Recovery from the infection is thought to confer lasting immunity.

Perinatal: Referring to the time near childbirth.

Zika virus also can be spread from an infected pregnant mother to her fetus. In 1% to 10% of these fetal infections, the most severe outcomes were the result of infection during the first or second trimester. By far, the most alarming complication is microcephaly. In Brazil, more than 2,000 cases of microcephaly or other fetal abnormalities of the central nervous system were reported during the 2016 epidemic. Researchers are unsure how the virus crosses the placental barrier and leads to microcephaly. Some evidence suggests the virus targets cells that make new neurons.

A small number of adults infected with the Zika virus might develop a paralyzing neurological condition called **Guillain-Barré syndrome**. This condition occurs if the immune system damages nerve cells, which can cause muscle weakness in the arms and legs. A temporary paralysis can affect breathing and might require the assistance of a breathing tube. The condition can last from a few weeks to several months.

Since the outbreak in 2016, disease cases have dropped in Latin America and the Caribbean. Scientists are not sure why ZIKV infections have disappeared. Much of the population in Latin America and the Caribbean now is immune to the virus, which would reduce the number of susceptible people. Some experts believe it will be at least a decade before there could be a large uptick in cases. Until there is a larger population of susceptible individuals, the virus will lie low and wait.

18.4 Viral Diseases of the Blood and Body Organs: Of Bugs and Foods

Viral diseases of the blood and such organs as the liver, spleen, and small and large intestines come from viruses introduced into the body tissues by insects or by contaminated food and drink.

Hemorrhagic Fevers

Hemorrhagic: Referring to blood escaping from the circulatory system.

Two of the most dangerous viral diseases causing **hemorrhagic** fevers are yellow fever and dengue fever.

Yellow Fever

Yellow fever is transmitted by mosquitoes infected with the yellow fever virus (YFV). Most infected individuals develop a flu-like illness and improve without medical treatment. However, about 15% of YFV-infected people develop a more severe form of the disease. In these individuals, infection of the liver causes an overflow of bile pigments into the blood, a condition called **jaundice** that causes the person's complexion to become yellow. These seriously ill patients experience bleeding gums, bloody stools and vomit, and delirium. Patients might die of internal bleeding, as the mortality rates are between 20% and 50%. There was a major outbreak of yellow fever in Brazil in 2016 through 2018 that killed more than 500 people.

Except for supportive therapy, no treatment exists for yellow fever. However, for people planning on traveling to or living in areas at risk of yellow fever, they can be protected by vaccination with a weakened YFV vaccine.

Dengue Fever

Dengue fever is a viral disease where the dengue fever virus (DFV) replicates in white blood cells. Like YFV, DFV also is transmitted by mosquitoes. Following a bite from an infected mosquito, high fever and severe exhaustion are early signs of infection. These are followed by sharp pains in the muscles and joints, and patients often report sensations that their bones are breaking. The name "dengue" comes from the Swahili word *dinga* that means "cramp-like attack" that describes some of the symptoms.

As many as 100 million people are infected every year. Death is uncommon, but if another strain of the virus later enters the body, a condition called **dengue hemorrhagic fever** might occur. Although there is no medical treatment or vaccine for either form of dengue, proper management and supportive care can keep the mortality rate below 1%.

Ebola Virus Disease

Since 1976, outbreaks of **Ebola virus (EV) disease** have appeared sporadically in central Africa. Most of these outbreaks were brought under control and eliminated within a few weeks of starting. Then, between 2014 and 2016, the largest-ever outbreak and epidemic of EV disease occurred in West Africa. Before the epidemic was brought under control, almost 29,000 reported cases and more than 11,000 reported deaths occurred in Guinea, Sierra Leone, and Liberia. In 2018–2019, another Ebola outbreak occurred in the Democratic Republic of Congo. As of June 2019, more than 1,500 people had died, making the outbreak the second deadliest in history.

EV disease is caused by infection with any of four infectious strains of EV (**FIGURE 18.11**). The virus is transmitted to an individual who has had direct contact with the blood or bodily fluids of a person or animal that is infected with EV. A person also can become infected though exposure to objects (such as needles) that have been contaminated with EV-containing secretions.

Initial symptoms can appear anywhere from 2 to 21 days after exposure (average is 8 to 10 days) and include fever, severe headache, joint and muscle pain, sore throat, and weakness. This is followed in 2 to 3 days by diarrhea, vomiting, and stomach pain. A rash, red eyes, and hiccups might be seen in some patients and visible hemorrhaging occurs in less than half the cases.

No antiviral drugs are available yet for EVD. However, an experimental EVD vaccine is being used as a preventative vaccine in the Ebola outbreak in the Democratic Republic of Congo. Preliminary studies suggest the vaccine is 97.5% effective in preventing EBV infections.

Other Viral Diseases of Body Organs

A few other viruses are associated with some well-known infectious diseases.

FIGURE 18.11 The Ebola Virus. False-color scanning electron microscope image of Ebola viruses (blue) being released from an infected cell (yellow). (Bar = 2 μm.)

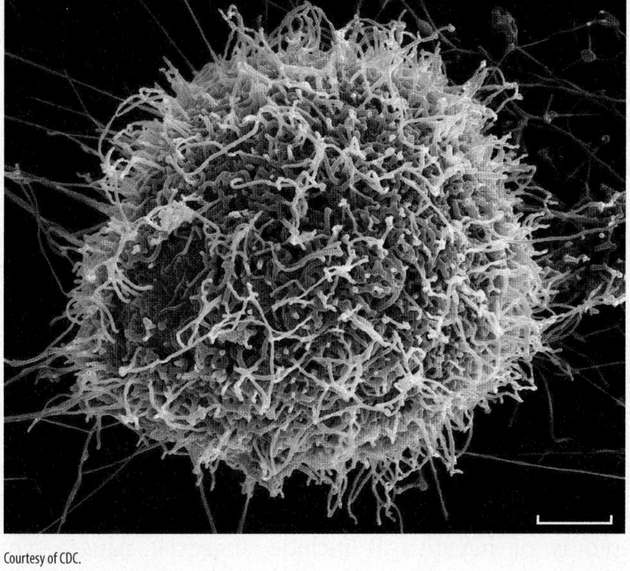

Courtesy of CDC.

Infectious Mononucleosis

Infectious mononucleosis or "**mono**" is common in young adults. It is sometimes called the "kissing disease" because it can be spread by contact with saliva containing the Epstein-Barr virus (EBV). EBV is one of the most common human viruses and occurs worldwide, and most people become infected with EBV sometime during their lives.

Infectious mononucleosis is a blood disease, especially affecting the antibody-producing B cells of the immune system. Infection of the lymph nodes, where B cells reside, causes an enlargement of the nodes. These so-called "swollen glands" can be accompanied by a sore throat, fever, and enlarged spleen. After recovery, individuals still can be carriers for several months and shed the viruses in their saliva.

There is no specific treatment for mono and there are no antiviral drugs or vaccines available. Recent reports have suggested that people who have had mono are more predisposed to develop neurodegenerative conditions, such as multiple sclerosis, later in life. Like the herpes viruses, EBV can lie dormant in B cells. If activated later, the virus can invade the spine and brain, attacking the protective lining around the nerves. Research studies on this association are ongoing.

Hepatitis

Hepatitis is an acute virus disease of the liver caused primarily by three different hepatitis viruses.

- **Hepatitis A**. The hepatitis A virus (HAV) is a small RNA-containing virus commonly transmitted through contaminated food or water, or by infected food handlers. HAV also can be transmitted by raw shellfish because clams and oysters filter and concentrate the viruses along with their food from contaminated seawater.

  A hepatitis A infection usually is not life threatening. However, in late 2017 and continuing into 2019, Kentucky has experienced the largest hepatitis A outbreak ever in the United States. As of early May, 2019, more than 4,400 outbreak-associated cases and 53 deaths of hepatitis A have been reported in Kentucky. Most of the cases were spread by drug users. Their deaths are the result of other health issues, including hepatitis C infections causing liver disease. There also has been a surge of hepatitis A cases reported in California and Arizona, mostly in homeless individuals.

  There are two vaccines available that contain inactivated HAV. The vaccines are recommended for individuals over 1 year of age and for those who are at increased risk for infection.
- **Hepatitis B.** Another form of hepatitis is caused by the hepatitis B virus (HBV), which is a DNA virus. The disease is a global health problem, accounting for 780,000 deaths every year. Two billion people, representing almost one-third of the world's population, have been exposed to HBV, and some 240 million have long-term HBV infections. About 20% of these individuals are at risk of dying from HBV-related liver disease.

  Transmission of HBV usually involves direct or indirect contact with infected body fluids such as blood. For example, transmission can occur by contact with blood- contaminated needles, such as hypodermic syringes or those used for tattooing, acupuncture, or ear piercing (**FIGURE 18.12**). HBV also can be transmitted from an infected partner through vaginal, anal, or oral sex.

  Symptoms of hepatitis B include anorexia, nausea, vomiting, and low-grade fever. Discomfort in the abdomen follows, as the liver enlarges.

FIGURE 18.12 Some Methods for the Transmission of Hepatitis B. Some form of blood contact is involved with all the transmission methods. Most result from improperly sterilized equipment and instruments.

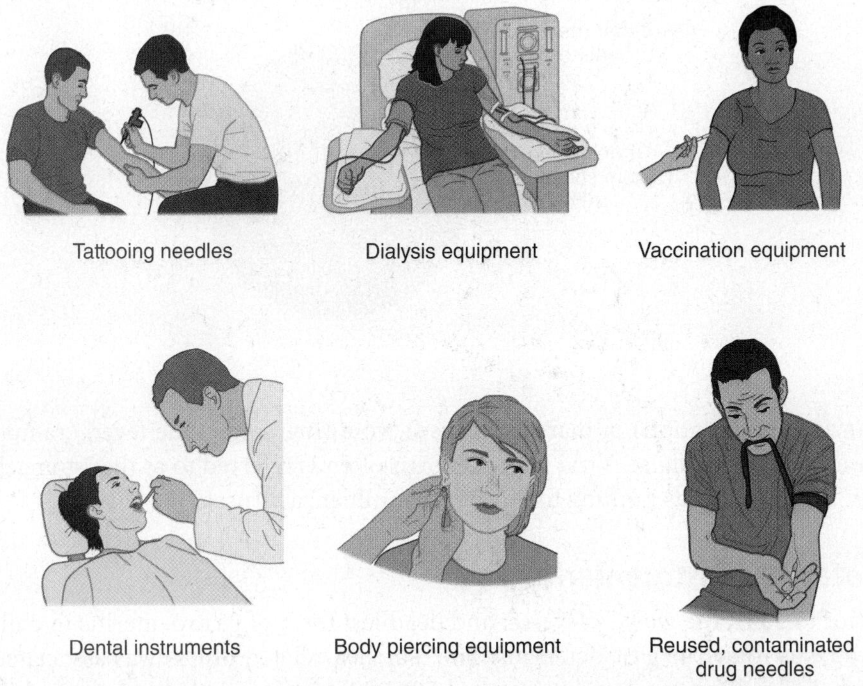

Considerable jaundice usually follows the onset of symptoms and the urine darkens.

The hepatitis B vaccine contains only protein fragments from HBV. Getting the hepatitis B vaccine is important because infection with HBV can lead to long-term liver damage and perhaps liver cancer. The CDC reports that there has been an 82% drop in hepatitis B cases since the introduction of the vaccine in 1991. For individuals already infected, several antiviral drugs are available and regular monitoring of the patient is necessary to detect if any liver damage has occurred or if liver cancer has developed.

- **Hepatitis C**. Another RNA virus, the hepatitis C virus (HCV), causes infections that are the most common type of long-term bloodborne infection in the United States. More than 3 million Americans are chronically infected. HCV is not efficiently transmitted sexually, so most infections come from injecting drugs using dirty needles (**FIGURE 18.13**).

  Up to 75% of infected patients develop a chronic HCV illness and develop symptoms that are similar to those for hepatitis B. In fact, liver damage from HCV is the primary reason for liver transplants in the United States.

  No vaccine is available to prevent HCV infection. However, a new hepatitis C blood test can screen for, detect, and confirm an HCV infection more rapidly than previous tests. In addition, new antiviral medicines taken in combination for up to 12 weeks can cure (i.e., no detectable virus in the blood three months after treatment is completed) 95% of individuals with a chronic HCV infection. Currently, some of these drugs are extremely expensive, and access to diagnosis and treatment is low in much of the developing world.

Viral Gastroenteritis

Viral gastroenteritis is an inflammation condition of the stomach and the intestines caused by a variety of viruses. It usually has an explosive onset with

FIGURE 18.13 Hepatitis C Infections—United States, by Source. There has not been a reported case of transfusion-related hepatitis C in the United States since 1993.

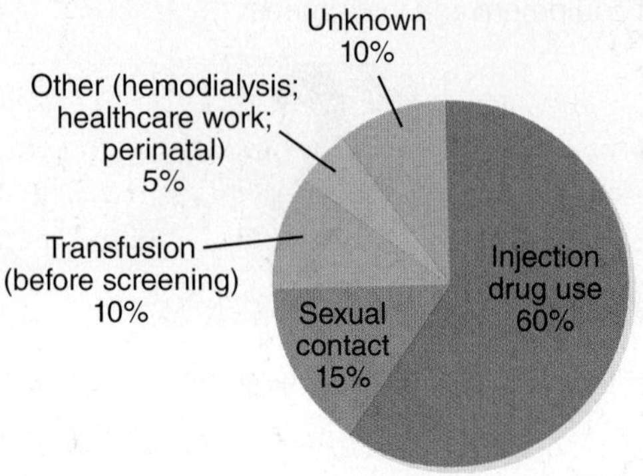

Data from: CDC.

varying combinations of diarrhea, nausea, vomiting, low-grade fever, cramps, headache, and malaise. Viral gastroenteritis often is referred to as the "stomach flu," although it has nothing to do with the influenza viruses.

Rotavirus Gastroenteritis

Prior to 2009, the world's severest and deadliest form of gastroenteritis in children was rotavirus gastroenteritis. The diarrhea-related illness was associated with 125 million cases and more than 500,000 deaths worldwide among children younger than 2 years of age. In the United States, there were an estimated 75,000 hospitalizations and 20 to 40 childhood deaths each year. However, since 2008, two vaccines have been available in the United States. With the use of these vaccines, hospitalizations for severe diarrhea have been reduced by almost 95%.

Rotavirus infections tend to occur in the cooler months (October–April) in the United States, and the infections are often referred to as "winter diarrhea." Transmission occurs by the fecal–oral route because the viruses are shed in great numbers in the stool. After an incubation period of two days, the virus infects cells in the small intestine and disrupts normal absorption. Thus, the patient can exhibit diarrhea, vomiting, and chills. The disease normally runs its course in 3 to 8 days. Recovery from the infection does not guarantee immunity; indeed, many children have multiple rounds of reinfection. Severe water and electrolyte loss due to diarrhea can be life threatening.

Norovirus Gastroenteritis

The noroviruses are the most common cause of viral gastroenteritis. The CDC reports that norovirus gastroenteritis causes 21 million infections, 71,000 hospitalizations, and some 800 deaths each year.

Noroviruses are transmitted by consumption of virus-contaminated food or water, or by direct person-to-person spread. Virus contamination of surfaces also can be a source of infection because the viruses are extremely stable in the environment.

In an infection, the symptoms of norovirus gastroenteritis include fever, diarrhea, abdominal pain, and vomiting. These symptoms can last about 24 to 48 hours. Although recovery is complete, dehydration is the most common complication. As with rotavirus gastroenteritis, the only treatment for norovirus gastroenteritis is fluid and electrolyte replacement. Washing hands and having safe food and water are important prevention measures.

18.5 HIV Infection and AIDS: A Global Epidemic

It has been over 35 years since what became known as **acquired immunodeficiency syndrome** (**AIDS**) was first reported and the causative virus, the **human immunodeficiency virus** (**HIV**), was identified. Today, HIV infections and AIDS remain a persistent problem for the United States and countries around the world.

Some Current Statistics

More than 1 million people in the United States are living with an HIV infection, and one in 7 infected individuals do not know they are infected. Consequently, there are more than 38,000 new HIV infections every year. The CDC estimates that approximately 15,000 people with an AIDS diagnosis die each year, and approximately 636,000 people in the United States with an AIDS diagnosis have died since the epidemic began in 1981.

Globally, there are about 1.5 million new cases of HIV each year and almost 37 million people are living with an HIV infection or AIDS. An estimated 1 million people die each year from AIDS-related illnesses.

Transmission

HIV transmission occurs in any of several ways that involve the blood or semen. Intimate sexual contact, including anal intercourse, is a common method of transmission. If rectal tissues bleed, a portal of entry gives the virus access to the bloodstream. Unprotected vaginal intercourse also is a high-risk sexual activity. Again, lesions, cuts, or abrasions of the vaginal tissues provide an entry point for the virus. The sharing of blood-contaminated needles by injection drug users also transmits HIV. Lastly, there is the possibility of HIV transfer from an infected mother to fetus.

Virus Infection

HIV is an RNA virus with an envelope that contains spikes (**FIGURE 18.14**). The spikes recognize the cell surface of **helper T cells**. As described in the chapter on Disease and Resistance, these cells are essential for the immune system to destroy infected cells and to produce antibodies.

Following virus entry, the RNA is released in the T-cell cytoplasm. Using the enzyme **reverse transcriptase**, the viral RNA serves as a template to synthesizes double-stranded DNA (see Figure 18.14). The DNA molecule called a **provirus**, then integrates into the host cell's DNA, where it can remain silent indefinitely.

Once active, the provirus transcribes its genetic message into new HIV particles. The viral particles then "bud" from the host cell and infect more helper T cells.

Disease Progression

Illness caused by HIV can be separated into three progressive stages, assuming the individual is not on antiviral treatment.

Acute HIV Infection

An infected person starts with an **acute HIV infection**. This person might not experience any early symptoms, or the individual might suffer flu-like symptoms. In either case, the amount of HIV in the body is increasing, and the virus can be easily spread to other individuals during this period.

FIGURE 18.14 The Replication Cycle of the Human Immunodeficiency Virus (HIV). Replication of HIV is dependent on the presence and activity of the reverse transcriptase enzyme. The black rectangles represent places in the replication cycle where antiviral drugs block entry or replication of the virus.

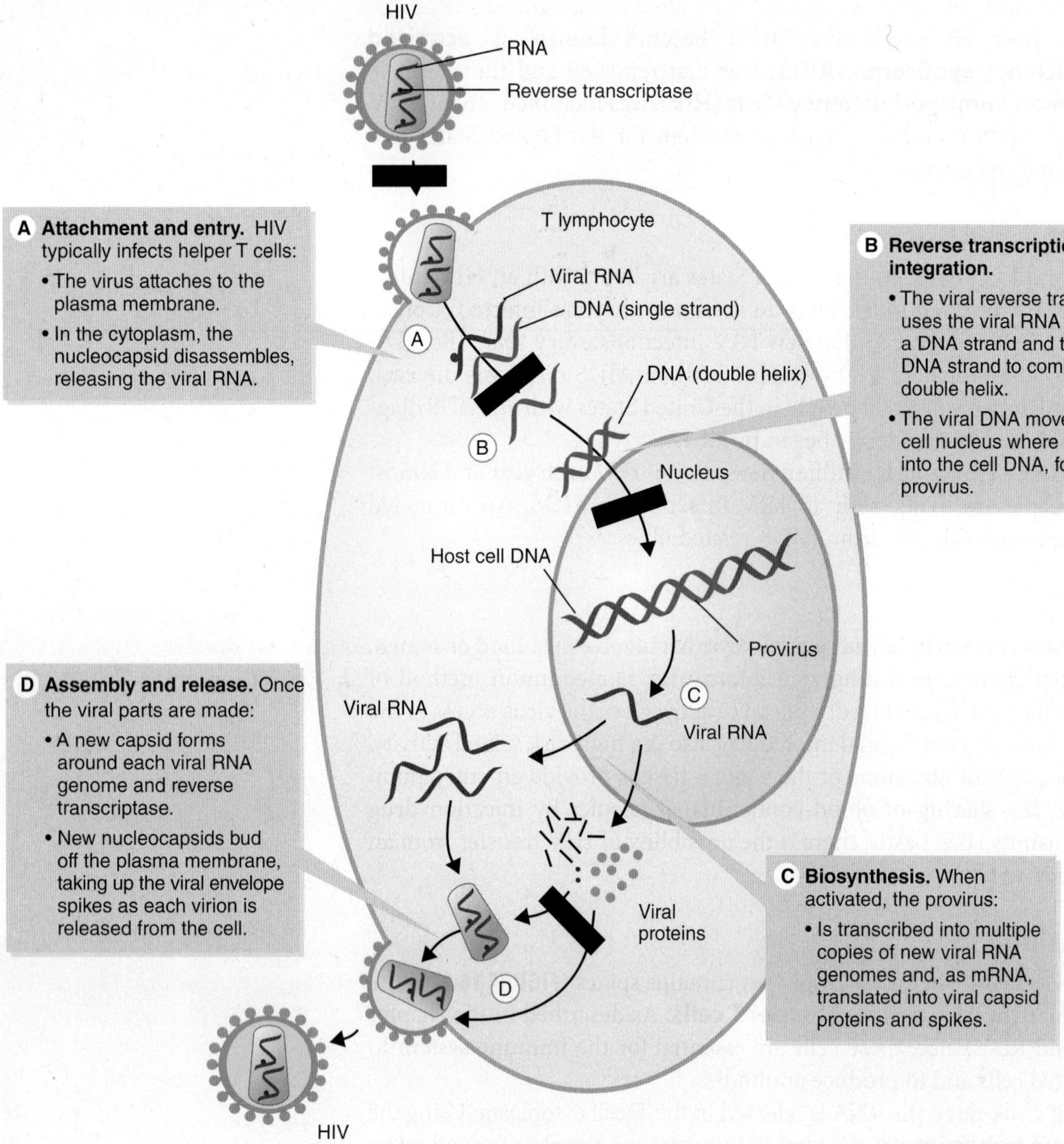

Asymptomatic HIV Infection

Most infected individuals then enter a stage called an **asymptomatic HIV infection**, which can last for many years. Although the virus level stays somewhat stable, HIV is still replicating at low levels. Without antiviral treatment (see below), as the years pass, the virus starts to get the upper hand, and the viral load in the body starts to increase as the helper T cell population declines. Now, the individual might start to have unusual illness symptoms.

AIDS

Once the number of helper T cells has fallen below 200 per microliter, the person is considered to have AIDS. The person has so few helper T cells that the immune system is incapable of combating other infectious agents. Often

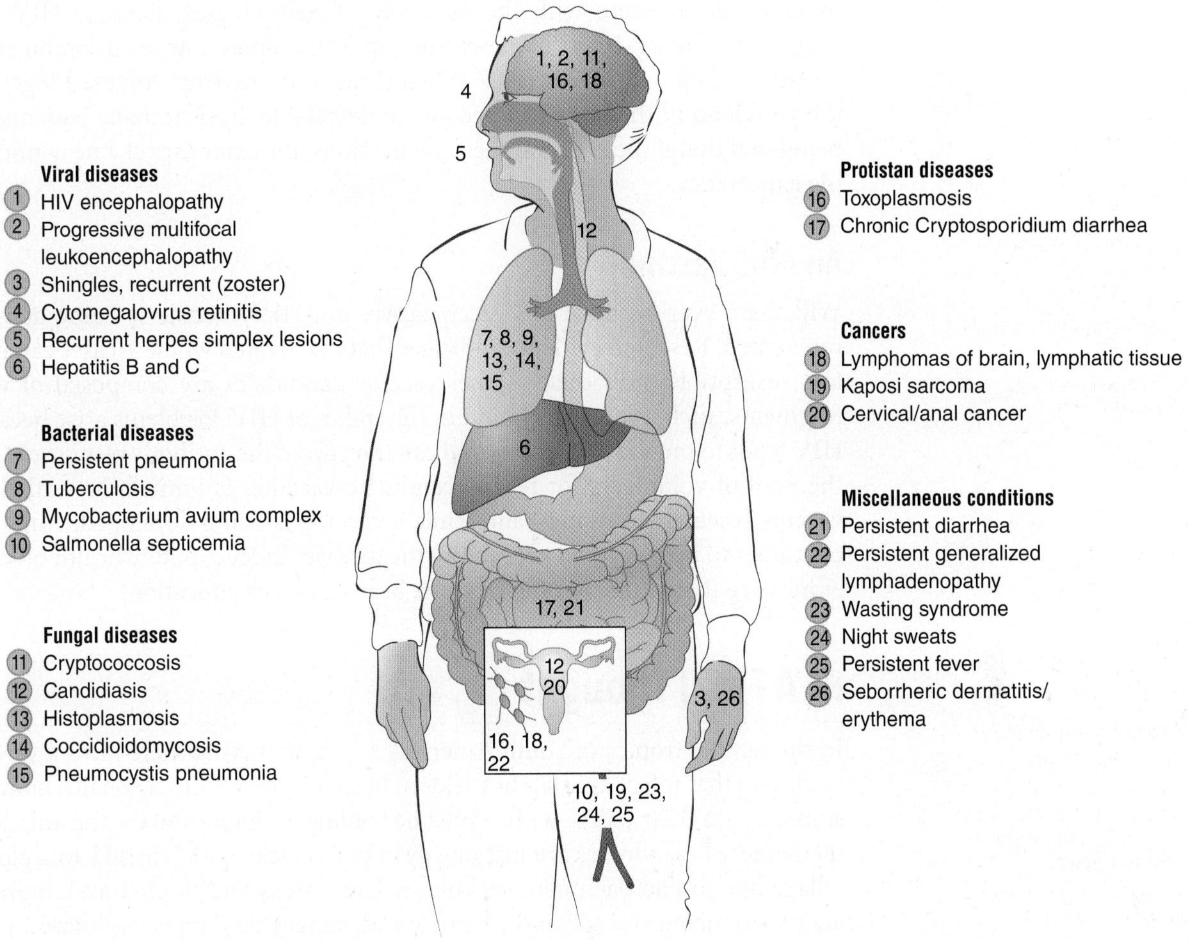

FIGURE 18.15 Defining Conditions in AIDS Patients. A variety of opportunistic illnesses can affect the body as a result of infection with HIV. Note the various systems that are affected, and the numerous microorganisms involved. Inset: female reproductive system.

opportunistic illnesses start to occur prior to the 200 level, so that too would represent defining conditions diagnosed as AIDS (**FIGURE 18.15**). With the onslaught of opportunistic diseases and without treatment, an AIDS patient has a life expectancy of three years or less.

Opportunistic: Referring to a situation where pathogens take advantage of a weakened immune system to generate an infection.

Virus Detection

There are several diagnostic tests a doctor can order to determine if a person has been infected with HIV. The simplest involves the detection of HIV antibodies by a blood test. This can be done in the clinic. In addition, testing kits can be purchased at local drugstores or pharmacies—or even on the Internet. A drop of blood is put on a paper disk and sent to the designated laboratory. Within a few days, results are available by contacting the lab and giving them the confidential code number that came with the test kit.

Although the blood tests are extremely accurate, if a person who bought a test kit does test positive for HIV, the individual should see a doctor as soon as possible to have another test run to detect the presence of actual virus in the blood. If positive, antiviral treatment can be started.

Antiviral Treatment

First, there is no cure for AIDS. Once a person is infected with HIV, the virus remains in that person for life, even with antiviral therapy.

Antiretroviral therapy (**ART**) is intended to control virus levels in the body by blocking a step of virus replication (see Figure 18.14). Some antiretroviral drugs interfere with reverse transcriptase activity or integration, while other drugs interfere with the assembly of new viruses. Because HIV can become resistant to many of these drugs, most therapies now use a combination of drugs called a drug "cocktail." When three or more drugs are used together, the viral load in the body can become undetectable. In fact, many patients are being told that if they stay on their medications, they can expect a near-normal life expectancy.

An AIDS Vaccine

Will there ever be an AIDS vaccine? An anti-HIV vaccine is being actively researched. Research evidence indicates that inactivated whole viruses can protect susceptible individuals. Other vaccine candidates are composed of viral fragments, such as proteins found in the spikes of HIV. Problems arise because HIV tends to mutate in the body and can then avoid the antibody response. Also, the pool of volunteers for testing candidate vaccines is limited. Nevertheless, vaccine research continues, and while a vaccine is being developed, emphasis continues to be placed on preventing the disease. Indeed, public health officials emphasize that at the present time, the best vaccine is education.

A Final Thought

In the remote tropics of South America, Africa, and Asia, there lurks a variety of viruses that infect animals but seldom bother humans. Occasionally, humans stumble into their paths, with results horrifying enough to mark the annals of medicine. Ebola viruses are just one example. As the world shrinks to a global village due to encroachment on wild, tropical areas and to air travel, humans are experiencing and transmitting new and sometimes dangerous microorganisms, especially viruses. Today, it is imperative that health agencies around the world monitor for such outbreaks and be prepared to snuff out any dangerous microbes before they can cause an epidemic or pandemic. The task is challenging but not insurmountable.

Chapter Discussion Questions

What Was He Thinking?

From your reading of this chapter, identify and discuss five major points about viral diseases that the author was trying to get across to you.

Questions to Consider

1. The CDC reports an outbreak of measles at an international gymnastics competition. A total of 700 athletes and numerous coaches and managers from 51 countries are involved. What steps would you take to avert a disastrous international epidemic and quell the spread of the disease?
2. Thomas Sydenham was a London physician who, in 1661, differentiated measles from scarlet fever, smallpox, and other fevers, and set down the foundations for studying these diseases. How would you go about distinguishing the variety of look-alike skin diseases discussed in this chapter?
3. An airliner bound for Kodiak, Alaska, developed engine trouble and was forced to land. While the airline located and brought in another aircraft,

the passengers sat for four hours in the unventilated cabin on the runway. One passenger was coughing heavily. The replacement aircraft eventually arrived, and all the passengers made it to their destination. However, within four days, 38 of the 54 passengers on the stranded plane had come down with the flu. What lessons about infectious disease does this incident teach?

4. As a state health inspector, you are suggesting all restaurant workers should be immunized with the hepatitis A vaccine. Do you believe restaurant owners will agree or disagree with your suggestion? Explain.
5. Sicilian barbers are renowned for their skill and dexterity with razors (and sometimes their singing voices). French researchers studied a group of 37 Sicilian barbers and found that 14 had antibodies against hepatitis C, despite never having been sick with the disease. By comparison, when a random group of 50 blood donors was studied, none had the antibodies. Propose an explanation for the high incidence of exposure to hepatitis C among these Sicilian barbers.
6. With many diseases, the immune system efficiently combats the infectious agent and the person recovers. With some other diseases, the infectious agent overcomes the immune system's defenses and death follows. Compare this broad overview of disease and resistance to what is taking place with AIDS and explain why AIDS is probably unlike any other known human disease.
7. During a blood donation drive, a young man arrives at the local blood donation bank to donate. The donation goes smoothly and the individual leaves. Because the blood would be used for transfusion purposes, it must be tested for several viruses. When this is done, it is discovered the blood is positive for HIV. Consequently, the blood will not be used. However, there is a lively debate as to whether the blood donor should be informed of the positive result. What is your opinion? Explain.

CHAPTER 19

Bacterial Diseases of Humans: Slate-Wipers and Current Concerns

The Courage to Stay

Each year, a group of pilgrims gather in the English countryside outside the village of Eyam (about 160 miles north of London) and pay homage to the townsfolk who, almost 355 years earlier, gave their lives so others might live. The pilgrims bow their heads and remember what happened that fateful year.

In the summer of 1665, a village tailor in Eyam received a shipment of cloth from London where plague was epidemic. Unknown to the cloth merchant, the shipment was infested with fleas, which today we know can carry the plague bacterium. An assistant opened the roll of cloth, and fleas quickly scattered about. Within one week, the assistant was dead. It was Eyam's first plague death—and not the last (**FIGURE 19.1**).

By December, 42 citizens of Eyam had died of the disease. Plague cases then dropped during the winter, but when spring arrived, plague erupted in Eyam. Some frightened townsfolk fled to the countryside, trying to escape death. But the village rector, William Mompesson, believed it was his duty to prevent the spread of the disease. Therefore, he suggested to the remaining citizens of Eyam that they would spread the disease to nearby villages if they fled. He and all his parishioners needed to isolate themselves by staying in the village.

After much soul-searching, everyone reluctantly decided to stay and take their chances. The people of Eyam marked off the village limits with a circle of boundary stones, where neighboring villagers set food and other supplies for the

CHAPTER 19 OPENER This painting shows the unloading of dead bodies into "plague pits" during the Plague of London in 1665. These pits were no more than mass graves used to bury the plague victims who had been gathered up from the streets on "dead carts." At least 70,000 deaths were recorded in the city, and other parts of the country, such as Eyam, also suffered.

FIGURE 19.1 The Plague Cottage. This is the place where Eyam's plague outbreak and first death occurred in 1665.

self-quarantined group. By summer, the plague spread unchecked throughout the village. Despite this, the people remained, even those who had been reluctant to stay.

By September, plague cases finally were dropping, and by November, no new cases were recorded. The quarantine had worked, as there were few plague cases in the surrounding towns and villages.

Sadly, in just over a year, 260 of the village's 320 residents, from some 76 different families, died from the bubonic plague. Their actions undoubtedly saved the lives of thousands in the surrounding villages.

In this final chapter of the textbook, we study some of the major bacterial diseases of humans according to their mode of transmission. Some of the diseases are of historical interest and are currently under control. But just as a garden always faces new onslaughts of weeds and pests, so too are the human body and society in general continually confronted with pathogens and newly emerging diseases.

LOOKING AHEAD

After reading and completing this chapter, you will be able to:

19.1 Identify several airborne bacterial diseases and the body systems that they affect.
19.2 Contrast a bacterial intoxication and a bacterial infection, giving examples.
19.3 Compare the soilborne bacterial diseases anthrax and tetanus.
19.4 Describe the symptoms of Lyme disease, and locate its geographic "hot spots."
19.5 Distinguish between the three major sexually transmitted infections.
19.6 Name several bacterial diseases associated with the skin, the oral cavity, and the urinary tract.
19.7 Explain what healthcare-associated infections (HAIs) are, and list the major sites of infection.

19.1 Airborne Bacterial Diseases: Some Major Players

Bacterial diseases of the respiratory tract can be severe. Moreover, the respiratory tract is a portal of entry to the blood, and, from there, a disease can spread and affect the more sensitive internal organs.

Streptococcal Diseases

The streptococci are a large and diverse group of gram-positive bacterial species occurring in chains (**FIGURE 19.2**). Various species can cause human diseases.

Strep Throat

Pharynx: The area of the throat behind the mouth and nasal cavity.

Streptococcus pyogenes: strep-toe-KOK-us pie-AHJ-en-eez

A sore throat, known medically as **pharyngitis**, is an inflammation of the **pharynx**. About 20% of pharyngitis cases are caused by *Streptococcus pyogenes.* If the normal microbiome in the pharynx has been reduced, *S . pyogenes* dominates and produces a potentially more dangerous form of pharyngitis popularly known as **strep throat**. Patients experience a high fever, coughing, and swollen lymph nodes and tonsils.

Scarlet fever is a disease arising in about 10% of children with strep throat. This occurs if *S. pyogenes* secretes toxins that cause a pink-red skin rash on the neck, chest, and soft-skin areas of the arms. Other symptoms include a fever and a very red, inflamed tongue. Most individuals recover within two weeks without treatment.

A rare but serious complication resulting from untreated strep throat is **rheumatic fever**. This post-infection condition involves antibodies that were produced against *S. pyogenes*. These antibodies mistakenly bind to (cross-react with) proteins on heart muscle. The ensuing immune response causes permanent scarring and distortion of the heart valves, producing a potentially deadly condition called **rheumatic heart disease**.

FIGURE 19.2 ***Streptococcus pyogenes.*** A light microscope image of a gram-stained throat specimen, showing the chain arrangement of *S. pyogenes* cells. (Bar = 20 μm.)

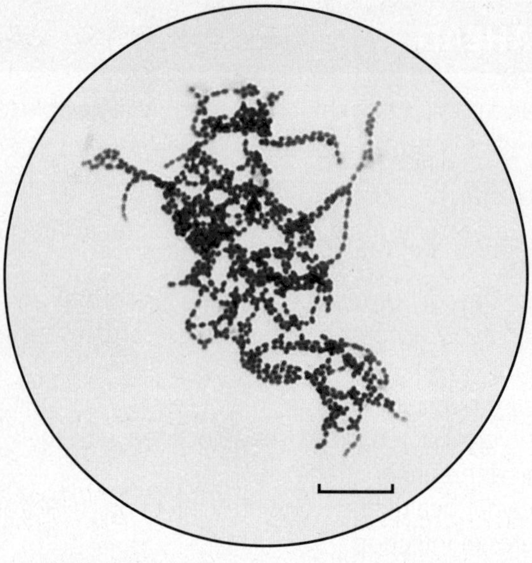

Courtesy of Dr. Jeffrey Pommerville.

FIGURE 19.3 Necrotizing Fasciitis. The extensive removal of infected tissue has been done to save the leg of this patient.

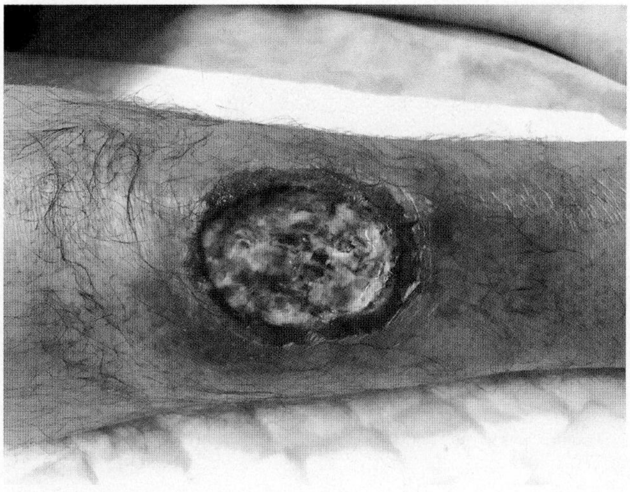

Necrotizing Fasciitis

S. pyogenes also is the causative agent of a rare but life-threatening infection resulting from a wound or some other trauma to the skin tissue. If a wound or cut becomes infected with *S. pyogenes*, the invasive streptococci infect the fatty tissue, called the **fascia**, lying over the muscles and beneath the skin. The toxins produced by *S. pyogenes* destroy the fascia so quickly that it appears the pathogen is "eating" the flesh. Tissue destruction causes cell death, which is called **necrosis**. Therefore, the medical term for the disease is **necrotizing fasciitis** (**FIGURE 19.3**). Removal of the dead and infected tissue through surgery often is required to stop the spread of this fast-moving infection. In severe cases, amputation might be the only recourse to remove the infection.

Pertussis

Pertussis is one of the most dangerous and highly contagious bacterial diseases in children under 5 years of age. Severe disease in the lungs most often occurs in infants only weeks to a few months old. Worldwide, the World Health Organization (WHO) reports more than 16 million pertussis cases and almost 200,000 deaths annually.

After infection by *Bordetella pertussis*, the toxins it produces cause cell death and mucus accumulation in the airways. This leads to labored breathing. Children then experience violent spells of rapid-fire coughing in one exhalation, followed by a forced inhalation over a partially closed windpipe. The rapid inhalation results in the characteristic "whoop"—and hence the name "whooping cough." The coughs can be so violent that facial injury occurs (**FIGURE 19.4**).

Bordetella pertussis: bore-deh-TEL-lah per-TUS-sis

Recently, there has been an upswing in reported pertussis cases. Major outbreaks have been reported in California, Michigan, and Washington, as well as in Latin America and Australia. The reason for the increased occurrence is not completely understood. Health experts believe that some infants and children are not being vaccinated on schedule (vaccine exemptions). These unvaccinated individuals therefore have not built up protective immunity. Another thought is that immunity in adolescents and adults is waning. If they get a mild infection, they could spread the bacterium to unvaccinated children. Also, the nature of the pathogen has changed. The recent outbreak strain of *B. pertussis* appears to be more resistant to vaccine-generated immunity.

FIGURE 19.4 A Child with Pertussis. The cough associated with pertussis was so violent that it caused broken blood vessels in the eyes and bruising on the face.

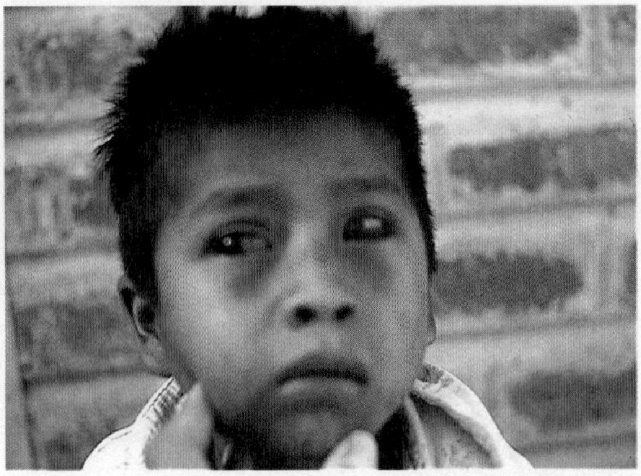

Courtesy of Thomas Schlenker, MD, MPH, Chief Medical Officer, Children's Hospital of Wisconsin.

Acute Bacterial Meningitis

The term **meningitis** refers to several diseases of the **meninges**. These are the three membranous layers covering the brain and spinal cord. The most common form of meningitis is **acute bacterial meningitis** (**ABM**). It is a rapidly developing infection affecting adults and children. Infection usually requires direct contact, such as kissing an infected individual, or being directly coughed or sneezed on by an infected person. Meningitis patients suffer pounding headaches, neck paralysis, and numbness in the extremities.

Meningococcal Meningitis

One of the most dangerous forms of ABM is **meningococcal meningitis**. This disease is caused by *Neisseria meningitidis*, a small, gram-negative bacterium. Because the cells form a diplococcus arrangement, the organism commonly is called the meningococcus. The disease begins as an influenza-like upper-respiratory infection (**FIGURE 19.5**). Once the infection has spread to the bloodstream (**meningococcemia**), the cells cross the blood-brain barrier. The meninges then become inflamed, putting pressure on the spinal cord and brain.

Neisseria meningitidis: nye-SEER-ee-ah meh-nin-jih-TIE-diss

To treat the infection, it is critical that doctors recognize symptoms of meningococcal meningitis. A principal criterion for diagnosis is the observation of gram-negative diplococci in samples of spinal fluid obtained by a spinal tap. If the infection is caught early, aggressive treatment with antibiotics can begin before irreversible nerve damage occurs.

Haemophilus Meningitis

Another form of ABM is *Haemophilus* meningitis. It is caused by *Haemophilus influenzae* type b (Hib) and the infection occurs primarily in children between the ages of 6 months and 2 years. Symptoms include stiff neck, severe headache, and other evidence of neurological involvement, such as listlessness, drowsiness, and irritability. There are antibiotics that can be used for treatment.

Haemophilus influenzae: hee-MAH-fill-us in-flew-EN-zeye

Haemophilus meningitis once was the most prevalent ABM in American children under 5 years of age. Then, in 1985, a Hib vaccine was developed to prevent Hib infections. The vaccine protects infants as young as 6 weeks old and those adolescents and adults who are at increased risk for *Haemophilus* meningitis.

FIGURE 19.5 Steps Leading to Meningitis.

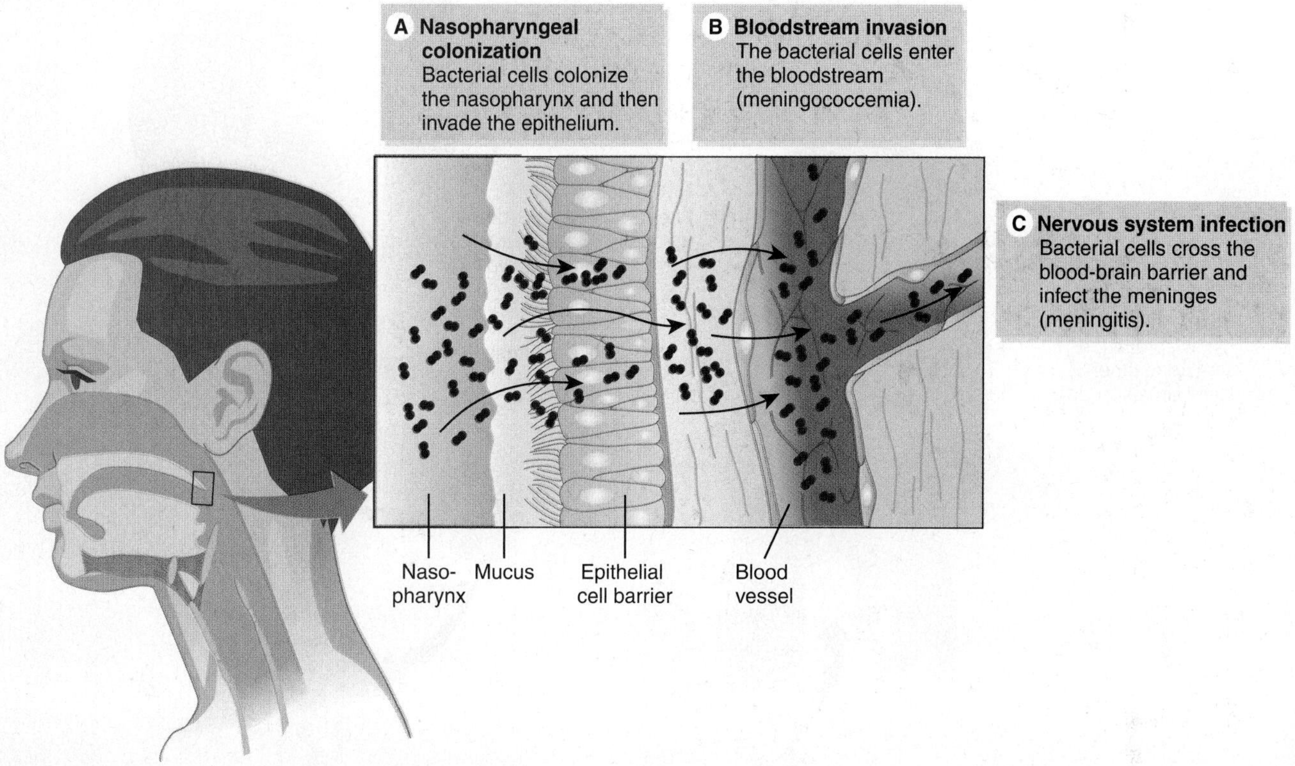

Tuberculosis

Tuberculosis (TB) is an infectious disease usually caused by *Mycobacterium tuberculosis*. The bacillus-shaped cells generally infect the lungs, but they also can spread to other parts of the body. The rod-shaped cells are carried in respiratory droplets spread through the cough of a contagious individual (**FIGURE 19.6**).

Mycobacterium tuberculosis: my-koh-back-TIER-ee-um too-ber-cue-LOH-sis

If inhaled by a susceptible person, the *M. tuberculosis* cells infect the **alveoli** (**sing. alveolus**) of the lungs. The immune system then attempts to form a wall of white blood cells and fibrous materials around the bacilli in the alveoli. As these materials accumulate, a hard nodule called a **tubercle** arises (hence the name "tuberculosis"). The cells in the tubercles produce no known toxins. However, in many cases, growth of the bacilli is so unrelenting that the immune system cannot contain them, and lung tissue is destroyed in the process. Infected patients experience chronic cough, chest pain, and high fever.

Alveolus: A hollow cavity forming the ends of the respiratory tract in the lungs.

A special staining procedure called the **acid-fast test** is an important screening tool used on the patient's **sputum**. Early detection of tuberculosis also is aided by the **tuberculin skin test**, a procedure that begins with the application of a purified protein derivative of *M. tuberculosis* to the skin. If the patient has been exposed to the TB bacillus, an immune reaction causes the skin to become thick, and a raised, red welt develops within 48 to 72 hours. An X-ray exam also can identify tubercles in the lungs.

Sputum: Respiratory mucus.

Physicians have treated tuberculosis patients with several antibiotics. However, *M. tuberculosis* is showing strong resistance to these drugs. Such patients are said to be infected with **multidrug-resistant tuberculosis (MDR-TB)** and must be given stronger drugs. Drug therapy is intensive and extends over a period of 6 to 9 months or more. And the scary part is, there now are strains of the pathogen that are resistant to almost *all* drugs used to treat TB. For centuries, TB has been a "slate-wiper"—a particularly deadly disease that has the potential to "wipe the slate clean" of humankind—and it continues to challenge the best medical treatment available today.

FIGURE 19.6 The Progress of Tuberculosis.

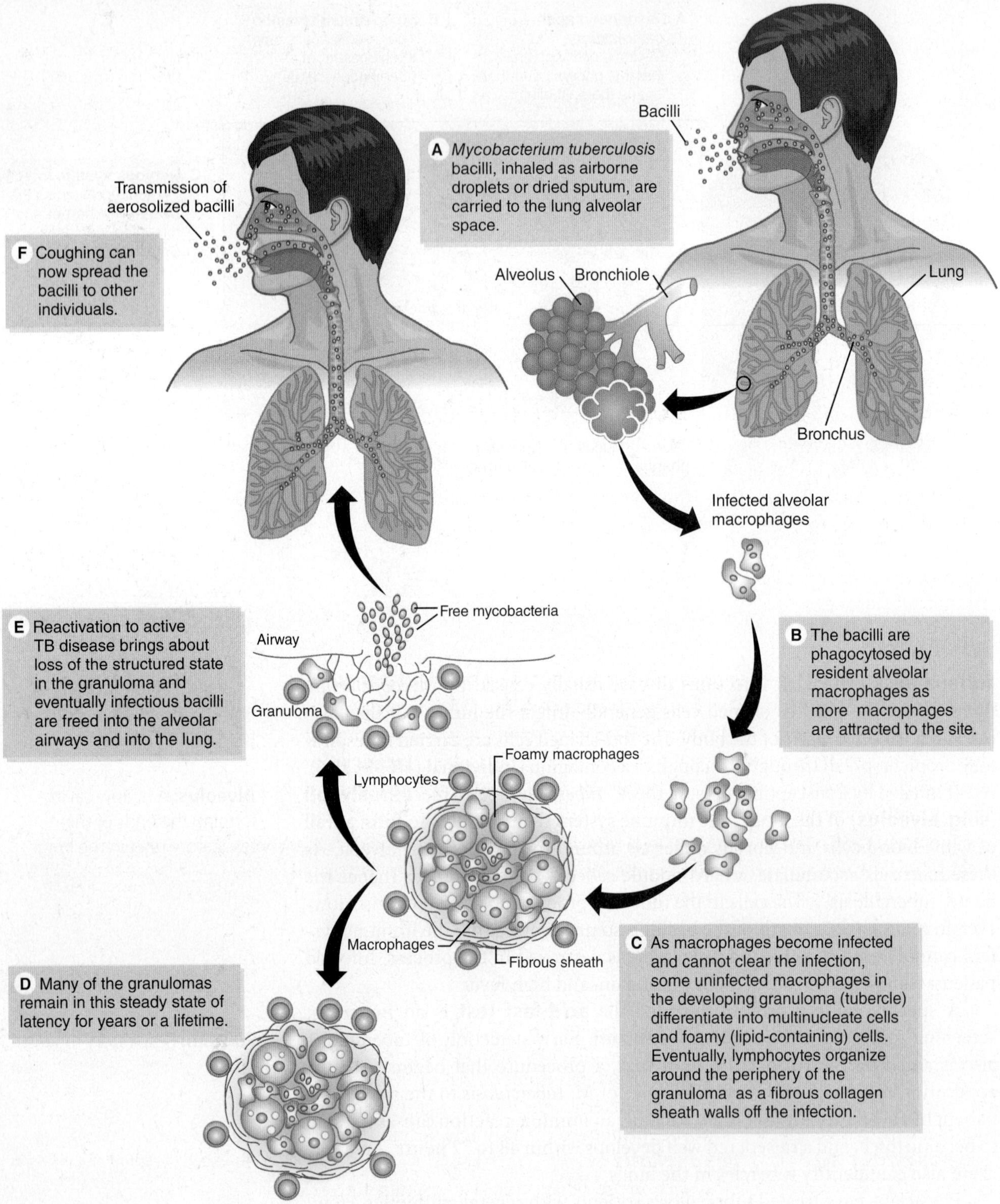

There is no effective, long-term TB vaccine for everyone in the population. Short-term protection for children is available with the bacillus Calmette-Guérin (BCG) vaccine. However, the vaccine does not provide life long immunity and is not effective against adult pulmonary TB. BCG generally is not recommended for use in the United States because of the low risk of infection with *M. tuberculosis*.

Bacterial Pneumonia

The term **pneumonia** refers to an inflammation in the lungs. Here, the **bronchioles** and alveoli in the lungs become inflamed and filled with fluid. In the United States, there are 2 to 3 million pneumonia infections resulting in 45,000 deaths each year. In adults, bacteria are the most common cause of pneumonia. Two of the more prevalent bacterial forms are described here.

Bronchiole: A passageway through which air passes in going from the nose or mouth to the alveoli (air sacs) of the lungs.

Pneumococcal Pneumonia

Pneumococcal pneumonia is caused by *Streptococcus pneumoniae*, a gram-positive chain of diplococci traditionally known as the pneumococcus. Patients experience high fever, sharp chest pains, difficulty breathing, and rust-colored sputum resulting from blood seeping into the air sacs of the lungs. The disease can occur in all age groups but primarily in infants, young children, older adults, and those with underlying medical conditions. Pneumonia in children can be particularly dangerous, as **A CLOSER LOOK 19.1** points out.

Streptococcus pneumoniae: strep-toe-KOK-us new-MOH-nee-eye

When *S. pneumoniae* multiplies in the alveoli, the bacterial cells damage the alveolar lining, which allows fluids and blood cells to enter the sacs. Pneumonia then develops as fluid fills the alveoli and reduces oxygen transfer.

Currently, there are two pneumococcal vaccines available. One, called Prevnar 13, is available for children under 5 years of age and adults 65 years of age and older. A second vaccine, called Pneumovax 23, is available for use in all adults who are older than 65 years of age.

Primary Atypical Pneumonia

A second type of bacterial pneumonia is termed **primary atypical pneumonia (PAP)**. It is called "primary" because it occurs in previously healthy individuals, whereas pneumococcal pneumonia usually develops in people who are already ill. It is referred to as "atypical" because the symptoms are unlike those in pneumococcal disease. With PAP, the patient experiences fever, fatigue, and a characteristic dry, hacking cough.

The agent of PAP is *Mycoplasma pneumoniae*, one of the smallest bacterial pathogens. Often, it is called "walking pneumonia," even though the term has no clinical significance. The disease is rarely fatal.

Mycoplasma pneumoniae: my-koh-PLAZ-mah new-MOH-nee-eye

Legionnaires' Disease (Legionellosis)

In 1976, another form of pneumonia appeared at an American Legion Convention in Philadelphia. The infection affected 182 conventioneers and 39 other people in or near the convention hotel. Those affected suffered from headaches, fever, coughing, and shortness of breath. Thirty-four individuals ultimately died of the disease or its complications. The bacterial species eventually identified with the disease was named *Legionella pneumophila*. The illness became known as **Legionnaires' disease** or **legionellosis**.

Legionella pneumophila: lee-ja-NEL-lah new-MAH-fil-lah

The cells of *L. pneumophila* exist where water collects. Cooling towers, industrial air-conditioning units, humidifiers, stagnant pools, and puddles of water have been identified as sources of the pathogen. If susceptible individuals breathe the contaminated aerosols into the respiratory tract, disease develops within a few days. The symptoms of Legionnaires' disease include fever, a dry cough with little sputum, and lung infection. Legionnaires' disease is not transmitted person to person. In 2015, two unrelated outbreaks in New York City affected more than 133 people and killed 16. The source appeared to be *Legionella*-contaminated cooling towers.

19.2 Foodborne and Waterborne Bacterial Diseases: Gastroenteritis

Food and water can become contaminated with microbial pathogens in many ways (**FIGURE 19.7**). Our intestinal microbiome helps block the infection process as described in the chapter on Microbial Crosstalk. Still, intestinal pathogens often either generate toxins or reproduce so quickly that the gut microbiome cannot contain them.

A CLOSER LOOK 19.1

The Killer of Children

Global Health Magazine recently reported the following: "Chitra Kumal knows the pain of losing a child. When her daughter, Sunita, was 15 months old, she developed a respiratory infection that quickly progressed into pneumonia. With no health facilities in her Nepalese village, Kumal depended on the advice and treatment of a traditional healer or shaman. After just 3 days of fever, fast breathing, and chest indrawing, her only daughter died."

Similar stories are reported everyday around the world. According to the World Health Organization (WHO), pneumonia kills 2 million children under 5 years of age each year. This is more than AIDS, malaria, and measles combined. In addition, pneumonia accounts for nearly one in five child deaths globally (see **FIGURE A**). However, because pneumonia often is misdiagnosed as malaria, this number might be an underestimate as nearly half of all pneumonia cases occur in parts of the world where malaria is endemic.

A major way to reduce childhood mortality is to prevent and treat childhood pneumonia. However, reducing the cases of pneumonia is not straightforward for several reasons. First, only about one in four caregivers know that the two key symptoms of pneumonia are fast breathing and difficulty breathing (indrawing). Second, many healthcare professionals have proposed that if antibiotics were available to children with pneumonia, around 600,000 lives could be saved each year. Although significant, this reduction would only represent about 25% of all pneumonia cases.

Therefore, several other prevention measures have been proposed to lower the number of childhood pneumonia cases further. Among the proposed measures are the promotion of balanced nutrition, reduction in environmental air pollution, and an increase in immunization rates. Vaccines are available to prevent pneumonia caused by *Streptococcus pneumoniae* (pneumococcus) and *Haemophilus influenzae* type b (Hib). However, only about 50% of pneumonia cases in Africa and Asia are caused by these two organisms. Consequently, other vaccines must be developed that target the other bacterial species (and viruses) that cause pneumonia. Lastly, like in all areas of infectious disease, handwashing can play an important role in reducing the transmission of pneumonia.

FIGURE A Pneumonia accounts for almost 20% of global childhood deaths.

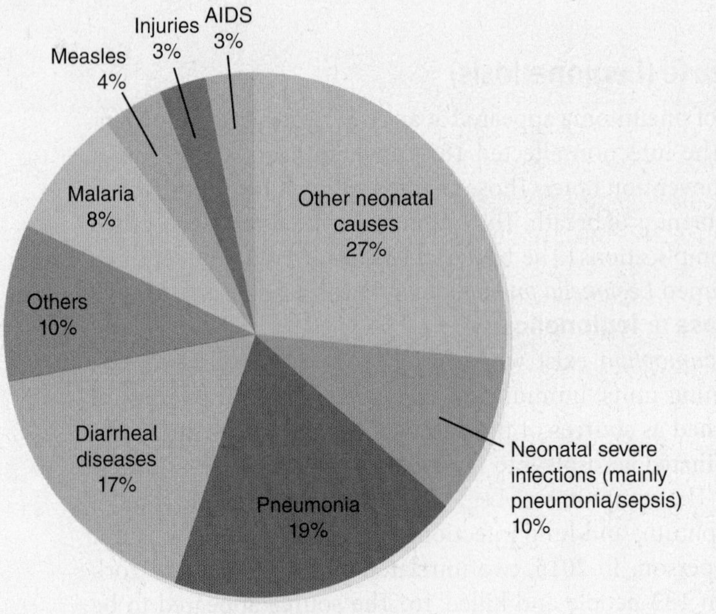

Source: Child Health Epidemiology Resources Group (CHERG), with additional data from UNICEF.

FIGURE 19.7 Foodborne Illness Surveillance.

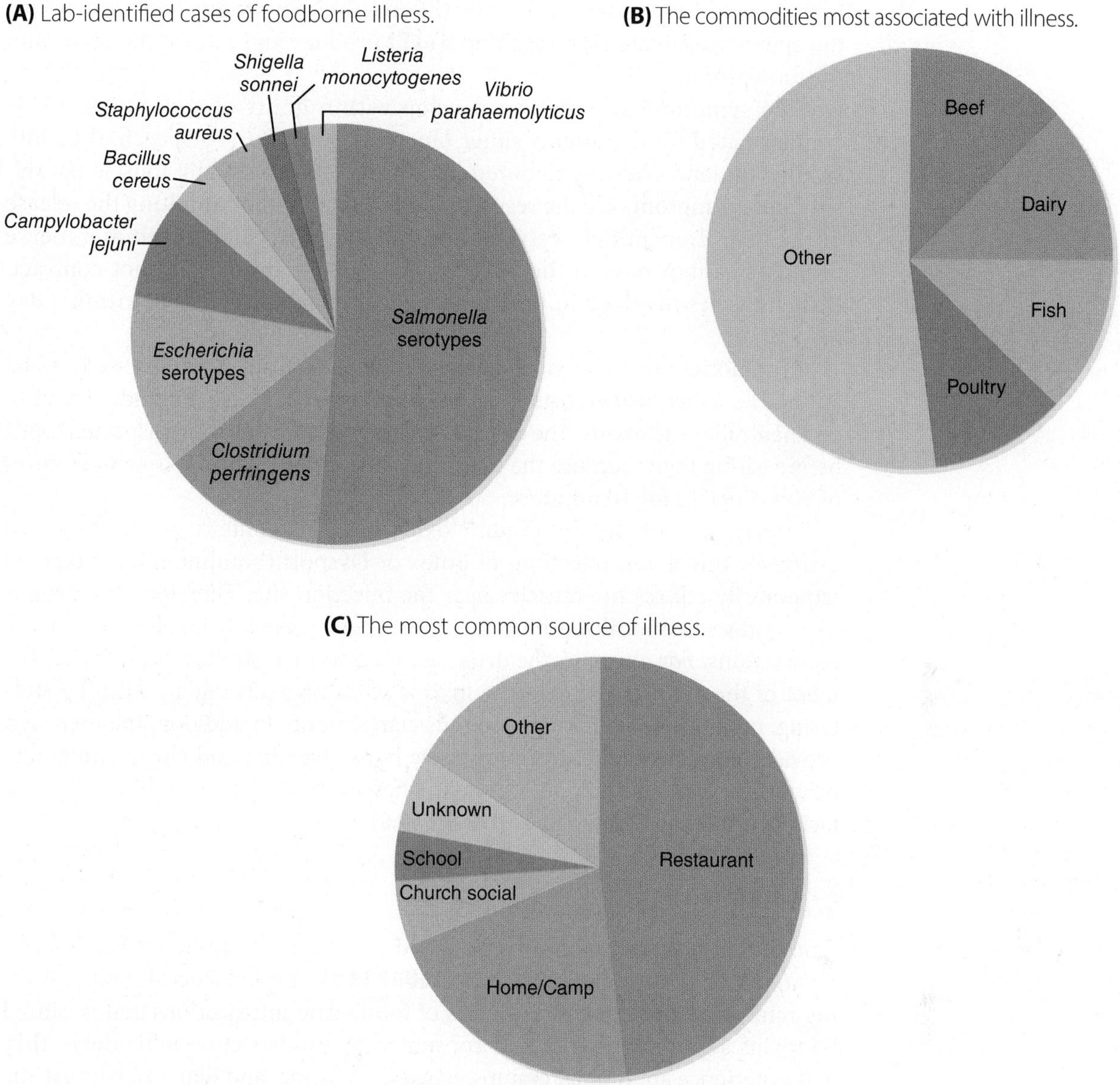

Data from CDC.

Most foodborne and waterborne illnesses affect the gastrointestinal (GI) tract and are introduced through the **fecal-oral route**. Thus, these illnesses represent some form of **gastroenteritis**, which is an inflammation of the stomach and the intestines, usually with vomiting and diarrhea. Some cause **food intoxications** (**poisoning**), which are illnesses that result from the ingestion of bacterial toxins in food or water. The appearance of symptoms usually has a short incubation period, and the illness resolves in a relatively brief time. Other foodborne and waterborne diseases are the result of **food infections**. In this case, live bacterial pathogens in food or water are ingested. For infections, the incubation period is longer, and the disease takes longer to run its course, than for an intoxication.

Fecal-oral route: A mode of disease transmission, where pathogens in fecal material pass from one person and are introduced through food or water into the oral cavity of another person.

The potential list of bacterial foodborne and waterborne diseases is extensive. Therefore, only a selection of the organisms and the diseases will be described in the following sections.

Botulism

Of all the foodborne intoxications in humans, none is more dangerous than **botulism**. Botulism is caused by *Clostridium botulinum*, a spore-forming,

Clostridium botulinum: kla-STRIH-dee-um bot-you-LIE-num

gram-positive rod that produces extremely poisonous toxins. The spores exist in the soil. If they inadvertently enter the anaerobic environment of cans or jars, the spores germinate. The resulting bacilli produce and release the toxin into the food product.

The symptoms of botulism develop within hours of consuming toxin-contaminated food. Patients suffer blurred vision, slurred speech, difficulty swallowing and chewing, labored breathing, and eventually muscle paralysis. These symptoms are the result of the botulism toxin inhibiting the release of the neurotransmitter acetylcholine. Without the neurotransmitter, nerve impulses cannot pass to the muscles. As a result, muscles do not contract. If the muscles involved in breathing weaken, death can follow within a day or two.

Antibiotics are of no value against a toxin. Instead, large doses of special antibodies called **antitoxins** must be administered. These antibodies bind to and neutralize the toxins. The disease can be avoided by heating suspected foods before eating them because the toxin is destroyed by exposure to temperatures of 90°C (194°F) for 10 minutes.

Today, one of the botulinum toxins has been put to practical use. In extremely tiny doses, injections of Botox or Dysport (botulinum toxin type A) temporarily relaxes the muscles near the injection site. Therefore, the drug is useful on several localized movement disorders caused by involuntary muscle contractions. For example, the drugs are used to treat strabismus, or misalignment of the eyes (cross-eye). The toxin also can be valuable in relieving stuttering, uncontrolled blinking, and musician's cramp. In addition, the toxin can provide temporary relief from excessive body sweating and chronic migraine headaches. Most commonly, the toxin has been used to temporarily relieve facial wrinkles and frown lines (**FIGURE 19.8**).

Staphylococcal Food Poisoning

Staphylococcus aureus: staff-ih-loh-KOK-us OH-ree-us

One of the most common forms of food poisoning is caused by *Staphylococcus aureus*, a gram-positive sphere (**FIGURE 19.9**). Staphylococcal food poisoning ranks first among reported cases of foodborne intoxications and is caused by toxins secreted into foods. If consumed by unsuspecting individuals, they will experience abdominal cramps, nausea, vomiting, and diarrhea because the toxin triggers the release of water. The symptoms last for several hours, and recovery is usually rapid and complete.

FIGURE 19.8 Cosmetic Injection of Botulinum Toxin. The botulinum toxin in controlled doses can be used to temporarily minimize wrinkles.

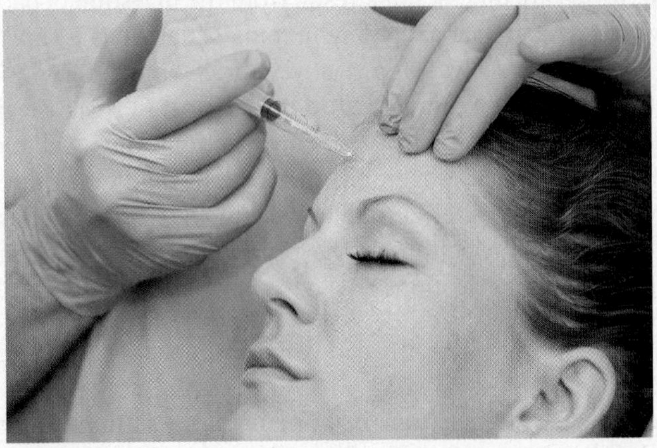

FIGURE 19.9 ***Staphylococcus aureus.*** A false-color scanning electron microscope image of *S. aureus* showing clusters of spherical cells. (Bar = 2 µm.)

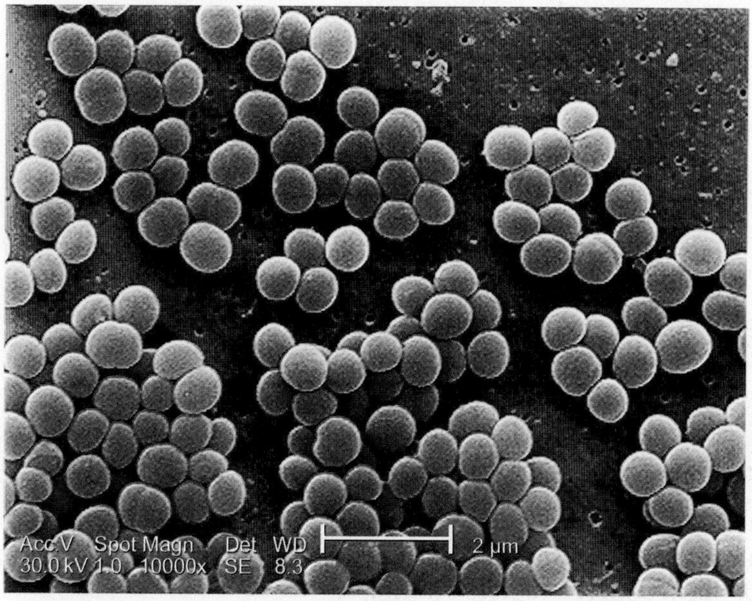

Courtesy of CDC.

Salmonellosis and Typhoid Fever

For *Salmonella enterica*, there are some 2,500 unique **serotypes** causing foodborne infections. Different serotypes of *S. enterica* are responsible for two disease conditions: salmonellosis and typhoid fever.

Salmonella enterica: sal-mon-EL-lah en-TAIR-eh-kah

Serotype: A group of closely related microbes distinguished by a characteristic set of antigens.

Salmonellosis

According to the Unites States Centers for Disease Control and Prevention (CDC), every year there are about 42,000 cases of **salmonellosis** reported in the United States. However, many cases are mild and may not be reported, so the actual number of total cases is probably well over one million.

Salmonellosis requires a relatively large infectious dose (>100,000 cells) to cause illness. Therefore, foods, most commonly beef, poultry, and eggs, must be heavily contaminated. Poultry products are particularly notorious because *Salmonella* species commonly infect chickens and turkeys (**FIGURE 19.10**). The microbes can be consumed directly from improperly cooked poultry. Salmonellosis also can occur after ingesting contaminated poultry products, such as chicken salad or cold cuts made from chicken. Cooking poultry properly and not allowing raw chicken and its fluids to encounter raw foods can prevent many salmonellosis outbreaks.

Typhoid Fever

Typhoid fever is one of the historical diseases that have ravaged human populations for centuries. *S. enterica* serotype Typhi (usually called S. Typhi) causes typhoid fever. The CDC estimates there are about 5,700 cases each year in the United States. About 75% of these cases are acquired during foreign travel. In fact, globally there are more than 21 million cases of typhoid fever every year.

The resistance to environmental conditions outside the body allows S. Typhi to remain alive for long periods in water, sewage, and certain foods. Thus, it is easily transmitted by flies, contaminated food, and the fecal-oral route. In the small intestine, S. Typhi causes deep ulcers and bloody stools. Blood invasion follows, and after a few days, the patient experiences mounting fever,

FIGURE 19.10 ***Salmonella*** **Contamination.** Several food types can be contaminated with *Salmonella*.

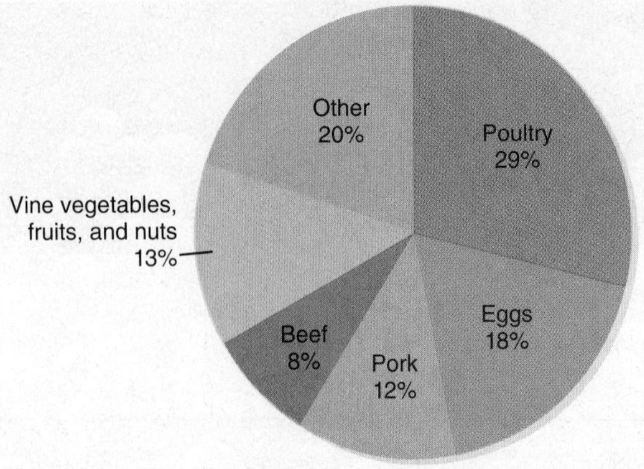

Courtesy of CDC.

tiredness, and delirium. The abdomen becomes covered with rose spots, an indication that blood is hemorrhaging in the skin. Often a grayish-/yellowish-coated tongue with red edges is evident.

Treatment of typhoid fever generally is successful with antibiotics, but about 5% of recoverees are carriers and continue to harbor and shed the organisms for a year or more. Consequently, people traveling to a country where typhoid fever is common should be vaccinated before traveling. However, immunity is lost after only a few years.

Shigellosis

Shigella sonnei: shih-GEL-lah SON-nee-ee

Another intestinal disease common among young children is **shigellosis**. Most cases of shigellosis are caused by *Shigella sonnei*. After ingesting the organisms in contaminated water or foods, the cells usually penetrate the intestinal lining. After 2 to 3 days, enough toxins have been produced to trigger water release. Shigellosis is manifested by waves of intense abdominal cramps and frequent passage of small-volume, bloody, mucoid stools. However, most infections go unreported, so the CDC estimates that there are about 500,000 diarrhea cases annually. About 130,000 of these cases are foodborne.

Most cases of shigellosis resolve within a week and usually produce few complications. Antibiotics are sometimes effective, but many strains of *Shigella* are becoming resistant to antibiotics.

Cholera

No diarrheal disease can compare with the extensive diarrhea associated with **cholera**. The WHO estimates there are more than 100,000 cases and over 1,900 deaths annually. The outbreak in Haiti that began after the 2010 earthquake is considered the worst epidemic of cholera in recent Haitian history. There have been more than 700,000 cases and some 8,500 deaths. More recently, between January and March 2019, there have been 110,000 suspected cases of cholera in Yemen. About one third of these cases have been in children under the age of five. At least 190 people have died.

Vibrio cholerae: VIB-ree-oh KAHL-er-eye

The causative agent, *Vibrio cholerae*, enters the intestinal tract in contaminated water or food. As the cells move along the intestinal wall, they secrete a toxin that stimulates an unrelenting loss of fluid. In the most severe cases, an infected patient might lose up to 1 liter of colorless, watery fluid every hour for several hours. The dehydration causes the patient's skin to wrinkle, dry, and feel cold to the touch.

Muscular cramps occur in the arms and legs. The blood thickens, urine production ceases, and the sluggish blood flow to the brain leads to shock and coma. In untreated cases, the mortality rate for cholera might reach 70%.

Antibiotics kill the bacterial cells, but the key treatment is to restore the body's water and **electrolyte** balance. For mild to moderate cases, this entails **oral rehydration therapy**. This involves drinking solutions of electrolytes and glucose designed to restore the normal water/salt balance in the body. For the most severe cases, rehydration requires intravenous injections.

Electrolyte: A mineral, like sodium or potassium in your blood and other body fluids, that carries an electric charge.

E. coli Diarrheas

Besides the harmless strains that are part of the human gut microbiome, there are other pathogenic strains of *Escherichia coli*. One such strain induces diarrhea in infants when it invades the intestinal lining and produces powerful toxins that cause water loss. Other strains cause **traveler's diarrhea**, a term usually applied to a disease in which a traveler experiences diarrhea within 2 weeks of visiting the affect area. The diarrhea lasts up to 10 days.

Escherichia coli: esh-er-EE-key-ah KOH-lee

It is possible to contract a very serious form of hemorrhagic diarrhea due to *E. coli* O157:H7. In the large intestine, the strain causes bloody diarrhea, a complication known as **hemorrhagic colitis**. When the disease involves the kidneys, it can lead to kidney failure and is called **hemolytic uremic syndrome** (**HUS**). Seizures, coma, colon perforation, and liver disorder have been associated with HUS.

Accounting for underdiagnosis, the CDC estimates there are over 96,000 cases of *E. coli* O157 annually. Contaminated foods have included ground meat, fresh spinach, romaine lettuce, hazelnuts, cheeses, and even cookie dough. It is uncertain how some of these foods get contaminated. The prevailing wisdom for ground meat contamination is that the O157:H7 cells exist in the intestines of cattle but cause no disease in the animals. Contamination during slaughtering brings *E. coli* to beef products. Excretion of the bacteria from cattle into the soil could account for transfer to plants and fruits.

Campylobacteriosis

Campylobacteriosis is one of the most common causes of diarrheal illness in Americans. Dairy products and water contaminated with *Campylobacter jejuni* are typical sources of infection. The symptoms of the disease range from mild diarrhea to severe gastrointestinal distress. Fever, abdominal pains, and bloody stools also are common. Most patients recover in less than a week, although the CDC estimates there are about 75 deaths every year from *Campylobacter* infections.

Campylobacter jejuni: kam-pill-oh-BAK-ter jeh-JU-nee

Listeriosis

In the fall of 2011, there was a multistate outbreak of **listeriosis** linked to cantaloupes. About 150 people were infected and 33 died. Listeriosis is caused by *Listeria monocytogenes*, a small, gram-positive rod. The disease primarily affects older adults, pregnant women, newborns, and individuals with a weakened immune system. It can occur as **listeric meningitis**, with headaches, stiff neck, delirium, and coma. Another form is a blood disease accompanied by high numbers of white blood cells. In pregnant women, an infection of the uterus might result in miscarriage.

Listeria monocytogenes: lis-TEH-ree-ah mah-no-sigh-TAH-jeh-neez

Listeria species commonly are found in the soil and in the intestines of many animals. Consequently, the bacterial cells can be transmitted to humans by food contaminated with fecal matter, as well as by the consumption of contaminated animal foods. Cold cuts, as well as soft cheeses (e.g., Brie, Camembert, and feta), have been associated with listeria outbreaks. Antibiotics are effective treatments.

Peptic Ulcer Disease

Helicobacter pylori: HE-lick-oh-bak-ter pie-LOW-ree

Today, peptic ulcers are considered an infectious disease. As crazy as it sounds, most cases of **peptic ulcer disease** are caused by a bacterium called *Helicobacter pylori*. This gram-negative, curved rod apparently is transmitted by the fecal-oral route from person-to-person. *H. pylori* can survive in the intense acidity of the stomach (**FIGURE 19.11**). The bacterial cells attach to the stomach wall and then secrete an enzyme that digests urea in the area, producing ammonia. The ammonia neutralizes the stomach acid in the vicinity of the infection and the organisms begin their destruction of the tissue, supplemented

FIGURE 19.11 The Progression of Gastric Ulcers. Most peptic ulcers are caused by *Helicobacter pylori*. Inset: false-color scanning electron microscope image of *H. pylori* cells. (Bar = 2 µm.)

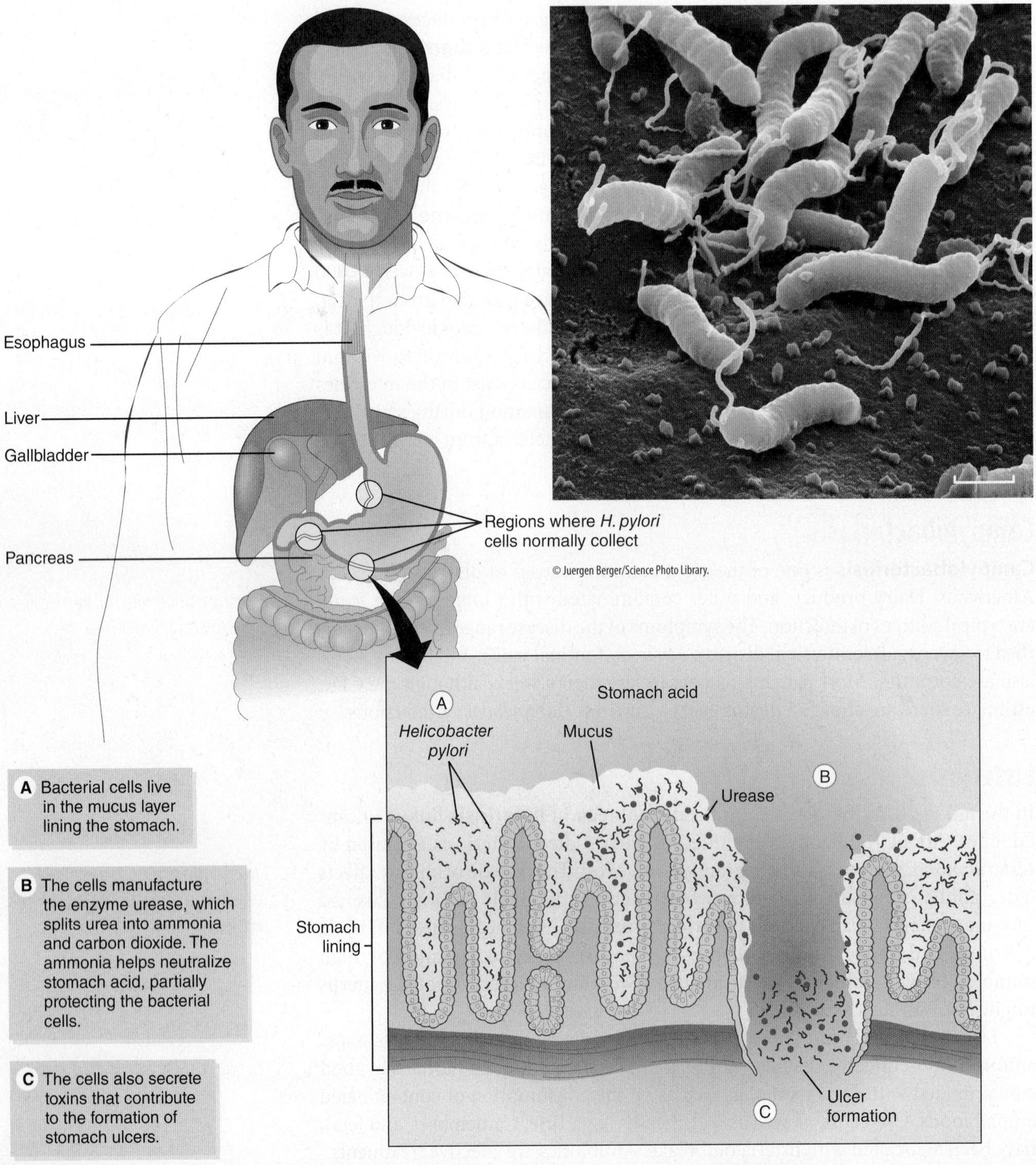
Esophagus
Liver
Gallbladder
Pancreas
Regions where *H. pylori* cells normally collect
A Bacterial cells live in the mucus layer lining the stomach.
B The cells manufacture the enzyme urease, which splits urea into ammonia and carbon dioxide. The ammonia helps neutralize stomach acid, partially protecting the bacterial cells.
C The cells also secrete toxins that contribute to the formation of stomach ulcers.
Stomach acid
A
Helicobacter pylori
Mucus
B
Urease
Stomach lining
C
Ulcer formation

by the stomach acid. Infection with *H. pylori* is also the major cause of gastric cancer, which represents the fourth most common cancer and the second leading cause of cancer deaths in the world—about 1 million deaths each year.

Doctors have revolutionized the treatment of ulcers by prescribing antibiotics. They have achieved cure rates of up to 94%. Relapses of ulcers are uncommon, and the cases of stomach cancer have been declining as *H. pylori* is eliminated from people's stomachs. However, the loss of *H. pylori* might not be all good news, as the cost-benefit analysis in **A CLOSER LOOK 19.2** reports.

A CLOSER LOOK 19.2

Helicobacter pylori: A Cost-Benefit Analysis

In the business community, a cost-benefit analysis allows a company to weigh its expected *costs* against expected *benefits* when determining the best (or most profitable) course of action for the company. We can do a similar analysis for infection with the bacterium *Helicobacter pylori*, which has been part of the human local stomach microbiome for more than 200,000 years. Specifically, what are the costs and benefits to humans of having an *H. pylori* infection? Let's do the analysis.

COST (of having *H. pylori*)		BENEFIT (of not having *H. pylori*)
The bacterium *H. pylori* infects 3 billion people around the world and 2% (60 million) contract **gastric (peptic) ulcer disease**. About 2% of the ulcer patients (1 million) will develop **stomach cancer** (in the US: 22,000 patients are diagnosed annually of whom 10,000 are expected to die).	↔	With the advent of antibiotic therapy to cure peptic ulcer disease, the incidence of *H. pylori* has dropped sharply. In developed nations, infection has dropped from 80% to just a few percent. That means, many fewer people are developing peptic ulcer disease and stomach cancer.
The presence of *H. pylori* might be linked to **adult type 2 diabetes**, which is the most common form of diabetes where the body does not respond properly to insulin.	↔	Elimination of *H. pylori* might lessen the chances of developing adult type 2 diabetes.
Several studies have suggested that people with **Parkinson's disease** (a brain disorder that leads to tremors and difficulty with walking and coordination) are more likely to have ulcers and to be infected with *H. pylori* than are healthy (non- Parkinson's) individuals.	↔	Eradication of *H. pylori* may lessen the chances of a person developing Parkinson's disease.

COST (of not having *H. pylori*)		BENEFIT (of having *H. pylori*)
Research suggests that people without *H. pylori* are at greater risk of **acid (esophageal) reflux disease** and **esophageal cancer**. These diseases have been rising dramatically as peptic ulcers have declined.	↔	Having *H. pylori* might lessen the chances of developing acid reflux disease and esophageal cancer. It would appear that *H. pylori* is in some way protecting the esophagus.
Immunologists and allergy experts report that people without *H. pylori* may be more prone to **allergy-induced asthma**.	↔	Infection with *H. pylori* provides protection from allergy-induced asthma. It appears that somehow *H. pylori* "trains" the immune system to lower its response to triggers causing asthma.

Discussion Point

So, as a cost-benefit analysis, is colonization by H. pylori *a good or bad situation? For example, if you had a peptic ulcer, would you submit to antibiotic treatment to eliminate the infection knowing that* H. pylori *eradication could have other unhealthy or healthy consequences?*

19.3 Soilborne Bacterial Diseases: Endospore Formers

Soilborne diseases are those whose bacterial agents are transferred from the soil to the unsuspecting individual. To remain alive in the soil, the bacterial cells must resist environmental extremes, and often they form **endospores**, as these two diseases illustrate.

Endospore: A dormant, tough, and non-reproductive structure produced by a few bacterial species when nutrients become limiting.

Bacillus anthracis: bah-SIL-lus an-THRAY-sis

Anthrax

Anthrax is a blood disease that occurs in cattle, sheep, goats, and rarely, in humans. The disease is caused by *Bacillus anthracis*, a gram-positive, spore-forming rod. Patients inhale the spores, ingest them in contaminated meat, or make skin contact with spores in the air (**FIGURE 19.12**). Violent dysentery with bloody stools accompanies the intestinal form.

There are antibiotics used for treatment after exposure or after infection. For treatment to be successful, the illness needs to be identified and treated early. In untreated cases, the mortality rate is more than 80%. There are few, if any, cases in the United States because of the testing of imported animal products. Of course, there was an anthrax bioterrorism incident in 2001 and anthrax remains a potential bioterror weapon, which was discussed in the chapter on Viral Diseases of Humans.

Clostridium tetani: kla-STRIH-dee-um TEH-tahn-ee

Tetanus

It is possible to get **tetanus** through a puncture wound from a soil-contaminated nail. However, the gram-positive organism, *Clostridium tetani*, is found everywhere in the environment. Its spores enter a wound and revert to multiplying bacilli. The growing cells produce the tetanus toxin that provokes sustained and uncontrolled contractions of the muscles, and spasms occur throughout the body.

FIGURE 19.12 Anthrax. This cutaneous lesion is a result of infection with anthrax bacilli. Lesions like this one develop when anthrax spores contact the skin, germinate and produce cells that multiply.

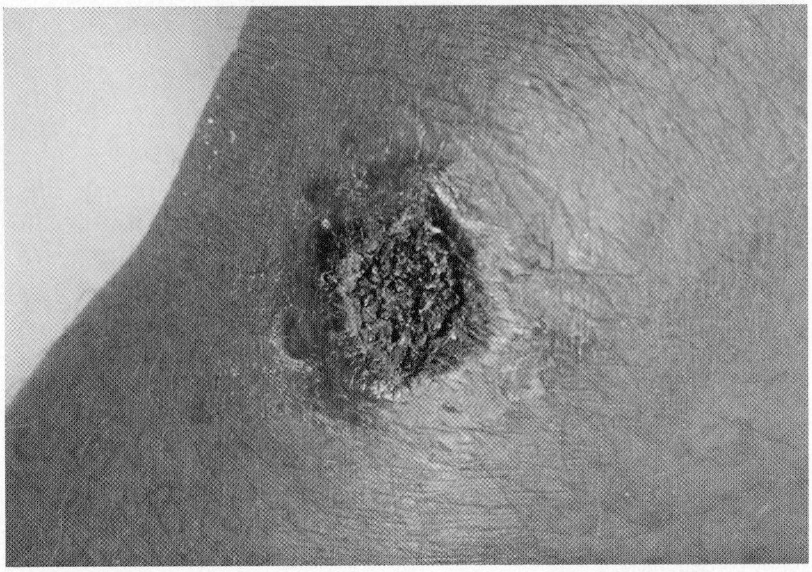

Tetanus patients are treated with sedatives and muscle relaxants and are placed in quiet, dark rooms. Physicians prescribe penicillin to destroy the bacterial cells and tetanus antitoxin to neutralize the toxin. The United States has had a steady decline in the incidence of tetanus due to immunization with the tetanus vaccine. There are only a few dozen cases reported to the CDC each year. Booster injections of tetanus toxoid in the "tetanus shot" are recommended every 10 years to keep the level of immunity high.

19.4 Arthropodborne Bacterial Diseases: The Bugs Bite

Fleas, lice, and ticks are examples of arthropods that transmit diseases to or among humans. Transmission usually happens by the arthropod taking a blood meal from an infected animal or person and themselves becoming infected. Then, these arthropods, called **vectors**, pass the microbes to another individual during the next blood meal. Arthropod-related diseases occur primarily in the bloodstream, and they often are accompanied by a high fever and a body rash.

Bubonic Plague

Few diseases have had a more terrifying history than **bubonic plague**. In fact, few can match the array of social, economic, and religious changes the disease created. The chapter opener is testament to its social significance. The pandemic in the 1300s was known as the "Black Death" because of the purplish-black splotches on victims. By some accounts, plague killed an estimated 40 million people in Europe, almost one-third of the population (**FIGURE 19.13**).

FIGURE 19.13 The Spread of Bubonic Plague in 14th-Century Europe. Over a period of 8 years, the Black Death spread clockwise from the Black Sea through Europe to Moscow.

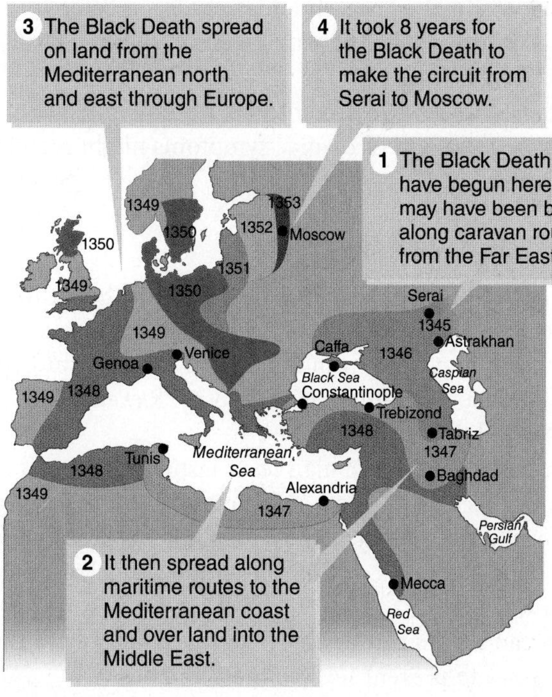

Bubonic plague first appeared in San Francisco in 1900, carried by rats on ships arriving from Asia. The disease spread to ground squirrels, prairie dogs, and other wild rodents. Today, it is endemic in the southwestern United States.

Yersinia pestis: yer-SIN-ee-ah PESS-tiss

Plague is caused by the gram-negative rod *Yersinia pestis*. The bacterial cells are transmitted by infected rat fleas when they take a blood meal. In humans, the bacterial cells localize in the lymph nodes, especially those of the armpits, neck, and groin. Hemorrhaging causes substantial swellings called **buboes**—hence the name bubonic plague. From there, the microbes can spread to the bloodstream, where they cause **septicemic plague** and then to the lungs, where they cause **pneumonic plague**. There is extensive coughing and hemorrhaging, and many patients suffer cardiovascular collapse. In the pneumonic form, the cells can be spread by respiratory droplets from person to person.

When detected early, plague can be treated with specific antibiotics, reducing mortality to less than 10%. A vaccine consisting of dead *Y. pestis* cells is available for high-risk groups. Without treatment, mortality rates for pneumonic plague approach 100%.

Lyme Disease

Lyme disease is named for Old Lyme, Connecticut, the suburban community where the first cluster of disease cases occurred in 1975. The disease was traced to deer ticks, and several years later, the causative agent was identified and named *Borrelia burgdorferi*. Lyme disease is currently the most commonly reported arthropodborne illness in the United States. About 95% of cases occur in the northeastern, mid-Atlantic, and north-central states (**FIGURE 19.14**).

Borrelia burgdorferi: bore-RELL-ee-ah burg-DOOR-fer-ee

About 20% of people infected with *B. burgdorferi* suffer nothing more than flu-like symptoms. For many of the rest, the illness starts with a rash at the site of the tick bite. It expands slowly, eventually forming an intense red border and a red center, resembling a bull's eye (**FIGURE 19.15**). Left untreated, some cases enter a second stage where the patient experiences pain, swelling, and arthritis in the large joints, especially the knee, shoulder, ankle, and elbow. In some individuals, a third stage occurs, where the arthritis is complicated by damage to the cardiovascular and nervous systems.

In the rash stage, effective treatment can be rendered with antibiotics. Patients developing neurological or cardiac symptoms might require treatment with intravenous antibiotics.

Rocky Mountain Spotted Fever

Rocky Mountain spotted fever (**RMSF**) is a serious tickborne illness that can be deadly if not treated early. RMSF cases occur throughout the United States but are most commonly reported from North Carolina, Tennessee, Missouri, Arkansas, and Oklahoma. The illness is caused by *Rickettsia rickettsii* and is carried by dog and wood ticks.

Rickettsia rickettsii: rih-KET-sea-ah rih-KET-sea-ee

One of the features of RMSF is a high fever lasting for several days. A skin rash also occurs. It begins as pink spots and progresses to pink-red pimple-like spots that fuse to form a flat, red area on the skin. The rash generally begins on the palms of the hands and soles of the feet and progressively spreads to the trunk and face (**FIGURE 19.16**).

Because RMSF can be life threatening, early treatment with the antibiotic doxycycline is necessary to prevent severe illness and death.

FIGURE 19.14 Reported Cases of Lyme Disease—by County of Residence. Although cases of Lyme disease have been reported in most every state, most cases are reported from southern Maine to northern Virginia and from Wisconsin, northern Minnesota, and northern Illinois.

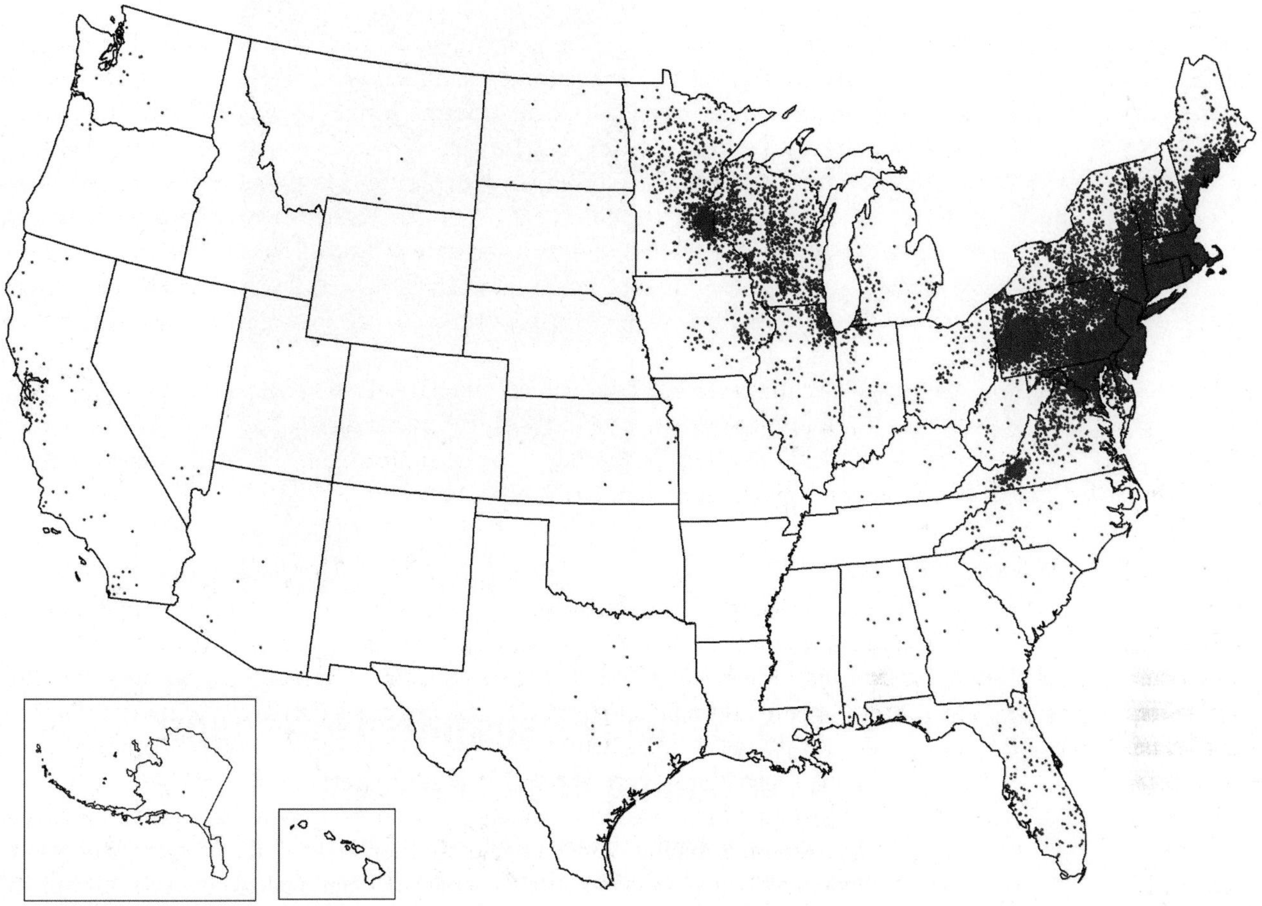

Courtesy of CDC.

FIGURE 19.15 The Bull's-Eye Rash. Lyme disease can start with a rash that consists of a large red center and an intense red border. It is usually hot to the touch, and it expands with time.

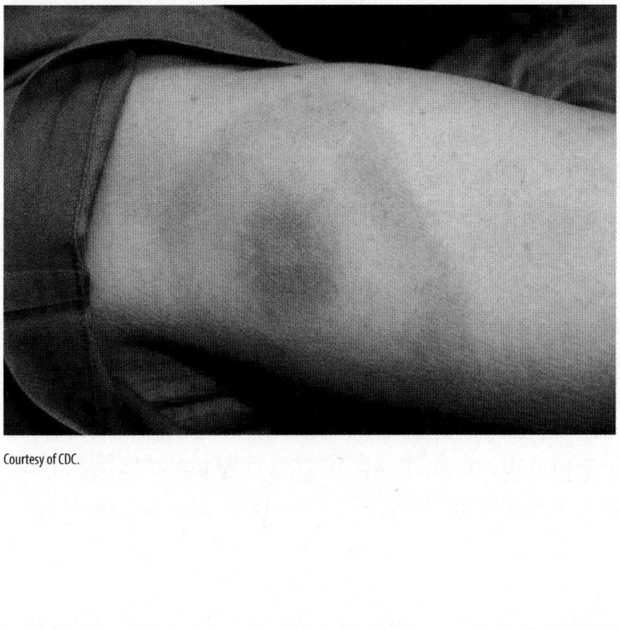

Courtesy of CDC.

FIGURE 19.16 Rocky Mountain Spotted Fever. A child's face displays the characteristic spotted rash of RMSF.

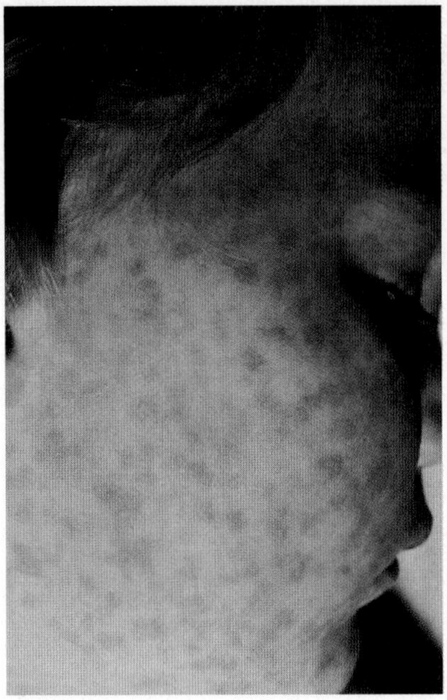

Courtesy of CDC.

19.5 Sexually Transmitted Infections: A Continuing Health Problem

The **sexually transmitted infections** (**STIs**) belong to a broad category of illnesses transmitted by direct contact. The contact in this case is with the reproductive organs. Person-to-person transmission is necessary for bacterial survival because the microbes usually cannot remain alive outside the body tissues. The major STIs continue to be a problem in the United States, as they make up four of the top ten reported infectious diseases (**FIGURE 19.17**). Incidentally, STIs have been referred to as **sexually transmitted diseases** (**STDs**). The term STI has become more informative because often these infections are without a definite disease outcome.

Syphilis

Treponema pallidum: treh-poh-NEE-mah PAL-eh-dum

Over the centuries, Europeans have had to contend with four pox diseases: chickenpox, cowpox, smallpox, and the Great Pox. The latter now is known as syphilis. **Syphilis** is caused by *Treponema pallidum*, whose spiral bacterial cells penetrate the skin surface and cause a disease that can progress through three stages.

- **Primary syphilis.** This stage is characterized by a **chancre**, a painless, hard, circular, purplish ulcer often on the genital organs. It persists for 2 to 6 weeks, and then disappears.
- **Secondary syphilis.** This second stage occurs several weeks later. The patient experiences lesions over the entire body surface, fever, rash, and a patchy loss of hair on the head. Recovering patients bear pitted scars from the lesions, and they remain "pockmarked."

FIGURE 19.17 Reported Cases of Notifiable Diseases in the United States, 2017. Four of the top six most reported microbial diseases in the United States are sexually transmitted.

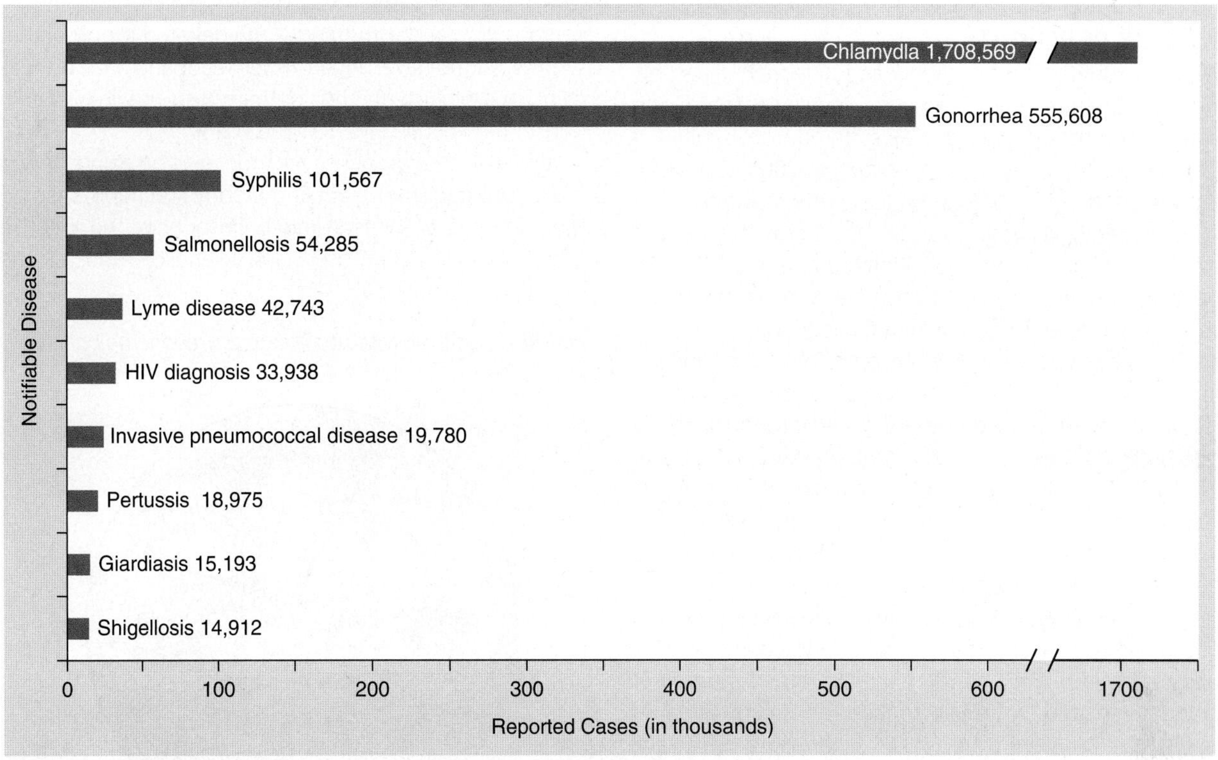

Data from CDC, Summary of Notifiable Diseases, 2017.

In terms of primary and secondary syphilis, the CDC has reported a 167% increase in the number of reported cases between 2007 and 2017.

- **Tertiary syphilis.** In about 40% of syphilis patients, a third stage of the disease occurs. Its hallmark is the **gumma**. This is a soft, gummy, granular lesion that weakens the blood vessels, causing them to bulge and burst. In the brain, they might alter the patient's personality and judgment and cause insanity so intense that for many generations, people with tertiary syphilis were confined to mental institutions.

However, some people believe the infection sometimes can alter a person's personality and judgment in other ways, as **A CLOSER LOOK 19.3** examines.

Penicillin is the drug of choice for the primary and secondary stages of syphilis. However, antibiotics are ineffective in tertiary syphilis. The cornerstone of syphilis control is the identification and treatment of the sexual contacts of patients. Syphilis also is a serious problem in pregnant women because the bacterial cells penetrate the placental barrier and cause **congenital syphilis**, which is a severe, disabling, and often life-threatening infection.

Gonorrhea

Gonorrhea is another common STI that infects both males and females. In fact, it is the second most frequently reported notifiable disease in the United States (see Figure 19.17). The number of reported cases has increased by 67% since 2013 (**FIGURE 19.18**). The STI is caused by *Neisseria gonorrhoeae*, a small, gram-negative diplococcus commonly known as the gonococcus. The great majority of cases of gonorrhea are transmitted by person-to-person contact during sexual intercourse.

In women, the gonococci invade the **cervix** and the **urethra**. Patients often report a discharge, abdominal pain, and a burning sensation on urination. In some

Neisseria gonorrhoeae: nye-SEER-ee-ah gah -nor-REE-eye

Cervix: The entrance to the womb and leading to the vagina.

Urethra: The tube in mammals carrying urine from the bladder out of the body.

A CLOSER LOOK 19.3

A Spark of Vision?

What do Abraham Lincoln, Adolf Hitler, Friedrich Nietzsche, Oscar Wilde, Ludwig van Beethoven, and Vincent van Gogh have in common? Very likely, all suffered from syphilis, if Deborah Hayden's research is correct. In 2003, she wrote a book entitled *Pox: Genius, Madness, and the Mysteries of Syphilis* (New York: Basic Books). In the book, she looks at 14 eminent figures from the 15th to 20th centuries whose behavior, careers, or personalities were more than likely shaped by this sexually transmitted infection.

There was no cure for syphilis prior to the introduction of penicillin in 1943. Patients suffered a long-lasting and relapsing disease as the bacterium spread through the body (see **FIGURE A**). Then, years later, the infection would reappear as tertiary syphilis. In this most dangerous form, patients suffered excruciating headaches and gastrointestinal pains. In many cases, the afflicted individuals eventually developed deafness, blindness, paralysis, and insanity.

Yet sometimes, ecstasy and fierce creativity were part of the "symptoms" of tertiary syphilis. As Hayden says, "one of the 'warning signs' of tertiary syphilis is the sensation of being serenaded by angels." In fact, Danish writer Karen Blixen (Isak Dinesen), who wrote *Out of Africa*, once said, "Syphilis sold [my] soul to the devil for the ability to tell stories." Hayden believes it is just such emotions that provided much of the creative spark for many of the notable historical figures she describes in her book. Perhaps the most intriguing is the debated proposal that syphilis might have driven Hitler mad, and that he was dying of syphilis when he committed suicide in his Berlin bunker in the final days of World War II.

If Hayden's arguments are true, it is amazing how a bacterial organism can affect the body and mind, shaping the thoughts of writers and philosophers, the creative genius of artists, composers, and scientists—and yes, the madness of dictators.

FIGURE A Light microscope image of *Treponema pallidum* (dark spirals).

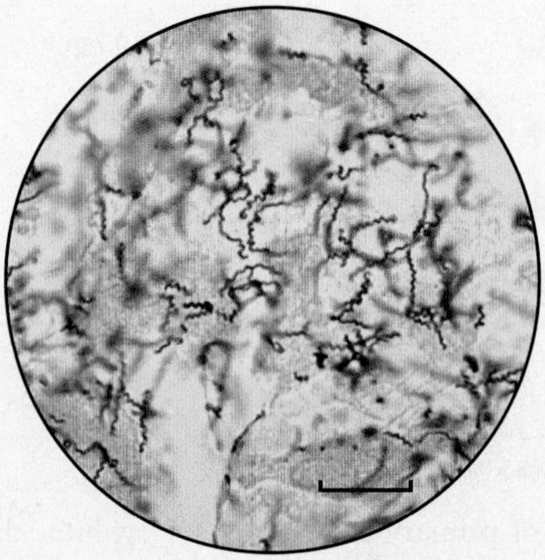

Courtesy of CDC.

FIGURE 19.18 Reported Cases of Chlamydia, Gonorrhea, and Syphilis—United States and U.S. Territories, 1995–2017. In 2017, the CDC reported the largest ever number of cases of chlamydia and gonorrhea.

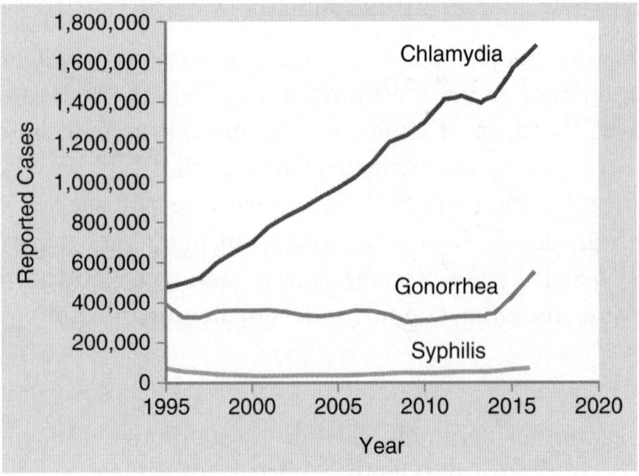

Fallopian tube: One of a pair of long, narrow ducts that transport sperm cells to the egg and transport the egg from the ovary to the uterus; also called an oviduct.

women, gonorrhea also spreads to the **fallopian tubes**, and these thin passageways become riddled with adhesions, causing a difficult passage for egg cells. In men, gonorrhea occurs primarily in the urethra. Onset is usually accompanied by a tingling sensation in the penis, followed in a few days by pain when urinating. There is also a thin, watery discharge at first, and later a whitened, thick fluid that resembles semen. In adults, gonorrhea therapy consists of antibiotic treatment.

The gonococci do not restrict themselves to the urogenital organs. **Gonococcal pharyngitis** from oral sex can develop in the pharynx. In infants born to infected women, gonococci might cause a disease of the eyes called **gonococcal ophthalmia**. To preclude the blindness that might ensue, most states have laws requiring the eyes of newborns be treated with antibiotics.

Chlamydia

Chlamydia is the most prevalent bacterial STI reported to the CDC (see Figure 19.17). In fact, for 2017 the CDC reported all-time highs in the number of cases of chlamydia, representing a 28% increase in reported cases since 2013 (see Figure 19.18). Chlamydia is caused by small bacterium called *Chlamydia trachomatis* and is most commonly spread through vaginal, oral, and anal sex.

Chlamydia trachomatis: kla-MIH-dee-ah trah-KO-mah-tiss

Women suffering from chlamydia often note a slight vaginal discharge as well as inflammation of the cervix. Burning pain can be experienced on urination. In complicated cases, the disease might spread to the fallopian tubes and block the passageways. In men, chlamydia is characterized by painful urination and a discharge that is waterier and less copious than that of gonorrhea. Tingling sensations in the penis are generally evident, and inflammation of the **epididymis** might result in sterility.

Epididymis: A coiled tube attached to the back and upper side of each male testicle.

A newborn might contract *C. trachomatis* from an infected mother. Contact with the pathogen during childbirth can lead to disease of the eyes known as **chlamydial ophthalmia**. **Chlamydial pneumonia** might also develop in a newborn. Health officials estimate that each year in the United States, over 75,000 newborns suffer chlamydial ophthalmia and 30,000 newborns experience chlamydial pneumonia. The disease can be treated successfully with antibiotics.

19.6 Contact and Miscellaneous Bacterial Diseases: Still More Pathogens

Several human diseases are transmitted by contact. Usually, some form of skin contact takes place, as these diseases will illustrate.

Staphylococcal Skin Disease

Staphylococcus aureus, the grapelike cluster of gram-positive cocci, is a species involved with several skin diseases.

Abscesses

The hallmark of an *S. aureus* skin infection is the **abscess**, a confined, pus-filled lesion. A boil is an example of a skin abscess. Deeper skin abscesses, called **carbuncles**, develop when the staphylococci work their way into the tissues below the skin (**FIGURE 19.19A**).

Impetigo

Impetigo represents a more superficial infection and involves patches of epidermis just below the outer skin layer (**FIGURE 19.19B**).

Both skin conditions are commonly treated with penicillin, but resistant strains of *S. aureus* have become a real treatment problem, both in hospital-acquired and community-acquired infections. Therefore, it is important that the doctor identifies the strain causing the infection, so an appropriate antibiotic can be selected. The chapter on Controlling Microbes discusses antibiotics and the antibiotic resistance problem.

FIGURE 19.19 Staphylococcal Skin Diseases.

(A) A severe carbuncle on the back of the head and neck.

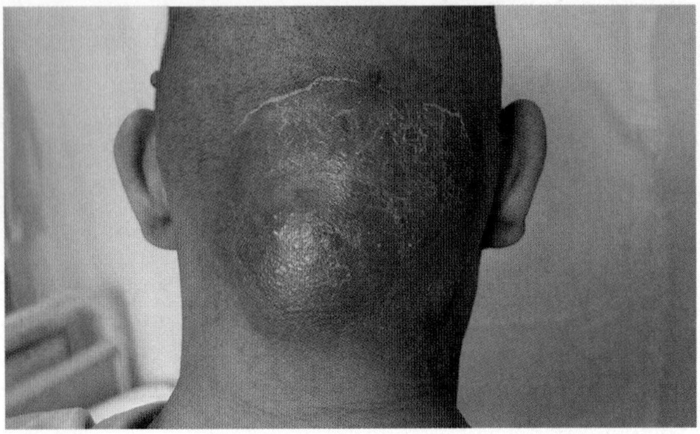

(B) A patient with impetigo on the cheeks of the face.

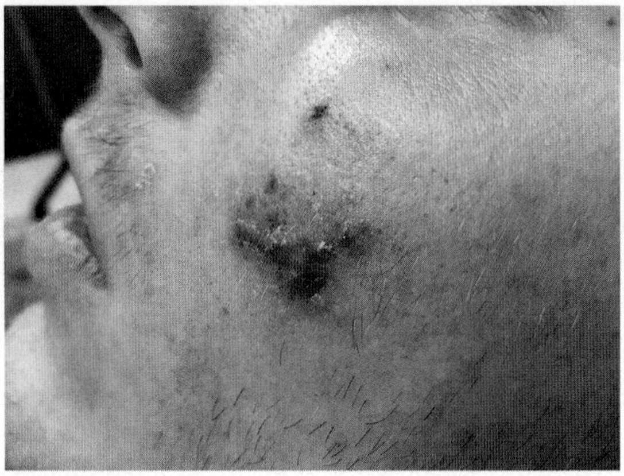

(C) A patient with toxic shock syndrome.

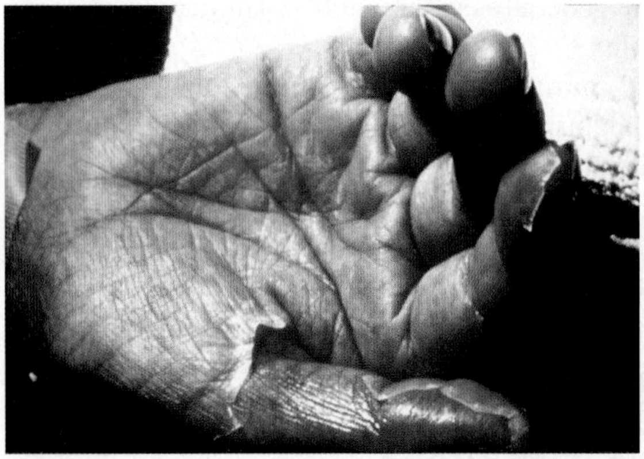

Toxic Shock Syndrome

Toxic shock syndrome (**TSS**) is caused by a toxin-producing strain of *S. aureus*. The earliest symptoms of TSS include a rapidly rising fever, accompanied by vomiting and watery diarrhea. Patients then experience a sore throat and severe muscle aches. A sunburn-like rash with peeling of the skin, especially on the soles of the feet and palms of the hands also occurs (**FIGURE 19.19C**). Clinical symptoms can progress rapidly, leading to a sudden drop in blood platelets, renal failure, shock, and death. Antibiotics can be used to inhibit bacterial growth.

Dental Diseases

Scientists estimate that there are between 50 billion and 100 billion bacterial cells in the adult mouth at any one time. Some can cause infections under the right conditions.

Dental Caries

Streptococcus mutans: strep-toe-KOK-us MEW-tanz

One of the primary bacterial pathogens of **dental caries**, or tooth decay, is *Streptococcus mutans*. This gram-positive coccus can attach to the smooth surfaces, pits, and fissures of a tooth. If they remain in the dental plaque as a biofilm on the tooth surface, they can ferment sugars to lactic acid and other acids.

The acids eventually dissolve the calcium compounds of the tooth enamel, and protein-digesting enzymes break down any remaining organic materials. Preventing dental caries means combating plaque buildup by normal brushing and flossing of the teeth along with dental cleaning to remove the plaque. Dietary modifications that minimize sucrose in foods also helps.

Periodontal Disease

Caries is not the only form of dental disease. The teeth are surrounded by periodontal tissues that provide essential support. Infection in these tissues can lead to **periodontal disease**, which affects some 80% of American adults. One of the most common forms of early-stage periodontal disease is **gingivitis** (**FIGURE 19.20A**). This condition develops when bacterial cells in the plaque multiply and build up between the teeth and gums. Left untreated, gingivitis can progress to what is called **periodontitis**, a serious disease of the soft tissue and bone supporting the teeth (**FIGURE 19.20B**). Untreated, periodontitis can result in loosening of the teeth and tooth loss.

Urinary Tract Infections

It has been estimated that up to 50% of humans will suffer a **urinary tract infection (UTI)** at some time during their lives. In the United States, the CDC estimates there are about 4 million ambulatory-care visits each year to treat UTIs. This represents about 1% of all outpatient visits and accounts for about 10 million doctor visits each year.

The major bacterial agents causing UTIs are *E. coli* and *Staphylococcus saprophyticus*. Sufferers report abdominal discomfort, burning pain on urination, and frequent urges to urinate. Most infections develop in the bladder, and women are apparently infected more often than men.

Staphylococcus saprophyticus: staff-ih-loh-KOK-us sah-pro-FI-tih-kus

Such practices as avoiding tight-fitting clothes and urinating soon after sexual intercourse can reduce the possibility of infection. Unfortunately, if UTIs are left untreated, the infections eventually can involve the entire bladder or spread to the kidneys.

FIGURE 19.20 Periodontal Disease.

(A) Gingivitis is an inflammation of the gums (gingiva) around the.

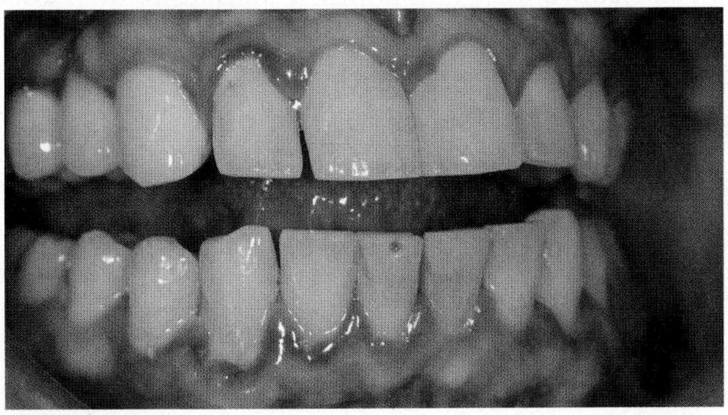

(B) Periodontitis refers to conditions when gum disease affects the structures supporting the teeth. In this example, gum recession and bone loss has occurred, which can lead to loosening of the teeth and even tooth loss.

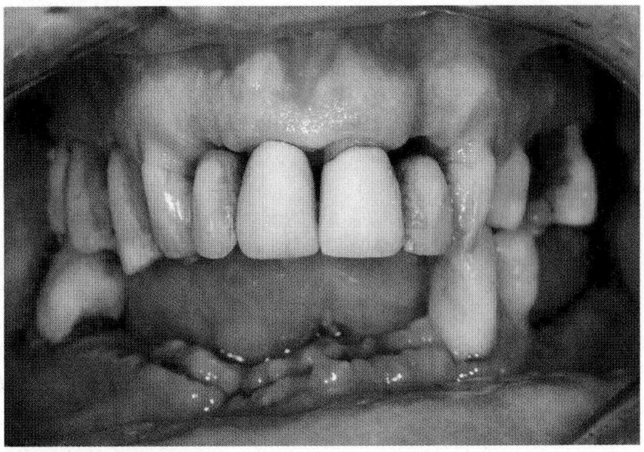

19.7 Healthcare-Associated Infections: Treatment Threats

Healthcare-associated infections (HAIs), also called nosocomial infections, are illnesses that are acquired in a healthcare setting. The infection results when a patient gets infected with a microbial pathogen as a result of receiving treatment for another condition or illness. Sources of the infectious agents include caregivers, hospital staff, and the surfaces and environment of the healthcare facility. Infections also can be associated with contaminated devices used in medical procedures, such as the use of catheters or ventilators, and through infection at surgical sites.

Until recently, HAIs accounted for nearly 1.7 million infections and as many as 99,000 deaths each year in the United States. Some estimates of the annual cost in treating HAIs range from $4.5 billion to $11 billion. However, between 2011 and 2015, a CDC study of inpatients in American hospitals indicates that HAIs have fallen by 16%.

Clostridioides difficile: kla-strih-dee-OY-deez DIF-fih-sil-ee

There are many different bacterial organisms capable of causing HAIs. The most common are *Clostridioides difficile*, *S. aureus*, and *E. coli*. The most common sites affected are the urinary and respiratory tracts and surgical sites. In most cases, HAIs can be prevented with increased diligence in handwashing and surface disinfection. In the above-mentioned survey, the reduction in HAI infections was due, in part, to the decrease in use of urinary catheters.

A Final Thought

As we conclude this chapter of *Microbes and Society*, you should take a moment to reflect on the vast body of knowledge you have acquired over the many pages of this book. You have undoubtedly become better aware of the microbes and their profound influence on your life. The last two chapters have focused on the negative aspects, but you will certainly also remember the numerous ways in which they affect your life for the better. You have learned an alphabet of microbes and a language of microbiology that will allow you to keep learning the rest of your life. No longer will you bypass a news headline referring to an emerging disease, a biotechnology breakthrough, or a new drug. Now you can read and digest the contents of the story. Then, in using and sharing your new knowledge, you will become a better citizen. And that's what education is all about. Congratulations! You now are a citizen microbiologist!

Chapter Discussion Questions

What Was He Thinking?

From your reading of this chapter, identify and discuss five major points about bacterial diseases that the author was trying to get across to you.

Questions to Consider

1. The CDC estimates 40,000 people in the United States die annually from pneumococcal pneumonia. Despite this high statistic, only 30% of older adults who could benefit from the pneumococcal vaccine are vaccinated

(compared to over 50% who receive an influenza vaccine yearly). As an epidemiologist in charge of bringing the pneumonia vaccine to a greater percentage of older Americans, what would you do to convince older adults to be vaccinated?

2. A children's hospital in a major American city reported a dramatic increase in the number of rheumatic fever cases. Doctors were alerted to start monitoring sore throats more carefully. Why do you suppose this prevention method was recommended?
3. The story is told of a doctor in New York City in the early 1800s who was an expert at diagnosing typhoid fever even before the symptoms of disease appeared. He would go up and down the rows of hospital beds, feeling the tongues of patients and announcing which patients were in the early stages of typhoid. Sure enough, a few days later the symptoms would surface. What do you think was the secret to his success?
4. You read in the newspaper that botulism was diagnosed in 11 patrons of a local restaurant. The disease was subsequently traced to mushrooms bottled and preserved in the restaurant. What special cultivation practice enhances the possibility that mushrooms will be infected with the spores of *Clostridium botulinum*?
5. A classmate plans to travel to a tropical country for spring break. To prevent traveler's diarrhea, she was told to take 2 ounces or two tablets of Pepto-Bismol four times a day for 3 weeks before travel begins. Short of turning pink, what better measures can you suggest she use to prevent traveler's diarrhea?
6. A column by Ann Landers carried the following letter: "I am a 34-year-old married woman who is trying to get pregnant, but it doesn't look promising.... When I was in college, I became sexually active. I slept with more men than I care to admit.... Somewhere in my wild days, I picked up an infection that left me infertile.... The doctor told me I have quite a lot of scar tissue inside my fallopian tubes." The woman went on to implore readers to be careful in their sexual activities. What advice do you think the woman gave to readers?
7. A frozen-food manufacturer recalls thousands of packages of jumbo stuffed shells and cheese lasagna after a local outbreak of salmonellosis. Which parts of the pasta products would attract the attention of inspectors as possible sources of salmonellosis? Why?

Appendix

Pronouncing Microorganism Names and Taxonomic Terms

This pronunciation guide is to help students master the pronunciation of microbe names and taxonomic terms. The syllable in all CAPS is emphasized in the pronunciation.

Prokaryotic Genera	**Pronunciation or Species**
Acetobacter aceti	a-SEA-toh-bak-ter a-SET-ee
Acinetobacter baumannii	a-sih-NEH-toe-bak-ter bow-MAHN-nee
Aeromonas	AIR-oh-mo-nass
Alcaligenes	all-kah-LIH-jen-eez
Aliivibrio fischeri	alee-ee-VIB-ree-oh FISH-er-ee
Amycoletopis orientalis	AH-my-coh-leh-toh-pis oh-ree-en-TAL-iss
Anabaena	an-nah-BEE-nah
Asticcacaulis excentricus	as-tik-ka-CAW-liss ek-SEN-trih-kus
Azotobacter	a-ZOE-toe-bak-ter
Bacillus anthracis	bah-SIL-lus an-THRAY-sis
B. cereus	SEH-ree-us
B. licheniformis	lie-ken-ih-FOR-miss
B. polymyxa	pawl-ee-MIX-ah
B. sphaericus	SFEH-rih-kus
B. subtilis	SUH-til-iss
B. thuringiensis	thur-in-je-EN-sis
Bacteroides	BAK-teh-roy-deez
Beijerinckia	bi-yeh-RINK-ee-ah
Bifidobacterium lactis	bi-fih-doe-back-TIER-ee-um LACK-tiss
Bordetella pertussis	bore-deh-TEL-lah per-TUS-sis
Borrelia burgdorferi	bore-RELL-ee-ah burg-DOOR-fer-ee
Campylobacter jejuni	KAM-pill-oh-bak-ter jeh-JU-nee
Chlamydia trachomatis	kla-MIH-dee-ah trah-KO-mah-tiss
Clostridioides difficile	kla-strih-dee-OY-deez DIF-fih-sil-ee
Clostridium botulinum	kla-STRIH-dee-um bot-you-LIE-num
C. tetani	TEH-tahn-ee
Corynebacterium glutamicum	KOH-ree-nee-back-tier-ee-um glu-TAH-meh-cum
Deinococcus radiodurans	DIE-no-kok-kus ray-dee-oh-DUR-anz
Diplococcus pneumoniae	DIP-loh-kok-us new-MO-nee-eye
Enterobacter aerogenes	en-teh-roh-BACK-ter air-RAH-jen-eez
Enterococcus faecalis	en-teh-roh-KOK-us FEE-kah-liss
E. faecium	FEE-cee-um
Erwinia	err-WIH-nee-ah

Escherichia coli	esh-er-EE-key-ah KOH-lee
Ferroplasma acidarmanus	fair-roh-PLAZ-mah ah-sid-ARE-mah-nuss
Flavobacterium	flay-voh-bak-TIER-ee-um
Francisella tularensis	FRAN-sis-el-lah too-lah-REN-sis
Haemophilus influenzae	hee-MAH-fill-us in-flew-EN-zeye
Helicobacter pylori	HE-lick-oh-bak-ter pie-LOW-ree
Klebsiella pneumoniae	kleb-sea-EL-lah new-MO-nee-eye
Lactobacillus acidophilus	lack-toe-bah-SIL-lus ah-sid-OFF-ill-us
L. bulgaricus	bull-GAIR-ee-kus
L. casei	KAY-sea-ee
L. curvatus	KUR-vah-tuss
L. plantarum	plan-TAR-um
Lactococcus lactis	lack-toe-KOK-us LACK-tiss
Legionella pneumophila	lee-ja-NEL-lah new-MAH-fil-lah
Leuconostoc citrovorum	lou-koh-NOS-tock sit-roh-VOR-um
L. mesenteroides	mez-en-ter-OY-deez
Listeria monocytogenes	lis-TEH-ree-ah mah-no-sigh-TAH-jeh-neez
Micrococcus luteus	my-kroh-KOK-us lu-TEE-us
Micromonospora purpurea	my-kroh-moh-NOS-poh-rah per-POO-ree-ah
Mycobacterium tuberculosis	my-koh-back-TIER-ee-um too-ber-cue-LOH-sis
M. vaccae	VAK-keye
Mycoplasma genitalium	my-koh-PLAZ-mah jen-ih-TAY-lee-um
M. pneumoniae	new-MOH-nee-eye
Myxococcus xanthus	micks-oh-KOK-us ZAN-thus
Neisseria gonorrhoeae	nye-SEER-ee-ah gah-nor-REE-eye
N. meningitidis	meh-nin-jih-TIE-diss
Nostoc	NOS-tock
Pelagibacter ubique	peh-LAJ-eh-back-ter u-BEAK
Photorhabdus luminescens	fo-tow-RAB-dus lu-mih-NES-senz
Propionibacterium acnes	pro-pea-OHN-ee-bak-tier-ee-um AK-nees
Proteus	PROH-tee-us
Pseudomonas aeruginosa	sue-doh-MOH-nahs ah-rue-gih-NO-sah
Rhizobium	rye-ZOH-bee-um
R. radiobacter	ray-de-oh-BACK-ter
Rickettsia rickettsii	rih-KET-sea-ah rih-KET-sea-ee
Salmonella enterica	sal-mon-EL-lah en-TAIR-eh-kah
S. enterica (Typhi)	TIE-fee
Serratia marcescens	ser-RAH-tee-ah mar-SES-senz
Shewanella	shoo-ah-NELL-ah
Shigella dysenteriae	shih-GEL-lah dis-en-TEH-ree-eye
S. sonnei	SON-nee-ee
Staphylococcus aureus	staff-ih-loh-KOK-us OH-ree-us
S. epidermidis	eh-peh-DER-mih-dis
S. equorum	eh-KWOR-um
S. saprophyticus	sah-pro-FI-tih-kus
Streptococcus lactis	strep-toe-KOK-us LAK-tiss
S. mutans	MEW-tanz
S. pneumoniae	new-MOH-nee-eye
S. pyogenes	pie-AHJ-en-eez
S. salivarius	sal-ih-VAIR-ee-us
S. thermophilus	ther-MOH-fill-us
Streptomyces cattleya	strep-toe-MY-seas KAT-lee-ah

S. erythreus	err-ih-THRAY-us
S. griseus	GRIH-sea-us
S. rimosus	rih-MOH-sus
S. venezuelae	veh-neh-zoo-EH-leye
Thermotoga maritima	ther-moh-TOE-gah mar-ih-TEA-mah
Thiomargarita namibiensis	thi-oh-mar-gah-REE-tah nah-mih-bee-EN-sis
Treponema pallidum	treh-poh-NEE-mah PAL-eh-dum
Vibrio cholerae	VIB-ree-oh KAHL-er-eye
Yersinia pestis	yer-SIN-ee-ah PESS-tiss
Xanthomonas	zan-tho-MO-nass
Zoogloea ramigera	ZO-oh-glee-ah rah-mih-JER-ah

Eukaryotic Genera or Species	**Pronunciation**
Fungi	
Alternaria	all-ter-NARE-ee-ah
Ashbya gossypii	ASH-be-ah gos-SIP-ee-ee
Aspergillus flavus	a-sper-JIL-lus FLAY-vus
A. fumigatus	few-mih-GAH-tus
A. niger	NYE-jer
A. oryzae	OH-rye-zeye
Batrachochytrium dendrobatidis	bah-trah-koh-KIH-tree-um den-dro-bah-TIE-diss
Botrytis	bow-TRI-tiss
Candida albicans	KAN-did-ah AL-bih-kanz
Cephalosporium acremonium	sef-ah-low-SPOH-ree-um ak-reh-MOH-nee-um
Claviceps purpurea	KLA-vi-seps pur-POO-ree-ah
Coccidioides immitis	kok-sid-ee-OY-deez IM-mi-tiss
Penicillium camemberti	pen-ih-SIL-lee-um kam-am-BER-tee
P. chrysogenum	cry-SAH-gen-um
P. notatum	know-TAH-tum
P. roqueforti	row-ko-FOR-tee
Rhizopus	rye-ZOH-puss
Saccharomyces carlsbergensis	sack-ah-roe-MY-seas ka-ruls-ber-GEN-sis
S. cerevisiae	seh-rih-VIS-ee-eye
S. ellipsoideus	ee-lip-SOY-dee-us
S. kefir	KEY-fur
Trichoderma	trick-oh-DER-mah
Protists	
Acanthamoeba	a-kan-thah-ME-bah
Botryococcus braunii	bow-tree-oh-KOK-kus BRAWN-ee-ee
Chlamydomonas	klam-ih-do-MO-nahs
Entamoeba histolytica	en-tah-MEE-bah hiss-toe-LIH-tih-kah
Giardia intestinalis	gee-ARE-dee-ah in-tes-tin-AL-iss
Paramecium	pair-ah-ME-sea-um
Phytophthora infestans	fı-TOF-tho-rah in-FES-tanz
Plasmodium falciparum	plaz-MOH-dee-um fal-SIP-arr-rum
Toxoplasma gondii	toxs-oh-PLAZ-mah GONE-dee-ee
Trichomonas vaginalis	trick-oh-MOAN-as vah-gin-AL-iss
Trypanosoma brucei	trih-PAH-no-soe-mah BREW-sea-ee
T. cruzi	CREWZ-ee
Volvox	VOLE-vox

Taxonomic Terms	Pronunciation
Actinobacteria	ack-TIN-oh-bak-tier-ee-ah
Archaea	arr-KEY-ah
Bacteroidetes	BAK-teh-roy-deh-teez
Chlamydiae	kla-MIH-dee-eye
Crenarchaeota	cren-are-key-OH-tah
Cyanobacteria	SIGH-ann-oh-bak-tier-ee-ah
Eukarya	YOU-care-ee-ah
Euryarchaeota	ur-ee-are-kee-OH-tah
Firmicutes	fir-mih-CUE-teez
Prokarya	pro-care-EE-ah
Proteobacteria	pro-TEE-oh-bak-tier-ee-ah
Rickettsiae	rih-KET-sea-eye
Spirochaetes	spy-row-KEY-teez

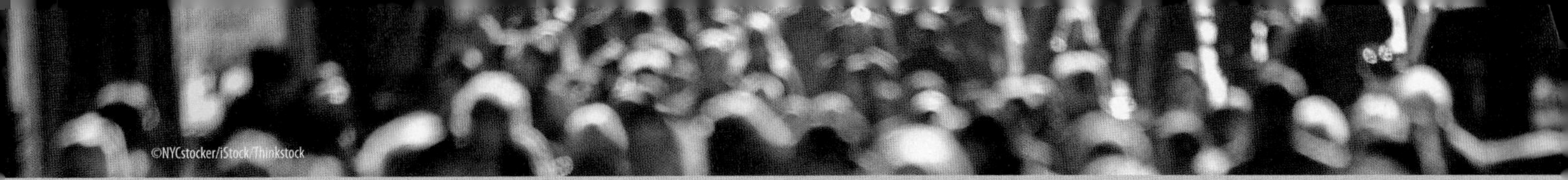
©NYCstocker/iStock/Thinkstock

Glossary

This glossary contains concise definitions for microbiological terms and concepts only. Please refer to the index for specific infectious agents, infectious diseases, anatomical terms and conditions, taxa, and specific antimicrobial drugs.

A

abscess A confined, pus-filled lesion.

acid A substance that releases hydrogen ions (H^+) in solution.

acid-fast test A staining process in which certain bacteria resist decolorization with acid-alcohol after being stained with a dye; used to visualize the tuberculosis bacteria.

acidophile A microorganism that grows at acidic pHs.

activated sludge The aerated sewage containing microorganisms to help break down the sewage.

active site The region of an enzyme where the substrate binds.

acute period (climax) The phase of a disease during which specific symptoms occur and the disease is at its height.

adaptive immunity The immune response that that is acquired as a result of experiencing an infectious agent or its products.

adenine (A) One of the nucleobases in nucleic acids.

adenosine diphosphate (ADP) A molecule in cells that is the product of ATP hydrolysis.

adenosine triphosphate (ATP) A molecule in cells that provides most of the energy for metabolism.

aerobe (aerobic) An organism (or referring to an organism) that uses oxygen gas (O_2) for growth and metabolism.

aerobic respiration A set of metabolic reactions and processes that take place in cells to convert biochemical energy from nutrients into cellular energy in the form of ATP.

aflatoxin A fungal toxin that is cancer causing in vertebrates.

agar A polysaccharide used as a solidifying agent in many microbiological culture media.

airborne transmission The spread of pathogens through respiratory droplets.

alcohol An antiseptic that works by disrupting protein structure and dissolving lipids, thereby inhibiting the growth of many microorganisms.

alcoholic fermentation A catabolic process that forms ethyl alcohol as a product of the fermentation process.

alga (pl. algae) A type of protist that carries out photosynthesis.

amino acid An organic acid containing one or more amino groups that build proteins in all living cells and viruses.

amoeba (pl. amoebae) A nonphotosynthetic protist that moves about using pseudopods.

amylase An enzyme that breaks down starch.

anabolic pathway (anabolism) A chemical process requiring an input of energy to build larger molecules from simpler ones.

anaerobe (anaerobic) An organism that does not require or cannot use oxygen gas (O_2) for growth and metabolism.

anaerobic respiration The production of ATP where the final electron acceptor is an inorganic molecule other than oxygen gas (O_2); examples include nitrate and sulfate.

animalcule A tiny, microscopic organism observed by Antony van Leeuwenhoek.

antibiotic A substance naturally produced by bacteria or fungi that inhibits or kills bacteria.

antibiotic resistance (ABR) The survival of a bacterium in the presence of an antibiotic drug meant to kill or inhibit the growth of the bacterium.

antibody A protein found in the blood that is produced by the body's immune system in response to a foreign agent, such as a bacterium or virus, invading the body.

antibody-mediated response The form of adaptive immunity conferred to an individual through the activity of B cells and the production of antibodies in the body fluids.

anticodon A three-nucleobase sequence on the tRNA molecule that binds to the codon on the mRNA molecule during translation.

antigen A foreign substance in the body that stimulates the production of antibodies into the blood.

antigen binding site The region on an antibody that binds to a portion of an antigen. *See also* epitope.

antimicrobial agent (drug) A chemical that inhibits or kills the growth of microorganisms.

antimicrobial resistance (AMR) The ability of a microbe to resist the effects of medication that once could successfully treat the pathogen.

antiretroviral therapy (ART) The combination of several (typically three or four) antiviral drugs for the treatment of infections caused by retroviruses, especially the human immunodeficiency virus.

antisepsis The use of chemical methods for eliminating or reducing microorganisms on the skin.

antiseptic A chemical used to reduce or kill pathogenic microorganisms on a living object, such as the surface of the human body.

antitoxin An antibody produced in the bloodstream to neutralize toxins produced by microorganisms.

applied science The use of the knowledge gained from basic science to develop more practical applications for, or to help solve, society's needs. *See also* basic science.

aqueous solution One or more substances dissolved in water.

Archaea The domain including small, single-celled organisms that excludes the Bacteria and Eukarya.

ART *See* antiretroviral therapy.

asexual reproduction The form of multiplication that maintains genetic constancy while increasing cell numbers.

aspartame An artificial sweetener found in diet soft drinks and many other dietary foods.

assembly The building of new virus particles from virus parts.

assimilation The process by which cells or organisms obtain nutrients.

associative learning The process in which a new response becomes associated with a stimulus.

atom The smallest portion into which an element can be divided and still take part in a chemical reaction.

atomic nucleus The positively charged core of an atom, consisting of protons and neutrons that make up most of the mass.

ATP *See* adenosine triphosphate.

ATP synthase The enzyme involved in generating ATP from ADP during electron transport.

attachment Connection of a virus to the cell surface of a host cell.

attenuated vaccine A preparation of bacteria or viruses with a reduced ability to reproduce or replicate.

autoclave An instrument used to sterilize microbiological materials by means of high temperature using steam under pressure.

avirulent Not likely to cause disease.

B

B lymphocyte (B cell) A small white blood cell that helps the body defend itself against infection.

bacillus (pl. bacilli) **1.** Any rod-shaped, prokaryotic cell. **2.** As *Bacillus*, a genus of aerobic or facultatively anaerobic, rod-shaped, endospore-producing, gram-positive bacterial cells.

Bacteria The domain of life that includes all small, single-celled organisms not classified as Archaea or Eukarya.

bacterial chromosome A closed loop of double-stranded DNA.

bacteriophage (phage) A virus that infects and replicates within bacterial cells.

bacterium (pl. bacteria) A tiny, single-celled organism lacking a cell nucleus and membrane-enclosed compartments.

bacteroid A modified cell formed by a symbiotic bacterium in a root nodule of a leguminous plant.

baculovirus A virus used to carry a foreign gene into insect and mammalian cells.

base-pair substitution A type of mutation involving the replacement or substitution of a single nucleotide base with another in DNA

basic (pure) science The field of science that describes and provides information and knowledge to explain phenomena in the natural world. *See also* applied science.

batch (vat) pasteurization A treatment in which milk is heated at 63°C for 30 minutes and then cooled rapidly to eliminate pathogens.

benign Referring to a tumor that usually is not life-threatening or likely to spread to another part of the body.

beta-lactam An antibiotic that has a beta-lactam ring at the core of the structure.

binary fission An asexual reproduction process in prokaryotic cells by which a cell divides to form two new cells while maintaining genetic constancy.

binomial nomenclature The method of naming organisms using two words (genus and specific epithet).

bioaugmentation The form of bioremediation in which bacterial cultures are added to speed up the rate of degradation of a compound.

biocrime An intentional introduction of biological agents into food or water, or by injection, to harm or kill groups of individuals.

biofilm A complex community of microorganisms that form a protective and adhesive matrix that attaches to a living or nonliving surface.

biofuel A material that stores potential energy and that is produced from living organisms.

biogeochemical cycle Any of the natural circulation pathways of the essential elements of living matter (e.g., carbon, nitrogen, phosphorus).

biological vector An infected arthropod, such as a mosquito or tick, that transmits disease-causing organisms between hosts. *See also* mechanical vector.

bioluminescent (bioluminescence) Emission of visible light by a living organism.

biomass The total weight of all organisms in a defined area or environment.

bioreactor A large vessel used to culture and grow large quantities of microbes under controlled conditions; also called a fermentor.

bioremediation The use of microorganisms to degrade toxic compounds in the environment into nontoxic substances.

biostimulation The form of bioremediation that modifies the environment to stimulate existing microbes to reproduce and better degrade a chemical.

biosynthesis 1. The production of complex molecules within living organisms or cells. 2. Manufacture of virus parts during virus replication.

biotechnology The use of microbes and their chemistry to manufacture products that will improve the quality of human life.

bioterrorism The intentional or threatened use of biological agents to cause fear in or actually inflict death or disease upon a large population.

blanching A process of putting food in boiling water for a few seconds to destroy enzymes.

bloom An excessive growth of bacteria or algae on or near the surface of water, often resulting from an oversupply of nutrients from organic pollution.

booster shot A repeat dose of a vaccine given periodically to maintain a high level of immunity.

broad-spectrum antibiotic An antimicrobial drug useful for treating many gram-positive and gram-negative bacteria. *See also* narrow-spectrum antibiotic.

broth A liquid containing nutrients for the growth of microorganisms.

Bt toxin A crystalline bacterial protein that is poisonous to insects.

budding An asexual reproduction process in fungi, in which a new cell forms as a swelling at the border of the parent cell and then breaks free to live independently.

C

callus An unorganized mass of plant cells formed in response to various living and nonliving stimuli.

cancer A disease characterized by the radiating spread of malignant cells that reproduce at an uncontrolled rate.

canning A food preservation method in which the food contents are processed and sealed in an airtight, sterile container.

capsid The protein coat that encloses the genome of a virus.

carbohydrate An organic molecule consisting of carbon, hydrogen, and oxygen that is an important source of carbon and energy for all organisms.

carbolic acid *See* phenol.

carbon cycle A series of interlinked processes involving carbon compound exchange between living organisms and the nonliving environment.

carbon-trapping reactions The stage of photosynthesis in which electrons and ATP are used to convert carbon dioxide gas (CO_2) to sugars.

carcinogen A chemical or physical substance capable of causing a tumor.

carrier An individual who has recovered from a disease but retains and continues to shed the infectious agent.

casein The major protein in milk.

catabolic pathway (catabolism) A chemical process that breaks down molecules and usually releases energy.

caterpillar The larva of a butterfly or moth.

catheter A thin, flexible tube that carries fluids into or out of the body.

CCP *See* critical control point

cell envelope The cell wall and cell membrane that surround a prokaryotic cell.

cell-mediated response The arm of adaptive immunity that attempts to resist infection through the activity of T lymphocytes.

cell membrane A thin bilayer of phospholipids and proteins that surrounds the prokaryotic cell cytoplasm. *See also* plasma membrane.

cell motility The ability of a cell to move spontaneously and actively in a watery or damp environment.

cell nucleus A membrane-enclosed structure in eukaryotic cells that contains most of the cell's genetic material.

cellular respiration The process of converting chemical energy into cellular energy in the form of ATP.

cellulase An enzyme that breaks down cellulose.

cellulose A polysaccharide composed of glucose subunits that forms the primary structural component of algal (and plant) cell walls.

cell wall A carbohydrate-containing structure surrounding fungal, algal, and most prokaryotic cells.

cesspool A concrete cylindrical ring with pores in the walls that is used to collect human waste.

chemical bond A force between two or more atoms that tends to hold those atoms together.

chemical element Any substance that cannot be broken down into a simpler one by a chemical reaction.

chemical reaction A process that changes the molecular composition of a substance by reorganizing atoms or groups of atoms without altering the number of atoms.

chitin A polysaccharide that provides rigidity to the cell walls of fungi.

chlorination The process of treating water with chlorine to kill harmful organisms.

chlorophyll A green or purple pigment in algae and some bacterial cells that functions in capturing light for photosynthesis.

chloroplast A double membrane-enclosed compartment in algae (and plant cells) that carries out photosynthesis.

chromosome A structure in the prokaryotic nucleoid or eukaryotic cell nucleus of a cell that carries hereditary information in the form of genes.

cilium (pl. cilia) A hair-like projection on some eukaryotic cells that, along with many others, assist in the movement of the cells.

citric acid cycle A series of chemical reactions essential for ATP production during cellular respiration.

class A category of related organisms consisting of one or more orders.

classification The cataloging of organisms into groups based on shared similarities. *See also* taxonomy.

climate change A change in the statistical distribution of weather patterns (e.g., temperature, rainfall) when that change lasts for an extended period.

climax *See* acute period

coccus (pl. cocci) A spherical-shaped prokaryotic cell.

codon A three-nucleobase sequence on the mRNA molecule that specifies a specific amino acid insertion for producing a polypeptide.

coenzyme A small, organic molecule that forms the nonprotein part of an enzyme molecule.

colony A visible mass of microorganisms of one type growing on agar.

commercial sterilization A canning process to eliminate all pathogens from the product.

communicable disease A disease that is readily transmissible between hosts.

community A group of organisms that live in the same area.

community immunity *See* herd immunity.

comparative genomics The process of analyzing the similarities of DNA sequences between organisms.

compost Organic matter that has been decomposed by microbes and recycled as a fertilizer.

compound A substance made by the combination of two or more different chemical elements.

conjugation A unidirectional transfer in prokaryotes of genetic material from a live donor cell into a live recipient cell during a period of cell contact.

conjugation pilus A hollow, tube-like projection for DNA transfer between the cytoplasms of donor and recipient bacterial cells.

consumer An organism that is the final user of a product.

contagious Referring to a disease in which the infectious agent passes with ease among hosts.

convalescence period Recovery from a disease when the body's systems return to normal.

coprophagy The activity of some animals, such as rabbits, to instinctively eat their fecal pellets, thereby giving the food a second pass through their intestine.

CRISPR A segment of prokaryotic DNA containing short repetitions of base sequences.

critical control point (CCP) A place in industrial food processing where contamination of the food product could occur.

culture medium (pl. culture media) A mixture of nutrients in which microorganisms can grow.

curd The result of curdling the milk protein casein.

cyanobacterium (pl. cyanobacteria) An oxygen-producing, pigmented bacterium occurring in unicellular and filamentous forms that carries out photosynthesis.

cytochrome A compound containing protein and iron that plays a role as an electron carrier in cellular respiration and photosynthesis; *see also* electron transport.

cytoplasm The jellylike solution containing a complex of chemicals and structures within a cell.

cytosine (C) One of the nucleobases in nucleic acids.

cytoskeleton An interconnected network of cytoplasmic protein fibers and threads that extends throughout the cytoplasm.

cytotoxic T cell (CTC) A type of T lymphocyte that searches out and destroys infected cells.

D

decline period The time when the signs and symptoms of a disease begin to subside.

decline (death) phase The final portion of a growth curve in which environmental factors adversely affect a population and result in cell death.

decomposer A microbe that breaks down dead plant and animal matter.

dehydration synthesis reaction A process of bonding two molecules together by removing a water molecule and joining the resulting open bonds.

denaturation A process caused by heat or pH in which proteins lose their function due to changes in their three-dimensional molecular structure.

denitrification The process of reducing nitrate and nitrite into gaseous nitrogen.

denitrifying bacterium A bacterial cell that carries out denitrification.

dental plaque The biofilm found on the tooth surface.

deoxyribonucleic acid (DNA) The genetic material of all cells and many viruses.

dermotrophic An illness that is attracted to, is localized in, or has entered by way of the skin.

desiccation The process of removing moisture from a substance.

diagnosis The process of identifying a disease, illness, or problem by examining an individual.

diatomaceous earth Filtering material composed of the remains of diatoms.

diplobacillus (pl. diplobacilli) A pair of rod-shaped prokaryotic cells.

diplococcus (pl. diplococci) A pair of spherical-shaped prokaryotic cells.

direct-contact transmission The form of disease spread involving close association between hosts; *see also* indirect-contact transmission.

disaccharide A sugar formed from two single sugar molecules.

disappearing microbe hypothesis The concept that the misuse and abuse of antibiotics and changes in the human diet are leading to alterations in the gut microbiome, resulting in human health challenges.

disease Any change from the general state of good health.

disinfectant A chemical used to kill or inhibit pathogenic microorganisms on a lifeless object such as a tabletop.

disinfection The process of killing or inhibiting the growth of pathogens.

DNA *See* deoxyribonucleic acid.

DNA double helix *See* double helix

DNA ligase In genetic engineering, an enzyme that seals a segment of DNA containing a gene of interest into the DNA of a plasmid.

DNA polymerase An enzyme that catalyzes DNA replication by combining complementary nucleotides to an existing strand.

DNA probe A known segment of single-strand DNA that is complementary to a desired DNA sequence of a bacterial species.

DNA replication The process of copying the genetic material.

DNA (gene) sequencing The process of determining the order of nucleobases (adenine, guanine, cytosine, and thymine) in a genome, a DNA segment, or an individual gene.

DNA virus A virus that has its genetic information in the form of DNA.

domain 1. The most inclusive taxonomic level of classification. 2. A loop of DNA consisting of about 10,000 bases.

double helix The structure formed by the two polynucleotide strands of DNA.

dry weight The weight of the materials in a cell after all the water is removed.

dysbiosis The imbalance in, or disturbance to, the normal microbial community (microbiome).

E

ecosystem A geographic area where plants, animals, and microbes, as well as weather and landscape, work together as a system.

edible vaccine In theory, a genetically modified plant that contains a vaccine that could be consumed when inoculations were needed.

electrolyte A mineral or salt in blood and other body fluids that regulates nerve and muscle function.

electron (e^-) A negatively charged particle with a small mass that moves around the atomic nucleus.

electron microscope An instrument that uses electrons and a system of electromagnetic lenses to produce a greatly magnified image of an object. *See also* transmission electron microscope; scanning electron microscope.

electron transport A series of electron carrier molecules that transfer electrons in cellular respiration to generate ATP.

element The single, smallest entity of a pure substance that can take part in chemical reaction.

emerging viral disease A new viral disease or changing disease that is seen within a population for the first time.

endemic Referring to the continual presence of disease or persistence of an infectious agent at a low level in a population.

endogenous retrovirus (ERV) Virus genes that are incorporated into human chromosomes and comprise up to 10% of the human genome.

endogenous viral element (EVE) A viral nucleic acid sequence, other than endogenous retroviruses, found in the human genome

endomembrane system A collection of membrane-enclosed subcompartments in the cytoplasm of eukaryotic cells.

endoplasmic reticulum (ER) A membrane network consisting of flat membranes with or without ribosomes in the cytoplasm of eukaryotic cells.

endospore An extremely resistant, dormant cell produced by a few gram-positive bacterial species.

endosymbiont One organism (partner) living inside another.

endosymbiont theory An explanation for the origins of mitochondria and chloroplasts in eukaryotic cells.

endotoxin A metabolic poison composed of lipid–polysaccharide–peptide complexes that is part of the bacterial cell wall of gram-negative bacteria.

enema An injection of fluid into the lower bowel by way of the rectum.

energy The ability to do work or the capacity to cause change.

energy pyramid A representation of the transfer of energy between feeding levels.

energy-trapping reactions The stage of photosynthesis in which light energy is locked in ATP and NADPH molecules.

enriched medium A growth medium in which special nutrients are added to stimulate a species to grow.

envelope The flexible membrane of protein and lipid that surrounds many types of viruses.

enveloped virus A virus in which the genome and capsid are surrounded by a membrane-like covering.

enzyme A reusable protein molecule that brings about a chemical reaction while itself remaining unchanged.

epidemic The occurrence of more cases of a disease than expected in population within a geographic area.

epidemiology The scientific study of the source, cause, and transmission of disease within populations.

epitope A section of an antigen molecule that stimulates antibody formation and to which the antibody binds.

ERV *See* endogenous retrovirus.

ethylene oxide A sterilizing gas that kills microbes and bacterial endospores.

Eukarya The taxonomic domain that includes protists, fungi, plants, and animals.

eukaryotic cell (eukaryote) A cell (organism) containing a cell nucleus and membrane-enclosed subcompartments; *see also* prokaryotic cell.

EVE *See* endogenous viral element.

exoenzyme An enzyme that is secreted by a cell and functions outside of that cell.

exotoxin A bacterial protein toxin secreted into the environment or body by living bacterial cells.

expiration date The last date that a food product should be used before it is considered spoiled.

extreme acidophile An archaeal organism living at an extremely acidic pH of 1 to 2.

extreme halophile An archaeal organism that grows at very high salt concentrations.

extremophile A microorganism that lives in an extreme environment, such as high temperature, high acidity, or high salt.

extremozyme A thermostable enzyme found in some archaeal species.

extrinsic factor An environmental characteristic that influences the growth of food microbes. *See also* intrinsic factor.

F

F factor (plasmid) A loop of DNA containing genes for its replication and the bacterial conjugation process.

facultative Referring to an organism that grows in the presence or absence of oxygen gas (O_2).

FAD *See* flavin adenine dinucleotide.

fat A type of lipid made up of a glycerol attached to three fatty acids that usually is solid at room temperature.

fatty acid A molecule composed of a long hydrophobic hydrocarbon chain and a hydrophilic carboxyl functional group.

fecal gut microbiome The microbes found in fecal material originating from the gut.

fecal microbiome transplantation (FMT) The addition of a fresh or frozen stool (microbiome) sample from a healthy donor into a patient's colon.

fecal-oral route A route of disease transmission, where pathogens in fecal material pass from one person and are introduced into the oral cavity of another person.

fermentation A metabolic process that breaks down carbohydrates to acid, gases, and/or alcohol using microbes.

fermentor *See* bioreactor.

fever An abnormally high body temperature that is usually caused by a bacterial or viral infection.

filtration A mechanical method to remove microorganisms by passing a liquid or air through a filter and trapping the organisms on the filter.

flagellum (pl. flagella) A long, hair-like appendage composed of protein and responsible for motion in many microorganisms.

flash pasteurization method A treatment in which milk or other liquid is heated at 72°C (161°F) for 15 seconds and then cooled rapidly to eliminate pathogens.

flavin adenine dinucleotide (FAD) A coenzyme derived from vitamin B2 (riboflavin) that transfers and transports electrons during electron transport.

floc A jelly-like mass that forms in a liquid and is made up of coagulated particles.

fomite An inanimate object, such as clothing or a utensil, that carries disease organisms.

food fermentation The process of using microbes to produce a broad range of ingredients resulting from the breakdown of carbohydrates and other large organic compounds in foods.

food infection The contamination of food by microorganisms carried by those preparing food or by other contaminated foods or food-processing equipment.

food intoxication (poisoning) Illness caused by consuming foods that contain bacterial toxins.

food preservation The application of a variety of methods and procedures to keep foods from spoiling.

food spoilage The result of food deteriorating to the point that it is not edible to humans or its quality of edibility becomes reduced, which can be due to microbial action.

fortified wine A wine that has had additional alcohol added.

frameshift mutation The deletion or insertion of a base in a DNA sequence such that it shifts the way the sequence is read.

freeze drying *See* lyophilization.

fungus (pl. fungi) A member of a large group of eukaryotic organisms that includes the yeasts, molds, and mushrooms.

G

gasohol Gasoline blended with ethanol.

gene A segment of a DNA molecule that provides the biochemical information for a functional product.

gene expression The processes by which the information in a gene is transcribed and translated into a protein.

genetic code The specific order of nucleobase sequences in DNA or RNA that encode specific amino acids for protein synthesis.

genetic diversity The range (or genetic variability) within species.

genetic engineering The use of bacterial and microbial genetics to isolate, manipulate, recombine, and express genes.

genetic recombination The process of bringing together different segments of DNA into one molecule.

genetically modified (GM) food A food that is the result of changes introduced into its DNA through genetic engineering.

genetically modified organism (GMO) An organism that has been genetically engineered to contain a foreign gene in its genome.

genetics The branch of biology concerned with the study of genes, genetic variation, and heredity in living organisms.

genome The complete set of genetic information in an organism or virus.

genomics The identification and study of gene sequences in an organism's DNA.

genus (pl. genera) A rank in the classification system of organisms composed of one or more species; a collection of genera constitutes a family. Also, the first of the two scientific words for a species. *See also* specific epithet.

germ *See* pathogen.

germ theory of disease The principle, formulated by Pasteur and proved by Koch, that microorganisms play a significant role in the development of infectious diseases.

global health emergency A serious public health event that, according to the World Health Organization, endangers international public health and global populations.

glucose A simple sugar that is an energy source for living organisms and that is a component of many carbohydrates.

gluten A substance in cereal grains consisting of two proteins that add elastic texture to dough.

glycocalyx A sticky polysaccharide matrix covering many prokaryotic cells to assist in attachment to a surface and impart resistance to desiccation.

glycolysis A series of chemical reactions in which glucose is broken down into two molecules of pyruvate with a net gain of two ATP molecules.

glyphosate A broad-spectrum systemic herbicide.

gnotobiotic Referring to an animal that is microbe-free or in which only certain known species are present.

Golgi apparatus A group of independent stacks of flattened membranes and vesicles in the cytoplasm of eukaryotic cells.

gram-negative Referring to a bacterial cell that stains red after Gram staining.

gram-positive Referring to a bacterial cell that stains purple after Gram staining.

Gram stain technique A staining procedure used two contrasting dyes to identify bacterial cells as gram positive or gram negative.

growth curve The plotted or graphed measurement of the size of a population of bacterial cells as a function of time.

guanine (G) One of the nucleobases in nucleic acids.

gut-brain axis (GBA) The pathway that connects the body's central nervous system with the nervous system of the gastrointestinal tract.

gut-brain-microbiome axis The addition of the gut microbiome to the pathway that connects the body's central nervous system with the nervous system of the gastrointestinal tract.

H

habitat The place where a particular species lives and grows.

HACCP *See* Hazard Analysis Critical Control Point.

halogen An antimicrobial chemical agent (iodine and chlorine) used for disinfection purposes.

halophile An organism that lives in environments with high concentrations of salt.

Hazard Analysis Critical Control Point (HACCP) A set of federally enforced regulations to ensure the dietary safety of seafood, meat, and poultry.

healthcare-associated infection (HAI) An invasion of microbes resulting from treatment for another condition while in a hospital or healthcare facility. Also called a nosocomial infection.

heavy (H) chain The larger polypeptide in an antibody molecule.

heavy metal An antimicrobial chemical element (mercury, copper, and silver) that often is toxic to microorganisms,

helicase An enzyme involved with the unwinding of the DNA double helix for DNA replication.

helper T lymphocyte (cell) A type of white blood cell that enhances the activity of B lymphocytes and stimulates cytotoxic T cells to search out pathogen-infected cells.

hemagglutinin (H spike) An enzyme composing one type of surface protein on influenza viruses that enables the viruses to bind to the host cell.

hemorrhagic Referring to blood escaping from the circulatory system.

herbicide-tolerant (HT) Referring to plants that are not killed by a substance that is toxic to plants.

herbivore An animal that survives on a diet of plants and grasses as the main component of its diet.

herd immunity The concept that if most of a population is immune to an infectious disease (through vaccination or prior illness), it is unlikely that the disease will spread person to person. Also called community immunity.

heredity The passing of genetic traits from parents to offspring.

high-efficiency particulate air (HEPA) filter A type of air filter that removes particles larger than 0.3 micrometers.

highly perishable Referring to foods that spoil easily.

homeostasis The maintaining of an internal steady state in a cell or organism.

horizontal gene transfer (HGT) The movement of genes from one organism to another within the same generation; also called lateral gene transfer.

host A cell or organism in which a microbe or virus can live, feed, and reproduce (replicate).

human genome The complete set a genetic information in a human cell.

Human Genome Project (HGP) An international scientific research project that sequenced the human DNA in a cell and is identifying and mapping all the genes.

human microbiome The community of microorganisms and viruses that normally resides on the surface of the skin and in the mouth, respiratory system, and gastrointestinal and urogenital tracts of the human body. *See also* microbiome.

Human Microbiome Project (HMP) An initiative that identified and characterized the microorganisms found in association with both healthy and diseased humans.

humus A complex organic substance resulting from the microbial breakdown of plant material.

hydrogen peroxide A household antiseptic that readily decomposes into water and oxygen gas (O_2) when in contact with damaged tissue.

hydrolysis reaction A process in which a molecule is split into two parts using a water molecule to effect the separation of a larger molecule into smaller ones.

hydrophilic Referring to a substance that dissolves in or mixes easily with water. *See also* hydrophobic.

hydrophobia A set of symptoms in the later stages of rabies in which the person has difficulty swallowing, shows panic when presented with liquids to drink, and cannot quench their thirst.

hydrophobic Referring to a substance that does not dissolve in or mix easily with water. *See also* hydrophilic.

hyperthermophile A prokaryote that has an optimal growth temperature above 80°C/176°C.

hypha (pl. hyphae) A microscopic filament representing the growing portion of a fungus. *See also* mycelium.

I

icosahedral Referring to a symmetrical figure composed of 20 triangular faces and is one of the major shapes of some viral capsids.

idiopathic Referring to a disease or disorder that has no known cause.

IgA (immunoglobulin A) The class of antibodies found in respiratory and gastrointestinal secretions that help neutralize pathogens.

IgD (immunoglobulin D) The class of antibodies found on the surface of B cells that act as receptors for binding antigen.

IgE (immunoglobulin E) The class of antibodies responsible for allergies.

IgG (immunoglobulin G) The class of antibodies abundant in serum that are major disease fighters.

IgM (immunoglobulin M) The first class of antibodies to appear in helping to fight pathogens.

immune system The complex interplay of cells and molecules in the body responsible for identifying, fighting, and defending against foreign substances (bacteria, viruses, fungi, parasites).

immunity The body's ability to resist infectious disease.

immunization The process by which an individual becomes protected against a particular disease. *See also* vaccination.

immunoglobulin (Ig) The class of proteins that react with an antigen; an alternate term for antibody.

immunology The scientific study of how the immune system works and responds to pathogens and other foreign agents.

inactivated vaccine A formulation containing bacteria or viruses that have been rendered inactive by physical or chemical processes.

inclusion body A cytoplasmic compartment in prokaryotic cells that concentrates and stores nutrients.

incubation period The time that elapses between the entry of a pathogen into the host and the appearance of signs and symptoms.

indicator organism A microorganism that when present signals fecal contamination of water.

indirect-contact transmission The mode of disease transmission involving nonliving objects. *See also* direct-contact transmission.

industrial fermentation Any large-scale industrial process, with or without oxygen gas (O_2), for growing microorganisms. *See also* fermentation.

industrial microbiology The field that uses microbes in the manufacturing of food and industrial products, including pharmaceuticals, beverages, and chemicals.

infection The entry, establishment, and multiplication of a pathogen in the host.

infectious disease A disorder arising from a pathogen invading a susceptible host and inducing medically significant symptoms.

infectious dose The number of microorganisms needed to bring about infection.

inflammation A nonspecific innate immune response to injury that is usually characterized by redness, warmth, swelling, and pain.

innate immunity Immune defenses that one is born with and that are nonspecific.

Integrative Human Microbiome Project (iHMP) The program to get a better understanding of the roles the human microbiome in human health and disease states.

interferon (IFN) An antiviral protein produced by body cells on exposure to viruses that signal the synthesis of antiviral proteins by neighboring cells.

intrinsic factor A characteristic of a food product that influences microbial growth. *See also* extrinsic factor.

ionizing radiation A type of radiation such as gamma rays and X rays that causes the separation of atoms or a molecule into electrically charged particles.

irradiation The process by which an object is exposed to radiation.

J

jaundice A condition in which bile seeps into the circulatory system, causing the skin to have a dull yellow color.

K

Koch's postulates A set of procedures by which a specific organism can be identified as the causative agent of a specific disease.

koji The common name for the fungus *Aspergillus oryzae*.

L

lactic acid bacteria (LAB) A group of gram-positive, acid-tolerant rods or cocci that are associated by their common metabolic and physiological characteristics.

lactose A milk sugar composed of one molecule of glucose and one molecule of galactose.

lag phase A portion of a growth curve encompassing the first few hours of the population's history during which no growth occurs.

lagering The secondary aging of beer.

latency A condition in which a virus integrates into a host chromosome without immediately causing a disease.

legume A plant that bears its seeds in pods.

leukocyte Any of several types of white blood cells.

lichen A symbiotic association between a fungal mycelium and a cyanobacterium or alga.

light (L) chain The smaller polypeptide in an antibody.

light microscope An instrument that uses visible light and a system of glass lenses to produce an enlarged image of an object.

lipid A organic energy compound composed of carbon, hydrogen, and oxygen; examples include animal fats, plant oils, and phospholipids in cell membranes.

lipopolysaccharide (LPS) A molecule composed of lipid and polysaccharide that is found in the outer half of the outer membrane of the gram-negative cell wall of bacterial cells.

logarithmic (log) phase The portion of a growth curve during which active growth leads to a rapid rise in cell numbers.

lymph node A bean-shaped organ located along lymph vessels that is involved in the immune response and contains phagocytes and lymphocytes.

lymphocyte A small white blood cell that helps the body defend against infection.

lyophilization (freeze drying) A process in which food or other material is deep frozen, after which the water is vaporized by vacuum pressure.

lysis The rupture of a cell and the loss of cell contents.

lysosome A cytoplasmic, membrane-enclosed structure containing digestive (hydrolytic) enzymes.

lysozyme An enzyme found in tears and saliva that digests the peptidoglycan of gram-positive bacterial cell walls.

M

macrophage A large white blood cell that is found within various tissues and helps the body defend itself against infection.

magnification *See* total magnification.

malaise An overall feeling of discomfort, illness, or lack of well-being.

malignant Referring to a tumor that invades the tissue around it and might spread to other parts of the body.

malting Referring to the process when barley begins to germinate by being soaked in water to produce simpler carbohydrates.

maltose A disaccharide sugar composed of two glucose molecules that is found in cereal grains.

manufacturer code A system used by manufacturers to identify products quickly.

mechanical vector A living organism, or an object, that transmits disease agents on its surface. *See also* biological vector

membrane filter technique A method to test water quality by identifying any indicator organisms trapped on a filter.

memory B cell A cell derived from B lymphocytes that helps the body defend itself against disease by remembering a prior exposure to a specific bacterium or virus.

memory T cell A cell derived from T lymphocytes that helps the body defend itself against disease by remembering a prior exposure to a specific bacterium or virus.

mesophile An organism that lives at ambient temperature ranges of 10°C (50°F) to 45°C (113°F).

messenger RNA (mRNA) An RNA transcript containing the information for synthesizing a specific polypeptide.

metabolism The sum of all biochemical processes taking place in a living cell or organism.

metabolite Any substance produced during metabolism.

metagenome The collective genomes from a population of organisms.

metagenomics The study of genes isolated directly from environmental samples.

metastasize The spread of tumor cells from the site of origin to other tissues in the body.

methanogen An archaeal organism that lives on simple compounds in anaerobic environments and produces methane gas during its metabolism.

microbe *See* microorganism.

microbial antagonism The process whereby resident microbes outcompete and inhibit the growth of pathogenic species.

microbial biotechnology The use of the techniques of genetic engineering to modify microbial genomes to produce substances the organisms would not produce naturally.

microbial forensics The discipline involved with the recognition, identification, and control of a pathogen.

microbial genomics The discipline of sequencing, analyzing, and comparing microbial genomes.

microbial load The total number of bacteria and fungi in a quantity of water or soil or on the surface of food.

microbiology The scientific discipline that studies microscopic organisms and viruses.

microbiome A specific environment characterized by a distinctive microbial community and its collective genetic material.

microbiostatic A chemical that slows the growth of microbes.

micrometer (μm) A unit of length equivalent to one thousandth of a millimeter.

microorganism (microbe) A microscopic form of life including bacterial, archaeal, fungal, and protistan cells.

mitochondrion (pl. mitochondria) A double membrane-enclosed compartment in eukaryotic cells that carries out cellular respiration.

mixed fermentation An anaerobic fermentation where the products are a complex mixture of acids that have numerous applications in biotechnology.

mold A type of fungus that grows as long filaments and appears as a fuzzy mass in culture.

molecule Two or more atoms held together by a sharing of electrons.

monosaccharide A simple sugar that cannot be broken down into simpler sugars.

mucous membrane A moist lining in the body passages of all mammals that contains mucus-secreting cells and is open directly or indirectly to the external environment.

mucus A sticky secretion of viscous fluid.

multi-drug resistance/resistant (MDR) Referring to microbes that are not killed or harmed by many different antimicrobial drugs.

mushroom A spore-bearing reproductive body of some fungi.

must The juice resulting from crushing grapes.

mutagen A physical agent or chemical substance capable of bringing about mutations in cells.

mutant An organism carrying a mutation. *See also* wild type.

mutation A permanent change in the genetic information in a DNA sequence.

mycelium (pl. mycelia) A mass of fungal filaments from which most fungi are built. *See also* hypha.

mycology The scientific study of the fungi.

mycorrhiza (pl. mycorrhizae) A close association between a soil fungus and the roots of many plants.

mycosis (pl. mycoses) A fungal infection of animals, including humans.

N

NAD^+ *See* nicotinamide adenine dinucleotide.

naked virus *See* nonenveloped virus.

nanometer (nm) A unit of length equivalent to one millionth of a millimeter.

narrow-spectrum antibiotic An antimicrobial drug that only works against a select group of bacteria. *See also* broad-spectrum antibiotic.

natural selection The process that results in the survival and reproductive success of individuals or populations best adjusted to their specific environment.

necrosis The death of most or all of the cells in an organ or tissue due to disease or injury.

neuraminidase (N spike) An enzyme composing one type of surface protein of influenza viruses that facilitates viral release from the host cell.

neutron (n) An uncharged particle in the atomic nucleus.

neutrophil A type of white blood cell involved in phagocytosis of pathogens.

niche A term that describes the "way of life" of a species or population in a habitat.

nicotinamide adenine dinucleotide (NAD^+) A coenzyme derived from vitamin B3 (niacin) that transfers and transports electrons during electron transport and fermentation reactions.

nicotinamide adenine dinucleotide phosphate (NADP) A coenzyme derived from vitamin B3 (niacin) that transfers and transports electrons during the energy trapping reactions of photosynthesis.

nitrification The biological process of converting ammonia (NH3) to nitrate (NO_2).

nitrifying bacterium A bacterial cell that carries out nitrification.

nitrogenase An enzyme that converts nitrogen gas (N_2) into ammonia (NH_3).

nitrogen cycle The processes that convert nitrogen gas (N_2) to nitrogen-containing substances in soil and living organisms.

nitrogen fixation The chemical process by which microorganisms convert nitrogen gas (N_2) into ammonia (NH_3).

nitrogen-fixing bacterium A bacterial cell that carries out nitrogen fixation.

nonenveloped (naked) virus A virus consisting of only the viral genome and a capsid.

nonperishable Referring to foods that are least likely to spoil.

nosocomial infection *See* healthcare-associated infection.

nucleic acid A organic compound consisting of nucleotide chains that convey genetic information and are found in all living cells and viruses. *See also* DNA; RNA.

nucleobase Any of five nitrogen-containing compounds found in nucleic acids.

nucleocapsid The combination of genome and capsid of a virus.

nucleoid The region of a prokaryotic cell containing the bacterial chromosome.

nucleotide A component of a nucleic acid consisting of a carbohydrate molecule, a phosphate group, and a nucleobase.

nutraceutical A food or part of a food that might provide medicinal or health benefits, including the prevention and treatment of disease.

nutrient agar A solidifying agent that contains nutrients for microbial growth.

O

obese Someone who is 30 pounds, or more, overweight (according to the National Institutes of Health).

objective lens The lens in a microscope that receives the first light rays from the object being observed.

obligate intracellular parasite An organism or virus that must get its nutrients from a host cell.

obligate symbiont A microbial species that must live in a host for its survival and reproduction.

ocean gyre A large system of circular ocean currents formed by global wind patterns and forces created by the Earth's rotation.

oil A type of lipid made up of a glycerol attached to three fatty acids that usually is liquid at room temperature.

oligosaccharide A carbohydrate that contains 3 to 10 simple sugars linked together.

oncogene A segment of DNA that can induce uncontrolled growth of a cell.

oncogenic virus A virus capable of causing a tumor or involved with a cancer.

oncology The scientific study of tumors and cancers.

operator A sequence of nucleobases in the DNA to which a repressor protein can bind.

operon The unit of bacterial DNA consisting of a promoter, operator, and a set of structural genes.

opportunistic Referring to pathogens that only cause disease when the person's immune system is weakened.

oral rehydration therapy A treatment that involves drinking solutions of electrolytes and glucose designed to restore the normal water/salt balance in the body.

order A category of related organisms consisting of one or more families.

organelle A specialized subcompartment in eukaryotic cells that has a specific function.

organic In chemistry, referring to chemicals [except carbon monoxide (CO) and carbon dioxide (CO_2)] that contain carbon atoms.

organic acid A small, carbon-containing compound with acidic properties.

organism An individual form of life composed of one or more cells.

osmosis The net movement of water molecules from an area of high concentration through a semipermeable membrane to a region of lower concentration.

outbreak A small, sudden, localized appearance of disease in a specific geographic area or population.

outer membrane A bilayer membrane forming part of the cell wall of gram-negative bacteria.

oxidation lagoon A large pond in which sewage can remain undisturbed so that digestion of organic matter will occur.

P

pandemic Referring to a disease occurring over a wide geographic area (worldwide) and affecting a substantial proportion of the global population.

parasite An organism or virus that depends on a host for reproduction or replication.

pasteurization A heating process that destroys disease-causing bacteria in a fluid such as milk and lowers the overall number of bacteria in the fluid.

pasteurizing dose The level of irradiation needed to eliminate all pathogens from a food product.

pathogen A disease-causing agent (bacterium, virus, fungus, or parasite).

pathway engineering The engineering of microbes to produce biochemicals or other products in a more sustainable or useful way.

PCR *See* polymerase chain reaction.

penetration The entry of a virus and its uncoating in a host cell during replication.

penicillinase An enzyme produced by certain microorganisms that breaks apart penicillin and thereby confers resistance against penicillin.

peptide bond A linkage between two amino acids.

peptidoglycan A complex molecule of the bacterial cell wall composed of alternating units of N-acetylglucosamine and N-acetylmuramic acid cross-linked by short peptide cross bridges.

periplasmic space A metabolic region between the cell membrane and outer membrane of gram-negative cells.

personalized medicine A medical procedure that separates patients into different groups—with medical treatment and/or products targeted to individual patients and their genetic makeup based on their predicted response or risk of disease.

pH A measure of the hydrogen ion (H^+) concentration of an aqueous solution. Solutions with a pH less than 7 are said to be acidic and solutions with a pH greater than 7 are alkaline (basic). Pure water has a pH of 7.

phage *See* bacteriophage.

phagocyte A white blood cell capable of engulfing and destroying foreign materials or cells, including bacteria and viruses.

phagocytosis A process by which foreign material, cells, or viruses are taken into a white blood cell and destroyed.

phenol (carbolic acid) An antimicrobial agent whose derivatives are used as an antiseptic or disinfectant.

phenotype The visible (physical) appearance of an organism resulting from the interaction between its genetic makeup and the environment.

phospholipid A water-insoluble compound containing glycerol, two fatty acids, and a phosphate group; forms part of the membrane in all cells.

phosphorus cycle The biogeochemical cycle that describes the movement of phosphorus through the water and soil.

photobiont The photosynthetic partner in a symbiotic relationship in a lichen.

photosynthesis A biochemical process in which light (solar) energy is converted to chemical energy in the form of sugars.

phylum (pl. phyla) A category of organisms consisting of one or more classes.

phytoplankton Microscopic photosynthetic communities of cyanobacteria and unicellular algae.

pilus (pl. pili) A hair-like extension from the cell membrane that is found on the surface of many bacterial cells and that is used for cell attachment and anchorage.

plasma cell An antibody-producing cell derived from B lymphocytes.

plasma membrane The phospholipid bilayer with proteins that surrounds the eukaryotic cell cytoplasm. *See also* cell membrane.

plasmid A small, closed-loop molecule of DNA apart from the chromosome and which replicates independently and carries nonessential genetic information.

point mutation The replacement, loss, or gain of one base in a DNA strand with another base.

polymerase chain reaction (PCR) A technique used to replicate a fragment of DNA millions of times.

polypeptide A chain of linked amino acids.

polysaccharide A complex carbohydrate made up of sugar molecules linked into a branched or chain structure.

portal of entry The site at which a pathogen enters the host.

portal of exit The site at which a pathogen leaves the host.

potable Referring to water that is safe to drink.

prebiotic A high fiber food or pill supplement that might trigger the growth or activity of the gut microbiome.

primary consumer An animal feeding on plants.

primary metabolite A small molecule essential to the survival and growth of an organism; *see also* secondary metabolite.

primary producer An organism that is part of the foundation of an ecosystem and form the foundation of the food web by creating food through photosynthesis.

primary structure The sequence of amino acids in a polypeptide.

primary waste treatment The removal (sedimentation) of large solids from the wastewater via physical settling, screening, or filtration.

primordial soup A pond or body of water rich in substances that could provide favorable conditions for the emergence of life.

prion An infectious, self-replicating protein involved in human and animal diseases of the brain.

probiotic Living microbes that might help reestablish or maintain the human microbiome of the gut.

prodromal phase The phase of a disease during which general symptoms occur in the body.

producer An organism in an ecosystem that produces biomass from inorganic compounds.

product A substance resulting from an enzyme reaction.

productive infection The active assembly and maturation of viruses in a host cell.

prokaryotic cell (prokaryote) A cell (organism) having a single chromosome but no cell nucleus or other membrane-enclosed compartments. *See also* eukaryotic cell.

promoter The region of a DNA strand or operon to which RNA polymerase binds.

protease An enzyme that uses water to hydrolyze and break peptide bonds between amino acids of proteins.

protein A chain or chains of linked amino acids used as a structural material or enzymes in living cells.

protein synthesis The process of forming a polypeptide or protein through a series of chemical reactions involving amino acids.

protist A member of a very large and diverse group of single-celled eukaryotic microorganisms that includes the protozoa and algae.

proton (p^+) A positively charged particle in the atomic nucleus.

protozoan (pl. protozoa) An informal term for a single-celled protist that lacks a cell wall and usually feeds on organic matter.

provirus The viral DNA that has integrated into a eukaryotic host chromosome.

pseudopod A projection of the plasma membrane that allows movement of amoebae and some white blood cells.

psychrophile An organism that lives at cold temperature ranges of 0°C (32°F) to 20°C (68°F) .

psychrotroph An organism that lives at cold temperature ranges of 4°C (39°F) to 35° (95°F) .

pure science *See* basic (pure) science.

putrefaction The breakdown of proteins in meat, which often is detected as sliminess.

pyrogen A fever-producing substance.

pyruvate The three-carbon product of glycolysis.

Q

quaternary ammonium compound (quat) A detergent-type disinfectant agent for disinfection of industrial equipment and food utensils, as well as in hospitals.

quaternary structure Two or more polypeptides bonded together to form the final functional protein.

quorum The minimum number of "members" of a deliberative assembly necessary to conduct the business of that group.

quorum sensing (QS) The process of sensing cell numbers within a biofilm through chemical communication between the cells in the biofilm.

R

reactant A substance that interacts with another in a chemical reaction.

recombinant DNA A molecule containing DNA from two or more different sources.

recombinant DNA technology The process of joining together of DNA molecules from two or more different species and inserting the new DNA into a host organism to produce new genetic combinations.

red tide A brownish-red discoloration in seawater caused by increased numbers of dinoflagellates. *See also* bloom.

reemerging infectious disease A disease showing a resurgence in incidence or a spread in its geographical area.

regulatory gene A DNA segment that codes for a repressor protein.

release The exiting of a virus from a host cell after replication.

rem *See* Roentgen Equivalent in Man.

rennin A protein-digesting enzyme that curdles milk.

repressor protein A protein that, when bound to the operator in a bacterial DNA sequence, blocks transcription.

reproductive fitness The ability of an organism to transmit genes onto the next generation in a way that ensures that next generation can pass them on to their next generation.

reservoir The location or organism where disease-causing agents exist and maintain their ability for infection.

respiratory droplet Droplets of moisture expelled from the nose or throat tract through sneezing or coughing.

restriction enzyme (endonuclease) A type of enzyme that splits open a DNA molecule at a specific restricted point.

retrovirus An RNA virus that can reverse transcribe its RNA into DNA.

reverse transcriptase An enzyme that synthesizes a double-stranded DNA molecule from the code supplied by a single-stranded RNA molecule.

ribonucleic acid (RNA) The nucleic acid involved in protein synthesis and gene control; also, the genetic information in some viruses.

ribosomal RNA (rRNA) An RNA transcript that forms part of the ribosome's structure.

ribosome A cellular component that makes proteins.

ripened cheese The addition of salt and microbes to cheese curds.

RNA *See* ribonucleic acid.

RNA polymerase The enzyme that synthesizes an RNA polynucleotide from a DNA template.

RNA virus A virus that has its genetic information in the form of RNA.

Roentgen Equivalent in Man (rem) A measure of radiation dose related to biological effect.

rumen The first chamber in the digestive tract of ruminant animals.

ruminant Any hooved animal that digests its food and chews the cud regurgitated from the rumen.

S

sanitation The process of reducing the number of microbes to a safe level.

saturated Referring to a water-insoluble compound that cannot incorporate any additional hydrogen atoms; *see also* unsaturated.

sauerkraut Cabbage that has been fermented by various lactic acid bacteria.

scanning electron microscope (SEM) The type of electron microscope that allows electrons to scan across an object, generating a three-dimensional image of the object.

science The organized body of knowledge that is derived from observations and can be verified or tested by further investigation.

secondary consumer (carnivore) An animal that feeds on primary consumers.

secondary metabolite A small molecule not essential to the survival and growth of an organism; *see also* primary metabolite.

secondary structure The region of a polypeptide folded into a helix or sheet-like structure.

secondary waste treatment The removal of smaller solids and particles remaining in the wastewater through fine filtration aided by the use of membranes or through the use of microbes.

sedimentation In waste and water treatment, it refers to the tendency for particles in the liquid to settle out of the fluid by gravity and come to rest at the bottom of the container.

selective medium A growth medium that contains ingredients to inhibit unwanted microorganisms while encouraging the growth of desired species.

semiconservative replication The DNA copying process where each parent (old) strand serves as a template for a new complementary strand.

semiperishable Referring to foods that spoil less quickly.

semisynthetic Referring to a chemical that has been chemically modified from its natural form.

septic tank An enclosed, concrete box that collects waste from the home.

serotype A distinct variation within a species based on immunological characteristics.

serum (pl. sera) The fluid portion of the blood consisting of water, minerals, salts, proteins, and other organic substances, including antibodies; contains no clotting agents.

sexually transmitted infection (STI) An infection that normally is passed from one person to another through sexual activity. Also called a sexually transmitted disease (STD).

sexual reproduction The generation of new living organisms by combining genetic information from two individuals of different types (sexes).

sewage treatment The process of removing household sewage plus some industrial waste from municipal wastewater.

shelf life The length of time that a commodity, such as food, can be stored without becoming unfit for use or consumption.

shock A state of physiological collapse marked by a weak pulse, coldness, sweating, and irregular breathing due to too low a blood flow to the brain.

sign An indication of the presence of a disease, especially one observed by a doctor but not apparent to the patient; *see also* symptom.

silage Grass or other green fodder compacted and preserved through fermentation in airtight conditions and used as animal feed in the winter.

simple stain technique The use of a single dye to contrast cells.

sludge The solids in sewage that separate out during sewage treatment.

soft rot A softening of tissues in the skin of vegetables.

soil A mixture of organic matter, minerals, gases, liquids, and organisms that together support life.

solute The substance dissolved in a solution.

solvent The substance doing the dissolving in a solution.

soy sauce A condiment made from a fermented paste of boiled soybeans, roasted grain, salt, and a mold.

species (pl. species) The fundamental rank in the classification system of organisms that is composed of the genus and specific epithet words.

specific epithet The second of the two scientific words for a species. *See also* genus.

spike A protein projecting from the viral envelope or capsid that aids in attachment and penetration of a host cell.

spirillum (pl. spirilla) A bacterial cell shape characterized by twisted or curved rods.

spirochete A twisted bacterial rod with a flexible cell wall.

spontaneous mutation A change in the sequences of bases in DNA that arises naturally and not as a result of exposure to any physical or chemical agent.

spore A reproductive structure formed by a fungus.

sporulation The process of spore formation.

sputum Respiratory mucus.

staphylococcus (pl. staphylococci) 1. An arrangement of bacterial cells characterized by spheres in a grapelike cluster. 2. As *Staphylococcus*, a genus of facultatively anaerobic, nonmotile, non-spore-forming, gram-positive spheres in clusters.

starch An energy polysaccharide that is built from many glucose molecules.

stationary phase The portion of a growth curve in which population growth is arrested.

stem cell: A nonspecialized cell from which all other cells with specialized functions are generated.

sterile Free from living microorganisms, spores, and viruses.

sterilization The destruction or removal of all life forms, including bacterial spores and viruses.

sterol A lipid containing several carbon rings with side chains.

sticky end The unpaired nucleotides that extend beyond the paired nucleotides of a DNA fragment that has been cut with a restriction endonuclease (enzyme).

streptobacillus (pl. streptobacilli) 1. A chain of bacterial rods. 2. As *Streptobacillus*, a genus of facultatively anaerobic, nonmotile, gram-negative rods.

streptococcus (pl. streptococci) 1. A chain of bacterial cocci. 2. As *Streptococcus*, a genus of facultatively anaerobic, nonmotile, non- spore-forming, gram-positive spheres in chains.

structural gene A segment of a DNA molecule that provides the biochemical information for a polypeptide.

substrate The substance upon which an enzyme acts.

subunit vaccine A preparation that contains parts of microorganisms, such as virus spikes or purified pili.

sucrose A disaccharide sugar composed of glucose molecule and the monosaccharide fructose.

sulfite A food preservative used to control food spoilage in fruits and vegetables, wines, sausages, and fresh shrimp.

superbug A microbe that has become resistant to multiple antibiotic drugs.

superinfection The overgrowth (replacement) of susceptible strains by antibiotic resistant ones.

symbiont An organism that is very closely associated (living) with another, usually larger, organism.

symbiosis (symbiotic) An interrelationship (or referring to an interrelationship) between two populations of organisms where there is a close and permanent association.

symptom An indication of some disease or other disorder that is experienced by the patient. *See also* sign.

syndrome A collection of signs or symptoms that together are characteristic of a disease.

synergism A combination of drugs that together are of more benefit than either drug alone.

synthetically Referring to a chemical that is made in a research lab, rather than produced naturally.

synthetic drug A medicine developed and produced by a pharmaceutical company or research lab.

T

T lymphocyte (T cell) A type of white blood cell that matures in the thymus gland and is associated with cell-mediated responses of adaptive immunity.

Taq polymerase A thermostable DNA polymerase used for the polymerase chain reaction (PCR).

taxon A group organisms that form a common unit.

taxonomy The science dealing with the systematized arrangements of related organisms in categories. *See also* classification.

teichoic acid A molecule in the cell wall of gram-positive bacterial cells that strengthens the wall.

tertiary structure The folding of a polypeptide back on itself to form a unique three-dimensional structure.

tertiary waste treatment Involves the disinfection of the wastewater through chlorination.

tetrad An arrangement of four bacterial cells forming a cube shape.

Theory of Natural Selection Darwin's idea that when change occurs in an environment, those organisms best suited to the new circumstances (environment) will survive and succeed.

thermoacidophile An archaeal organism living under high temperature and high acid conditions.

thermophile An organism that lives at high temperature ranges of 40°C (104°F) and higher.

three-domain system The classification scheme that places all living organisms into one of three groups (domains) based on evolutionary relationships.

thymine (T) One of the nucleobases in DNA.

Ti plasmid A circular DNA molecule used to insert a gene into a chromosome of a plant.

tincture A low concentration of a chemical dissolved in alcohol.

topical Referring to a surface area, especially the skin.

total magnification The number of times an object has been enlarged.

toxin A poisonous chemical substance produced by an organism.

toxoid A preparation of a microbial toxin that has been rendered harmless by chemical treatment but that is capable of stimulating antibodies.

toxoid vaccine A vaccine produced from a toxin (poison) that has been made harmless but can still trigger an immune response against the toxin.

transcription The biochemical process in which RNA is synthesized according to a code supplied by the bases of a gene in the DNA molecule.

transduction The transfer of a few bacterial genes from a donor cell to a recipient cell via a bacterial virus.

transfer RNA (tRNA) A molecule of RNA that unites with amino acids and transports them to the ribosome in protein synthesis.

transformation The transfer and integration of DNA fragments from a dead and lysed donor cell to a recipient cell's chromosome.

transgenic Referring to an organism containing a gene or genes from another organism.

translation The biochemical process in which the code on the mRNA molecule is translated into a sequence of amino acids in a polypeptide.

transmission electron microscope (TEM) The type of electron microscope that allows electrons to pass through the object, resulting in a detailed view of the object's structure.

transposon A segment of DNA that moves from one site on a DNA molecule to another site, carrying information for protein synthesis. Also called jumping genes.

tree of life (TOL) A concept based on evolutionary relationships that is used to relate all known forms of life.

triclosan A phenol derivative incorporated as an antimicrobial agent into a wide variety of household products.

trophic level The position of an organism occupies in a food web.

truffle The reproductive body of a subterranean fungus.

tubercle A hard nodule that develops in tissue infected with the tuberculosis bacterium.

tuberculin skin test A procedure to establish if someone has been exposed to the tuberculosis bacterium.

tumor A mass of cells resulting from an abnormal, uncontrolled growth of cells.

tumor-causing virus *See* oncogenic virus.

tumor suppressor gene (TSG) A normal gene that inhibits tumor formation.

U

ultra-high temperature (UHT) pasteurization A treatment in which milk is heated at 140°C for 1-3 seconds to destroy all pathogens and most, if not all other microbes.

ultraviolet radiation (UV light) A type of electromagnetic radiation of short wavelengths that damages DNA.

uncoating Referring to the loss of the viral capsid once inside an infected eukaryotic cell.

unripened cheese The product of curd formation.

unsaturated Referring to a water-insoluble compound that can incorporate additional hydrogen atoms. *See also* saturated.

uracil (U) One of the nucleobases in RNA.

V

vaccination Inoculation with a weakened or dead infectious microbe or virus in order to generate immunity and prevent the disease. *See also* immunization.

vaccine A preparation containing weakened or dead microorganisms or viruses, treated toxins, or parts of microorganisms or viruses to protect the body from infectious disease.

vaporized Referring to a solid being converted directly into vapor without going through a liquid phase.

vector An arthropod that transmits the agents of disease from an infected host to a susceptible host.

vertical gene transfer The passing down of genetic information from one generation to the next.

vesicle A small structure within a cell, consisting of fluid enclosed by a lipid bilayer.

viable but noncultured (VBNC) Referring to microbes that are alive but not dividing and therefore cannot be cultured in any known growth medium.

vibrio 1. A prokaryotic cell shape occurring as a curved rod. 2. As *Vibrio*, a genus of facultatively anaerobic, gram-negative curved rods with flagella.

vinegar A liquid consisting mainly of acetic acid and water that results from the fermentation of ethanol by acetic acid bacteria.

viral genome The genetic material contained within a virus particle.

viroid An infectious RNA segment associated with certain plant diseases.

virosphere The places where viruses are found or interact with their hosts.

virulence The relative capacity of a pathogen to overcome the body's immune defenses.

virulence factor A molecule on or produced by a pathogen that increases its ability to invade or cause disease to a host.

virulent Referring to a virus or microorganism that can be extremely damaging when in the host.

virus An infectious, noncellular agent that replicates within living cells and causes disease.

W

water pollution The contamination of water with contaminants.

whey The clear liquid remaining after protein has curdled out of milk.

white blood cell *See* leukocyte.

whole agent vaccine A vaccine containing whole bacterial cells, viruses, or toxins.

wild type The normal, nonmutated characteristic of a cell or organism. *See also* mutant.

wort A sugary liquid produced from crushed malted grain and water to which is added yeast and hops for the brewing of beer.

Y

yeast 1. A type of unicellular, nonfilamentous fungus. 2. A term sometimes used to denote the unicellular form of pathogenic fungi.

Z

zoonotic disease (zoonosis) A disease spread from another animal to humans.

zooplankton Small animal-like protists that are a component of the aquatic food webs.

Index

Note: Page numbers followed by *b*, *f*, and *t* indicate materials in boxes, figures, and tables respectively.

A

D

E

H

I

N

O

P

Q

R